STATISTICAL PHYSICS

Fundamentals and Application to Condensed Matter

With Lectures, Problems and Solutions

STATISTICAL PHYSICS

Fundamentals and Application to Condensed Matter

With Lectures, Problems and Solutions

Hung T. Diep

University of Cergy-Pontoise, France

World Scientific

NEW JERSEY · LONDON · SINGAPORE · BEIJING · SHANGHAI · HONG KONG · TAIPEI · CHENNAI

Published by

World Scientific Publishing Co. Pte. Ltd.
5 Toh Tuck Link, Singapore 596224
USA office: 27 Warren Street, Suite 401-402, Hackensack, NJ 07601
UK office: 57 Shelton Street, Covent Garden, London WC2H 9HE

British Library Cataloguing-in-Publication Data
A catalogue record for this book is available from the British Library.

STATISTICAL PHYSICS
Fundamentals and Application to Condensed Matter
With Lectures, Problems and Solutions

ISBN 978-981-4696-13-5
ISBN 978-981-4696-25-8 (pbk)

In-house Editor: Rhaimie Wahap

Typeset by Stallion Press
Email: enquiries@stallionpress.com

Printed in Singapore

To my wife and our children Samuel, Tuan, Kim and Sarah.

To my mother.

Preface

Statistical mechanics provides general methods to study properties of systems composed of a large number of particles. It establishes general formulas connecting physical parameters to various physical quantities. When parameters of a system such as interaction between particles and temperature are known, one can deduce its macroscopic properties. In general, microscopic mechanisms leading to interactions are provided by quantum mechanics. The combination of statistical mechanics and quantum mechanics has provided an understanding of properties of matter leading to spectacular technological innovations and discoveries which have radically changed our daily life since the sixties.

This book is based on the author's lectures for students of the third year of the Bachelor's degree in Physics. The second part of the book treats selected advanced subjects taught at the Master's level. In each chapter, fundamental notions and techniques are presented and followed by applications chosen among frequently encountered phenomena. Demonstrations leading to main results are given in details.

In the first part, after an introduction of basic definitions and mathematical tools (chapter 1), the book covers the foundation of statistical physics at equilibrium: starting from the fundamental postulate, the book deals with systems under various situations going from isolated systems (chapter 2) and systems maintained at a constant temperature (chapter 3) to open systems (chapter 4). The main properties of free fermions (chapter 5) and free bosons (chapter 6) are studied to a great extent. The first part ends with chapter 7 where the method of second quantization is shown. This method, though conceptually more abstract than the Schrödinger equation, is technically less cumbersome to handle, and is very useful in the study of weakly interacting many-particle systems. A large number of applications of this method is found in the remaining chapters.

In the second part, advanced techniques and applications in condensed matter are presented. Selected topics in condensed matter include vibrations of atoms in crystals, conducting electrons in metals and superconductors, band structures in semiconductors, and magnetic properties of materials. Statistical physics contributes with quantum mechanics to the success of these fields in the last fifty years. In chapter 8 the crystalline symmetry is presented with all necessary notions for studying properties of electrons and atoms in crystals. In chapter 9 systems of interacting atoms in crystals are considered. Quantized atom vibrations are called phonons which dominate thermodynamic properties of solids. Systems of interacting conducting electrons are studied in chapter 10 along with general properties of Fermi liquids. The origin of energy bands of electrons in semiconductors is shown in chapter 11. Conducting electrons are at the heart of charge and spin transport phenomena with an enormous number of applications. The spin carried by an electron plays thus a very important role in condensed matter physics. Magnetic properties due to spins cannot be separated from other properties of matter. Note that systems of interacting spins constitute one of the most important subjects in statistical physics. They are studied in chapter 12 where collective excitations, called spin waves or magnons, are shown in details. The abundance of magnetic materials, natural or artificial compounds, provides an inexhaustible source of applications. Chapter 13 deals with the phase transition in spin systems where basic notions such as symmetry breaking and universality class are introduced. The mean-field approximation and the Landau-Ginzburg theory for second-order phase transitions are presented. The concept of the renormalization group is described. In chapter 14, the Ginzburg-Landau theory for the superconductivity is developed to explain properties of type I and type II superconductors. The microscopic Bardeen-Cooper-Schrieffer theory for conventional superconductors is presented in details in this chapter. The second part of the book ends with chapter 15 where basic notions on systems out of equilibrium and the Boltzmann's equation are introduced. As applications of the Boltzmann's equation, properties of electron transport in metals and semiconductors are studied to a great extent.

In the third part of the book, solutions of problems are given. These problems are conceived for self-training and to help the reader discover new related phenomena which are complementary to the lectures.

H. T. Diep
Professor of Physics, University of Cergy-Pontoise, France

Acknowledgments

I am grateful to my many colleagues at the University of Cergy-Pontoise for sharing uncountable stimulating moments in my professional life and for their precious friendship during the last 25 years. My sincere gratitude goes to all the people who have contributed in one way or another to forging the site into the education and research institution that it is today. Thanks to them, it is the place where I go to work every day with enthusiasm and eagerness.

I am in particular indebted to numerous administrative staffs who have been working with me over the years, for their generosity and for carrying out with me our collective duties in joy and mutual trust. I am proud of what we have achieved together. It was always a team effort which is key in overcoming obstacles and fighting adversity.

I would like to express here my deep affection for my former and current doctorate students with whom I shared innumerable wonderful moments not only in our research activities but also in discussions on many subjects of life.

H. T. Diep

Contents

Solutions of Problems   375

Appendices   **571**

List of Problems

Problem 2. Next-nearest-neighbor interaction in a centered cubic lattice

Problem 3. Next-nearest-neighbor interaction in a square lattice

Problem 4. System of two spins

Problem 5. Improved mean-field approximation: Bethe's approximation

Problem 6. Chain of Ising spins by micro-canonical method

Problem 7. Chain of Ising spins by canonical method

Problem 8. Mean-field approximation for antiferromagnets

Problem 9. Ferrimagnets by mean-field theory

Problem 10. Chain of Ising spins by exact method

Problems of chapter 14: Superconductivity

Problem 1. Demonstration of Ginzburg-Landau Eqs. (14.13) and (14.14)

Problem 2. Current density $\vec{J}$: gauge-invariance

Problem 3. Theory of Gorter-Casimir

Problem 4. Energy of a vortex

Problem 5. Electron gas in a strong magnetic field: Landau's levels

Problems of chapter 15: Transport in Metals and Semiconductors

Problem 1. Effect of magnetic field: Demonstration of Eq. (15.24)

Problem 2. Electrons in a strong electric field: an approximation

Problem 3. Semiconductors: effect of temperature on conductivity

Problem 4. Semiconductor: effect of magnetic field on the gap

Problem 5. Effect of doping in semiconductors

Problem 6. Swallow impurity states in semiconductors

Problem 7. Recombinations in semiconductors

Problem 8. Dielectric relaxation

Problem 9. Polarized $p - n$ junction: direct current

Problem 10. Transport in a superlattice

Problem 11. Hall effect - Magnetoresistance

PART 1
Fundamentals of Statistical Physics

Chapter 1

Basic Concepts and Tools in Statistical Physics

1.1 Introduction

Statistical mechanics provides general methods to study properties of systems composed of a large number of particles. It establishes general formulas connecting physical parameters to various physical quantities. From these formulas, one can deduce properties of a system if one knows the system's parameters. In general, microscopic mechanisms leading to interaction parameters are provided by quantum mechanics. Statistical mechanics and quantum mechanics are at the heart of the modern physics which has allowed a microscopic understanding of properties of matter and spectacular technological innovations and progress which have radically changed our daily life since 50 years.

While quantum mechanics searches for microscopic structures and mechanisms which are not perceptible at the macroscopic scale, statistical mechanics studies macroscopic properties of systems composed of a large number of particles using information provided by quantum mechanics on interaction parameters. The progress made in the 20-th century in the theory of condensed matter shows the power of methods borrowed from statistical physics.

Statistical physics of systems at equilibrium is based on a single postulate called "fundamental postulate" introduced in the case of an isolated system at equilibrium. Using this postulate, one studies in chapter 2 properties of isolated systems and recovers results from thermodynamics. Moreover, as one sees in chapter 3 and 4, one can also study properties of non isolated systems in some particular situations, such as systems maintained at a constant temperature and systems maintained at a constant temperature and a constant chemical potential, where the fundamental postulate

3

can be used in each case to calculate the probability of a microscopic state, or "microstate" for short.

In this chapter one recalls basic mathematical tools and definitions which are used throughout this book.

1.2   Combinatory analysis

In the following, one recalls some useful definitions in the combinatory analysis.

1.2.1   *Number of permutations*

Let N be the number of discernible objects. The number of permutations two by two of these objects is given by

$$P = N ! \tag{1.1}$$

To demonstrate this formula, let us consider an array of N "cases" and N objects numbered from 1 to N:

-there are N ways to choose the first object to put into the first case

-there are $N-1$ ways to choose the second object to put into the second case, etc.

The total number of different configurations of objets in the cases is thus $N(N - 1)(N - 2)...1 = N!$.

Example: The number of permutations of 3 objects numbered from 1 to 3 is $3! = 6$. They are $|1|2|3|$, $|1|3|2|$, $|2|1|3|$, $|2|3|1|$, $|3|1|2|$, $|3|2|1|$.

1.2.2   *Number of arrangements*

The number of arrangements of n objects taken among N objects is the number of ways to choose n objects among N objects, one by one, taking into account the order of sorting-out. This number is given by

$$A_N^n = \frac{N!}{(N - n)!} \tag{1.2}$$

Example: Consider 4 objects numbered from 1 to 4. One chooses 2 objects among 4. The possible configurations are (1,2), (1,3), (1,4), (2,1), (2,3), (2,4), (3,1), (3,2), (3,4), (4,1), (4,2) and (4,3) where the first and second numbers in the parentheses correspond respectively to the first and second sorting. If one takes into account the sorting order, then (1,2) and

(2,1) are considered to be different. There are thus 12 "arrangements", as given by $A_4^2 = 4!/(4-2)! = 12$.

1.2.3 *Number of combinations*

The number of combinations of n objects among N objects is the number of ways to choose n objects among N objects, without taking into account the sorting order. This number is given by

$$C_N^n = \frac{N!}{(N-n)!n!} \tag{1.3}$$

One sees that C_N^n is equal to A_N^n divided by the number of permutations of n objects: $C_n^n = A_N^n/n!$. In the example given above, if (1,2) and (2,1) are counted only once, and the same for the other similar pairs, then one has 6 combinations, namely C_4^2.

1.3 Probability

Statistical mechanics is based on a probabilistic concept. As one sees in the following, the value of a physical quantity experimentally observed is interpreted as its most probable value. In terms of probabilities, this value corresponds to the maximum of the probability to find this quantity. One sees below that the observed quantity is nothing but the mean value resulting from the statistical average over microstates.

Hereafter, one recalls the main properties of the probability and presents some frequently used probability laws.

1.3.1 *Definition*

One distinguishes two cases: discrete events and continuous events.

1.3.1.1 *Ensemble of discrete events*

Discrete events can be distinguished from each other without error or ambiguity. Examples include rolling a dice or flipping a coin in which the result of each event is independent of the previous one.

Let us realize N experiments. Each experiment gives a result. Let n_i be the number of times that an event of the type i occurs among N results. The probability of that event is defined by

$$P_i = \lim_{N \to \infty} \frac{n_i}{N} \tag{1.4}$$

Example: One flips N times a coin. One obtains n_1 heads and n_2 tails. If N is very large, one expects $n_1 = n_2 = \frac{N}{2}$. The ratios $\frac{n_1}{N}$ and $\frac{n_2}{N}$ are called "probabilities" to obtain "heads" and "tails", respectively.

Remark: For all i, $0 \le P_i \le 1$.

1.3.1.2 *Ensemble of continuous events — Density of probability*

Variables in continuous events are always given with an uncertainty (error).

Example: The length of a table is measured with an uncertainty. Let $\Delta P(x)$ be the probability so that the length of the table belongs to the interval $[x, x + \Delta x]$. Let $\Delta n(x)$ be the number of times among N measurements where x belongs to $[x, x + \Delta x]$. One has by definition

$$\Delta P(x) = \lim_{N \to \infty} \frac{\Delta n(x)}{N} \tag{1.5}$$

The density of probability is defined by

$$W(x) = \lim_{\Delta x \to 0} \frac{\Delta P(x)}{\Delta x} = \frac{dP(x)}{dx} \tag{1.6}$$

The probability of finding x in the interval $[x, x + dx]$ is thus $dP(x) = W(x)dx$.

Remarks:
1) $W(x) \ge 0$
2) For more than one variable one writes $dP(x, y, z) = W(x, y, z)dxdydz$ or $dP(\overrightarrow{r}) = W(\overrightarrow{r})d\overrightarrow{r}$.

1.3.2 *Fundamental properties*

1.3.2.1 *Normalization of probabilities*

From their definition, one has for discrete and continuous cases the following normalization relations

$$\sum_i P_i = \frac{\sum_i n_i}{N} = 1 \tag{1.7}$$

$$\int_{-\infty}^{\infty} W(x)dx = \int dP(x) = \frac{1}{N} \int dn(x) = 1 \tag{1.8}$$

1.3.2.2 *Addition rule*

Let e_1 and e_2 be two incompatible discrete events of probabilities $P(e_1)$ and $P(e_2)$, respectively. The probability of finding e_1 or e_2 is given by

$$P(e_1 \text{ or } e_2) = P(e_1) + P(e_2) \tag{1.9}$$

For continuous variables, the probability to find x between a and b is

$$P(a \leq x \leq b) = \int_a^b W(x)dx \tag{1.10}$$

If e_1 and e_2 are not incompatible in N experiments, then one should remove the number of times where e_1 and e_2 simultaneously take place. One has

$$P(e_1 \text{ or } e_2) = P(e_1) + P(e_2) - P(e_1 \text{ and } e_2) \tag{1.11}$$

1.3.2.3 *Multiplication rule*

Let e_1 and e_2 be two independent events. The probability to find e_1 and e_2 is given by

$$P(e_1 \text{ and } e_2) = P(e_1)P(e_2) \tag{1.12}$$

Example: One flips a dice twice. The probability to find face 1 or face 2 is $\frac{1}{6} + \frac{1}{6}$, and that to find face 1 and face 2 is $\frac{1}{6} \times \frac{1}{6}$.

1.3.3 *Mean values*

One defines the mean value of a quantity f by

$$\bar{f} = \sum_m P_m f_m \tag{1.13}$$

where f_m is the value of f in the microstate m of probability P_m. The sum is performed over all states.

If the states (or events) are characterized by continuous variables, one has

$$\bar{f} = \int_{-\infty}^{\infty} W(x)f(x)dx \tag{1.14}$$

where $f(x)$ is the value of f in the state x.

If a value A occurs many times with different events m (or different states), one can write

$$\overline{A} = \sum_{A_i} P(A_i)A_i \qquad (1.15)$$

where $P(A_i)$ is the probability to find A_i, namely

$$P(A_i) = \sum_{m, A_m = A_i} P_m \qquad (1.16)$$

where the sum is made only on the events m having $A_m = A_i$. In the case of a continuous variable, one has

$$\overline{A} = \int W(A)A dA \qquad (1.17)$$

where $W(A)$ is the density of probability to find A.

The most probable value of A_i (or A) corresponds to the maximum of $P(A_i)$ (or $W(A)$).

The variance is defined by

$$(\Delta f)^2 = \overline{\left(f - \overline{f}\right)^2} \qquad (1.18)$$

One can also write

$$
\begin{aligned}
(\Delta f)^2 &= \sum_m P_m \left(f_m - \overline{f}\right)^2 \\
&= \sum_m P_m \left[f_m^2 + (\overline{f})^2 - 2f_m \overline{f}\right] \\
&= \overline{f^2} + (\overline{f})^2 - 2\overline{f}\ \overline{f} \\
&= \overline{f^2} - (\overline{f})^2 \qquad (1.19)
\end{aligned}
$$

One calls $\Delta f = \sqrt{(\Delta f)^2}$ "standard deviation". This quantity expresses the dispersion of the statistical distribution. In the same manner, one has for the continuous case

$$(\Delta A)^2 = \int_{-\infty}^{+\infty} W(A)\left[A - \overline{A}\right]^2 dA = \overline{A^2} - (\overline{A})^2 \qquad (1.20)$$

1.4 Statistical distributions

1.4.1 *Binomial distribution*

When there are only two possible outcomes in an experiment, the distribution of the results in a series of experiments follows the binomial law.

Example: flipping a coin $\rightarrow$ two possible results, head or tail.

Let P_A and P_B be the probabilities of the two types of outcome. The normalization of probability imposes $P_A + P_B = 1$. If after N experiments one obtains n times A and $N - n$ times B, then the probability to find n times A and $(N - n)$ times B is given by the multiplication rule

$$P(N, n) = C_N^n P_A^n P_B^{N-n} \tag{1.21}$$

where the factor $C_N^n = \frac{N!}{n!(N-n)!}$ is introduced to take into account the number of combinations in which n times A and $(N - n)$ times B occur. One shows that $P(N, n)$ is normalized:

$$\sum_{n=0}^{N} P(N, n) = \sum_{n=0}^{N} C_N^n P_A^n P_B^{N-n}$$
$$= (P_A + P_B)^N$$
$$= 1 \tag{1.22}$$

where one has used the formula of the Newton binomial to go from the first to the second line.

One shows in the following some main results:

- the mean value of n is $\overline{n} = NP_A$:

$$\overline{n} = \sum_{n=0}^{N} nP(N, n) = \sum_{n=0}^{N} nC_N^n P_A^n P_B^{N-n}$$
$$= P_A \frac{\partial}{\partial P_A} \left[\sum_{n=0}^{N} C_N^n P_A^n P_B^{N-n} \right]$$
$$= P_A \frac{\partial}{\partial P_A} (P_A + P_B)^N$$
$$= P_A N (P_A + P_B)^{N-1}$$
$$= NP_A \tag{1.23}$$

where one has used in the last line $P_A + P_B = 1$.

- the variance:
  The variance is defined by $(\Delta n)^2 = \overline{n^2} - (\overline{n})^2$. One calculates $\overline{n^2}$ as follows:

$$\overline{n^2} = \sum_{n=0}^{N} n^2 P(N, n) = n^2 C_N^n P_A^n P_B^{N-n}$$

$$= (P_A \frac{\partial}{\partial P_A})^2 \left[\sum_{n=0}^{N} C_N^n P_A^n P_B^{N-n} \right]$$

$$= (P_A \frac{\partial}{\partial P_A})^2 (P_A + P_B)^N$$

$$= P_A \frac{\partial}{\partial P_A} \left[P_A N (P_A + P_B)^{N-1} \right]$$

$$= N P_A + N(N-1) P_A^2 \tag{1.24}$$

where one has used in the last equality $P_A + P_B = 1$. One finds

$$(\Delta n)^2 = \overline{n^2} - (\overline{n})^2 = N P_A + N(N-1) P_A^2 - (N P_A)^2$$

$$= N P_A (1 - P_A) = N P_A P_B \tag{1.25}$$

from which the standard deviation is $\Delta n = \sqrt{N P_A P_B}$.

- the relative uncertainty or relative error:
  The relative error on $\overline{n}$ is given by

$$\frac{\Delta n}{\overline{n}} \simeq \frac{\sqrt{N}}{N} = \frac{1}{\sqrt{N}} \tag{1.26}$$

The relative error (or uncertainty) decreases with increasing N. For $N = 1000$ (which is the standard sample for polls), $\Delta n / \overline{n} \simeq 3\%$.

1.4.2   *Gaussian distribution*

The Gaussian distribution applies in the case of continuous events. The density of probability $W(x)$ of the Gaussian distribution, or Gaussian law, is given by

$$W(x) = A \exp \left[-\frac{(x - x_0)^2}{2\sigma^2} \right] \tag{1.27}$$

where A is a constant determined by the normalization of $W(x)$, x_0 the central value of the distribution and 2σ the full width at half-maximum of $W(x)$ (see Fig.1.1).

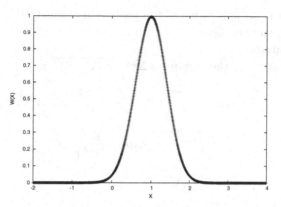

Fig. 1.1 Gaussian density of probability $W(x)$ shown with $x_0 = 1$, $\sigma = 0.4$.

To calculate A, one writes

$$1 = \int_{-\infty}^{\infty} W(x)dx = A \int_{-\infty}^{\infty} \exp\left[-\frac{(x-x_0)^2}{2\sigma^2}\right] = A\sqrt{2\pi\sigma^2} \qquad (1.28)$$

where $u = x - x_0$ and where one has used in the last equality the Gauss integral (see Appendix A):

$$\int_{-\infty}^{\infty} \exp\left[-au^2\right] du = \sqrt{\frac{\pi}{a}} \qquad (1.29)$$

One then has $A = \frac{1}{\sigma\sqrt{2\pi}}$.

One shows some important results in the following:

- the mean value of x:

  With (1.27), one has

$$\bar{x} = \int_{-\infty}^{\infty} xW(x)dx = A \int_{-\infty}^{\infty} x \exp\left[-\frac{(x-x_0)^2}{2\sigma^2}\right] dx$$

$$= A \int_{-\infty}^{\infty} (x-x_0) \exp\left[-\frac{(x-x_0)^2}{2\sigma^2}\right] dx$$

$$+ A \int_{-\infty}^{\infty} x_0 \exp\left[-\frac{(x-x_0)^2}{2\sigma^2}\right] dx$$

$$= 0 + x_0 A \int_{-\infty}^{\infty} \exp\left[-\frac{(x-x_0)^2}{2\sigma^2}\right] dx$$

$$= x_0 \qquad (1.30)$$

where the first integral of the second line is zero (integral of an odd function with two symmetrical opposite bounds), and one has used the normalization of the probability density $W(x)$ in the line before the last line.

- the variance:

Using $\bar{x} = x_0$, the variance $(\Delta x)^2 = \overline{(x - \bar{x})^2}$ is

$$
\begin{aligned}
(\Delta x)^2 &= A \int_{-\infty}^{\infty} (x - x_0)^2 \exp\left[-\frac{(x - x_0)^2}{2\sigma^2}\right] dx \\
&= A \int_{-\infty}^{\infty} y^2 \exp\left[-\frac{y^2}{2\sigma^2}\right] dy \\
&= A \frac{1}{2} \sqrt{\pi(2\sigma^2)^3}
\end{aligned}
$$

where one used in the last line a formula in Appendix A. Replacing A by $\frac{1}{\sigma\sqrt{2\pi}}$, one obtains

$$
(\Delta x)^2 = \sigma^2 \tag{1.31}
$$

from which the standard deviation is $\Delta x = \sigma$.

One can also show that (see Problem 1) the binomial law becomes the Gaussian law when $N \gg n \gg 1$. This equivalence is known as "central limit theorem".

1.4.3   *Poisson law*

When the probability of an event is very small with respect to 1, namely a rare event, the probability to find n events of the same kind is given by the Poisson law:

$$
P(n) = \frac{\mu^n \exp(-\mu)}{n!} \tag{1.32}
$$

where μ is a constant which is the mean value of n (see Problem 2). One shows that $P(n)$ is normalized:

$$
\sum_{n=0}^{\infty} P(n) = \sum_{n=0}^{\infty} \frac{\mu^n \exp(-\mu)}{n!} = \exp(-\mu) \sum_{n=0}^{\infty} \frac{\mu^n}{n!} = \exp(-\mu)\exp(\mu) = 1 \tag{1.33}
$$

One can also show that the binomial law becomes the Poisson law when $P_A \ll P_B$ and $N \gg n \gg 1$ (see Problem 2).

1.5   Microstates — Macrostates

1.5.1   *Microstates — Enumeration*

A microstate of a system is defined in general by the individual states of the particles which constitute the system. These individual states are characterized by the physical parameters which define the system. Let us mention some examples. The microstates of an atom are given by the states of the electrons of the atom: each microstate is defined by four quantum numbers (n, l, m_l, m_s). The microstates of an isolated system of energy E are given by the different distributions of E on the system particles. The microstates, at a given time, of a system of N classical particles of mass m are defined by the positions and the velocities of the particles at that time. The microstates of a system of independent quantum particles are given by the wave-vectors of the particles.

It is obvious that for a given system there are a large number of microstates which satisfy the physical constraints imposed on the system (volume, energy, temperature, ...). These microstates are called "realizable microstates". It is very important to know how to calculate the number of these states. Systems of classical and quantum independent particles are shown in details in chapter 2. In the following, one gives some examples and formulas to enumerate the microstates of simple systems. In general, one distinguishes the cases of indiscernible and discernible particles. Discernible particles are those one can identify individually such as particles with different colors or particles bearing each a number. Indiscernible particles are identical particles impossible to distinguish the one from the others such as atoms in a mono-atomic gas or conduction electrons in a metal.

Example: Consider an isolated system of total energy equal to 3 units. This system has 3 particles supposed to be discernible and bearing letters A, B and C. Each of the particles can occupy levels of energy at 0, 1, 2, or 3 units. With the constraint that the sum of energies of the particles is equal to 3, one has the 3 following categories of microstates (see Fig. 1.2):

-level 0: B, C; level 3: A (Fig. 1.2, left) $\rightarrow$ permutations of A with B and A with C give 3 microstates

-level 0: C; level 1: B ; level 2: A (Fig. 1.2, middle) $\rightarrow$ permutations of A, B and C give 6 microstates

-level 1: A, B, C (Fig. 1.2, right) $\rightarrow$ 1 microstate.

Remark: Permutations between particles of the same level do not give rise to new microstates.

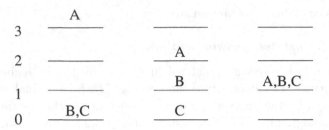

Fig. 1.2   Three categories of microstates (left, middle, right) according to occupation numbers of energy levels.

In the example given above, one can use the following formula to calculate the number of microstates of each category in Fig. 1.2

$$\omega = \frac{N!}{n_0! n_1! n_2! \dots} \tag{1.34}$$

where n_i $(i = 0, 1, 2, \dots)$ is the number of particles occupying the i-th level and N the total number of particles. One can obviously verify that this formula gives the number of microstates enumerated above in each category. Furthermore, one sees that the total number of microstates (all categories) is 10. This number can be calculated in the general case with arbitrary N and E by

$$\Omega(E) = \frac{(N + E - 1)!}{(N - 1)! E!} \tag{1.35}$$

For $E = 3$ and $N = 3$, one recovers $\Omega = \frac{(3+3-1)!}{(3-1)!3!} = 10$.

The demonstrations of Eqs. (1.34) and (1.35) are done in Problem 3.

Remark: In the above example, if the particles are indiscernible, the number of microstates is reduced to 3 because permutations of particles on the different levels do not yield new microstates.

It should be emphasized that the knowledge of the number of microstates allows one to calculate principal physical properties of isolated systems as one will see in chapter 2.

The microstates having the same energy are called "degenerate states". The number of degenerate states at a given energy is called "degeneracy". In the above example, the degeneracy is 10 for $E = 3$.

Each microstate l has a probability P_l. If one knows P_l, one can calculate the mean values of physical quantities as will be seen in the following chapters. That is why the main objective of statistical mechanics is to find a way to determine P_l according to the conditions imposed on the system.

Remark: Quantum particles, by their nature, are indiscernible. One distinguishes two kinds of particles: bosons and fermions. Bosons are particles having integer spins $(0, 1, 2, ...)$ and fermions are those having half-integer spins $(1/2, 3/2, 5/2, ...)$. The symmetry postulate in quantum mechanics states that wave functions of bosons are invariant with respect to permutation of two particles in their states, while wave functions of fermions do change their sign at each permutation. A consequence of this postulate is that a quantum microstate can contain any number of bosons but only zero or one fermion. This affects obviously the enumeration of microstates as one will see in chapters 5 and 6.

1.5.2 *Macroscopic states*

A macroscopic state , or macrostate for short, is an ensemble of microstates having a common macroscopic property. An observed macroscopic property is considered as the mean value of a statistical mixing of microstates of the ensemble. So, its definition depends on which macroscopic property one wants to observe: there are thus many ways to define a macrostate. In the example shown in Fig. 1.2, one can define a macrostate by the occupation numbers of the energy levels: there are thus 3 macrostates corresponding to the following occupation numbers in the first 4 levels $(2,0,0,1)$, $(1,1,1,0)$, $(0,3,0,0)$ (Fig. 1.2, from left to right). One can also define a macrostate as the one having an energy $E = 3$. In this case, the macrostate contains the ensemble of all 10 microstates.

1.5.3 *Statistical averages — Ergodic hypothesis*

When a system is at equilibrium, the mean values of its macroscopic variables do not vary anymore with time evolution. However, often the system fluctuates between different microstates giving rise to small variations around the mean values of the macroscopic variables. One defines the time average of a variable f by

$$\tilde{f} = \lim_{\tau \to \infty} \frac{1}{\tau} \int_{t_0}^{t_0+\tau} f(t)dt \tag{1.36}$$

where τ is the averaging time.

On the other hand, as seen above, the mean value of a quantity can be calculated over the ensemble of microstates. Let P_l be the probability of the microstate l in which the variable f is equal to f_l. One defines the so-called "ensemble average" of f by

$$\bar{f} = \sum_{(l)} P_l f_l \tag{1.37}$$

The ergodic hypothesis of statistical mechanics at equilibrium introduced by Boltzmann in 1871 states that the time and ensemble averages are equal. The validity of this principle has been experimentally verified for systems at equilibrium. Of course, for systems with extremely long relaxation times such as glasses, this equivalence is not easy to prove.

1.6   Statistical entropy

In statistical mechanics, the fundamental quantity which allows one to calculate principal properties of a system is the statistical entropy introduced by Boltzmann near the end of the 19th century. It is defined by

$$S = -k_B \sum_l P_l \ln P_l \tag{1.38}$$

where P_l is the probability of the microstate l of the system and k_B the Boltzmann constant.

One will see in the following chapters that S coincides with the thermodynamic entropy defined in a macroscopic manner in the second principle of the classical thermodynamics. The advantage of the definition (1.38) is that the physical consequences associated with the entropy are clear as seen below:

- As $0 \le P_l \le 1$, one has $S \ge 0$
- When one of the events is sure, namely one probability is equal to 1 (other probabilities are zero by normalization), one sees that S is zero (mimimum). When all probabilities are equal, namely when

the uncertainty on the outcome is maximum, one can show that S is maximum (see Problem 8). In other words, S represents the uncertainty or the lack of information on the system. With the above definition, one understands easily that why in thermodynamics entropy S is said to express the "disorder" of the system.

Statistical entropy is additive: consider two independent ensembles of events $\{e_m; m = 1, ..., M\}$ and $\{e'_{m'}; m' = 1, ..., M'\}$ of respective probabilities $\{P_m; m = 1, ..., M\}$ and $\{P'_{m'}; m' = 1, ..., M'\}$. The probability to find e_m and $e'_{m'}$ is thus $P(m, m') = P_m P'_{m'}$. The entropy of these double events is then

$$S(e, e') = -k_B \sum_{m,m'} P_m P'_{m'} \ln(P_m P'_{m'}) = -k_B \sum_{m,m'} P_m P'_{m'} [\ln P_m + \ln P'_{m'}]$$

$$= -k_B \sum_m P_m \ln P_m \sum_{m'} P'_{m'} - k_B \sum_{m'} P'_{m'} \ln P'_{m'} \sum_m P_m$$

$$= S(e) + S(e') \tag{1.39}$$

using $\sum_{m'} P'_{m'} = 1$ and $\sum_m P_m = 1$. The total entropy is the sum of partial entropies.

1.7 Conclusion

Since statistical mechanics is based on a probabilistic concept, the enumeration of microstates and the use of probabilities are very important. In this chapter, basic definitions and fundamental concepts on these points have been presented. In particular, most frequently used probability laws such as binomial, Gaussian and Poisson distributions have been shown. Microstates and macrostates were defined using examples for better illustration. Finally, the definition of the statistical entropy which is the most important quantity in statistical mechanics was shown and discussed in connection with the thermodynamic entropy. The above definitions and tools will be used throughout this book.

1.8 Problems

Problem 1. Central limit theorem:

Show that the binomial law becomes the Gaussian law when $N \gg n \gg 1$.

Problem 2. Poisson law (1.32):

a) Calculate $\bar{n}$.

b) Show that the binomial law becomes the Poisson law when $P_A \ll P_B$ and $N \gg n \gg 1$.

Problem 3. Demonstrate the formulas (1.34) and (1.35).

Problem 4. Application of the binomial law:

Calculate the probability to find the number of heads between 3 and 6 (boundaries included) when one flips 10 times a coin.

Problem 5. Random walk in one dimension:

Consider a particle moving on the x axis with a constant step length l on the right or on the left with the same probability.

a) Calculate the probability for the particle to make n steps to the right and n' steps to the left after N steps.

b) Calculate the averages $\bar{n}$ and $\bar{n'}$. Calculate the variance and relative uncertainty on $\bar{n}$.

c) Calculate the probability to find the particle at the position $x = ml$ from the origin. Calculate $\bar{x}$ and $(\Delta x)^2$.

d) Suppose that the step length x_i is not constant. What is the density of probability to find the particle at X after N steps with $N \gg 1$?

Problem 6. Random walk in three dimensions:

Consider a particle moving in three dimensions with variable discrete step lengths l and a density of probability independent of the direction of the step. Each step corresponds to a change of direction of the particle motion which is due to a collision with another particle. This is a simple model of the Brownian motion.

a) Calculate the Cartesian components (X, Y, Z) of the particle position after N steps. Calculate the averages $(\bar{X}, \bar{Y}, \bar{Z})$ and the corresponding variances $(\Delta X)^2$, $(\Delta Y)^2$ and $(\Delta Z)^2$. Deduce the average $\overline{l^2}$.

b) What is the density of probability to find the particle at (X, Y, Z) after N steps?

c) Putting $(\Delta X)^2 = 2Dt$ where D is the diffusion coefficient and t the lapse of time of the motion of the particle, demonstrate $D = \frac{1}{6}\frac{\overline{l^2}}{\tau}$ where τ is the average time between two particle collisions.

Problem 7. Exchange of energy:

Consider two isolated systems. The first one has $N_1 = 2$ particles and an energy equal to $E_1 = 3$ units. The second one has $N_2 = 4$ particles and $E_2 = 5$ units. Each particle can have 0, 1, 2, ... energy units.

a) Calculate the number of microstates of the total system composed of systems 1 and 2 separated by an insulating wall.

b) Remove the wall. The total system is kept isolated. Calculate the number of microstates of the total system in this new situation. Comment.

c) In the situation of the previous question, calculate the number of microstates in which there are at least 2 particles having 2 energy units each. Deduce the probability of this macrostate.

Problem 8. Statistical entropy:

Show that the statistical entropy $S = -k_B \sum_l P_l \ln P_l$ is maximum when all probabilities P_l are equal.

Chapter 2

Isolated Systems: Micro-Canonical Description

2.1 Introduction

To study properties of a system composed of a very large number of particles requires methods which take into account simultaneously the individual states of all particles. One cannot study properties of particles one by one except when they are independent. For a small system, one can solve the Schrödinger equation to obtain its physical properties. However, exact solutions are limited to systems of a few particles in quantum mechanics. Therefore, one should look for methods which are able to treat macroscopic systems of particles such as electrons in crystals with 10^{22} atoms per cm^3. Statistical physics provides methods which are based on the probability distribution of microstates of a given system defined by external parameters. Therefore, the first task is to find out this probability distribution as a function of external parameters imposed on a system.

The foundation of any field of science is based on a small number of postulates and principles that one should admit. Using these postulates and principles, subsequent results are demonstrated and their validity is experimentally verified. Statistical physics for systems at equilibrium is no exception: it is based on a single postulate which admits the probability of a microstate of an isolated system. In this chapter, this postulate is presented and its consequences on the fundamental physical properties of isolated systems are demonstrated.

2.2 Fundamental postulate

By definition, an isolated system does not have any interaction with the remaining universe. This definition imposes that interactions of any kind

21

are simply neglected because they are so small that no significant effects can be detected on the system under study. The external parameters such as energy E, volume V and number of particles N are constant. However, "internal variables" are free to fluctuate. These fluctuations follow a statistical law to be determined.

Accessible microstates of an isolated system are states which satisfy external constraints. Let Ω be the number of such microstates. The fundamental postulate of statistical mechanics states that *all accessible microstates of an isolated system at equilibrium have the same probability.*

According to the above postulate, the probability P_l of the microstate l is independent of l and is given by

$$P_l = \frac{1}{\Omega} \tag{2.1}$$

This probability is called "micro-canonical probability". The microstates satisfying (2.1) constitute the micro-canonical ensemble. The description of properties of a system using the micro-canonical ensemble is called "micro-canonical description".

The statistical entropy S of an isolated system is

$$S = -k_B \sum_l P_l \ln P_l = k_B \sum_l \frac{1}{\Omega} \ln \Omega$$

$$= k_B \frac{1}{\Omega} \ln \Omega \sum_l 1 = k_B \ln \Omega \tag{2.2}$$

Expression (2.2) is called "micro-canonical entropy".

The micro-canonical temperature T is defined by

$$\frac{1}{T} = \left(\frac{\partial S}{\partial E} \right)_{N,V} \tag{2.3}$$

The micro-canonical pressure p is defined by

$$\frac{p}{T} = \left(\frac{\partial S}{\partial V} \right)_{E,N} \tag{2.4}$$

The micro-canonical chemical potential μ is defined by

$$\frac{\mu}{T} = -\left(\frac{\partial S}{\partial N} \right)_{E,V} \tag{2.5}$$

One shows in the following the physical meaning of each of the above "mathematical" definitions by examining the spontaneous evolution of an isolated system when one modifies an external constraint. One then shows that the above definitions correspond to quantities of the same names in thermodynamics.

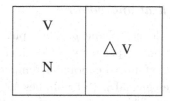

Fig. 2.1 Experimental setup for a Joule expansion.

2.3 Properties of an isolated system

2.3.1 *Spontaneous evolution of an isolated system toward equilibrium*

Consider an isolated system at equilibrium. If one removes one of the external constraints, the system evolves toward a new equilibrium. An example is the well-known Joule expansion. Consider a container having two compartments separated by a wall. In the initial state, one compartment is occupied by a gas, the other is empty (see Fig. 2.1). One removes the wall: the gas makes an expansion to occupy both compartments. It is obvious that the number of microstates increases because the system has new states thanks to the supplementary volume. One has

$$\Omega(E, N, V + \Delta V) > \Omega(E, N, V) \tag{2.6}$$

Thus,

$$\Delta S = S(E, N, V + \Delta V) - S(E, N, V) > 0 \tag{2.7}$$

This relation is demonstrated in Problem 1.

With the same argument, when the energy E is increased, the number of microstates Ω increases because the supplementary energy allows for new microstates which were forbidden before. S is thus an increasing function of E. It is interesting to note that, according to the definition (2.3), the micro-canonical temperature T is always positive.

It is concluded that when an external constraint is removed, the system acquires new microstates so that Ω increases. The spontaneous evolution of the system is thus simultaneously accompanied by an increase of the micro-canonical entropy [see (2.2)]. The new equilibrium is attained when S is maximum.

2.3.2   *Exchanges of heat and volume*

One has mathematically defined T and p. Their physical meaning is better understood by the following experiment. Consider two systems initially isolated from each other at micro-canonical temperatures T_1 and T_2, of micro-canonical pressures p_1 and p_2, of volumes V_1 and V_2, separated by a fixed and adiabatic wall. The wall becomes now diathermic and mobile. There are exchanges of heat and pressure. Since the total system is isolated, one has

$$E = E_1 + E_2^* = \text{constant} \rightarrow dE_1 + dE_2 = 0$$

$$V = V_1 + V_2 = \text{constant} \rightarrow dV_1 + dV_2 = 0$$

One can choose E_1 and V_1 as independent variables. The entropy is additive so that

$$S = S_1(E_1, V_1) + S_2(E_2, V_2)$$

The variation of S during the spontaneous evolution is written as

$$
\begin{aligned}
dS &= \frac{\partial S}{\partial E_1} dE_1 + \frac{\partial S}{\partial V_1} dV_1 \\
&= \frac{\partial S_1}{\partial E_1} dE_1 + \frac{\partial S_2}{\partial E_1} dE_1 + \frac{\partial S_1}{\partial V_1} dV_1 + \frac{\partial S_2}{\partial V_1} dV_1 \\
&= \frac{\partial S_1}{\partial E_1} dE_1 - \frac{\partial S_2}{\partial E_2} dE_1 + \frac{\partial S_1}{\partial V_1} dV_1 - \frac{\partial S_2}{\partial V_2} dV_1 \\
&= \frac{1}{T_1} dE_1 - \frac{1}{T_2} dE_1 + \frac{p_1}{T_1} dV_1 - \frac{p_2}{T_2} dV_1 \\
&= \left(\frac{1}{T_1} - \frac{1}{T_2} \right) dE_1 + \left(\frac{p_1}{T_1} - \frac{p_2}{T_2} \right) dV_1
\end{aligned}
\tag{2.8}
$$

During the evolution, one knows that the total entropy S increases, namely $dS > 0$. Let us first fix $dV_1 = 0$. If $T_1 > T_2$, then dE_1 should be negative because $dS > 0$. This means that the subsystem of higher temperature looses heat to the other subsystem. The new equilibrium is attained when S is maximum, i. e. when $dS = 0$. This corresponds to $T_1 = T_2$. Thus, at the new equilibrium, the two micro-canonical temperatures are equal. This is what one observes when one puts in contact two systems of different temperatures. The micro-canonical temperature is thus the temperature as defined in thermodynamics. Let us consider the volume exchange. If $p_1 > p_2$, one has $dV_1 > 0$ because $dS > 0$. One concludes that the subsystem of higher pressure expands its volume. The new equilibrium is attained when the pressures are equal, namely when $dS = 0$ (S is maximum).

Remark: In the heat exchange, the subsystem having higher temperature looses heat even if its energy is smaller than that of the other subsystem. It is the difference in temperatures that determines the direction of the evolution to new equilibrium.

2.3.3 *Exchange of particles*

In the previous experiment, if the wall is permeable to particles, an exchange of particles takes place. Since the total number of particles, N, is constant, the only variable is N_1 ($N_2 = N - N_1$). One writes

$$dS = \left[\frac{\partial S_1}{\partial N_1} + \frac{\partial S_2}{\partial N_1} \right] dN_1$$

$$= \left[\frac{\partial S_1}{\partial N_1} - \frac{\partial S_2}{\partial N_2} \right] dN_1$$

$$= \left[-\frac{\mu_1}{T_1} + \frac{\mu_2}{T_2} \right] dN_1 > 0 \tag{2.9}$$

Consider the system at thermal equilibrium, namely $T_1 = T_2$. One supposes now that $\mu_1 > \mu_2$. In a spontaneous evolution $dS > 0$, so one should have $dN_1 < 0$. This means that the subsystem of higher chemical potential looses particles. This is what observed in thermodynamics. Thus, in a spontaneous evolution to equilibrium, particles go from regions of high chemical potential to regions of low chemical potential. The new equilibrium is established when all chemical potentials are equal, namely when $dS = 0$ (S is maximum).

In summary, a spontaneous evolution of an isolated system composed of subsystems is accompanied by an increase of its micro-canonical entropy S. The new equilibrium is established when S is maximum. At the new equilibrium, the micro-canonical temperatures of all subsystems are equal. The same is for micro-canonical chemical potentials and pressures.

2.3.4 *Statistical distribution of an internal variable*

Consider an isolated system at equilibrium characterized by external parameters $\{E, x\}$ where x represents the ensemble of external parameters

other than energy E. The density of probability to find the value of an
internal variable y lying between y and $y + dy$ is written as

$$W^*(E, x, y) = \frac{\Omega^*(E, x, y)}{\Omega(E, x)} \tag{2.10}$$

where $\Omega^*(E, x, y)$ is the number of microstates having the value y and
$\Omega(E, x)$ the total number of microstates. One has

$$\int_{-\infty}^{+\infty} \Omega^*(E, x, y)dy = \Omega(E, x) \tag{2.11}$$

The normalization of $W^*(E, x, y)$ is

$$\int_{-\infty}^{+\infty} W^*(E, x, y)dy = 1 \tag{2.12}$$

One rewrites (2.10) as

$$\ln W^*(E, x, y) = \ln \Omega^*(E, x, y) - \ln \Omega(E, x) \tag{2.13}$$

By analogy with the definition of entropy S, one defines the "partial micro-
canonical entropy" dependent on y by

$$s^*(E, x, y) \equiv k_B \ln \Omega^*(E, x, y) \tag{2.14}$$

Replacing (2.14) and (2.2) in (2.13), one obtains

$$W^*(E, x, y) = \exp\left[\frac{1}{k_B}(s^*(E, x, y) - S(E, x))\right] \tag{2.15}$$

The dependence on y is in the first term of the exponential argument. One
rewrites it as

$$W^*(E, x, y) = A \exp\left[\frac{s^*(E, x, y)}{k_B}\right] \tag{2.16}$$

where A is a constant to be determined by the normalization of W^*.

The most probable value y_0 of y corresponds to the maximum of the density of probability W^*, namely to the maximum de $s^*(E, x, y)$ [see (2.16)].
One has at $y = y_0$,

$$\left[\frac{\partial s^*(E, x, y)}{\partial y}\right]_{y=y_0} = 0 \tag{2.17}$$

One supposes that there is only one solution y_0 for this equation. To
study the fluctuations of y around y_0, one expands s^* around y_0:

$$\ln W^*(y) \cong \ln A + \frac{1}{k_B}s^*(y_0) + \frac{1}{2}(y - y_0)^2 \frac{1}{k_B}\left(\frac{\partial^2 s^*}{\partial y^2}\right)_{y=y_0} + \dots \tag{2.18}$$

Putting

$$\frac{1}{\Delta^2} = -\frac{1}{k_B} \left(\frac{\partial^2 s^*}{\partial y^2} \right)_{y=y_0} \tag{2.19}$$

and neglecting terms of higher order in Eq. (2.18), one obtains

$$W^*(y) = W^*(y_0) \exp \left[-\frac{(y - y_0)^2}{2\Delta^2} \right] \tag{2.20}$$

where $W^*(y_0) = A \exp \left[\frac{1}{k_B} s^*(y_0) \right]$.

The statistical distribution of the internal variable y is thus Gaussian within this approximation which is valid for large systems.

2.4 Phase space — Density of states

In chapter 1, a number of examples on how to enumerate the number of microstates has been shown. Once the total number of microstates $\Omega(E)$ for an isolated system of energy E is known, one can calculate by Eq. (2.2) the micro-canonical entropy from which one obtains other physical quantities such as temperature, pressure and chemical potential.

In this section, one shows how to calculate the density of microstates, or "density of states" for short, in the phase space. The phase space is defined as a space of dimension equal to the number of degrees of freedom of the system. These degrees of freedom determine the microstates. One will see a few examples below. In particular, one will see how to use the density of states to calculate physical properties of a system.

2.4.1 *Density of states*

One wishes to calculate the mean value of A taken over all microstates

$$\overline{A} = \sum_l P_l A_l \tag{2.21}$$

where $P_l A_l$ depends in general on the energy E_l of the state l. One puts

$$P_l A_l = f(E_l) \tag{2.22}$$

where $f(E_l)$ is a function of E_l. If for a value E_l, there are $g(E_l)$ distinct microstates (degenerate states), one can replace the sum on l in (2.21) by

$$\overline{A} = \sum_l f(E_l) = \sum_{E_l} g(E_l) f(E_l) \tag{2.23}$$

In general, for a small system, energy E_l is a discrete variable. However, for systems of macroscopic dimension, namely systems at the thermodynamic limit, the distance between two successive energy levels is so small that E can be considered as a continuous quantity. In this case, one can replace the sum $\sum_{E_l}$ by an integral. Let $\Delta n(E)$ be the number of microstates having energy between E and $E + \Delta E$. The density of states $\rho(E)$ is defined by

$$\rho(E) = \lim_{\Delta E \to 0} \frac{\Delta n(E)}{\Delta E} = \frac{dn(E)}{dE} \tag{2.24}$$

For a large system, one writes

$$\overline{A} = \sum_{E_l} g(E_l) f(E_l) = \int_{E_0}^{\infty} f(E)\rho(E)dE \tag{2.25}$$

where E_0 is the lowest energy, namely energy of the ground state. One sees that $\rho(E)dE$ replaces the degeneracy $g(E)$ when E is considered as a continuous variable.

One calculates $\rho(E)$ as follows. Let $\phi(E)$ be the number of states of energy lower or equal to E. The number of states of energy lying between E and $E + \Delta E$ is thus

$$\Delta n(E) = \phi(E + \Delta E) - \phi(E) \tag{2.26}$$

The density of states is therefore

$$\rho(E) = \lim_{\Delta E \to 0} \frac{\Delta n(E)}{\Delta E} = \frac{d\phi(E)}{dE} \tag{2.27}$$

With this expression, it suffices to calculate $\phi(E)$ and take its derivative with respect to E to obtain $\rho(E)$. One considers in the following two important examples of which the results will be used in the next chapters.

2.4.2   *Density of states of free quantum particles*

Consider a free particle of mass m inside a box of dimensions $L_x, L_y, L_z = L$. The time-independent Schrödinger equation is

$$-\frac{\hbar^2 \Delta}{2m} \varphi(\vec{r}) = E\varphi(\vec{r}) \tag{2.28}$$

where E is the eigen-energy of the particle in the eigen-state $\varphi(\vec{r})$. For a free particle, $\varphi(\vec{r})$ is a plane wave:

$$\varphi_{\vec{k}}(\vec{r}) = \frac{1}{\sqrt{V}} e^{i\vec{k}\cdot\vec{r}} \tag{2.29}$$

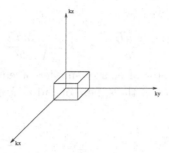

Fig. 2.2 Space of wave vectors $\vec{k}$. The volume of a microstate is represented by the cube of linear dimension $\frac{2\pi}{L}$.

where $V = L^3$ is the system volume. Replacing $\varphi(\vec{r})$ in (2.28) one finds $E_k = \hbar^2 k^2 / (2m)$. The values of $\vec{k}$ are determined by the boundary conditions. Using the periodic boundary conditions, one has for the x direction

$$\varphi_{\vec{k}}(x, y, z) = \varphi_{\vec{k}}(x + L, y, z) \tag{2.30}$$

thus

$$e^{ik_x x} = e^{ik_x(x+L)} \tag{2.31}$$

$$k_x L = 2\pi n_x \tag{2.32}$$

$$k_x = \frac{2\pi n_x}{L} \qquad n_x = 0, \pm 1, \pm 2, \pm 3, \ldots \tag{2.33}$$

In the same manner, one has $k_y = \frac{2\pi n_y}{L}$, $k_z = \frac{2\pi n_z}{L}$. The energies E_k constitute thus discrete levels

$$E_k = E(n_x, n_y, n_z) = \frac{\hbar^2 k^2}{2m} = \frac{2\pi^2 \hbar^2}{m L^2}(n_x^2 + n_y^2 + n_z^2) \tag{2.34}$$

For $L = 1$ cm and $m = m_e$ (electron mass), the distance between successive energy levels is

$$\Delta E = \frac{\hbar^2}{2m}\left(\frac{2\pi}{L}\right)^2 \approx \frac{10^{-68}}{2 \times 9.1 \times 10^{-31}}\left(\frac{2\pi}{10^{-2}}\right)^2 \text{ SI} \simeq 10^{-14}\text{eV} \tag{2.35}$$

This separation is so small that E can be considered as continuous at this macroscopic system size.

In the space (k_x, k_y, k_z) each point $(\frac{2\pi}{L} n_x, \frac{2\pi}{L} n_y, \frac{2\pi}{L} n_z)$ represents a microstate of energy $E(n_x, n_y, n_z)$. The elementary volume of a state is thus

$$\omega = \frac{(2\pi)^3}{L^3}$$

For a given E, k is given by

$$k^2 = \frac{2mE}{\hbar^2} \quad \Rightarrow \quad k = \sqrt{\frac{2mE}{\hbar^2}}$$

The number of microstates of energy smaller or equal to E, called $\phi(E)$ above, is thus the volume of the sphere of radius k divided by the volume of the microstate. One has

$$\phi(E) = \frac{\frac{4\pi k^3}{3}}{(\frac{2\pi}{L})^3} = \frac{V}{6\pi^2}\left[\frac{2m}{\hbar^2}\right]^{3/2} E^{3/2}$$

Using (2.27), one obtains the following density of states

$$\rho(E) = \frac{V}{4\pi^2}\left[\frac{2m}{\hbar^2}\right]^{3/2} E^{1/2} \tag{2.36}$$

Note that this relation is obtained for a single particle. For a system of N free particles, the phase space has $3N$ degrees of freedom, instead of 3. The volume of a hyper-sphere of radius k in this space is equal to Ak^{3N}. The calculation of the coefficient A is given in section A.1 of Appendix A. Using this result, one obtains for a system of N free particles:

$$\phi(E) \propto E^{3N/2} \tag{2.37}$$

Thus,

$$\rho(E) \propto E^{3N/2-1} \tag{2.38}$$

One should emphasize that the above densities of states are calculated without the spin degeneracy. It is known that when the particle has a spin of magnitude S, each point $\vec{k}$ of the phase space represents $(2S+1)$ states with spin components $-S, -S+1, ..., S-1, S$ [see (5.9)]. In the case of a single free particle of spin S, one has

$$\rho(E) = (2S+1)\frac{V}{4\pi^2}\left[\frac{2m}{\hbar^2}\right]^{3/2} E^{1/2} \tag{2.39}$$

2.4.3   *Density of states of free classical particles*

The state of a classical particle at a given time is defined by its position $\vec{r}$ and its momentum $\vec{p}$. For a system of N classical particles in three dimensions, the number de degrees of freedom is thus $6N$ (each particle has 6 degrees of freedom: x, y, z, p_x, p_y, p_z). To calculate the density of states in such a continuous space, one makes the following device: one associates a "volume" to each microstate in the phase space. This volume should be as small as possible. The choice is to associate to each degree of freedom a linear size $\sqrt{h}$ (h: Planck constant) representing the linear size of a state. This choice is consistent with the Heisenberg uncertainty principle. The elementary volume of a microstate is thus $(\sqrt{h})^{6N} = h^{3N}$. The number of microstates in a portion of the phase space is equal to the volume of this portion divided by the volume of a microstate. It is obvious that one recovers the results of classical mechanics by letting h tend to zero at the end. The sum over discrete microstates becomes integrals in the classical phase space:

$$\sum_l \cdots = \frac{1}{h^{3N}} \int d\vec{p}_1 \int d\vec{p}_2 \cdots \int d\vec{p}_N \int d\vec{r}_1 \int d\vec{r}_2 \cdots \int d\vec{r}_N \cdots \tag{2.40}$$

One now calculates $\phi(E)$:

$$\phi(E) = \frac{1}{h^{3N}} \int d\vec{p}_1 \int d\vec{p}_2 \cdots \int d\vec{p}_N \int d\vec{r}_1 \int d\vec{r}_2 \cdots \int d\vec{r}_N \tag{2.41}$$

where the integration domain of $\vec{r}_i$ is V and that of $\vec{p}_i$ is limited by the energy E as follows:

$$\frac{1}{2m}(p_1^2 + p_2^2 + \ldots + p_N^2) \leq E \tag{2.42}$$

The integrals over $\vec{p}_i (i = 1, \ldots, N)$ yield the volume of a sphere of radius $\sqrt{2mE}$ which is $A(\sqrt{2mE})^{3N}$ where A is a constant. The integrals over $\vec{r}_i$ give V^N. One finally has:

$$\phi(E) = \frac{V^N}{h^{3N}} A(2mE)^{3N/2} \tag{2.43}$$

One recovers here the same power of E as in the quantum case shown above [see (2.37)]. One obtains thus the following density of states

$$\rho(E) = \frac{3NV^N}{2h^{3N}} A(2m)^{3N/2} E^{3N/2-1} \tag{2.44}$$

When $N \gg 1$, one can take $\rho(E) \propto E^{3N/2}$.

2.5 Applications of the micro-canonical method

To study an isolated system, the first quantity to calculate is the number of microstates Ω as a function of E. Using Ω, one can calculate micro-canonical entropy S by (2.2). The micro-canonical temperature, pressure and chemical potential can be then calculated by (2.3)-(2.5).

As said above, for large systems at the thermodynamic limit, instead of using $\Omega(E)$ one uses the density of states $\rho(E)$ to calculate S. This is justified as follows. Since the energy E is always given with an uncertainty $\pm \Delta E/2$, the number of microstates having energy lying in the interval ΔE around E is equal to $\Omega(E)$. Using the definition of $\rho(E)$ one writes $\Omega(E) = \rho(E)\Delta E$. Micro-canonical entropy S is then

$$S = k_B \ln \Omega(E) = k_B \ln \rho(E) + k_B \ln \Delta E \qquad (2.45)$$

For a system of size N, the first term is proportional to N [see (2.38)] while the second term is proportional to $\ln N$ (because $\Delta E/E \simeq 1/\sqrt{N}$). When N becomes large, the second term is negligible with respect to the first one. One then has

$$S \simeq k_B \ln \rho(E) \qquad (2.46)$$

One studies in the following two frequently encountered examples. To explicit the steps to follow, one presents each example under the form question-answer.

2.5.1   *Example 1: two-level systems*

Consider a system of N independent and discernible particles, with $N \gg 1$. Each particle has two energy levels ϵ_1 and ϵ_2. The system is isolated with a total energy equal to E. Using the micro-canonical description,

a) Calculate the total number of accessible microstates $\Omega(E)$.
b) Calculate micro-canonical entropy $S(E)$.
c) Deduce micro-canonical temperature T as a function of ϵ_1 and ϵ_2. Find the number of particles on each level as functions of T, ϵ_1 and ϵ_2.
d) Find the limits at high and low temperatures of these numbers taking $\epsilon(> 0) = -\epsilon_1 = \epsilon_2$.
e) Find a relation between E and T.

Answer:

a) Energy of the system: $E = N_1\epsilon_1 + N_2\epsilon_2 = $ constant
Since $N = N_1 + N_2 = $ constant, one has $E = N_1\epsilon_1 + (N - N_1)\epsilon_2$.
As E is constant, N_1 is determined, so as N_2. The number of microstates Ω is equal to the number de ways to choose N_1 particles among N particles for level ϵ_1. One has $\Omega = \dfrac{N!}{N_1!N_2!} = \dfrac{N!}{N_1!(N-N_1)!}$.

b) Micro-canonical entropy is thus $S = k_B \ln \Omega = k_B[\ln N! - \ln N_1! - \ln(N - N_1)!$. Using the Stirling formula for $N \gg 1$:

$$\ln N! \simeq N \ln N - N \qquad (2.47)$$

one obtains

$$S \simeq k_B \left[N \ln N - N - N_1 \ln N_1 + N_1 - (N - N_1) \ln(N - N_1) + (N - N_1) \right]$$

$$= k_B \left[N \ln N - N_1 \ln \frac{N_1}{N - N_1} - N \ln(N - N_1) \right] \qquad (2.48)$$

where $N_1 = (E - N\epsilon_2)/(\epsilon_1 - \epsilon_2)$.

c)

$$T^{-1} = \left(\frac{\partial S}{\partial E} \right)_{V,N} = \left(\frac{\partial S}{\partial N_1} \frac{\partial N_1}{\partial E} \right)_{V,N}$$

$$= \frac{1}{\epsilon_1 - \epsilon_2} k_B \left[-\ln N_1 - 1 + \ln(N - N_1) + 1 \right]$$

$$= \frac{k_B}{\epsilon_1 - \epsilon_2} \ln \frac{N - N_1}{N_1} \qquad (2.49)$$

so that

$$\frac{\epsilon_1 - \epsilon_2}{k_B T} = \ln \frac{N - N_1}{N_1}$$

$$\frac{N_1}{N - N_1} = \exp[-\beta(\epsilon_1 - \epsilon_2)] \qquad (2.50)$$

where $\beta = (k_B T)^{-1}$. Therefore,

$$N_1 = N \frac{\exp[-\beta(\epsilon_1 - \epsilon_2)]}{1 + \exp[-\beta(\epsilon_1 - \epsilon_2)]}$$

$$= N \frac{1}{1 + \exp(-2\beta\epsilon)} \qquad (2.51)$$

d) At low temperatures ($\beta \to \infty$), $N_1 \to N$ and $N_2 \to 0$: all the particles occupy the lower level at $T = 0$. At high temperatures, $N_1 \to N/2$ so that $N_2 \to N/2$: the particles are on the two levels, half and half.

e) Energy E is connected to N_1 and N_2 by $E = -N_1\epsilon + N_2\epsilon = -N_1\epsilon + (N - N_1)\epsilon$. Replacing N_1 by (2.51), one obtains

$$E = -\epsilon(2N_1 - N) = -\epsilon \left[2N \frac{\exp(2\beta\epsilon)}{1 + \exp(2\beta\epsilon)} - N \right]$$

$$= -N\epsilon \left[\frac{2\exp(2\beta\epsilon) - 1 - \exp(2\beta\epsilon)}{1 + \exp(2\beta\epsilon)} \right]$$

$$= -N\epsilon \left[\frac{\exp(2\beta\epsilon) - 1}{1 + \exp(2\beta\epsilon)} \right]$$

$$= -N\epsilon \frac{\exp(\beta\epsilon) - \exp(-\beta\epsilon)}{\exp(\beta\epsilon) + \exp(-\beta\epsilon)}$$

$$= -N\epsilon \tanh(\beta\epsilon) \qquad (2.52)$$

2.5.2   Example 2: Classical ideal gas

Consider a classical ideal gas of N particles, of volume V. The gas is isolated with a total energy E. Using the micro-canonical description,

a) Show that the number of microstates having an energy $\leq E$ is given by $AV^N E^{\frac{3N}{2}}$ where A is a constant (not to be determined). Deduce the number of microstates having energy equal to E, provided $N \gg 1$.

b) Calculate the micro-canonical entropy. Deduce temperature T. Show that $E = \frac{3}{2} N k_B T$.

c) Calculate pressure p. Find the following equation of a classical ideal gas $pV = N k_B T$.

Answer:

a) The classical phase space has $6N$ dimensions ($3N$ for positions of particles and $3N$ for their momenta).

For a given E, the number of microstates having an energy $\leq E$ is

$$\phi(E) = \frac{1}{h^{3N}} \int_V d\vec{r}_1 \int_V d\vec{r}_2 ... \int_{p_1^2+p_2^2+....+p_N^2 \leq 2mE} d\vec{p}_1 d\vec{p}_2 ... d\vec{p}_N$$

$$= \frac{V^N}{h^{3N}} \int_{p_1^2+p_2^2+....+p_N^2 \leq 2mE} d\vec{p}_1 d\vec{p}_2 ... d\vec{p}_N \qquad (2.53)$$

where $\vec{p}_i$ is the momentum of the i-th particle. The integrals over $\vec{p}_i$ give the volume of the sphere in $3N$ dimensions which is proportional to its radius to the power of $3N$, namely $A(2mE)^{3N/2}$ where A is a constant [see (2.43)-(2.44) and A.1].

The number of microstates of energy equal to E, $\Omega(E)$, is the number of microstates on the surface of the sphere, namely the derivative of $\phi(E)$ with respect to E. One has

$$\Omega(E) = A \frac{3NV^N}{2h^{3N}} (2m)^{3N/2} E^{3N/2-1} \simeq A \frac{3NV^N}{2h^{3N}} (2m)^{3N/2} E^{3N/2} \qquad (2.54)$$

because $N \gg 1$.

b) Micro-canonical entropy is $S = k_B \ln \Omega(E) = k_B \ln V^N E^{3N/2} + B$ where B is a constant independent of E and V. The temperature is given by

$$T^{-1} = \frac{\partial S}{\partial E} = k_B \frac{3N}{2E} \qquad (2.55)$$

from which, one has $E = 3N k_B T/2$.

c) With (2.4), one obtains

$$\frac{p}{T} = \frac{\partial S}{\partial V} = \frac{k_B N}{V} \tag{2.56}$$

from which one gets $pV = Nk_B T$.

One recovers here the expression of the energy and the equation of state of a classical ideal gas found in thermodynamics.

2.6 Conclusion

In this chapter, the fundamental postulate of statistical mechanics has been introduced for isolated systems at equilibrium. This is the only postulate needed to study properties of such systems. Using this postulate, a relation between the statistical entropy S and the number of microstates Ω was established [Eq. (2.2)]. From S, micro-canonical temperature, pressure and chemical potential have been defined [Eqs. (2.3)-(2.5)]. These relations allow one not only to determine the direction of spontaneous evolutions but also to calculate principal properties of isolated systems which have been found almost empirically in thermodynamics. Some applications have been presented. Others are found in the list of problems below.

It is noted that for an isolated system, its energy is a parameter imposed from outside. However, its temperature is an internal parameter which is calculated using the micro-canonical entropy. One shows in the following chapters that the fundamental postulate can be used in a number of situations to find appropriate probability distributions for systems which are not isolated. One of such situations is the case of systems in contact with a very large heat reservoir which imposes on the system its temperature.

2.7 Problems

Problem 1. Joule expansion:

Consider the following experiment of Joule expansion of a classical ideal gas of N molecules. The gas initially occupies compartment A of volume V_0. After the removal of the wall, the gas occupies both compartments $A + B$ of total volume $2V_0$.

a) Describe the final state of the gas.

b) Determine the probability to find a molecule in the compartment A in the final state. Determine the probability to find all N molecules in A. What is the variation of the statistical entropy by the expansion?

Problem 2. Exchange of heat:

Consider an isolated system composed of two identical subsystems separated by a wall impermeable to particles and initially insulating. The energy levels of a particle in each subsystem are given by $\epsilon_{n_x, n_y} = \epsilon_0(n_x^2 + n_y^2)$, where n_x, n_y are integers ≥ 1 [this is the case of an electron in two dimensions (2.34), but taking only positive integers for simplicity].

a) Determine the energy levels of each particle and their degeneracy, namely the number of microstates for each level.

b) At the time $t = 0$ each of the two subsystems contains two discernible particles. The first subsystem has the energy $E_1 = 15\epsilon_0$ and the second one $E_2 = 10\epsilon_0$. Calculate the number of accessible microstates of subsystem 1, of subsystem 2 and of the total system.

c) The separating wall becomes now diathermic. It allows for an exchange of heat between subsystems. The total system evolves toward a new equilibrium.

- Which quantity is conserved during the spontaneous evolution?
- Determine the accessible microstates of the total system and their number.
- Calculate the possible energies of subsystem 1. Calculate the numbers of microstates of these energies.

Remark: Problem 14 is the version in three dimensions of the above Problem.

Problem 3. Distribution of energy on particles:

Consider an isolated system of N identical, independent, indiscernible particles of a total energy E. The i-th particle has an energy $\epsilon_i = n_i u$ where n_i is a positive integer or zero, and u a positive constant.

a) Calculate the number of accessible microstates.

b) Calculate micro-canonical entropy S as a function of E, u and N. Assuming that E/u and N are very large with respect to 1, show that S is approximatively given by $S = k[(E/u+N)\ln(E/u+N) - (E/u)\ln(E/u) - N\ln N]$. One uses the Stirling formula (2.47).

c) Calculate micro-canonical temperature T. Deduce E as a function of N, T and u. Find the limits of E/N at low and high temperatures. Plot E/N versus $k_B T/u$.

Problem 4. System of magnetic moments in a field:

Consider an isolated system of $N(\gg 1)$ free discernible particles having each an intrinsic magnetic moment $\vec{\mu}$. Under an applied magnetic field $\vec{B}$, if the magnetic moment of a particle is parallel to $\vec{B}$ its energy is equal to $-\mu B$, if it is antiparallel to $\vec{B}$, its energy is $+\mu B$. The system energy is thus $E = -(n_1 - n_2)\mu B$ where n_1 and n_2 are the numbers of moments parallel and antiparallel to $\vec{B}$, respectively.

a) The system is isolated with an energy equal to E. Calculate the total number of microstates $\Omega(E)$.

b) Find the expression of $\ln \Omega(E)$ using the Stirling formula (2.47).

c) Calculate micro-canonical entropy S and temperature T. Deduce n_1 as a function of T. Find the limits of n_1 at low and high temperatures.

Problem 5. Density of states in one and two dimensions:

Calculate the density of states $\rho(E)$ of a free particle in one and two dimensions.

Problem 6. Classical ideal gas in one and two dimensions:

Calculate the energy of a classical ideal gas of N particles in one and two dimensions.

Problem 7. Classical ideal gas:

Show that if one uses (2.46) for a molecule of a classical ideal gas, the obtained result is wrong. Explain why.

Problem 8. Classical harmonic oscillator:

Consider an isolated classical harmonic oscillator of energy E in three dimensions. Calculate its density of states. Using (2.46) to calculate its micro-canonical entropy and temperature T. Find the expression of E as a function of T. Is this expression correct ? Justify the answer.

Problem 9. System of classical harmonic oscillators:

Calculate the density of states of an isolated system composed of N classical harmonic oscillators of total energy E, using the following formula of the volume of a sphere of radius r in n dimensions: $V_n(r) = \pi^{n/2} r^n / \Gamma(n/2 + 1)$ where $\Gamma(m + 1) = m\Gamma(m)$, $\Gamma(1/2) = \sqrt{\pi}$, $\Gamma(m + 1) = m!$ (see Appendix A.1).

Find micro-canonical entropy S and temperature T. Calculate E as a function of T.

Problem 10. System of quantum harmonic oscillators:

Consider an isolated system composed of N independent quantum harmonic oscillators of total energy E in three dimensions with $N \gg 1$. Each oscillator has an energy given by $\epsilon_i = (n_i^x + n_i^y + n_i^z + 3/2)h\nu$ where h is the Planck constant, ν the frequency and n_i^α $(\alpha = x, y, z)$ a positive integer or zero.

a) Calculate the number of microstates of the system.

b) Find micro-canonical entropy S and temperature T, using the Stirling formula (2.47).

c) Find the expression of E as a function of T. Calculate the calorific capacity C_v. Plot $E(T)$ and $C_v(T)$ versus T.

d) Show that at high temperatures, one recovers the results of a system of classical harmonic oscillators found in Problem 9.

Problem 11. System composed of subsystems of quantum harmonic oscillators:

Consider an isolated system composed of two subsystems of N_1 and N_2 quantum harmonic oscillators of respective energies $E_1 = N_1 h\nu/2 + M_1 h\nu$ and $E_2 = N_2 h\nu/2 + M_2 h\nu$ $(N_1, N_2 \gg 1)$. where M_1 and M_2 are integers. One recalls that the energy of each oscillator is given by $e_i = h\nu/2 + n_i h\nu$ where ν is the frequency and n_i an integer (positive or zero).

a) Initially the two subsystems are separated by an insulating and immobile wall. Find the number of microstates of each subsystem. Find the total number of microstates and the micro-canonical entropy S_0 of the system.

b) One makes the wall diathermic but still impermeable to oscillators. The system is isolated with an energy E.

- Find the number of microstates as a function of E, N_1 and N_2.

- Calculate micro-canonical entropy S of the total system. Show that $S > S_0$. Comment.

- Find the energies of the two subsystems at equilibrium. Show that the energy per oscillator is the same in the two subsystems.

Fig. 2.3   Model of Frenkel's defects (Problem 12): Black and white sites represent ordered sites and interstices (defects), respectively.

Problem 12. Frenkel's defects:

Consider a crystal containing N regular sites ("ordered" sites) and N' defects ("interstices") shown respectively by black and white circles in Fig. 2.3. In the ground state, atoms occupy ordered sites and there are no atoms on the interstices. Let E_0 be the energy of the ground state. When an atom goes from an ordered site to an interstice, its energy increases by $\epsilon > 0$.

The crystal is isolated with an energy $E > E_0$. Let n be the number of defects corresponding to this energy.

a) Express n as a function of E, E_0 and ϵ

b) Calculate the number of microstates of the crystal.

c) Calculate micro-canonical entropy S and temperature T, using the Stirling formula (2.47) when $N, N' \gg 1$.

d) Find n as a function of T. Show that $n \simeq \sqrt{NN'} \exp[-\frac{\epsilon}{2k_B T}]$. Find the limits of n at low and high temperatures.

Problem 13. Schottky's defects:

Consider an isolated crystal of energy E, of $N + n$ sites among which N sites are occupied each by an atom and n sites are empty $(N, n \gg 1)$. This is an imperfect crystal. One assumes that each empty site (defect) has an energy $\epsilon > 0$ and that an occupied site has a zero energy.

a) Calculate micro-canonical entropy S. Deduce micro-canonical temperature T of the crystal.

b) Calculate the number of the empty sites n as a function of T for the case where $n \ll N$. Find the limits of n at low and high temperatures.

Problem 14. Exchange of heat in three dimensions:

Consider an isolated system composed of two identical cubic compartments separated by an insulating wall which is impermeable to particles. The energy levels of a particle in a compartment are given by $\epsilon_{n_x, n_y, n_z} = \epsilon_0(n_x^2 + n_y^2 + n_z^2)$ where n_x, n_y, and n_z are

Fig. 2.4 Schottky's defects (Problem 13): Black and white circles represent occupied and empty sites respectively.

integers ≥ 1 [see (2.34) for the origin of this expression, but for simplicity one takes only positive values of $n_i (i = x, y, z)$].

a) At the time $t = 0$, each compartment contains two discernible particles. The energies of the two compartments are $E_1 = 9\epsilon_0$ and $E_2 = 12\epsilon_0$. Calculate the number of microstates of each compartment and of the total system.

b) The wall between compartments 1 and 2 is now diathermic: there is exchange of heat. The total system evolves toward a new equilibrium:

- Indicate the quantity which is conserved during the evolution.
- Describe the microstates of the system and calculate its number.
- Indicate the direction (increase or decrease) of the evolution of the number of microstates. Estimate its variation.

c) One considers the total system at the new equilibrium.

- Give the probability of a given microstate.
- Determine the probability to find the compartment 1 with an energy of i) $6\epsilon_0$, ii) $9\epsilon_0$, and iii) $12\epsilon_0$.
- Calculate the average energies of compartments 1 and 2 at equilibrium. Determine the most probable energy for each of them.
- Determine the energy increase of compartment 1.

Problem 15. Binary alloy:

Consider a crystal of centered square lattice having N sites as shown in Fig. 2.5 where there are two types of sites: sites of type I (white circles) and of type II (black circles). Assume that there are two kinds of atom, A and B, which occupy the sites of this lattice. The total number of each kind is $N/2$. The interaction energies

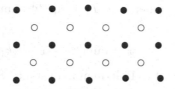

Fig. 2.5 Binary alloy (Problem 15): White and black circles represent sites of type I and II, respectively.

between two nearest neighbors are given by: ϵ if the atoms are of the same kind, ϕ if they are of different kinds. One assumes that $\epsilon > \phi$.

In a completely disordered state, half of A atoms occupy randomly black sites and the other half white sites. The same is for B atoms.

a) Explain why an atom is surrounded by atoms of the other kind in the ground state.

b) The crystal is isolated at an energy E. Let $N_{A,I}$ be the number of A atoms occupying sites of type I. One defines the variable x by

$$N_{A,I} = N(1+x)/4 \qquad (2.57)$$

- Determine the domain of variation of x. In which state is the crystal when $x = 0$? Calculate as a function of x the number of A atoms occupying the sites of type II ($N_{A,II}$). The same question is for B atoms.
  Assume $x > 0$ in the following.
- Calculate the probabilities as functions of x so that an A atom is found on a site of type I and on a site of type II, assuming that the probability for an atom to occupy a site is independent of the occupation of neighboring sites. The same question is for B atoms.
- Let $N_{A,A}$, $N_{B,B}$, and $N_{A,B}$ be the numbers of pairs AA, BB and AB, respectively. Calculate these quantities as functions of x. Show that $N_{A,A} = N(1 - x^2)/2$, $N_{B,B} = N(1 - x^2)/2$, $N_{A,B} = N(1 + x^2)$.
- Calculate E as a function of x, ϵ and ϕ. Show that E can be written as

$$E = N(\epsilon + \phi) - N(\epsilon - \phi)x^2 \qquad (2.58)$$

- Calculate $\Omega(E)$ the number of microstates of energy E. Find micro-canonical entropy S as a function of x.

- Calculate temperature T at equilibrium. Show the following relation

$$x = \tanh[2(\epsilon - \phi)x/(k_B T)] \tag{2.59}$$

where k_B is the Boltzmann constant. Show that x tends to 1 at low temperatures. Show that x tends to 0 (disordered state) when T tends to $T_c = 2(\epsilon - \phi)/k_B$.

Chapter 3

Systems at a Constant Temperature: Canonical Description

Most of experiments to study properties of materials are conducted at a given temperature. The system under study is maintained at a constant temperature by a thermostat. The thermostat is considered as a heat reservoir very large with respect to the system so that its temperature T does not vary by the contact with the system. The thermostat imposes its temperature on the system. This situation is called "canonical situation". One shows in this chapter that in such a situation, one can use the fundamental postulate to calculate the probability of a microstate of the system in contact with a thermostat. Microstates obeying such a "canonical" probability law form a canonical ensemble. This ensemble is used to calculate properties of the system at a constant temperature. There is of course an exchange of heat between the thermostat and the system but due to the huge energy reservoir of the thermostat this heat exchange does not change the temperature of the thermostat as said earlier. The description using the canonical ensemble is called "canonical description".

3.1 Canonical probability

When a system is in interaction with other systems, the fundamental postulate can be used if one considers the total system composed of all interacting subsystems as an isolated system. In the general case, it is impossible to study such a global system because there are too many involved parameters. However, in some cases one needs just a smaller number of parameters to calculate desired properties of a system interacting with outside. One of such situations is called "canonical situation" described in the Introduction. The knowledge of the temperature suffices to determine the main properties of the system as will be seen below.

One determines first of all the probability of a microstate l in the canonical situation described above. One assumes that the total system, composed of a heat reservoir τ and the system ζ one wishes to study, is isolated. The total energy is written as

$$E_{tot} = E_L + E_l \tag{3.1}$$

where E_L is the energy of the heat reservoir in its microstate L, and E_l the energy of the system in its microstate l. The microstate of the total system is thus characterized by two indices L and l. Let P_l be the probability of the microstate l of energy E_l. This probability is, by construction, equal to the probability of the state L of energy E_L of the reservoir. One writes

$$P_l = \frac{\Omega_\tau(E_L = E_{tot} - E_l)}{\Omega_{tot}} \tag{3.2}$$

where $\Omega_\tau(E_L = E_{tot} - E_l)$ is the number of microstates of the thermostat and Ω_{tot} the total number of microstates of the total system. The microcanonical entropy of the thermostat is written as

$$S_\tau(E_L = E_{tot} - E_l) = k_B \ln \Omega_\tau(E_L = E_{tot} - E_l) \tag{3.3}$$

In the canonical situation $E_l << E_L$: an expansion of $S_\tau(E_L = E_{tot} - E_l)$ around E_{tot} gives

$$S_\tau(E_L = E_{tot} - E_l) \simeq k_B \left[\ln \Omega_\tau(E_{tot}) + \frac{\partial \ln \Omega_\tau}{\partial E_L}(E_L - E_{tot}) \right] \tag{3.4}$$

$$= k_B \left[\ln \Omega_\tau(E_{tot}) - \frac{E_l}{k_B T} \right] \tag{3.5}$$

where (2.3) and (3.1) have been used in the last equality. From (3.2), one has

$$\ln P_l = \ln \Omega_\tau(E_L = E_{tot} - E_l) - \ln \Omega_{tot} \tag{3.6}$$

$$= \frac{S_\tau(E_L = E_{tot} - E_l)}{k_B} - \ln \Omega_{tot} \tag{3.7}$$

$$\simeq - \ln \Omega_{tot} + \ln \Omega_\tau(E_{tot}) - \frac{E_l}{k_B T} \tag{3.8}$$

The first two terms are independent of l, one thus has

$$P_l = A e^{-\frac{E_l}{k_B T}} \tag{3.9}$$

where A is a constant to be determined by the probability normalization:

$$\sum_l P_l = A \sum_l e^{-\frac{E_l}{k_B T}} = 1$$

Thus,

$$\frac{1}{A} = \sum_l e^{-\frac{E_l}{k_B T}} \tag{3.10}$$

The sum of the right-hand side is called "partition function" Z:

$$Z = \sum_l e^{-\frac{E_l}{k_B T}} \tag{3.11}$$

Note that this function depends on the external variables imposed on the system such as T, V (system volume) and N (number of particles of the system). One will see that this function plays a very important role in the calculation of fundamental properties of the system. Equation (3.9) is rewritten as

$$P_l = \frac{e^{-\beta E_l}}{Z(T, V, N)} \tag{3.12}$$

where $\beta = \frac{1}{k_B T}$. P_l is called canonical probability of the microstate l. The ensemble of the microstates obeying the canonical probability is called "canonical ensemble". The description of the properties of a system using (3.12) is called "canonical description".

3.2 Partition function

The partition function Z depends on T, V and N: $Z = Z(T, V, N)$. As Z plays an important role in the calculation of physical properties of a system, one presents in the following several practical forms of Z frequently used:

- The sum in (3.11) is performed over the microstates l, not on the energies E_l. If one wishes to sum on E_l, one has to take into account the degeneracy $g(E_l)$ of each energy level E_l, namely

$$Z = \sum_l e^{-\frac{E_l}{k_B T}} = \sum_{E_l} g(E_l) e^{-\frac{E_l}{k_B T}} \tag{3.13}$$

- When the dimension of the system is large, i. e. at the thermodynamic limit, the energy can be considered as a continuous variable. One can replace the sum on E_l by an integral in which the degeneracy is replaced by the density of states $\rho(E)$ (see chapter 2). One writes

$$Z = \sum_{E_l} g(E_l) e^{-\frac{E_l}{k_B T}} = \int_{E_0}^{\infty} \rho(E) e^{-\frac{E_l}{k_B T}} \, dE \qquad (3.14)$$

where E_0 is the minimum energy (ground-state energy) of the system.

- In the case of a system of classical particles [see section 2.4.3], the sum on the microstates is replaced by an integral in the phase space. For example, in three dimensions, one writes

$$Z = \sum_l e^{-\frac{E_l}{k_B T}}$$

$$= \frac{1}{h^{3N}} \int_V d\vec{r}_1 \int_V d\vec{r}_2 \ldots \int_V d\vec{r}_N \int d\vec{p}_1 \int d\vec{p}_2 \ldots \int d\vec{p}_N e^{-\frac{E}{k_B T}}$$

$$(3.15)$$

Note that the partition function is multiplicative. Consider two independent systems 1 and 2, maintained at the same temperature T. The energy of the total system in the microstate M is $E_M = E_l + E_k$ where l and k are the microstates of 1 and 2, respectively. The partition function of the total system is thus

$$Z = \sum_M e^{-\frac{E_M}{k_B T}} = \sum_l \sum_k e^{-\frac{E_l + E_k}{k_B T}} = \sum_l e^{-\frac{E_l}{k_B T}} \sum_k e^{-\frac{E_k}{k_B T}} = Z_1 Z_2 \quad (3.16)$$

One sees that the total partition function is the product of the partition functions of the two subsystems.

3.3 Properties of a system at a constant temperature

One shows in the following how to calculate various physical quantities using Z:

- *Average energy and calorific capacity:*

The average energy $\overline{E}$ of a system at T is given by

$$\overline{E}(T,N,V) = \sum_l E_l P_l = \sum_l \frac{E_l e^{-\frac{E_l}{k_B T}}}{Z}$$

$$= -\frac{1}{Z}\frac{\partial}{\partial \beta}\sum_l e^{-\beta E_l} = -\frac{1}{Z}\frac{\partial Z}{\partial \beta}$$

$$= -\left(\frac{\partial \ln Z}{\partial \beta}\right)_{V,N} \tag{3.17}$$

The calorific capacity is

$$C_V = \left(\frac{\partial \overline{E}}{\partial T}\right)_{V,N} = \frac{\partial}{\partial T}\left[-\frac{\partial \ln Z}{\partial \beta}\right]$$

$$= -\frac{\partial^2 \ln Z}{\partial \beta^2}\left(\frac{d\beta}{dT}\right) = \frac{1}{k_B T^2}\frac{\partial^2 \ln Z}{\partial \beta^2} \tag{3.18}$$

- *Canonical entropy:*

Replacing P_l in the statistical entropy (1.38) by the canonical probability (3.12), one obtains the canonical entropy

$$S(T,N,V) = -k_B \sum_l \frac{e^{-\beta E_l}}{Z}(-\beta E_l - \ln Z) \tag{3.19}$$

$$= k_B \left[\beta \overline{E} + \ln Z \sum_l \frac{e^{-\beta E_l}}{Z}\right] \tag{3.20}$$

$$= \frac{\overline{E}}{T} + k_B \ln Z \tag{3.21}$$

- *Free energy:*

The free energy F is defined by

$$F = -k_B T \ln Z \tag{3.22}$$

F is a function of T, V, and N, just as Z. Thus, the canonical entropy can be rewritten as

$$S = \frac{\overline{E}}{T} - \frac{F}{T} \tag{3.23}$$

or, more frequently,

$$F = \overline{E} - TS \tag{3.24}$$

Note that F is additive (extensive): from (3.16), one has

$$F = -k_B T \ln Z = -k_B T(\ln Z_1 + \ln Z_2) = F_1 + F_2 \tag{3.25}$$

- *Canonical pressure:*
  The canonical pressure p is defined by

$$p = -\left(\frac{\partial F}{\partial V}\right)_{T,N} \tag{3.26}$$

- *Canonical chemical potential:*
  The canonical chemical potential μ is defined by

$$\mu = \left(\frac{\partial F}{\partial N}\right)_{T,V} \tag{3.27}$$

3.4   Statistical distribution of an internal variable

The probability to find a property x is written as

$$P(x) = \frac{1}{Z}\sum_{j\in x}\exp(-\beta E_j) \tag{3.28}$$

where the sum runs over the microstates having the property x. One defines the "partial partition function" $Z^*(x)$ by

$$Z^*(x) = \sum_{j\in x}\exp(-\beta E_j) \tag{3.29}$$

from which, one has

$$P(x) = \frac{Z^*(x)}{Z}$$
$$\ln P(x) = \ln Z^*(x) - \ln Z \tag{3.30}$$

One defines the "partial free energy" $F^*(x)$ by

$$F^*(x) \equiv -k_B T \ln Z^*(x) \tag{3.31}$$

At the most probable value of x ($x = x_0$), the corresponding probability $P(x)$, namely $\ln P(x)$, is maximum. One has therefore

$$\left[\frac{\partial \ln P(x)}{\partial x}\right]_{x=x_0} = 0 \tag{3.32}$$

From (3.30), on obtains at $x = x_0$,

$$\left[-\beta\frac{\partial(F^*(x) - F)}{\partial x}\right]_{x=x_0} = 0 \tag{3.33}$$

As F is independent of x, one finally has

$$\left[\frac{\partial F^*(x)}{\partial x}\right]_{x=x_0} = 0 \tag{3.34}$$

This relation allows one to determine the most probable value of an internal variable x: one expresses F^* as a function of x and then looks for the value of x which minimizes $F^*(x)$.

An expansion of $\ln P(x)$ around x_0 gives

$$\ln P(x) = \ln P(x_0) + \left[\frac{\partial \ln P(x)}{\partial x}\right]_{x=x_0} (x - x_0)$$

$$+ \frac{1}{2}\left[\frac{\partial^2 \ln P(x)}{\partial x^2}\right]_{x=x_0} (x - x_0)^2 + \dots$$

$$\simeq \ln P(x_0) - \frac{1}{2\sigma^2}(x - x_0)^2 \tag{3.35}$$

where (3.32) has been used and one has put

$$[\sigma^2]^{-1} \equiv -\left[\frac{\partial^2 \ln P(x)}{\partial x^2}\right]_{x=x_0} \tag{3.36}$$

From (3.35), one has

$$P(x) = P(x_0) \exp\left[-\frac{1}{2\sigma^2}(x - x_0)^2\right] \tag{3.37}$$

Under this form, one sees that x fluctuates around its most probable value x_0 according to a Gaussian law. Note that this result has been demonstrated for a large system.

3.5 Spontaneous evolution of a canonical system

3.5.1 *Criterion for equilibrium*

One shows in the following that during a spontaneous evolution of a system in the canonical situation after a modification of an external constraint, the system free energy F diminishes. The new equilibrium is reached when F is minimum.

Consider a total system composed of a system in contact with a thermostat. One modifies now one of the external parameters of the system (volume, number of particles, ...), then one leaves the system to evolve.

The thermostat will not change due to its huge dimension. However, since the total system is isolated, the micro-canonical entropy of the total system should increase during the evolution (see chapter 2). The new equilibrium is reached when that entropy is maximum. One gives the details of the calculation: let S_τ and S be the micro-canonical entropies of the thermostat and of system, L and l their microstates. The probability of the microstate l, L of the total system is

$$P_{l,L} = P_l P_L \simeq P_l(E_l) P_L^*(E_L = E_{total} - E_l) \tag{3.38}$$

where P_L of the thermostat is replaced by a micro-canonical probability because the thermostat is so large that it is considered as an isolated system. In this hypothesis, one writes

$$\ln P_{l,L} = \ln P_l(E_l) + \ln P_L^*(E_L = E_{total} - E_l) \tag{3.39}$$

so that the micro-canonical entropy of the total system is

$$\begin{aligned}
S_{total} &= -k_B \sum_{l,L} P_{l,L} \ln P_{l,L} = -k_B \sum_{l,L} P_l(E_l) P_L^*(E_L = E_{total} - E_l) \\
&\quad \times [\ln P_l(E_l) + \ln P_L^*(E_L = E_{total} - E_l)] \\
&= -k_B \sum_l P_l \ln P_l(E_l) \sum_L P_L^*(E_L = E_{total} - E_l) \\
&\quad - k_B \sum_l P_l \sum_L P_L^*(E_L = E_{total} - E_l) \ln P_L^*(E_L = E_{total} - E_l) \\
&= S + \sum_l S_\tau(E_L = E_{total} - E_l) P_l \tag{3.40}
\end{aligned}$$

where one has used in the line before the last line $\sum_L P_L^*(E_L = E_{total} - E_l) = 1$, for any l. Now, one expands $S_\tau(E_L = E_{total} - E_l)$ around E_{total}:

$$S_\tau(E_L = E_{total} - E_l) \simeq S_\tau(E_{total}) + \frac{\partial S_\tau(E_{total})}{\partial E_L}(E_L - E_{total})$$

$$= S_\tau(E_{total}) - \frac{1}{T} E_l \tag{3.41}$$

Using this relation in (3.40), one has

$$S_{total} = S_\tau(E_{total}) + S - \sum_l \frac{1}{T} E_l P_l = S_\tau(E_{total}) + S - \frac{\overline{E}}{T}$$

$$S_{total} = S_\tau(E_{total}) - \frac{F}{T} \tag{3.42}$$

where one has used (3.24) in the last line. In a spontaneous evolution, S_{total} increases: since the first term of the right-hand side of (3.42) is a constant, the free energy F (second term) should decrease. The new equilibrium is attained when F is minimum.

3.5.2 *Direction of spontaneous evolution*

Consider two systems initially separated by a fixed wall impermeable to particles. The two systems are maintained at the same temperature T by a thermostat (see Fig.3.1). One now makes the wall mobile and permeable to particles.

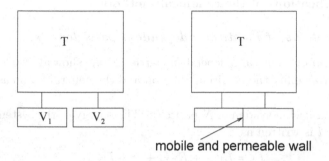

Fig. 3.1 Left: Two independent systems, maintained at the temperature T by a thermostat. Right: Exchange of volume and of particles between the two systems via a mobile and permeable wall.

The two systems exchange their volume and particles but conserving the total volume $V = V_1 + V_2$ and the total number of particles $N = N_1 + N_2$. During the exchange, they evolve toward a new equilibrium state: one determines in the following the direction of this evolution.

One has seen above that during a spontaneous evolution, the system free energy F decreases. Since there are only two independent variables V_1 and N_1 (because $dV_2 = -dV_1$ and $dN_2 = -dN_1$), one writes

$$0 > dF$$

$$0 > \left(\frac{\partial F}{\partial V_1}\right)_{N_1,T} dV_1 + \left(\frac{\partial F}{\partial N_1}\right)_{V_1,T} dN_1$$

$$0 > \frac{\partial(F_1 + F_2)}{\partial V_1} dV_1 + \frac{\partial(F_1 + F_2)}{\partial N_1} dN_1$$

$$0 > \left[\left(\frac{\partial F_1}{\partial V_1}\right)_{N_1,T} - \left(\frac{\partial F_2}{\partial V_2}\right)_{N_2,T}\right] dV_1$$

$$+ \left[\left(\frac{\partial F_1}{\partial N_1}\right)_{V_1,T} - \left(\frac{\partial F_2}{\partial N_2}\right)_{V_2,T}\right] dN_1$$

$$0 > [-p_1 + p_2] \, dV_1 + [\mu_1 - \mu_2] \, dN_1 \tag{3.43}$$

To satisfy this relation, each term should be negative. If $p_1 < p_2$, then $dV_1 < 0$: the system with smaller canonical pressure diminishes its volume, as experimentally expected. If $\mu_1 < \mu_2$, then $dN_1 > 0$: the system with larger canonical chemical potential looses particles.

3.6   Applications of the canonical method

3.6.1   *Systems of identical independent particles*

For a system of identical independent particles, one shows below that one can greatly simplify the partition function and the general formulas of section 3.3.

Consider such a system of N particles. The energy of the system in the microstate l is written as

$$E_l = E_{1,\lambda_1} + E_{2,\lambda_2} + ... + E_{N,\lambda_N} \tag{3.44}$$

where E_{i,λ_i} $(i = 1, ..., N)$ is the energy of the i-th particle which occupies the state λ_i. Since all particles are identical, it is obvious that each particle takes its state on the same "list of states". λ_i is an "individual state" belonging to the list. Therefore, the total partition function Z can be factorized:

$$
\begin{aligned}
Z &= \sum_l \exp(-\beta E_l) = \sum_{\lambda_1,...,\lambda_N} \exp\left[-\beta(E_{1,\lambda_1} + ... + E_{N,\lambda_N})\right] \\
&= \sum_{\lambda_1} \exp(-\beta E_{1,\lambda_1}) ... \sum_{\lambda_N} \exp(-\beta E_{N,\lambda_N}) \\
&= z_1 ... z_N = z^N
\end{aligned}
\tag{3.45}
$$

where z_i $(i = 1, ..., N)$ is the partition function of the i-th particle. Since the particles have the same list of states, their partition functions are identical. One has used $z_i \equiv z$ for all i.

In the expression (3.45) one has assumed that the particles are discernible: for each of them, the sum runs over all individual states. However, if the particles are indiscernible, then one has to exclude in the sum the "equivalent" states. This is understood by the following argument. If one defines a microstate of the system by the numbers of the particles in each individual states (briefly called hereafter "occupation numbers") $\{N_{\lambda_i}\}$ of $\{\lambda_i\}$ with $i = 1, ..., N$, then each repartition of the particles in individual states is one microstate of the system. Permutations among particles in a given repartition do not create new microstates because the occupation numbers do not change. One did not exclude these permutations in the calculation leading to (3.45): one has "circulated" each particle on all individual states: in doing so one counts G times each repartition, where G is given by the formula (1.34) which is the number of permutations of discernible particles in a repartition $(N_{\lambda_1}, N_{\lambda_2}, ...)$. One has

$$G = \frac{N!}{N_{\lambda_1}! N_{\lambda_2}! ...} \qquad (3.46)$$

Therefore, one should divide Z by G in the case of indiscernible particles:

$$Z = \frac{z^N}{G} \qquad (3.47)$$

Let us examine the following cases

- The case of bosons (particles having integer spins): In this case $N_{\lambda_1} = 0, 1, 2, ...$ with the constraint $\sum_{\lambda_i} N_{\lambda_i} = N$. The formula (3.47) is valid with G given by (3.46).
- The case of fermions (particles having half-integer spins): In this case, $N_{\lambda_i} = 0$ or 1. The formula (3.47) becomes

$$Z = \frac{z^N}{N!} \qquad (3.48)$$

- The case of dilute systems: When there are very few particles to occupy individual states λ_i (dilute systems), one has $\overline{N}_{\lambda_i} \ll 1$. Then, $N_{\lambda_i}! \simeq 1$. Equation (3.47) becomes, for both cases of bosons and fermions,

$$Z \simeq \frac{z^N}{N!} \qquad (3.49)$$

This approximation is called "Maxwell-Boltzmann approximation". Note that the condition $\overline{N}_{\lambda_i} \ll 1$ means a dilute density, i. e. the average distance between particles is large so that interaction between them can be neglected. This corresponds to the case of a classical ideal gas. One shows in Problem 2 that the condition $\overline{N}_{\lambda_i} \ll 1$ also corresponds to a system, including boson and fermion cases, at high temperatures.

Let us come back to the representation of microstates by occupation numbers of individual states λ_i discussed above. The energy of the microstate l in a repartition of particles on the individual states is written by

$$E_l = \sum_{\lambda_i} N_{\lambda_i} \epsilon_{\lambda_i} \qquad (3.50)$$

where ϵ_{λ_i} is the energy of the individual state λ_i. Replacing E_l in Z one obtains

$$Z = \sum_l \exp(-\beta E_l) = \sum_{\{N_{\lambda_i}\}} \exp\left(-\beta \sum_{\lambda_i} N_{\lambda_i} \epsilon_{\lambda_i}\right) \qquad (3.51)$$

where the sum in the argument of the exponential is performed over all levels λ_i for a given particle repartition, while the sum $\sum_{\{N_{\lambda_i}\}}$ is made over all particle repartitions. The average number of particles in the individual state λ_j is

$$
\begin{aligned}
\overline{N}_{\lambda_j} &= \frac{1}{Z} \sum_{\{N_{\lambda_i}\}} N_{\lambda_j} \exp\left[-\beta \sum_{\lambda_i} N_{\lambda_i} \epsilon_{\lambda_i}\right] \\
&= -\frac{1}{Z} \frac{1}{\beta} \frac{\partial}{\partial \epsilon_{\lambda_j}} \sum_{\{N_{\lambda_i}\}} \exp\left[-\beta \sum_{\lambda_i} N_{\lambda_i} \epsilon_{\lambda_i}\right] \\
&= -k_B T \frac{1}{Z} \frac{\partial Z}{\partial \epsilon_{\lambda_j}} \\
&= -k_B T \frac{\partial \ln Z}{\partial \epsilon_{\lambda_j}} \qquad (3.52)
\end{aligned}
$$

3.6.2   *Classical ideal gas*

Consider an ideal gas of N classical particles, of volume V, maintained at the temperature T by a thermostat. Using the canonical description presented in this chapter, one recovers easily the well-known equation of the average energy $\overline{E} = 3N k_B T / 2$ and the equation of state $pV = N k_B T$ as seen below.

Since the particles are identical, independent, indiscernible, the partition function is given by $Z = z^N / N!$ where z is the partition function of a particle [see (3.49)]. In three dimensions, z is given by [see (3.15)]

$$z = \frac{1}{h^3} \int_V d\vec{r} \int d\vec{p}\,\exp[-\beta E] = \frac{1}{h^3} \int_V d\vec{r} \int d\vec{p}\,\exp[-\beta p^2/(2m)]$$

$$= \frac{V}{h^3} \int_{-\infty}^{\infty} \exp[-\beta p_x^2/(2m)]dp_x \int_{-\infty}^{\infty} \exp[-\beta p_y^2/(2m)]dp_y$$

$$\times \int_{-\infty}^{\infty} \exp[-\beta p_z^2/(2m)]dp_z$$

$$= \frac{V}{h^3} \left[\sqrt{\frac{2\pi m}{\beta}}\right]^3$$

$$= \frac{V}{h^3} \left[\frac{2\pi m}{\beta}\right]^{3/2} \tag{3.53}$$

where one has used the Gauss integral for integrations over p_x, p_y and p_z (see Appendix A). Using (3.49) and (3.17), one obtains

$$\overline{E} = -\left(\frac{\partial \ln Z}{\partial \beta}\right)_{N,V} = -\frac{\partial}{\partial \beta}[N \ln z - \ln N!]$$

$$= \frac{3N}{2\beta} = \frac{3}{2}Nk_BT \tag{3.54}$$

Using (3.24) and (3.26), one gets the canonical pressure

$$p = -\left(\frac{\partial F}{\partial V}\right)_{N,T} = \frac{\partial k_B T \ln Z}{\partial V}$$

$$= Nk_BT\frac{\partial}{\partial V}\left[\ln V + \ln \frac{1}{h^3}\left(\frac{2\pi m}{\beta}\right)^{3/2}\right]$$

$$= \frac{Nk_BT}{V} \tag{3.55}$$

One recovers here by the canonical method the well-known results of the classical ideal gas.

3.6.3   Two-level systems

One has studied a two-level system with the micro-canonical method in section 2.5.1. One shows here that one can also use the canonical method to find those results.

Consider a system of N identical and independent particles. Each particle has two energy levels $\epsilon_1 = -\epsilon$ and $\epsilon_2 = \epsilon$ with $\epsilon > 0$. The system is in contact with a thermostat at temperature T. As always in the canonical method, first one tries to calculate the partition function Z from which one obtains other physical properties.

If the particles are indiscernible (for example a gas of free electrons of spin 1/2 in an applied magnetic field as seen in Problem 4 of chapter 2), the partition function of the system is written as $Z = z^N/N!$ where z is the partition function of a particle [see (3.49)]. If the particles are discernible (for example, atoms localized on a lattice), one then uses (3.45). One has

$$z = \sum_l e^{-\beta E_l} = e^{-\beta \epsilon_1} + e^{-\beta \epsilon_2}$$

$$= e^{\beta \epsilon} + e^{-\beta \epsilon}$$

$$= 2 \cosh(\beta \epsilon) \tag{3.56}$$

Assuming that the particles are indiscernible, one has

$$\overline{E} = -\frac{\partial \ln Z}{\partial \beta} = -\frac{\partial}{\partial \beta}[N \ln z - \ln N!]$$

$$= -N \frac{\partial}{\partial \beta} \ln[2 \cosh(\beta \epsilon)] = -N\epsilon \frac{2 \sinh(\beta \epsilon)}{2 \cosh(\beta \epsilon)}$$

$$= -N\epsilon \tanh(\beta \epsilon) \tag{3.57}$$

One recovers here with the canonical method the result (2.52) obtained by the micro-canonical method. It is obvious that physical properties of a system do not depend on the method of investigation.

At low temperatures, one sees that $\overline{E} \rightarrow -N\epsilon$: all particles occupy the low energy level $-\epsilon$. At high temperatures, $\overline{E} \rightarrow 0$ so that half of the system particles occupy the level $-\epsilon$ and the other half the level ϵ.

3.6.4   *Theorem of equipartition of energy*

In thermodynamics, the temperature is defined as a quantity which is proportional to the kinetic energy of a particle. One writes $E = AT$ where A is the constant of proportionality which is equal to $\frac{1}{2}k_B$ for each degree of freedom of the particle. This yields immediately that the energy of a free particle in three dimensions is equal to $E = \frac{3}{2}k_BT$, and that of N particles is $E = \frac{3}{2}Nk_BT$ which is the energy of a classical ideal gas. One shows below that the above definition of T in thermodynamics is justified by the theorem of equipartition of energy. As will be seen, this theorem is valid not only for the kinetic energy but also for the potential energy if this potential depends on the square of any degree of freedom (position for instance) of the particle.

The theorem is stated as follows:

If the energy of a particle is a sum of terms each of which is quadratic in a degree of freedom of the particle, then each term gives rise to a contribution equal to $\frac{1}{2}k_BT$ to the average energy regardless of the physical origin of that term.

Examples:

- Kinetic energy of a free particle in three dimensions (3D) $E = \frac{1}{2m}p^2$
  $\rightarrow$ average energy at T is $\overline{E} = \frac{3}{2}k_BT$
- Energy of a classical 3D harmonic oscillator $E = \frac{1}{2m}p^2 + \frac{K}{2}r^2$ (K: coupling, r: distance to the equilibrium position) $\rightarrow$ average energy at T is $\overline{E} = \frac{3}{2}k_BT + \frac{3}{2}k_BT = 3k_BT$ (3 degrees of freedom for p and 3 for r) $\rightarrow$ the same contribution for p and r, independent of coefficients $\frac{1}{2m}$ and $\frac{K}{2}$.

Demonstration: One assumes that the energy E has the following form

$$E = \sum_{i=1}^{N} A_i v_i^2 \tag{3.58}$$

where v_i is a parameter associated with the degree of freedom i of a physical quantity. It can be a component of the momentum $\vec{p}$ or of the position $\vec{r}$. A_i is the corresponding coefficient. One calculates now $\overline{E}$ as follows:

$$\overline{E} = \sum_{i=1}^{N} A_i \overline{v_i^2} \tag{3.59}$$

$$A_i \overline{v_i^2} = \frac{1}{z} \int \frac{dv_i}{\sqrt{h}} A_i v_i^2 \exp(-\beta A_i v_i^2)$$

where z is the partition function for a degree of freedom [see (3.15)] given by

$$z = \frac{1}{\sqrt{h}} \int_{-\infty}^{\infty} dv_i \exp(-\beta A_i v_i^2) = \frac{1}{\sqrt{h}} \sqrt{\frac{\pi}{\beta A_i}} \tag{3.60}$$

where integral I_0 of Appendix A has been used. One has, on the other hand,

$$\int_{-\infty}^{\infty} \frac{dv_i}{\sqrt{h}} A_i v_i^2 \exp(-\beta A_i v_i^2) = \frac{A_i}{\sqrt{h}} \frac{1}{2\sqrt{\beta A_i}} \sqrt{\frac{\pi}{\beta A_i}} \tag{3.61}$$

where integral I_2 of Appendix A has been used. From the above results, one obtains

$$A_i \overline{v_i^2} = \frac{1}{2\beta} = \frac{k_B T}{2} \tag{3.62}$$

Replacing in (3.59), on gets the average energy $\overline{E} = k_B T / 2 \sum_{i=1}^{N} 1 = N k_B T / 2$ which is independent of the coefficients A_i.

More applications of the canonical method are found in the list of Problems given below.

3.7   Conclusion

Using the fundamental postulate for an isolated system composed of a system in contact with a huge thermostat which imposes its temperature T on the system, one has found the "canonical probability" of a microstate of the system. The difference with the micro-canonical case is that for an isolated system the energy is a constant and the temperature is an internal variable which fluctuates, while for a canonical system the temperature is constant and the energy is an internal quantity which fluctuates around its average value.

Using the canonical probability one calculates physical properties of the system via a quantity called "partition function" Z. One has defined the free energy F which plays an important role in the determination of the the system equilibrium: F decreases during a spontaneous evolution and is at its minimum when the new equilibrium is reached. Note that for an isolated system, it is the maximum of the micro-canonical entropy which indicates the system equilibrium (see section 2.3).

Let us give the last remark on the micro-canonical and canonical methods. The difference in their description of a system comes from the condition in which the system is found: isolated or in contact with thermostat. However, physical properties of the system are equivalent in the thermodynamic limit. At this limit, fluctuations of T in an isolated system tend to zero, namely T is "constant". If now this system is put into contact with a thermostat of temperature T, nothing happens for the system, because it is already at the thermostat temperature, namely already in an equilibrium canonical situation. Consider now a canonical system of thermodynamic size: fluctuations in energy become zero so that its energy is considered as a constant E. It is therefore equivalent to an isolated system of energy E. The equivalence of canonical et micro-canonical descriptions allows one to choose the description which is convenient to study a system according to the condition in which the system is found.

3.8 Problems

Problem 1. Show that the calorific capacity C_V is given by

$$C_V = \frac{\overline{(E - \overline{E})^2}}{k_B T^2} \tag{3.63}$$

where T, E and $\overline{E}$ are the temperature, energy and average energy of the system.

Problem 2. Maxwell-Boltzmann approximation:

Consider a classical ideal gas. Show that the condition $\overline{N}_{\lambda_i} \ll 1$ for the validity of the Maxwell-Boltzmann approximation (3.49) corresponds to the high-temperature regime.

Problem 3. Classical ideal gas in the gravitational field:

Consider a classical ideal gas of N particles of mass m contained in a cylinder of section A with an infinite height in the gravitational field. The gas is at equilibrium. Calculate the partition function, the average energy, the calorific heat, the canonical entropy and the free energy.

Problem 4. Bi-dimensional classical ideal gas:

Consider a bi-dimensional classical ideal gas of N particles on a surface S, maintained at a constant temperature T.

a) Show that the density of states ρ is equal to a constant A to be determined.

b) Calculate the partition function z for a particle. Find the partition function for N particles.

c) Calculate the average energy of the gas.

d) Find the following equation of state: $pS = Nk_B T$.

Problem 5. Classical harmonic oscillators:

Consider a system of independent classical harmonic oscillators of mass m, in three dimensions, maintained at the temperature T. Calculate the partition function. Find the average energy and the calorific capacity.

Problem 6. Quantum harmonic oscillators:

Consider a system of independent quantum harmonic oscillators of the same pulsation ω in three dimensions, maintained at the temperature T. Calculate the partition function. Find the average energy and the calorific capacity.

Note: This model is called the "Einstein model" which was initially used to study vibrations of interacting atoms on a lattice (see chapter 9).

Problem 7. Three-level system:

Consider a system of N independent particles, of volume V, maintained at the temperature T. Each particle has three non degenerate states of energy $-\epsilon, 0, \epsilon$ $(\epsilon > 0)$.

a) Calculate the partition function of the system and its average energy.

b) Find the probability of each of the three energy levels. Calculate the numbers of particles N_1 , N_2 and N_3 of these levels as functions of T, ϵ and N.

c) Find the limits at low and high temperatures of N_1 , N_2 and N_3. Comment.

Problem 8. Frenkel's defects:

One considers again Problem 12 of chapter 2 but assumes now that the system is maintained at the temperature T. One supposes in addition $E_0 = 0$ for simplicity (this will not alter the final result).

a) Find the probability for the system to have the energy $E_n = n\epsilon$.

b) Calculate E_M, the most probable energy of the system. Compare the result with that obtained by the micro-canonical method used in Problem 12 of chapter 2.

c) Is it possible to calculate the partition function Z for N and N' large but finite ? In the case where $N, N' \to \infty$, calculate Z and $\overline{E}$. Do the results correspond to the initial problem ?

Problem 9. Schottky's defects:

One considers again Problem 13 of chapter 2 but assumes now that the system is maintained at the temperature T (canonical situation).

a) Find the probability to have the energy of vacancies equal to $E_n = n\epsilon$.

b) Calculate E_M, the most probable energy of the system. Compare the result with that obtained by the micro-canonical method used in Problem 13 of chapter 2.

Problem 10. Velocity distribution in a classical ideal gas:

Let n be the density of molecules per volume unit of a classical ideal gas maintained at the temperature T.

a) Write down the expression of $dN(v_x, v_y, v_z)$, the number of molecules whose velocity Cartesian components are found between v_x and $v_x + dv_x$, v_y and $v_y + dv_y$, and v_z and $v_z + dv_z$. Calculate $dN(v)$ the number of molecules whose velocity modulus is found between v and $v + dv$.

b) Find for a Neon gas ($M = 20.2$ g) at $T = 300$ K

- the most probable velocity v_p
- the average velocity $\bar{v}$
- $\sqrt{\bar{v^2}}$

c) Determine the number of atoms of a Neon gas whose velocities are inside the interval $v_p \pm 10\%$.

Problem 11. System of spins in an applied magnetic field:

Consider a solid composed of N independent atoms each of which bears a spin $\vec{S}_i$ ($i = 1, .., N$). One applies a magnetic field $\vec{B}$ parallel to the Oz axis. One assumes that the z-component of $\vec{S}_i$ has four values $S_i^z = -3/2, -1/2, 1/2, 3/2$. The Zeeman energy of the i-th atom is given by $E_i = g\mu_B \vec{B} \cdot \vec{S}_i$ where g and μ_B are the Landé factor and the Bohr magneton, respectively. The solid is at equilibrium at the temperature T.

a) Calculate the partition function Z of the solid. Show that

$$Z = \frac{\sinh^N(2\alpha)}{\sinh^N(\alpha/2)} \tag{3.64}$$

where $\alpha = \frac{g\mu_B B}{k_B T}$ and k_B is the Boltzmann constant.

b) Calculate the average energy and the specific heat C_V of the solid.

c) Calculate the canonical entropy S. Define physically the low- and high-temperature limits. Comment on the values of S at these limits.

d) Find the average magnetic moment per atom. Find its value at the limit $\alpha << 1$. Comment.

Problem 12. Equilibrium of a vapor-solid system:

Consider a solid in three dimensions composed of atoms assimilated to independent quantum harmonic oscillators. In the Einstein model, all oscillators have the same frequency ν. One supposes that the energy needed to transform an atom at the surface into a free atom is ϕ (> 0). The solid is maintained at the temperature T. Due to thermal agitation, a number of atoms quits the surface to form a vapor above the surface. Assume that this vapor is an ideal gas.

a) Calculate the partition function Z_s of the solid as a function of T, ν, ϕ, and N_s (number of atoms of the solid).

b) Calculate the partition function Z_g of the gas as a function of T and N_g (number of atoms of the gas).

c) Suppose that the global system (solid and gas) conserves the total number of atoms. Calculate N_g when the global system is at equilibrium at T.

d) Calculate the pressure of the vapor at equilibrium.

Problem 13. Harmonic oscillators:

Consider an ensemble of N independent harmonic oscillators of pulsations ω_j ($j = 1, .., N$) in three dimensions, maintained at the temperature T.

a) Calculate the partition function of an oscillator of pulsation ω_j in the following cases: i) the oscillator is a classical one with a mass m, ii) the oscillator is a quantum one. Show that these results are equivalent at high temperatures.

b) Suppose in this question that all pulsations are identical and equal to ω_E. Calculate in the quantum case the partition function of the system. Find the average energy, the calorific capacity, the free energy and the canonical entropy of the system.

c) In this question, the pulsations ω_j are different.

- Calculate the free energy F of the system. Show that F is given by

$$F = 3k_BT \sum_j \ln[2\sinh(\hbar\omega_j/2k_BT)] \qquad (3.65)$$

where k_B is the Boltzmann constant.

- For $N >> 1$, one can write

$$F = F(T, V, N) = 3k_BT \int_0^\infty \ln[2\sinh(\hbar\omega/2k_BT)]g(\omega)d\omega \qquad (3.66)$$

where $g(\omega)$ is the number of pulsations between ω and $\omega + d\omega$, V the system volume. Show that the average energy is given by

$$\overline{E} = \int_0^\infty \epsilon(\omega, T)g(\omega)d\omega \qquad (3.67)$$

where $\epsilon(\omega, T)$ is to be determined. Give the meaning of $\epsilon(\omega, T)$. Find the integral which gives the calorific capacity C of the system.

d) Suppose that $g(\omega) = 3\omega^2/\omega_D$ if $\omega < \omega_D$ and $g(\omega) = 0$ if $\omega > \omega_D$ where ω_D is a constant (called "Debye pulsation"). Calculate the integral of C in the following cases: i)$\hbar\omega_j/k_BT \ll 1$, ii) $\hbar\omega_j/k_BT \gg 1$.

Chapter 4

Open Systems at Constant Temperature: Grand-Canonical Description

4.1 Introduction

In this chapter, one studies the case of a system in contact with a reservoir of heat and particles which is very large with respect to the size of the system. By its large dimension, the reservoir imposes its temperature T and its chemical potential μ on the system during the exchange of heat and particles between them. In such a situation, one can use the fundamental postulate for the total system composed of the reservoir and the system to calculate the probability of a microstate of the system. This situation is called "grand-canonical situation" and the microstates obeying such a probability law form the so-called "grand-canonical ensemble". The description using the grand-canonical ensemble is called "grand-canonical description".

One shows in the following that properties of a system in the grand-canonical situation can be calculated as functions of two parameters T and μ.

4.2 Grand-canonical probability

Assuming that the total system composed of the reservoir (called τ) and the system under study (called ζ) is isolated, one has

$$E_{tot} = E_L + E_l = \text{constant} \qquad (4.1)$$

$$N_{tot} = N_L + N_l = \text{constant} \qquad (4.2)$$

where E_L and N_L are respectively the energy and the number of particles of τ in its microstate L, while E_l and N_l are the energy and the number of particles of ζ in its microstate l. Since the total system is isolated, the probability P_l to find the system with an energy E_l and N_l particles is equal to the probability to find the reservoir with an energy $E_L = E_{tot} - E_l$ and a number of particles $N_L = N_{tot} - N_l$. One writes

$$P_l = A\Omega_\tau(E_L = E_{tot} - E_l, N_L = N_{tot} - N_l) \qquad (4.3)$$

where $\Omega_\tau(E_L = E_{tot} - E_l, N_L = N_{tot} - N_l)$ is the number of microstates of τ having $E_L = E_{tot} - E_l$ and $N_L = N_{tot} - N_l$. A is a constant to be determined by the normalization of P_l. One expresses (4.3) as a function of the reservoir micro-canonical entropy S_τ [see (2.2)]:

$$\begin{aligned}
\ln P_l &= \ln A + \ln \Omega_\tau(E_L = E_{tot} - E_l, N_L = N_{tot} - N_l) \\
&= \ln A + S_\tau(E_L = E_{tot} - E_l, N_L = N_{tot} - N_l)/k_B \\
P_l &= A\exp[S_\tau(E_L = E_{tot} - E_l, N_L = N_{tot} - N_l)/k_B] \qquad (4.4)
\end{aligned}$$

Since the reservoir is very large with respect to the system, one can expand S_τ around E_{tot} and N_{tot} as follows:

$$\begin{aligned}
S_\tau(E_L = E_{tot} - E_l, N_L = N_{tot} - N_l) &\simeq S_\tau(E_{tot}, N_{tot}) + \frac{\partial S_\tau}{\partial E_L}(E_L - E_{tot}) \\
&\quad + \frac{\partial S_\tau}{\partial N_L}(N_L - N_{tot}) \\
&= S_\tau(E_{tot}, N_{tot}) - \frac{E_l}{T} + \frac{\mu N_l}{T} \qquad (4.5)
\end{aligned}$$

where (2.3), (2.5) , (4.1) and (4.2) have been used. In the above expression μ and T are the chemical potential and the temperature of the reservoir. Replacing (4.5) in (4.4) one obtains

$$P_l = Ce^{-\beta(E_l - \mu N_l)} \qquad (4.6)$$

where C contains all constant factors independent of l. C is determined by the normalization as follows:

$$\sum_l P_l = C\sum_l e^{-\beta(E_l - \mu N_l)} = 1$$

One gets

$$\frac{1}{C} = \sum_l e^{-\beta(E_l - \mu N_l)} \qquad (4.7)$$

One defines the "grand partition function" $\mathcal{Z}$ by

$$\mathcal{Z} = \sum_l e^{-\beta(E_l - \mu N_l)} \qquad (4.8)$$

$\mathcal{Z}$ is a function of T, μ and V. The grand-canonical probability P_l is then given by

$$P_l = \frac{e^{-\beta(E_l - \mu N_l)}}{\mathcal{Z}} \qquad (4.9)$$

Microstates obeying this probability law form a "grand-canonical ensemble".

4.3 Grand partition function $\mathcal{Z}$ — Grand potential J

The grand partition function $\mathcal{Z}$ plays an important role in the calculation of average values of physical quantities as seen below. For practical reasons, one gives hereafter several frequently used equivalent expressions of $\mathcal{Z}$.

First, to perform the sum on microstates in (4.8), one can sum over all states l of energy E_l at a fixed $N_l = N$, then one makes the sum over N. One has

$$\mathcal{Z} = \sum_l e^{-\beta(E_l - \mu N_l)} = \sum_N e^{\beta \mu N} \sum_l e^{-\beta E_l} = \sum_N e^{\beta \mu N} Z(N, T, V) \quad (4.10)$$

where $Z(N, T, V)$ is the canonical partition function which depends on N, T and V as seen in the previous chapter. In the case of a very large system, one can replace the sum over N by an integral, and replace Z by an integral over E as follows:

$$\mathcal{Z} = \int_0^\infty dN \int_{E_0}^\infty \rho(E) e^{-\beta(E - \mu N)} dE \qquad (4.11)$$

Let us define a new function called "grand potential" J by

$$J = -k_B T \ln \mathcal{Z} \qquad (4.12)$$

One sees in the next section that one can express average values of physical quantities as functions of J or $\mathcal{Z}$.

One shows that $\mathcal{Z}$ is multiplicative: one considers a system composed of two independent subsystems. The energy of the system in the microstate l is the sum over the energies of the subsystems, and the number of particles of the system is the sum of the numbers of particles of the subsystems: $E_l = E_{l_1} + E_{l_2}$ and $N_l = N_{l_1} + N_{l_2}$ where $l = (l_1, l_2)$. One has

$$\mathcal{Z} = \sum_l e^{-\beta(E_l - \mu N_l)} = \sum_{l_1, l_2} e^{-\beta(E_{l_1} + E_{l_2} - \mu N_{l_1} - \mu N_{l_2})}$$

$$= \sum_{l_1} e^{-\beta E_{l_1} - \mu N_{l_1}} \sum_{l_2} e^{-\beta E_{l_2} - \mu N_{l_2}} = \mathcal{Z}_1 \mathcal{Z}_2 \qquad (4.13)$$

The grand potential is thus additive:

$$J = -k_B T \ln \mathcal{Z} = -k_B T \ln \mathcal{Z}_1 \mathcal{Z}_2 = -k_B T \ln \mathcal{Z}_1 - k_B T \ln \mathcal{Z}_2 = J_1 + J_2$$

$$(4.14)$$

4.4   General properties of grand-canonical systems

One can express average values of physical quantities as functions of J as follows:

- *Average number of particles:*

$$\overline{N} = \sum_l N_l P_l = \frac{1}{Z} \frac{1}{\beta} \frac{\partial}{\partial \mu} \sum_l e^{-\beta(E_l - \mu N_l)}$$

$$= k_B T \left(\frac{\partial \ln Z}{\partial \mu} \right)_{T,V} \tag{4.15}$$

$$= - \left(\frac{\partial J}{\partial \mu} \right)_{T,V} \tag{4.16}$$

- *Average energy:*

$$\overline{E} - \mu \overline{N} = \sum_l (E_l - \mu N_l) P_l = -\frac{1}{Z} \frac{\partial}{\partial \beta} \sum_l e^{-\beta(E_l - \mu N_l)}$$

$$\overline{E} - \mu \overline{N} = - \left(\frac{\partial \ln Z}{\partial \beta} \right)_{\mu,V} \tag{4.17}$$

$$\overline{E} - \mu \overline{N} = \frac{\partial}{\partial \beta} (\beta J)$$

$$\overline{E} - \mu \overline{N} = J + \beta \left(\frac{\partial J}{\partial \beta} \right)_{\mu,V} \tag{4.18}$$

One gets

$$\overline{E} = J + \beta \frac{\partial J}{\partial \beta} - \mu \frac{\partial J}{\partial \mu}$$

$$= J + \left(\beta \frac{\partial}{\partial \beta} - \mu \frac{\partial}{\partial \mu} \right) J \tag{4.19}$$

- *Grand-canonical pressure:*
  The grand-canonical pressure p is defined by

$$p = - \left(\frac{\partial J}{\partial V} \right)_{\mu,T} \tag{4.20}$$

- *Grand-canonical entropy:*

Using P_l of (4.9), one writes

$$S = -k_B \sum_l P_l \ln P_l$$

$$= -k_B \frac{1}{Z} \sum_l e^{-\beta(E_l - \mu N_l)} [-\beta(E_l - \mu N_l) - \ln Z]$$

$$= k_B \beta(\overline{E} - \mu \overline{N}) + k_B \ln Z = \frac{1}{T}(\overline{E} - \mu \overline{N}) - \frac{J}{T} \quad (4.21)$$

$$= \frac{1}{T}(J + \beta \frac{\partial J}{\partial \beta}) - \frac{J}{T} = \frac{1}{k_B T^2} \frac{\partial J}{\partial \beta}$$

$$= -\left(\frac{\partial J}{\partial T} \right)_{\mu,V} \quad (4.22)$$

- *Statistical distribution of an internal variable:*
  The probability to find a property x is given by

$$P(x) = \frac{1}{Z} \sum_{j \in x} \exp[-\beta(E_j - \mu N_j)] \quad (4.23)$$

where the sum is performed over the sub-ensemble of microstates j having the property x. One defines the "partial grand partition function" $Z^*(x)$ by

$$Z^*(x) = \sum_{j \in x} \exp[-\beta(E_j - \mu N_j)] \quad (4.24)$$

from which

$$P(x) = \frac{Z^*(x)}{Z}$$

$$\ln P(x) = \ln Z^*(x) - \ln Z \quad (4.25)$$

One defines the "partial grand potential" $J^*(x)$ by

$$J^*(x) \equiv -k_B T \ln Z^*(x) \quad (4.26)$$

The most probable value of x, called x_0, corresponds to the maximum of the probability $P(x)$, namely to the maximum of $\ln P(x_0)$, one has

$$\left[\frac{\partial \ln P(x)}{\partial x} \right]_{x=x_0} = 0 \quad (4.27)$$

From (4.25), one writes

$$\left[-\beta \frac{\partial(J^*(x) - J)}{\partial x} \right]_{x=x_0} = 0 \quad (4.28)$$

Since J is independent of x, one can write

$$\left[\frac{\partial J^*(x)}{\partial x}\right]_{x=x_0} = 0 \qquad (4.29)$$

The above relation is important in the determination of the most probable value of the internal variable x: it suffices to express J^* as a function of x and then look for the value x_0 which minimizes $J^*(x)$.

In order to examine the nature of the fluctuations of x around x_0, one expands $\ln P(x)$ around x_0:

$$\ln P(x) = \ln P(x_0) + \left[\frac{\partial \ln P(x)}{\partial x}\right]_{x=x_0} (x - x_0)$$

$$+ \frac{1}{2}\left[\frac{\partial^2 \ln P(x)}{\partial x^2}\right]_{x=x_0} (x - x_0)^2 + \dots$$

$$\simeq \ln P(x_0) - \frac{1}{2\sigma^2}(x - x_0)^2 \qquad (4.30)$$

where one has used (4.27) and one has put $1/\sigma^2 \equiv -[\frac{\partial^2 \ln P(x)}{\partial x^2}]_{x=x_0}$. From (4.30), one gets

$$P(x) = P(x_0)\exp\left[-\frac{1}{2\sigma^2}(x - x_0)^2\right] \qquad (4.31)$$

The variable x fluctuates thus around its most probable value x_0, namely around its average value, according to a Gaussian law for a large system.

4.5   Spontaneous evolution of a grand-canonical system

When an external parameter is modified, one shows that a system in the grand-canonical situation evolves toward a new equilibrium. During this spontaneous evolution , the grand potential J diminishes. The new equilibrium state is attained when J is minimum.

One considers a total system composed of the reservoir τ and the system under study. One modifies one of its external parameters: the system evolves spontaneously toward a new equilibrium. Since the total system is isolated, the total entropy increases during the evolution and attains its maximum at the new equilibrium (see chapter 2). In the same manner used in section 3.5, let S_τ and S be the micro-canonical entropies of the reservoir and the system, L and l be their microstates. The probability of the microstate defined by l, L of the total system is given by

$$P_{l,L} = P_l P_L \simeq P_l(E_l, N_l) P_L^*(E_L = E_{total} - E_l, N_L = N_{total} - N_l) \quad (4.32)$$

where P_L of the reservoir was replaced by its micro-canonical probability because the reservoir is considered approximatively as isolated due to its huge size. In this hypothesis, one writes

$$\ln P_{l,L} = \ln P_l(E_l, N_l) + \ln P_L^*(E_L = E_{total} - E_l, N_L = N_{total} - N_l) \quad (4.33)$$

from which one gets the following micro-canonical entropy of the total system

$$
\begin{aligned}
S_{total} &= -k_B \sum_{l,L} P_{l,L} \ln P_{l,L} \\
&= -k_B \sum_{l,L} P_l(E_l) P_L^*(E_L = E_{total} - E_l, N_L = N_{total} - N_l) \\
&\quad \times [\ln P_l(E_l) + \ln P_L^*(E_L = E_{total} - E_l, N_L = N_{total} - N_l)] \\
&= -k_B \sum_l P_l \ln P_l(E_l) \sum_L P_L^*(E_L = E_{total} - E_l, N_L = N_{total} - N_l) \\
&\quad - k_B \sum_l P_l \sum_L P_L^*(E_L = E_{total} - E_l, N_L = N_{total} - N_l) \\
&\quad \times \ln P_L^*(E_L = E_{total} - E_l, N_L = N_{total} - N_l) \\
&= S + \sum_l S_\tau(E_L = E_{total} - E_l, N_L = N_{total} - N_l) P_l \quad (4.34)
\end{aligned}
$$

where one has replaced in the line before the last line $\sum_l P_l = 1$ and $\sum_L P_L^*(E_L = E_{total} - E_l, N_L = N_{total} - N_l) = 1$ (normalization of P_L^*, for any l). An expansion of $S_\tau(E_L = E_{total} - E_l, N_L = N_{total} - N_l)$ around (E_{total}, N_{total}) gives

$$
\begin{aligned}
S_\tau(E_L = E_{total} - E_l, N_L = N_{total} - N_l) &\simeq S_\tau(E_{total}, N_{total}) \\
&\quad + \frac{\partial S_\tau(E_{total}, N_{total})}{\partial E_L}(E_L - E_{total}) \\
&\quad + \frac{\partial S_\tau(E_{total}, N_{total})}{\partial N_L}(N_L - N_{total}) \\
&= S_\tau(E_{total}, N_{total}) - \frac{1}{T} E_l + \frac{\mu}{T} N_l \quad (4.35)
\end{aligned}
$$

Replacing this relation in (4.34) one has

$$S_{total} = S_\tau(E_{total}, N_{total}) + S + \sum_l \left[-\frac{1}{T}E_l + \frac{\mu}{T}N_l \right] P_l$$

$$= S_\tau(E_{total}, N_{total}) + S - \frac{\overline{E}}{T} + \frac{\mu \overline{N}}{T}$$

$$= S_\tau(E_{total}, N_{total}) - \frac{J}{T} \tag{4.36}$$

where (4.21) and (4.12) have been used in the last line. In a spontaneous evolution, one knows that S_{total} increases: the expression (4.36) thus implies that the grand potential J of the system should decrease. The new equilibrium is reached when S_{total} is maximum, namely when J is minimum.

One considers two systems in contact with the same reservoir. Initially the two systems are not in contact with each other. They have the same T and μ fixed by the reservoir. They are now put in contact with each other: in the same manner as in section 3.5.2 but with the free energy F replaced by the grand potential J in the calculation, one obtains during the evolution, using $dV_2 = -dV_1$,

$$dJ < 0$$

$$\frac{\partial J}{\partial V_1}dV_1 < 0$$

$$\left(\frac{\partial J_1}{\partial V_1} - \frac{\partial J_2}{\partial V_2} \right) dV_1 < 0$$

$$(-p_1 + p_2)dV_1 < 0 \tag{4.37}$$

One sees that if $p_1 < p_2$, one has $dV_1 < 0$: the subsystem of smaller grand-canonical pressure diminishes its volume as it is experimentally observed.

4.6   Systems of identical, independent particles

One considers a system of identical, independent and indiscernible particles. In such a system, it is obvious that each particle has the same list of microstates as the others. One calls these microstates "individual microstates" to distinguish with the microstates of the system which can be defined as follows: a microstate l of the system is defined by the set composed of the numbers of particles occupying the individual microstates of the list. Let k be the individual microstate of energy ϵ_k occupied by n_k particles. The system microstate l is written as $l = \{n_1, n_2, ..., n_k, ...\}$.

The total energy and the total number of particles of the system in the microstate l are thus

$$E_l = \sum_k n_k \epsilon_k \tag{4.38}$$

$$N_l = \sum_k n_k \tag{4.39}$$

4.6.1 *Factorization of $\mathcal{Z}$*

The grand partition function is

$$\mathcal{Z} = \sum_l e^{-\beta(E_l - \mu N_l)}$$

$$= \sum_{n_k} e^{-\beta \sum_k n_k (\epsilon_k - \mu)} \tag{4.40}$$

where the sum over l is made in two steps as follows: for a given l (particle repartition) $l = \{n_1, n_2, ..., n_k, ...\}$ one performs the sum in the argument of the exponential, one then makes it again for another repartition $l' = \{n_1', n_2', ...\}$ and repeats until all repartitions, namely all system microstates, have been used. Note that, due to the fact that the total particle number is not fixed (grand-canonical situation), each n_i takes all possible values regardless of other numbers n_j ($j \neq i$). One can therefore factorize $\mathcal{Z}$ as follows

$$\mathcal{Z} = \prod_k z_k \tag{4.41}$$

where

$$z_k = \sum_{n_k} e^{-\beta n_k (\epsilon_k - \mu)} \tag{4.42}$$

Note that the sum in z_k is made over all possible values of n_k which can occupy the individual state k of energy ϵ_k: one distinguishes two following cases:

- The case of bosons (particles of integer spin, including spin 0): In this case, $n_k = 0, 1, 2, ..., \infty$. One has

$$z_k^{BE} = \sum_{n_k=0}^{\infty} e^{-\beta n_k (\epsilon_k - \mu)} \tag{4.43}$$

$$= \frac{1}{1 - e^{-\beta(\epsilon_k - \mu)}} \tag{4.44}$$

where the superscript BE indicates the Bose-Einstein case.

- The case of fermions (particles of half-integer spin):
  In this case, according to the Pauli's principle, $n_k = 0, 1$. One gets

$$z_k^{FD} = \sum_{n_k=0}^{1} e^{-\beta n_k(\epsilon_k - \mu)} \tag{4.45}$$

$$= 1 + e^{-\beta(\epsilon_k - \mu)} \tag{4.46}$$

where the superscript FD indicates the Fermi-Dirac case.

It is interesting to calculate the average number of particles $\overline{n}_k$ of the state of energy ϵ_k: using (4.16), (4.39) and (4.41) one writes

$$\overline{N} = \sum_k \overline{n}_k = -\left(\frac{\partial J}{\partial \mu}\right)_{T,V}$$

$$= k_B T \frac{\partial}{\partial \mu} \sum_k \ln z_k = k_B T \sum_k \frac{\partial \ln z_k}{\partial \mu} \tag{4.47}$$

from which

$$\overline{n}_k = k_B T \frac{\partial \ln z_k}{\partial \mu} \tag{4.48}$$

4.6.2   Bose-Einstein distribution

Using (4.44), one obtains in the boson case:

$$\overline{n}_k^{BE} = \frac{1}{e^{\beta(\epsilon_k - \mu)} - 1} \tag{4.49}$$

This statistical distribution is called "Bose-Einstein distribution": $\overline{n}_k^{BE}$ is the average number of particles occupying the microstate k of energy ϵ_k, at the temperature T.

4.6.3   Fermi-Dirac distribution

In the fermion case, using (4.46) one obtains

$$\overline{n}_k^{FD} = \frac{1}{e^{\beta(\epsilon_k - \mu)} + 1} \tag{4.50}$$

This expression is called "Fermi-Dirac distribution": $\overline{n}_k^{FD}$ is the average number of fermions occupying the microstate k of energy ϵ_k at the temperature T.

Properties of systems of free bosons and free fermions are studied in the next two chapters. Numerous applications of Bose-Einstein and Fermi-Dirac distributions for various systems of quantum particles or quasi-particles will be shown in Part 2 of this book.

4.6.4   *Maxwell-Boltzmann distribution*

In the case where the average number of particles in the microstate of energy ϵ_k is very small, namely $\overline{n_k} \ll 1$, one has $e^{\beta(\epsilon_k - \mu)} \gg 1$. In such a situation, one can neglect 1 in the denominators of $\overline{n}_k^{BE}$ and $\overline{n}_k^{FD}$: the cases of bosons and fermions become equivalent

$$\overline{n}_k^{BE} \simeq \overline{n}_k^{FD} \simeq e^{-\beta(\epsilon_k - \mu)} \tag{4.51}$$

The above approximation is called "Maxwell-Boltzmann approximation" which is valid when $e^{\beta(\epsilon_k - \mu)} \gg 1$. This condition corresponds to the high-temperature region and/or to dilute systems, as shown in section 3.6.1 and in Problem 2 of chapter 3. In these situations, the quantum nature is lost: the system behaves as a system of classical particles.

4.7   Applications of the grand-canonical method

4.7.1   *Classical ideal gas*

One considers a classical ideal gas of volume V connected to a reservoir of heat and particles at temperature T and chemical potential μ.

Using the grand-canonical description, one calculates the grand partition function $\mathcal{Z} = \sum_N e^{N\beta\mu} Z(T, N, V)$ where Z is the partition function (of the canonical description). One has $Z = z^N/N!$, z being the partition function of a particle which is given by (3.53). Thus,

$$\mathcal{Z} = \sum_{N=0}^{\infty} e^{N\beta\mu} z^N/N! = \sum_{N=0}^{\infty} (\lambda z)^N/N! = \exp(\lambda z) \tag{4.52}$$

where $\lambda \equiv e^{\beta\mu}$ is called "fugacity".

Knowing $\mathcal{Z}$, the average number of particles of the system is calculated using (4.15) as follows:

$$\overline{N} = \frac{1}{\beta} \frac{\partial \ln \mathcal{Z}}{\partial \mu} = \frac{1}{\beta} \frac{\partial \lambda z}{\partial \mu}$$

$$\overline{N} = \frac{1}{\beta} z \frac{\partial e^{\beta\mu}}{\partial \mu} = z e^{\beta\mu} = \lambda z \tag{4.53}$$

The average energy is calculated by (4.17):

$$\overline{E} - \mu\overline{N} = -\frac{\partial \ln \mathcal{Z}}{\partial \beta} = -\frac{\partial \lambda z}{\partial \beta}$$

$$\overline{E} - \mu\overline{N} = -\frac{\partial(e^{\beta\mu}z)}{\partial \beta} = -\mu\lambda z - \lambda\frac{\partial z}{\partial \beta}$$

$$\overline{E} - \mu\overline{N} = -\mu\overline{N} - \lambda\frac{3V}{2h^3}(2\pi m)^{3/2}\beta^{-5/2}$$

$$\overline{E} = \frac{3\lambda z}{2\beta}$$

$$\overline{E} = \frac{3\overline{N}k_B T}{2} \tag{4.54}$$

where (3.53) has been used.

The grand-canonical pressure is

$$p = -\frac{\partial J}{\partial V} = k_B T\frac{\partial \ln \mathcal{Z}}{\partial V}$$

$$p = k_B T\lambda\frac{\partial z}{\partial V} = \frac{k_B T\lambda z}{V} \tag{4.55}$$

where z is given by (3.53). Using (4.53), one recovers the well-known result $pV = \overline{N}k_B T$ obtained before by both the micro-canonical and canonical methods, for a classical ideal gas.

4.7.2 *Two-level systems*

Two-level systems have been studied by micro-canonical and canonical methods. Here, one shows that the grand-canonical description is also very convenient for studying these systems. What one should bear in mind is whatever the method is to be used, physical results of a given system should be the same at the thermodynamic limit.

One considers a system of identical, independent particles maintained at the temperature T and at chemical potential μ by a reservoir. Each particle has two energy levels $\pm\epsilon$ with $\epsilon > 0$. One calculates the grand partition function $\mathcal{Z}$ using (4.10): $\mathcal{Z} = \sum_N \exp(\beta\mu N)Z(N,T,V)$. Since the particles are independent and identical, the partition function $Z(N,T,V)$ is written as $Z(N,T,V) = z^N/N!$ where z is the partition function of a particle. One has

$$z = \sum_l \exp(-\beta\epsilon_l) = \exp(\beta\epsilon) + \exp(-\beta\epsilon) = 2\cosh(\beta\epsilon) \tag{4.56}$$

from which

$$Z = \sum_{N=0}^{\infty} \exp(\beta\mu N)\frac{2^N \cosh^N(\beta\epsilon)}{N!} = \exp[\exp(\beta\mu)2\cosh(\beta\epsilon)] \quad (4.57)$$

One then obtains

$$\ln Z = \exp(\beta\mu)2\cosh(\beta\epsilon) = \lambda z \quad (4.58)$$

where $\lambda \equiv \exp(\beta\mu)$. The average number of particles of the system is given by (4.15):

$$\overline{N} = k_B T\frac{\partial \ln Z}{\partial \mu} = \exp(\beta\mu)2\cosh(\beta\epsilon) = \lambda z \quad (4.59)$$

The average energy is calculated by (4.17):

$$\overline{E} - \mu\overline{N} = -\frac{\partial \ln Z}{\partial \beta}$$

$$\overline{E} - \mu\overline{N} = -\mu\exp(\beta\mu)2\cosh(\beta\epsilon) - \exp(\beta\mu)2\epsilon\sinh(\beta\epsilon)$$

$$\overline{E} - \mu\overline{N} = -\mu\overline{N} - \exp(\beta\mu)2\epsilon\sinh(\beta\epsilon)$$

$$\overline{E} = -\exp(\beta\mu)2\epsilon\sinh(\beta\epsilon)$$

$$\overline{E} = -\frac{\overline{N}}{z}2\epsilon\sinh(\beta\epsilon)$$

$$\overline{E} = -\overline{N}\epsilon\tanh(\beta\epsilon) \quad (4.60)$$

This relation is identical to that obtained by the canonical method (3.57), and to that by the micro-canonical one (2.52).

One has

$$\overline{E} = -\overline{N}_1\epsilon + \overline{N}_2\epsilon = -\overline{N}_1\epsilon + (\overline{N} - \overline{N}_1)\epsilon = (\overline{N} - 2\overline{N}_1)\epsilon$$

Thus, the average number of particles occupying level 1 is

$$\overline{N}_1 = \frac{1}{2}(\overline{N} - \frac{\overline{E}}{\epsilon}) = \frac{\overline{N}}{2}[1 + \tanh(\beta\epsilon)]$$

$$\overline{N}_1 = \overline{N}\frac{\exp(2\beta\epsilon)}{1 + \exp(2\beta\epsilon)} \quad (4.61)$$

For level 2, one has $\overline{N}_2 = \overline{N} - \overline{N}_1$. These results are the same as (2.51). At low temperatures ($\beta \to \infty$), $N_1 \to N$ and $N_2 \to 0$: all the particles occupy the lower level at $T = 0$, and at high temperatures, $N_1 \to N/2$ so that $N_2 \to N/2$: the particles equally occupy the two levels.

4.8   Conclusion

The grand-canonical method has been presented in this chapter to study properties of a system maintained at a constant temperature and a constant chemical potential by a large reservoir of heat and particles. The energy and the number of particles of the system fluctuate around their average values according to a Gaussian law. These fluctuations tend to zero at the thermodynamic limit (infinite system size) so that the energy and the number of particles can be considered as constants. The microcanonical, canonical and grand-canonical descriptions then give equivalent results. Note that, using the grand-canonical description the demonstration of the Bose-Einstein and Fermi-Dirac distributions was straightforward. These distributions will be used in the following chapters to study properties of quantum ideal gases.

4.9   Problems

Problem 1. Fluctuations of N:

Show that the variance of the fluctuations of the number of particles N of a system in the grand-canonical situation is given by

$$\Delta N^2 = \overline{(N - \overline{N})^2} = k_B T \frac{\partial \overline{N}}{\partial \mu} \qquad (4.62)$$

Problem 2. Classical ideal gas in the gravitational field:

One supposes that the atmosphere is a classical ideal gas. Calculate, using the grand-canonical description, the variation of the pressure of the atmosphere as a function of the altitude y at the constant temperature T.

Problem 3. Two-level system:

One studies the two-level system shown in section 4.7.2, using (4.41) and (4.42).

a) Explain the difference between the partition function z of a particle and the function z_k in the formula (4.42).

b) Calculate z_k. Find the grand partition function $\mathcal{Z}$.

c) Calculate the average energy of the system and the average numbers of particles of the two levels.

Problem 4. Degeneracy in the case of fermions:

Show that in the case of fermions, if the energy level ϵ_k in (4.45) is g_k-fold degenerate, then the average number of particles of that level is given by

$$\overline{n}_k^{FD} = \frac{1}{\frac{1}{g_k}e^{\beta(\epsilon_k - \mu)} + 1} \tag{4.63}$$

This relation is often used in the study of semiconductors.

Problem 5. Particle trap:

Consider a crystal of volume V containing N sites ($N \gg 1$). The crystal is in contact with a reservoir of heat and particles which provides a constant temperature T and a chemical potential μ. Each site of the crystal can capture at most a particle. A captured particle has an energy equal to either $-\epsilon_1$ or $-\epsilon_2$, with $\epsilon_2 > \epsilon_1 > 0$. One neglects interaction between captured particles.

a) Calculate the grand partition function of the crystal.
b) Calculate the average number of captured particles $\overline{n}$, and their average energy, as functions of μ and T.
c) Calculate $\overline{n_1}$ and $\overline{n_2}$, the average numbers of particles captured on levels $-\epsilon_1$ and $-\epsilon_2$, as functions of μ and T. Calculate $\overline{n_0}$, the average number of the crystal empty sites.
d) Suppose now that the captured particles are electrons of magnetic moment $\vec{m}$ and the crystal is under an applied magnetic field $\vec{B}$. The energy levels $-\epsilon_1$ and $-\epsilon_2$ are given by: $-\epsilon_1 = -\epsilon_0 + mB$, $-\epsilon_2 = -\epsilon_0 - mB$ where $\epsilon_0 > 0$ is a constant. $-\epsilon_1$ is the energy of an electron having a spin antiparallel to $\vec{B}$ (Zeeman energy is thus $+mB$), while $-\epsilon_2$ is that of an electron having a spin parallel to $\vec{B}$. Using the results obtained in the previous questions, calculate the average magnetic moment $\overline{M}$ of the occupied sites. Find the low- and high-temperature limits of $\overline{M}$. Comment on the results.

Problem 6. Poisson law by the grand-canonical description:

Consider a classical ideal gas of N particles, of volume V maintained at temperature T. One examines a very small portion of volume v of V. The portion outside v, of volume $V - v$, plays the role of a reservoir of particles for the volume v. Using the grand-canonical description, show that the number of particles inside v obeys the Poisson probability law.

Problem 7. System of interacting electrons:

Conducting electrons of a metal can be considered as a gas of free particles. Suppose that the metal is maintained at temperature T. One introduces now N impurities into the metal. Each impurity can capture zero, one or two electrons. An electron captured by an impurity has an energy equal to ϵ_0 (< 0), regardless of its spin state. If two electrons are captured on the same impurity, their spins should be antiparallel (Pauli's principle). The energy of the spin pair is equal to $\epsilon_0 + g$ where g represents the on-site interaction between the two electrons which takes into account the combination of two degenerate spin states $\uparrow\downarrow$ and $\downarrow\uparrow$. One supposes that g is positive and $g < |\epsilon_0|$. In addition, one assumes that there is no interaction between electrons captured on different impurities. The electron gas is considered as a reservoir of particles for the impurities.

a) Calculate the average number of electrons $\bar{n}$ captured by impurities, and their average energy.

b) If the number of captured electrons is equal to N, calculate the chemical potential μ of the electrons.

c) Calculate the numbers of impurities having zero, one and two electrons in the case where $\bar{n} = N$. Find the low- and high-temperature limits of these numbers.

d) The whole system is now placed in a magnetic field $\vec{B}$. Calculate the magnetic moment due to captured electrons.

Problem 8. Lattice model for an ideal gas:

Consider a system at equilibrium with a reservoir of heat and particles of temperature T and chemical potential μ.

a) Recall the definition of the grand partition function $\mathcal{Z}$. Show that for a simple fluid (i. e. fluid composed of one kind of molecules) the grand potential is given by $J = -pV$ where p is the pressure and V the system volume.

b) In the lattice model for a gas, the volume V is divided into small cubes of volume v of the order of a volume occupied by an atom in a crystal. Each cube can contain at most one molecule of the gas. In addition, one assumes the gas is ideal and mono-atomic (i. e. one kind of atoms or molecules). Calculate the grand partition function $\mathcal{Z}_i$ of a cube. Find the grand partition function $\mathcal{Z}$ of the gas.

c) Calculate the average number of particles $\overline{N}$ and the pressure. By eliminating μ, find the equation of state of the gas. What is the limit of this equation at strong dilution ($\overline{N}v << V$) ?

Problem 9. Adsorption:

Consider the surface of a solid having N_0 sites. This surface is in contact with an ideal gas considered as a reservoir of heat and particles of temperature T and chemical potential μ. Each site of the surface can capture a molecule of the gas. This phenomenon is called "adsorption". If the bonding between the molecule and the surface is due to a physical interaction such as the van der Waals force or the Lennard-Jones potential, the bonding energy is relatively weak, the adsorption is called "physisorption". If the bonding is from a chemical mechanism, the bonding energy is much stronger, one has a chemisorption. Assume that the energy of an adsorbed molecule is equal to $-\epsilon$ ($\epsilon > 0$).

a) Calculate the grand partition function of molecules adsorbed on the surface.

b) Calculate the average number $\overline{N}_c$ of adsorbed molecules and their energy.

c) Calculate the grand potential. Find the entropy of adsorbed molecules as a function of $\overline{N}_c$. Compare this entropy to the micro-canonical entropy in the case where N_c molecules are captured by N_0 surface sites.

d) Express the chemical potential μ as a function of pressure p, of temperature T and of Broglie wavelength of the gas. Find the adsorption rate $\theta = \overline{N}_c/N_0$ as a function of p and T. Plot schematically curves of isotherms (called "Langmuir isotherms").

Problem 10. Adsorption of an ideal gas on the surface of a solid: Consider an ideal gas of N molecules in contact with the surface S of a solid. The surface plane is a xy plane. The total system composed of the gas and the surface S is maintained at temperature T. Assume that molecules of the gas can be adsorbed on the surface. The energy of an adsorbed molecule is written as $E = -\epsilon_0 + \frac{p^2}{2m}$ where $\epsilon > 0$ is the capture energy by the surface. The second term is the kinetic energy of the adsorbed molecule where only the components p_x and p_y of its momentum are conserved after adsorption.

a) Calculate the average number of molecules adsorbed on the surface as a function of T. Express this number as a function of the grand-canonical pressure of the ideal gas which serves as a reservoir of molecules.

b) Calculate the average energy of adsorbed molecules. Find the total energy of the system. Find the calorific capacity of the total system.

c) Find the above results using the canonical method.

Chapter 5

Free Fermi Gas

5.1 Introduction

A "free Fermi gas", also called "ideal Fermi gas", is an ensemble of a large number of independent fermions, namely fermions without interaction. In the previous chapter, the Fermi-Dirac distribution, Eq. (4.50), has been demonstrated for the case of free fermions. In this chapter, properties of a free Fermi gas are shown using this statistics. In particular, quantum behaviors of the Fermi gas at low temperatures are shown with an emphasis on the difference with respect to the results for the classical ideal gas. The free Fermi gas studied in this chapter is the first step toward more complicated quantum systems which can be treated by quantum statistical physics. Applications of quantum effects are numerous in our daily life going from electronic devices using semiconductors, such as in computers, smart phones and data storage, to plasmas and liquid crystals used for digital display. In spite of the fact that interaction between fermions is not taken into account in a free Fermi gas, the results presented in this chapter bear essential features which allow for a clear understanding of the quantum nature of particles and their fundamental effects.

5.2 Fermi-Dirac distribution

The Fermi-Dirac distribution, or Fermi-Dirac statistics, has been demonstrated in the previous chapter. It is given by (4.50):

$$f(E_\lambda, T, \mu) \equiv \overline{n}_\lambda^{FD} = \frac{1}{e^{\beta(E_\lambda - \mu)} + 1} \qquad (5.1)$$

with $\overline{n}_k^{FD}$ replaced by the widely used notation $f(E_\lambda, T, \mu)$ which represents the average number of particles occupying the microstate λ of energy E_λ at temperature T. In the following, one shows some fundamental properties of this function.

At $T = 0$, namely $\beta = \infty$, one has

$$f(E_\lambda, T, \mu) = 1 \ \text{ if } \ E_\lambda < \mu$$
$$= 0 \ \text{ if } \ E_\lambda > \mu \tag{5.2}$$

Thus, only microstates of energy smaller than μ are occupied at $T = 0$, while microstates of energy higher than μ are empty.

One defines the Fermi level E_F by $E_F = \mu_0$ where μ_0 is the value of μ at $T = 0$, namely the highest occupied energy level at $T = 0$.

For large systems, E_λ can be considered as a continuous variable E. The function $f(E, T, \mu)$ is shown in Figures 5.1 and 5.2 at $T = 0$ and for $T \neq 0$, respectively. Note that when $T \neq 0$, all energy levels are occupied up to $E = \infty$ with a decreasing occupation number.

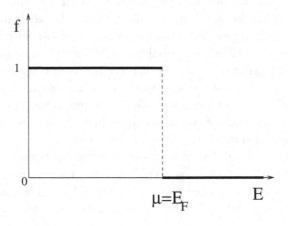

Fig. 5.1   Fermi-Dirac distribution $f(E, T, \mu)$ versus energy E at $T = 0$.

5.3   General properties of a free Fermi gas

5.3.1   *General formulas*

- Number of particles:
  The total number of fermions of a free Fermi gas is given by

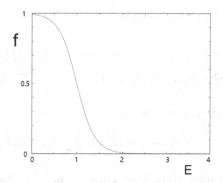

Fig. 5.2 Fermi-Dirac distribution $f(E, T, \mu)$ as a function of energy E at $T \neq 0$, with $\mu = 1$.

$$N = \sum_\lambda f(E_\lambda, T, \mu) = \sum_\lambda \frac{1}{e^{\beta(E_\lambda - \mu)} + 1} \qquad (5.3)$$

- Average energy:

  The average energy of the gas is written as

$$\overline{E} = \sum_\lambda f(E_\lambda, T, \mu) E_\lambda = \sum_\lambda \frac{E_\lambda}{e^{\beta(E_\lambda - \mu)} + 1} \qquad (5.4)$$

- Grand potential:

  Using (4.46) and (4.50) one writes

$$z_\lambda = \frac{1}{1 - f(E_\lambda, T, \mu)} \qquad (5.5)$$

with the notation $f(E_\lambda, T, \mu) \equiv \overline{n}_\lambda^{FD}$. The grand potential (see chapter 4) is thus

$$J = -k_B T \ln \mathcal{Z} = -k_B T \sum_\lambda \ln z_\lambda = k_B T \sum_\lambda \ln[1 - f(E_k, T, \mu)]$$

$$= -k_B T \sum_\lambda \ln[1 + x_\lambda] \qquad (5.6)$$

where $x_\lambda = e^{-\beta(E_\lambda - \mu)}$.

- Entropy:

  The entropy of the gas is

$$S = -\frac{\partial J}{\partial T} = k_B \sum_\lambda \{\ln[1 - f(E_\lambda, T, \mu)] + \beta(E_\lambda - \mu) f(E_\lambda, T, \mu)\}$$

$$= -\sum_\lambda \{(1 - f(E_\lambda, T, \mu)) \ln[1 - f(E_\lambda, T, \mu)]$$

$$+ f(E_\lambda, T, \mu) \ln f(E_\lambda, T, \mu)\} \qquad (5.7)$$

where the following relation has been used in the last equality:

$$\exp[\beta(E_\lambda - \mu)] = [1 - f(E_\lambda, T, \mu)]/f(E_\lambda, T, \mu)$$

which results from taking the logarithm of (5.1).

- Pressure:

The pressure is given by

$$p = -\frac{\partial J}{\partial V} = -k_B T \frac{\partial}{\partial V} \sum_\lambda \ln[1 - f(E_\lambda, T, \mu)] \qquad (5.8)$$

In general, the microstate λ of a free fermion is defined by the wave vector $\vec{k}$ (see section 2.4.2) and the spin state σ of the particle:

$$\lambda = (\vec{k}, \sigma)$$

The sum over λ in the above formulas is decomposed into two sums as follows:

$$\sum_\lambda ... = \sum_{\sigma=-S}^{S} \sum_{\vec{k}} ... \qquad (5.9)$$

One recalls that the energy of a free quantum particle (fermion or boson) is written as (see section 2.4.2):

$$E_k = \frac{\hbar^2 k^2}{2m} \qquad (5.10)$$

where m is the particle mass and k the modulus of the wave vector $\vec{k}$. The allowed values of $\vec{k}$ are determined by the periodic boundary conditions (see section 2.4.2). The energy of a free fermion does not depend on its spin σ. The sum on σ gives thus a factor $(2S+1)$ since $\sigma = -S, -S+1, ..., S$ where S is the spin amplitude. This factor is important, it shows that each "wave-vector" state $\vec{k}$ can be occupied by $(2S+1)$ fermions of different spin states. If $S = 1/2$, as in the case of electrons, each wave-vector state $\vec{k}$ can be occupied by two particles of spin $\pm 1/2$.

5.3.2 Formulas for large systems

For a very large system, E_k can be considered as a continuous variable. One can replace the sums in the previous subsection by integrals performed either in the wave-vector space $\vec{k}$ or in the energy space E. One has

$$\sum_k ... = \frac{L^3}{(2\pi)^3} \int d\vec{k}... = \int_{E_0}^{\infty} dE \rho(E)... \qquad (5.11)$$

where the factor in front of the integral on $\vec{k}$ results from the fact that the elementary volume of a microstate is equal to $(2\pi)^3/L^3$ [see (2.36) or section 2.4.2]. In the second equality, $\rho(E)$ is the density of states given by (2.36), and E_0 the lowest energy which is set to be zero in the following.

The exact formulas (5.3) and (5.4) become for a large system:

$$N = (2S+1)\frac{L^3}{(2\pi)^3}\int d\vec{k}\,\frac{1}{e^{\beta(E_k-\mu)}+1} \tag{5.12}$$

or

$$N = (2S+1)\int_0^\infty dE\rho(E)\frac{1}{e^{\beta(E-\mu)}+1} \tag{5.13}$$

and

$$\overline{E} = (2S+1)\frac{L^3}{(2\pi)^3}\int d\vec{k}\,\frac{E_k}{e^{\beta(E_k-\mu)}+1} \tag{5.14}$$

or

$$\overline{E} = (2S+1)\int_0^\infty dE\rho(E)\frac{E}{e^{\beta(E-\mu)}+1} \tag{5.15}$$

One sees that if N is fixed, the chemical potential μ is determined by (5.13) for a given value of T. Using the so-determined value of μ, one calculates $\overline{E}$ and other physical quantities. One shows now how to determine μ at temperature T by solving (5.13). The simplest way to do this is to solve this equation graphically: replacing $\rho(E)$ by (2.36), one writes

$$N = (2S+1)\frac{V}{4\pi^2}(\frac{2m}{\hbar^2})^{3/2}\int_0^\infty dE E^{1/2}\frac{1}{e^{\beta(E-\mu)}+1} \tag{5.16}$$

from which

$$\frac{N}{V}\frac{4\pi^2}{2S+1}(\frac{\hbar^2\beta}{2m})^{3/2} = \int_0^\infty dx x^{1/2}\frac{1}{e^x/\phi+1} \tag{5.17}$$

where $x \equiv \beta E$ and $\phi \equiv e^{\beta\mu}$ (fugacity). One plots the curve y_1 of the right-hand side versus ϕ in Fig. 5.3. The left-hand side, function y_2, is a constant with respect to ϕ, but its value depends on T: y_2 is parallel to the ϕ axis in Fig. 5.3. The intersection of y_1 and y_2 yields the solution of ϕ from which one obtains μ. For varying T, y_2 moves up or down, the intersection point varies, giving rise to the temperature-dependent μ. There is always an intersection so that a solution of μ is found whatever the temperature is.

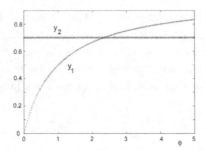

Fig. 5.3   Graphical determination of the value of μ at a given temperature T: intersection of y_1 and y_2 gives the solution ϕ which is used to find μ by the relation $\phi \equiv e^{\beta\mu}$.

One shows in Problem 1 that the grand potential J and the pressure of a free Fermi gas are given by

$$J = -\frac{2}{3}\overline{E}$$

$$p = \frac{2\overline{E}}{3V}$$

The last equation is different from the equation of state of a classical ideal gas because for the free Fermi gas $\overline{E}$ is given by (5.15) while for a classical ideal gas $\overline{E} = 3Nk_BT/2$.

5.4   Properties of a free Fermi gas at $T = 0$

At $T = 0$, $f = 1$ for $E < \mu_0 = E_F$, and $f = 0$ for $E > \mu_0$. One replaces therefore the upper limit in the integrals (5.13) and (5.15) by E_F and one replaces $f(E) = 1/[\exp(\beta(E - \mu)) + 1]$ by 1.

5.4.1   *Fermi energy*

Equation (5.13) becomes

$$N = (2S + 1)\int_0^{E_F} \rho(E)dE \tag{5.18}$$

Using $\rho(E) = AE^{1/2}$ [see (2.36)], one obtains

$$N = \frac{2}{3}BE_F^{3/2} \tag{5.19}$$

where

$$B = (2S + 1)A = (2S + 1)\frac{V}{4\pi^2}\left(\frac{2m}{\hbar^2}\right)^{3/2} \tag{5.20}$$

from which,

$$E_F = \frac{\hbar^2}{2m}\left(\frac{6\pi^2}{2S+1}\frac{N}{V}\right)^{2/3} \tag{5.21}$$

using B given by (5.20).

This result shows that the Fermi energy depends on the density of fermions $n = \frac{N}{V}$ of the gas. Since the Fermi energy is the chemical potential at $T = 0$, the chemical potential in general is closely related to the particle density.

5.4.2 Total average kinetic energy

The total average kinetic energy at $T = 0$ is

$$\overline{E}_0 = (2S+1)\int_0^{E_F} E\rho(E)dE \tag{5.22}$$

from which, one has

$$\overline{E}_0 = (2S+1)A\int_0^{E_F} E^{3/2}dE = \frac{2}{5}BE_F^{5/2} \tag{5.23}$$

Using (5.19) one gets

$$\overline{E}_0 = \frac{3}{5}NE_F \tag{5.24}$$

One sees that the energy of a free Fermi gas is not zero $T = 0$, in contrast to the case of a classical ideal gas.

5.5 Properties of a free Fermi gas at low temperatures

5.5.1 Sommerfeld's expansion

One considers the following integral

$$I = \int_0^\infty h(E)f(E)dE \tag{5.25}$$

where $h(E)$ is a function with finite derivative with respect to E at any order. At low temperatures, one can show that this integral can be expanded in powers of T as follows (see Appendix B):

$$I = \int_0^\mu h(E)dE + \frac{\pi^2}{6}(k_BT)^2 h^{(1)}(E)|_{E=\mu} + \frac{7\pi^4}{360}(k_BT)^4 h^{(3)}(E)|_{E=\mu} + \dots \tag{5.26}$$

where $h^{(n)}(E)|_{E=\mu}$ is the $n-th$ derivative of $h(E)$ with respect to E, taken at $E = \mu$.

5.5.2   Chemical potential, average energy and calorific capacity

Using (5.26) for the integral in (5.13) one obtains to the order of T^2:

$$N = \frac{2}{3}B\mu^{3/2} + \frac{\pi^2}{6}(k_BT)^2\frac{1}{2}B\mu^{-1/2} + O(T^4) \tag{5.27}$$

where B is given by (5.20). Since N does not vary with T, by equalizing (5.19) and (5.27) one obtains

$$\mu = E_F\left[1 - \frac{\pi^2}{12}\left(\frac{k_BT}{E_F}\right)^2 - O(T^4)\right] \tag{5.28}$$

This equation shows that the chemical potential μ is a function of temperature.

In the same manner, using (5.26) for (5.15) with $h(E) = E\rho(E)$, one gets

$$\overline{E} = \overline{E}_0\left[1 + \frac{5\pi^2}{12}\left(\frac{k_BT}{E_F}\right)^2 + O(T^4)\right] \tag{5.29}$$

where μ has been replaced by (5.28). The average energy thus increases as T^2 at low temperatures.

The calorific capacity of a free Fermi gas at low T is therefore

$$C_V = \frac{d\overline{E}}{dT} = \frac{1}{3}\pi^2\rho(E_F)k_B^2T = \gamma T \tag{5.30}$$

One sees that C_V is linear in T. Note that for a classical ideal gas $C_V = 3Nk_B/2$, independent of T.

5.6   Free Fermi gas at the high-temperature limit

When T is very high, one shows that the free Fermi gas becomes a classical ideal gas: the quantum nature disappears at high temperatures.

The energy (5.15) becomes at high T

$$\overline{E} \simeq (2S+1)\int_0^\infty dE\rho(E)Ee^{-\beta(E-\mu)} \tag{5.31}$$

where one has neglected 1 in the denominator because $e^{\beta(E-\mu)} \gg 1$ at high T. Replacing $\rho(E)$ by (2.36) and putting $u = \beta E$, one has

$$\overline{E} = (2S+1)A\int_0^\infty dEE^{3/2}e^{-\beta(E-\mu)} = (2S+1)e^{\beta\mu}A\frac{1}{\beta^{5/2}}\int_0^\infty duu^{3/2}e^{-u}$$

$$= (2S+1)e^{\beta\mu}A\frac{1}{\beta^{5/2}}\Gamma(5/2) \tag{5.32}$$

where one has used the definition of the Γ function (see Appendix A). Similarly, Eq. (5.13) becomes at high T

$$N \simeq (2S+1) \int_0^\infty dE \rho(E) e^{-\beta(E-\mu)} = (2S+1)e^{\beta\mu} A \frac{1}{\beta^{3/2}} \int_0^\infty du\, u^{1/2} e^{-u}$$

$$= (2S+1)e^{\beta\mu} A \frac{1}{\beta^{3/2}} \Gamma(3/2) \tag{5.33}$$

From these two equations, one finds

$$\frac{E}{N} = \frac{\Gamma(5/2)}{\beta\Gamma(3/2)} = \frac{3k_BT}{2} \tag{5.34}$$

This is the equation obtained in the previous chapters for a classical ideal gas. The free Fermi gas loses thus the quantum nature at high temperatures.

5.7 Applications

5.7.1 *Paramagnetism of conducting electrons in metals*

A metal can be approximatively viewed as a lattice of positively charged ions and an ensemble of conducting electrons. The latter is a Fermi gas because the electronic spin is $1/2$. In the first approximation, the ions are assumed to be immobile so that the Coulomb interaction between immobile positive ion charges yields a constant which is omitted. The electrons are considered as free electrons because the interaction energy between electrons are approximatively canceled by the interaction energy between electrons and ions of the metal (see for example [50]). Conducting electrons of a good metal form thus a free Fermi gas shown in the previous sections. In the following, one shall use the above results to study the effect of an applied magnetic field on conducting electrons in metals. Other applications are found in the list of problems given at the end of the chapter.

When a magnetic field $\vec{H}$ is applied to an electron gas, according to the Zeeman effect, the energy of an electron with spin parallel to $\vec{H}$ diminishes by $\mu_B H$, and that of an electron with spin antiparallel to $\vec{H}$ increases by the same amount. μ_B is the Bohr magneton defined by

$$\mu_B = \frac{e\hbar}{2m} \tag{5.35}$$

where e is the electron charge, m the electron mass and $\hbar = h/(2\pi)$ with h being the Planck constant. One writes

$$E_\uparrow = E - \mu_B H \tag{5.36}$$

$$E_\downarrow = E + \mu_B H \tag{5.37}$$

The magnetic moment of the electron gas is given by

$$M = \mu_B(N_\uparrow - N_\downarrow) \tag{5.38}$$

where $N_\uparrow$ and $N_\downarrow$ are the numbers of up and down spins ($\uparrow$ and $\downarrow$), respectively. Using the density of states for each kind of spin (2.36), one writes

$$M = \mu_B \int_0^\infty (\rho(E_\uparrow)f(E_\uparrow) - \rho(E_\downarrow)f(E_\downarrow))dE \tag{5.39}$$

In the case where H is weak, one can assume $\rho(E_\uparrow) \simeq \rho(E_\downarrow) \simeq \rho(E)$ because $\rho(E)$ is a smooth function. In the same spirit, one can replace $f(E_\uparrow)$ and $f(E_\downarrow)$ by their expansion around E:

$$f(E_\uparrow) \simeq f(E) + \frac{\partial f}{\partial E}(E_\uparrow - E) \tag{5.40}$$

$$f(E_\downarrow) \simeq f(E) + \frac{\partial f}{\partial E}(E_\downarrow - E) \tag{5.41}$$

Equation (5.39) becomes

$$M = \mu_B^2 H \int_0^\infty \rho(E)\left(-\frac{\partial f}{\partial E}\right) dE \tag{5.42}$$

At low temperatures, $\frac{\partial f}{\partial E}$ is important only around E_F (see Fig. 5.2), therefore,

$$M \simeq 2\mu_B^2 H\rho(E_F) \int_0^\infty \left(-\frac{\partial f}{\partial E}\right) dE = 2\mu_B^2 H\rho(E_F)[f(0) - f(\infty)]$$
$$= 2\mu_B^2 H\rho(E_F) \tag{5.43}$$

Ii is noted that $\rho(E_F)$ used here is for one kind of spin as defined by (2.36) (without the spin degeneracy).

The magnetic susceptibility is thus

$$\chi = \frac{dM}{dH} = 2\mu_B^2 \rho(E_F) \tag{5.44}$$

χ is called "susceptibility of Pauli paramagnetism". It is independent of temperature.

To calculate M at a higher order in T, one can use first the integration by parts for (5.42) in order to put it in the form (5.25), then the Sommerfeld's expansion. One will find that χ is the constant given by (5.44) plus a term of the order T^2 (see Problem 9).

5.7.2 *Thermo-ionic emission*

Under the effect of the temperature T, electrons can leave the surface of a metal. One calculates the current density in the vicinity of the surface.

Let $\phi = E_0 - E_F$ be the "work function" where E_0 is the energy level of the vacuum near the surface. The current density in the x direction perpendicular to the surface is written as

$$
\begin{aligned}
j &= -2e\frac{1}{h^3}\int\int\int n(E)v_x dp_x dp_y dp_z \\
&= -2e\frac{m^3}{h^3}\int\int\int \frac{v_x dv_x dv_y dv_z}{\exp[\beta(E-E_F)]+1} \\
&= -\frac{2m^3 e}{h^3}\int_{-\infty}^{\infty} dv_y \int_{-\infty}^{\infty} dv_z \int_{v_x^0}^{\infty} \\
&\quad \times \frac{v_x dv_x}{\exp[\beta(\frac{m}{2}(v_x^2 + v_y^2 + v_z^2) - E_F)]+1}
\end{aligned}
\tag{5.45}
$$

where one has used:

- integration in the phase space

$$
\frac{1}{h^3}\int dp_x \int dp_y \int dp_z ... = m^3 \int dv_x \int dv_y \int dv_z ...
$$

  [see (2.40)]
- factor 2 is for spin degeneracy
- $v_x^0 = \sqrt{2E_0/m} = [\frac{2}{m}(\phi + E_F)]^{1/2}$ = minimum velocity in order to be able to escape to the vacuum.

One supposes $\beta(E - E_F) \gg 1$ (high temperatures). One has then

$$
\begin{aligned}
j &\simeq -\frac{2m^3 e}{h^3}e^{\beta E_F}\int_{-\infty}^{\infty} dv_y \exp[-\beta m v_y^2/2]\int_{-\infty}^{\infty} dv_z \exp[-\beta m v_z^2/2] \\
&\quad \times \int_{v_x^0}^{\infty} dv_x v_x \exp[-\beta m v_x^2/2] \\
&= -\frac{2m^3 e}{h^3}e^{\beta E_F}\frac{2\pi k_B T}{m}\int_{v_x^0}^{\infty} dv_x v_x \exp[-\beta m v_x^2/2] \\
&= -\frac{2m^3 e}{h^3}e^{\beta E_F}\frac{2\pi k_B T}{m}\left[\frac{k_B T}{m}\exp\left(-\frac{m(v_x^0)^2}{2k_B T}\right)\right] \\
&= -\frac{4\pi m e}{h^3}(k_B T)^2 e^{-\beta\phi}
\end{aligned}
\tag{5.46}
$$

This relation is called "Richardson's equation" which has been experimentally verified.

5.8   Conclusion

In this chapter, properties of a free Fermi gas have been demonstrated at zero, low and high temperatures. The difference with a classical ideal gas at low temperatures was emphasized. The obtained results allow for an understanding of the behaviors of conducting electrons in metals. Though simple, the free-fermion theory presented here provides results qualitatively in agreement with experiments for example on the linear temperature-dependence of the specific heat and the almost temperature-independent susceptibility for good metals at low temperatures. The results of this chapter provide a necessary background for improvements by taking into account, for example, interaction between electrons (see chapter 10) or interaction with the crystalline potential (see chapter 11).

5.9   Problems

Problem 1. Free Fermi gas at thermodynamic limit:

   Consider a free Fermi gas of volume V which is assumed to be of macroscopic dimension.

   a) Calculate the grand potential J. Show that $J = -2\overline{E}/3$
   b) Calculate the grand-canonical entropy S
   c) Calculate the pressure p. Show that $p = 2\overline{E}/(3V)$.

Problem 2. Fermi gas at low temperatures:

   Consider the free Fermi gas of the previous problem at low temperatures.

   Calculate J, S and p at the lowest order in T. Compare these results to those of a classical ideal gas.

Problem 3. Free Fermi gas in one and two dimensions:

   Calculate the Fermi energy and average kinetic energy per electron at $T=0$ Kelvin in a metal in one and two dimensions as functions of the electron density n.

Problem 4. Free Fermi gas in two dimensions:

   a) Find the relation between the density of electrons and the Fermi wave vector in two dimensions.
   b) In two dimensions, the density of states for free electrons $\rho(E)$ is a constant D. Show that in this case all terms in the Sommerfeld's expansion for the number of electrons N vanish except the $T = 0$

term. Show that μ (chemical potential)=E_F (Fermi energy) for all temperature T.

c) Calculate the average energy $\overline{E}$ at low temperatures. Find the calorific capacity.

d) Calculate μ and $\overline{E}$ at high temperatures. Comment.

e) Calculate μ exactly and show that μ is given by

$$\mu = k_B T \ln\left[\exp(\beta\mu_0) - 1\right] \tag{5.47}$$

Problem 5. Pressure of a free Fermi gas:

Show that the pressure of a free Fermi gas is always stronger than that of a classical ideal gas.

Problem 6. Free Fermi gas with internal degrees of freedom:

Consider a free Fermi gas of N fermions of spin s, of volume V, maintained at temperature T. The particle i has a kinetic energy E_i and two energy levels 0 and $\epsilon > 0$ which are due to the internal degrees of freedom. One supposes that $\epsilon \ll \mu_0$ where μ_0 is the chemical potential at $T = 0$.

a) Establish a relation between N, T and μ.

b) Find a relation between μ_0 and ϵ. Solve the equation to determine μ_0.

c) Calculate at the lowest order in ϵ the energy of the gas at $T = 0$.

Problem 7. Ultra relativistic ideal gas:

Consider a gas of N free fermions ($N \gg 1$) of mass m and of spin $1/2$, contained in a recipient of volume V maintained at temperature T. One wishes to study some properties of this gas in the ultra relativistic limit where the energy of a particle with an impulsion $\vec{p}$ is equal to $E = cp$, p being the modulus of $\vec{p}$ and c the light velocity.

a) Write explicitly the sum over individual microstates according to the parameters given above.

b) Write down the expressions (integrals) of the following quantities as functions of T, V and m

- the number of particles N
- the average energy $\overline{E}$
- the grand potential J of the gas. Find a relation of the form $J = a\overline{E}$ where a is a constant to be determined.

c) One supposes now that the temperature of the gas is zero. Calculate as a function of N and V the chemical potential μ_0, the average energy $\overline{E}_0$ and the pressure P_0 of the gas.

d) Define the low-temperature limit. Calculate in this limit, to the lowest order in T, the following quantities as functions of T, μ_0, and $\overline{E}_0$: the chemical potential, the calorific capacity at constant volume and the entropy of the gas.

Problem 8. Electrons in Sodium:

a) Calculate the number of electrons occupying the states between E_F and $E_F - 0.5 \ eV$, per volume unit in metallic Sodium at 0 Kelvin. The mass per volume unit of Na is $m_v = 0.97$ g cm^{-3}, and its mass per mole is $m_m = 22.99$ g/mole.

b) What is the Fermi temperature T_F?

Problem 9. Pauli paramagnetism:

Consider a gas of free electrons under an applied magnetic field $\vec{B}$ in three dimensions. Calculate the susceptibility at low and high temperatures in the case where B is rather weak.

Problem 10. Electrons in Copper:

Consider a metal with a density of conducting electrons n per volume unit.

a) Calculate v_F, the velocity of electrons at the Fermi level, as a function of n.

Numerical application: Calculate v_F for Cu using $n = 8.45 \times 10^{22}$ cm^{-3}, $m = 9.1095 \times 10^{-28}$ g and $\hbar = 1.05459 \times 10^{-27}$ erg-s.

b) Calculate the average kinetic energy per electron at 0 K. Give it in the unit of Rydberg (1 Ry $= \frac{me^4}{2\hbar^2}$).

Chapter 6

Free Boson Gas

6.1 Introduction

In this chapter, one studies the fundamental properties of a free boson gas using the Bose-Einstein statistics demonstrated in chapter 4. In the boson case, each individual state λ can be occupied by an arbitrary number of particles. This leads to low-temperature physical properties completely different from those of a free fermion gas studied in chapter 5.

There is a large number of systems of particles or quasi-particles following the Bose-Einstein statistics. Bosons are particles of integer spins, while bosonic quasi-particles are quanta of energy of quantized elementary excitations of systems of interacting particles. Examples of bosonic quasi-particles includes phonons (quantized vibrations of coupled atoms in crystals), magnons (spin waves), plasmons (charge waves) and photons (light waves).

It is obvious that the simplest case of boson systems is a gas of free bosons. By definition, a free boson gas is a system of bosons without interaction. In this chapter, fundamental properties of such a system are presented. They allow one to understand spectacular quantum phenomena at low temperatures such as the Bose-Einstein condensation. They prepare the way for the investigation of properties of systems of interacting bosons which will be shown in the following chapters.

6.2 Bose-Einstein distribution

In chapter 4, one has demonstrated the Bose-Einstein statistics which gives the following expression of the average occupation number of the microstate λ of energy E_λ at the temperature T [see (4.49)]:

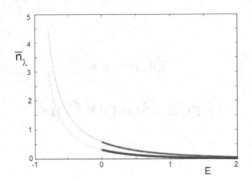

Fig. 6.1   Bose-Einstein distribution $\bar{n}_\lambda$ versus E for $T = 1$ (upper curve) and $T = 0.7$ (lower curve), with $\mu = -1$, $k_B = 1$. The discontinued curves are the continuations of $\bar{n}_\lambda$ in the non physical region $E < 0$.

$$\bar{n}_\lambda = \frac{1}{\exp[\beta(E_\lambda - \mu)] - 1} \qquad (6.1)$$

Since $\bar{n}_\lambda \geq 0$, one should have $\exp[\beta(E_\lambda - \mu)] \geq 1$. This constraint implies $E_\lambda - \mu \geq 0$. If the lowest energy E_0, namely the ground-state energy, satisfies the condition $E_0 - \mu \geq 0$, then all other energies will satisfy $E_\lambda - \mu > 0$. Note that $E_0 = 0$ for free bosons considered here, so that $\mu \leq 0$. Figure 6.1 shows $\bar{n}_\lambda$ versus energy E considered here as a continuous variable.

At $T = 0$, namely $\beta = \infty$, if $E_\lambda > E_0$, then $\bar{n}_\lambda = 0$ because $E_\lambda - \mu > 0$. If $E_\lambda = E_0$, then $\bar{n}_\lambda = \infty$ because $E_0 - \mu = 0$. This means that at $T = 0$, states other than the ground state are empty, and the ground state contains all the system particles.

6.3   General properties of a free boson gas

6.3.1   *General formulas*

One gives in the following principal general formulas of a free boson gas:

- Number of particles:
  The total number of particles of a boson gas is

$$N = \sum_\lambda \bar{n}_\lambda = \sum_\lambda \frac{1}{e^{\beta(E_\lambda - \mu)} - 1} \qquad (6.2)$$

- Energy:

The total energy is

$$\overline{E} = \sum_\lambda \overline{n}_\lambda E_\lambda = \sum_\lambda \frac{E_\lambda}{e^{\beta(E_\lambda - \mu)} - 1} \qquad (6.3)$$

- Grand potential:
  Using (4.46) and (4.50) one has

$$z_\lambda = 1 + \overline{n}_\lambda \qquad (6.4)$$

The grand potential (see chapter 4) is thus

$$J = -k_B T \ln \mathcal{Z} = -k_B T \sum_\lambda \ln z_\lambda$$

$$= -k_B T \sum_\lambda \ln[1 + \overline{n}_\lambda] \qquad (6.5)$$

or, by replacing $\overline{n}_\lambda$ by (6.1),

$$J = k_B T \sum_i \ln\left(1 - e^{-\beta(E_i - \mu)}\right) \qquad (6.6)$$

- Entropy:
  The entropy of the gas is given by

$$S = -\frac{\partial J}{\partial T} = k_B \sum_\lambda [\ln(1 + \overline{n}_\lambda) + \beta(E_\lambda - \mu)\overline{n}_\lambda]$$

$$= k_B \sum_\lambda \left[\ln(1 + \overline{n}_\lambda) + \overline{n}_\lambda \ln \frac{1 + \overline{n}_\lambda}{\overline{n}_\lambda}\right]$$

$$= k_B \sum_\lambda \left[(1 + \overline{n}_\lambda)\ln(1 + \overline{n}_\lambda^{BE}) - \overline{n}_\lambda \ln \overline{n}_\lambda\right] \qquad (6.7)$$

where one has taken the logarithm of the relation $\exp[\beta(E_\lambda - \mu)]$ $= [1 + \overline{n}_\lambda]/\overline{n}_\lambda$ deduced from (6.1).
- Pressure:
  The pressure of the gas is

$$p = -\frac{\partial J}{\partial V} = k_B T \frac{\partial}{\partial V} \sum_\lambda \ln[1 + \overline{n}_\lambda] \qquad (6.8)$$

As in the case of a free fermion, the microstate λ of a free boson is defined by its wave vector $\vec{k}$ and its spin state σ: $\lambda = (\vec{k}, \sigma)$. The sum over λ is decomposed into two sums $\sum_\lambda \cdots = \sum_{\sigma=-S}^{S} \sum_k \cdots$. If the particle energy does not depend on σ as in the case of a free particle, then the sum over σ gives a factor $(2S + 1)$ because $\sigma = -S, -S + 1, ..., S$ where S is the spin amplitude which is an integer (zero or positive) in the case of bosons.

6.3.2   *Formulas for large systems*

For systems of macroscopic sizes, namely at the so-called thermodynamic limit, the system energy can be considered as a continuous variable. If E_k depends only on the modulus k as in the free boson case, one can replace the sums on $\vec{k}$ by an integral over $\vec{k}$ or an integral over energy E. The exact formula (6.2) becomes

$$N = (2S+1)\frac{L^3}{(2\pi)^3}\int d\vec{k}\frac{1}{e^{\beta(E_k-\mu)}-1} \tag{6.9}$$

or

$$N = (2S+1)\int_0^\infty dE\rho(E)\frac{1}{e^{\beta(E-\mu)}-1} \tag{6.10}$$

Similarly, Eq. (6.3) becomes

$$\overline{E} = (2S+1)\frac{L^3}{(2\pi)^3}\int d\vec{k}\frac{E_k}{e^{\beta(E_k-\mu)}-1} \tag{6.11}$$

or

$$\overline{E} = (2S+1)\int_0^\infty dE\rho(E)\frac{E}{e^{\beta(E-\mu)}-1} \tag{6.12}$$

For a given value of N, the chemical potential μ is determined from the integral (6.10) for each value of T. Unlike the case of free fermions, there is no solution of Eq. (6.10) for μ at very low temperatures in the case of free bosons. In this case, one has to come back to the exact formula (6.2) and proceeds in the manner described in section 6.5. Above this low-temperature region, one can determine μ graphically as follows: replacing $\rho(E)$ by (2.36) in (6.10), one has

$$\frac{N}{V}\frac{4\pi^2}{2S+1}\left(\frac{\hbar^2\beta}{2m}\right)^{3/2} = \int_0^\infty dx x^{1/2}\frac{1}{e^x/\phi-1} \tag{6.13}$$

where $x \equiv \beta E$ and $\phi \equiv e^{\beta\mu}$ (fugacity). For a graphical solution, one plots in Fig. 6.2 the curve $y_1(\phi)$, representing the right-hand side of the above equation, versus ϕ. y_1 is bounded at $\phi = 1$ because $\mu \leq 0$. The left-hand side of Eq. (6.13), called y_2, is a constant with respect to ϕ. This constant depends however on T as $1/T^{3/2}$. y_2 is shown by the horizontal line in Fig. 6.2. The intersection of y_1 and y_2 gives the solution of ϕ which yields μ. For each given value of T, one obtains a solution for μ. One sees however that when T diminishes, μ increases and tends toward zero. At a low enough temperature where $y_2 > y_1(1)$ there is no more intersection. One shall treat this case in section 6.5.

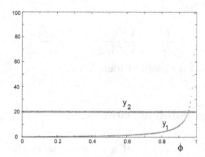

Fig. 6.2 Graphical determination of μ at a given temperature T: the intersection of $y_1(\phi)$ and y_2 gives the solution of $\phi = \exp(\beta\mu)$ from which one obtains μ. There is no solution when $y_2 > y_1(\phi = 1)$.

One shows below that the grand potential and the pressure of a free boson gas of large size are given by

$$J = -\frac{2}{3}\overline{E} \tag{6.14}$$

$$p = \frac{2\overline{E}}{3V} \tag{6.15}$$

These equations are the same as those of a free Fermi gas but the expression of $\overline{E}$ is different for the two cases.

6.4   High-temperature limit

As in a free Fermi gas, quantum effects in a free boson gas manifest at low temperatures. These effects disappear at high temperatures: the free boson gas behaves as a classical ideal gas as shown below.

One first calculates the energy of a free boson gas at high temperatures where $e^{\beta(E-\mu)} \gg 1$. Equation (6.12) becomes

$$\overline{E} = (2S + 1)\int_0^\infty dE\rho(E)Ee^{-\beta(E-\mu)} \tag{6.16}$$

One proceeds to a similar calculation as for (5.32): replacing $\rho(E)$ by (2.36) and putting $u = \beta E$, one obtains

$$\overline{E} = (2S + 1)e^{\beta\mu}A\frac{1}{\beta^{5/2}}\Gamma(5/2) \tag{6.17}$$

In the same manner as for (5.33), Eq. (6.10) becomes at high temperatures

$$N \simeq (2S + 1)e^{\beta\mu}A\frac{1}{\beta^{3/2}}\Gamma(3/2) \tag{6.18}$$

from which one has

$$\frac{\overline{E}}{N} = \frac{\Gamma(5/2)}{\beta\Gamma(3/2)} = \frac{3k_BT}{2} \tag{6.19}$$

This is the equation of a classical ideal gas.

6.5   Bose-Einstein condensation

The most remarkable property of a boson gas is the phenomenon called "Bose-Einstein condensation". One has seen in Fig. 6.2 that there is no solution of μ when $y_2 > y_1(1)$, namely when

$$\frac{N}{V}\frac{4\pi^2}{2S+1}(\frac{\hbar^2\beta}{2m})^{3/2} > \int_0^\infty dx x^{1/2}\frac{1}{e^x-1} \simeq \frac{\sqrt{\pi}}{2}2.612 \tag{6.20}$$

where one has used a formula in Appendix A. One deduces that no solution of μ can be found from the integral (6.10) for temperatures lower than T_B given by $y_2 = y_1(1)$, namely

$$T_B = \frac{1}{k_B}\frac{2\pi\hbar^2}{m}\left[\frac{N}{V}\frac{1}{2S+1}\frac{1}{2.612}\right]^{2/3} \tag{6.21}$$

T_B is called "Bose temperature".

For $T < T_B$, one goes back to the exact formula (6.2). In the ground state $E_0 = 0$, thus

$$N = N_0 = \frac{1}{\exp(-\beta\mu) - 1} \tag{6.22}$$

from which, one has

$$\exp(-\beta\mu) = 1 + \frac{1}{N_0} \tag{6.23}$$

One has seen above that the graphical solution presented in Fig. 6.2 shows that when T decreases, μ tends to 0, its upper limit. Replacing $\exp(-\beta\mu) \simeq 1 - \beta\mu$, one gets

$$\mu \simeq -\frac{k_BT}{N_0} \tag{6.24}$$

For $0 < T < T_B$, one puts

$$N = N_0 + N_e \tag{6.25}$$

where N_e is the number of bosons in excited states. Given the fact that μ is very small in this temperature region, one can calculate N_e using (6.10) by taking $\mu \sim 0$. One has

$$N_e \simeq (2S+1)\int_0^\infty dE\rho(E)\frac{1}{e^{\beta E} - 1} \tag{6.26}$$

Replacing $\rho(E) = AE^{1/2}$ by (2.36), one gets

$$N_e \simeq (2S+1)A \int_0^\infty \frac{dE E^{1/2}}{e^{\beta E} - 1} = (2S+1)A(k_B T)^{3/2} \int_0^\infty \frac{dx x^{1/2}}{e^x - 1} \quad (6.27)$$

where one has put $x = \beta E$. The last integral can be calculated as follows:

$$\int_0^\infty \frac{dx x^{1/2}}{e^x - 1} = \int_0^\infty \frac{dx x^{1/2} e^{-x}}{1 - e^{-x}} = \int_0^\infty dx x^{1/2} e^{-x} \sum_{n=0}^\infty e^{-nx}$$

$$= \int_0^\infty dx x^{1/2} \sum_{n=0}^\infty e^{-(n+1)x} = \sum_{m=1}^\infty \int_0^\infty dx x^{1/2} e^{-mx}$$

$$= \sum_{m=1}^\infty m^{-3/2} \int_0^\infty dy y^{1/2} e^{-y} = \zeta(\tfrac{3}{2})\Gamma(\tfrac{3}{2})$$

$$= 1.306\sqrt{\pi} \quad (6.28)$$

where one has used $\zeta(\tfrac{3}{2}) \simeq 2.612$ and $\Gamma(\tfrac{3}{2}) = \frac{\sqrt{\pi}}{2}$ (see Appendix A). Replacing (6.28) and A in (6.27), one obtains

$$N_e \simeq (2S+1)\frac{V}{4\pi^2} \left(\frac{2m}{\hbar^2}\right)^{3/2} (k_B T)^{3/2} 1.306\sqrt{\pi}$$

$$= (2S+1)\frac{1.306 V}{4} \left(\frac{2mk_B T}{\hbar^2 \pi}\right)^{3/2} \quad (6.29)$$

Using (6.21), one can rewrite N_e as

$$\frac{N_e}{N} = \left(\frac{T}{T_B}\right)^{3/2} \quad (6.30)$$

This relation is valid for $T < T_B$. When $T = T_B$, $N_e \simeq N$. Finally, one has

$$N_0 = N - N_e = N \left[1 - \left(\frac{T}{T_B}\right)^{3/2}\right] \quad (6.31)$$

This relation indicates that $N_0 = N$ when $T = 0$ as expected for the ground state.

6.6 Properties at temperatures higher than T_B

One calculates now general properties of a free boson gas at temperatures higher than T_B. One shall recover of course the result of a classical ideal gas at high temperatures found in section 6.4.

- Number of particles:
One has seen in the precedent section that the number of bosons in the ground state is zero for $T > T_B$, namely all bosons are in the excited states. One has

$$N_e = (2S+1) \int_0^\infty dE \rho(E) \frac{1}{\phi^{-1} e^{\beta E} - 1} \qquad (6.32)$$

where $\phi \equiv e^{\beta \mu}$ (fugacity). As $\mu \leq 0$, one has $\phi \leq 1$. Calculating in the same manner as for (6.27)-(6.29) by replacing $\rho(E)$ by $AE^{1/2}$ and by putting $x = \beta E$, one gets

$$N_e = (2S+1) A (k_B T)^{3/2} \int_0^\infty \frac{dx \, x^{1/2}}{\phi^{-1} e^x - 1} \qquad (6.33)$$

This integral can be calculated as follows:

$$\int_0^\infty \frac{dx \, x^{1/2}}{\phi^{-1} e^x - 1} = \int_0^\infty \frac{dx \, x^{1/2} \phi e^{-x}}{1 - \phi e^{-x}}$$

$$= \int_0^\infty dx \, x^{1/2} \phi e^{-x} \sum_{n=0}^\infty \phi^n e^{-nx}$$

$$= \int_0^\infty dx \, x^{1/2} \sum_{n=0}^\infty \phi^{(n+1)} e^{-(n+1)x}$$

$$= \sum_{m=1}^\infty \int_0^\infty dx \, x^{1/2} \phi^m e^{-mx}$$

$$= \sum_{m=1}^\infty \frac{\phi^m}{m^{3/2}} \int_0^\infty dy \, y^{1/2} e^{-y} = F(\phi) \Gamma(\frac{3}{2})$$

$$= F(\phi) \frac{\sqrt{\pi}}{2} \qquad (6.34)$$

where one has put $m = n+1$ and $y = mx$, and one has used the following definition:

$$F(\phi) = \sum_{m=1}^\infty \frac{\phi^m}{m^{3/2}} \qquad (6.35)$$

Finally, one has

$$N_e = (2S+1) A (k_B T)^{3/2} F(\phi) \frac{\sqrt{\pi}}{2} \qquad (6.36)$$

where

$$A = \frac{V}{4\pi^2} \left(\frac{2m}{\hbar^2} \right)^{3/2} \qquad (6.37)$$

- Average energy:
One has

$$\overline{E} = (2S+1)A \int_0^\infty \frac{dE E^{3/2}}{\phi^{-1}e^{\beta E} - 1}$$

$$= (2S+1)A(k_BT)^{5/2} \int_0^\infty \frac{dx x^{3/2}}{\phi^{-1}e^x - 1} \qquad (6.38)$$

where $x = \beta E$. The same calculation as that for N_e leads to

$$\overline{E} = (2S+1)A(k_BT)^{5/2}H(\phi)\Gamma(\frac{5}{2})$$

$$= (2S+1)A(k_BT)^{5/2}H(\phi)\frac{3\sqrt{\pi}}{4} \qquad (6.39)$$

where

$$H(\phi) = \sum_{m=1}^\infty \frac{\phi^m}{m^{5/2}} \qquad (6.40)$$

It is interesting to note that by using (6.36) and (6.39), one obtains

$$\overline{E} = \frac{3}{2}k_BTN_e\frac{H(\phi)}{F(\phi)} \qquad (6.41)$$

At high temperatures, $\phi \ll 1$. As a consequence, only the first term of $H(\phi)$ and the first term of $F(\phi)$ are relevant. One has then $H(\phi) \simeq F(\phi) \simeq \phi$: one recovers the result of the classical ideal gas $\overline{E} = \frac{3}{2}Nk_BT$.

- Grand potential:
Using (6.5), one writes

$$J = -k_BT \sum_\lambda \ln[1 + \overline{n}_\lambda]$$

$$= -k_BT \sum_\lambda \ln\left[1 + \frac{1}{e^{\beta(E_\lambda - \mu)} - 1}\right]$$

$$= k_BT \sum_\lambda \ln[1 - x_\lambda] \qquad (6.42)$$

where one has put $x_\lambda = e^{-\beta(E_\lambda - \mu)}$. For a system of macroscopic size, one has

$$J = k_BT(2S+1) \int_0^\infty dE\rho(E) \ln\left[1 - e^{-\beta(E-\mu)}\right]$$

$$= k_BT(2S+1)A \int_0^\infty dE E^{1/2} \ln\left[1 - \phi e^{-\beta E}\right]$$

$$= (2S+1)A(k_BT)^{5/2} \int_0^\infty dx x^{1/2} \ln\left[1 - \phi e^{-x}\right] \qquad (6.43)$$

This integral can be calculated by parts as follows:

$$\int_0^\infty dx\, x^{1/2} \ln\left[1 - \phi e^{-x}\right] = \left(\frac{2}{3}x^{3/2}\ln\left[1 - \phi e^{-x}\right]\right)_0^\infty$$

$$-\frac{2}{3}\phi\int_0^\infty dx\,\frac{x^{3/2}e^{-x}}{1 - \phi e^{-x}}$$

$$= 0 - \frac{2}{3}\int_0^\infty dx\,\frac{x^{3/2}}{\phi^{-1}e^x - 1}$$

$$= -\frac{2}{3}H(\phi)\Gamma(\frac{5}{2}) = -\frac{2}{3}H(\phi)\frac{3\sqrt{\pi}}{4}$$

$$(6.44)$$

where one has used the integral (6.38) in the last line. Finally,

$$J = -\frac{2}{3}(2S+1)A(k_BT)^{5/2}H(\phi)\frac{3\sqrt{\pi}}{4} \tag{6.45}$$

$$= -\frac{2}{3}(2S+1)\frac{V}{4\pi^2}\left(\frac{2m}{\hbar^2}\right)^{3/2}(k_BT)^{5/2}H(\phi)\frac{3\sqrt{\pi}}{4} \tag{6.46}$$

By comparing (6.39) to (6.45), one obtains

$$J = -\frac{2}{3}E \tag{6.47}$$

- Pressure:

The pressure of a free boson gas at $T > T_B$ is given by

$$p = -\frac{\partial J}{\partial V} = \frac{2}{3}(2S+1)\frac{1}{4\pi^2}(\frac{2m}{\hbar^2})^{3/2}(k_BT)^{5/2}H(\phi)\frac{3\sqrt{\pi}}{4} = -\frac{J}{V} \tag{6.48}$$

Replacing J by (6.47), one has

$$pV = \frac{2}{3}E \tag{6.49}$$

This is the equation of state of a free boson gas.

6.7   Applications

6.7.1   *Photons: black-body radiation*

"Black-body radiation" is a phenomenon observed in the cavity made inside a material when it is heated to a high temperature. Note however that this term indicates in general electromagnetic radiation inside or around a non-reflective object at thermodynamic equilibrium. The term "black-body was historically used because the radiation spectrum at room temperature is infra-red and cannot be perceived by the human eye. The spectrum depends only on the temperature of the object. At very high temperatures, the intensity of the spectrum increases and the colors range from dull red to blindingly brilliant blue-white. The observation of a perfect black-body radiation emitted by the wall of the cavity is made through a small hole. Other objects such as the sun make an approximate black-body radiation which allows an approximate estimation of their emitted energy.

One examines a perfect black-body radiation by a cavity in a material heated to the temperature T. At thermodynamic equilibrium, a number of atoms of the wall of the cavity absorbs thermal energies to go to excited states. When they come back to their initial states they emit absorbed energies as a radiation. They go and back between absorption and radiation. These electromagnetic radiations make multiple reflections on the cavity wall. Some of them escape through the hole which allows one to observe the emitted spectrum. At equilibrium, the numbers of absorptions and emissions are equal. Note that emitted electromagnetic waves in the cavity have different wave-lengths λ. One knows from the quantum mechanics that these waves are also particles called "photons" (wave-particle duality). A photon is a boson with zero mass and zero chemical potential. Its energy associated with the wave-length λ is written as $E = \hbar\omega = h\nu$ with $\nu = c/\lambda$ (c: light velocity).

One demonstrates in the following that when a body is heated to the temperature T, the intensity of the light emitted in the cavity (or the energy density radiated per frequency unit) undergoes a maximum at a certain frequency. This maximum moves to higher frequencies with increasing temperature as shown in Fig. 6.3.

One supposes that the photons inside the cavity form a free boson gas. The number of photons of energy $\hbar\omega$ is then

$$f(\omega) = \frac{1}{\exp(\beta\hbar\omega) - 1} = \frac{1}{\exp(\beta\hbar ck) - 1} \tag{6.50}$$

where one has replaced ω by ck because $\omega = 2\pi\nu = 2\pi c/\lambda = ck$ (k: modulus of the wave-vector $\vec{k}$). Using (6.9), the total number of photons in the cavity is given by

$$N = 2\frac{V}{(2\pi)^3} \int \frac{d\vec{k}}{\exp(\beta\hbar ck) - 1} \tag{6.51}$$

where the factor 2 is due to the degeneracy of the photon polarization, and V denotes the cavity volume. Using $d\vec{k} = 4\pi k^2 dk$ and putting $x = \beta\hbar ck$, one obtains

$$\frac{N}{V} = \frac{1}{(\pi)^2 c^3 \hbar^3}(k_B T)^3 \int_0^\infty \frac{x^2 dx}{\exp(x) - 1} \simeq 0.244 \left(\frac{k_B T}{\hbar c}\right)^3 \qquad (6.52)$$

where one has used an integral formula of Appendix A.

The energy of photons in the cavity is given by [see (6.11)]

$$\overline{E} = 2\frac{V}{(2\pi)^3} \int \frac{d\vec{k}\hbar ck}{\exp(\beta\hbar ck) - 1} \qquad (6.53)$$

As before, using $d\vec{k} = 4\pi k^2 dk$ and putting $x = \beta\hbar ck$, one gets

$$\overline{E} = V\hbar c \left(\frac{k_B T}{\hbar c}\right)^4 \int_0^\infty \frac{x^3 dx}{\exp(x) - 1} = \frac{8\pi^5 k_B^4 V}{15c^3 h^3}T^4 \qquad (6.54)$$

where a formula of Appendix A has been used. The relation (6.54) is often written as

$$\frac{\overline{E}}{V} = \sigma T^4 \qquad (6.55)$$

where $\sigma = \frac{8\pi^5 k_B^4}{15c^3 h^3}$ is called "Stefan constant". Note that the energy does not depend on the nature of the material. The expression (6.55), called "Stefan-Boltzmann law", is in an excellent agreement with experiments.

The relation (6.53) is in agreement with the Planck law which is stated as follows: the energy density (energy per frequency unit) of thermal radiation is given by

$$u(\lambda, T) = \frac{2hc^2}{\lambda^5} \frac{1}{\exp(\beta hc/\lambda) - 1} \qquad (6.56)$$

To see this agreement, it suffices to use the definition of the energy density given by (6.50):

$$d\overline{E} = u(\lambda, T)d\nu = \frac{h\nu}{\exp(\beta h\nu) - 1}d\nu = \frac{h\nu}{\exp(\beta h\nu) - 1}2\frac{V}{(2\pi)^3}d\vec{k} \qquad (6.57)$$

where $2\frac{V}{(2\pi)^3}d\vec{k}$ is the density of states in the $\vec{k}$ space. This equation is rewritten as

$$d\overline{E} = u(\lambda, T)d\nu = \frac{8\pi hV}{c^3} \frac{\nu^3}{\exp(\beta h\nu) - 1}d\nu \qquad (6.58)$$

using $k = 2\pi\nu/c$. One recovers here $u(\lambda, T)$ given by (6.56). At low frequencies, $\beta h\nu \ll 1$, one has

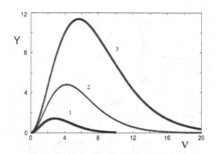

Fig. 6.3 Energy density $Y = u(\nu, T)$ (energy per frequency unit) of thermal radiation in the cavity of a material versus radiated frequency ν, at three temperatures. Curves 1, 2 and 3 correspond to temperatures T_1, T_2 and T_3 with $T_1 < T_2 < T_3$ in arbitrary unit.

$$u(\nu, T) = \frac{8\pi h V}{c^3} \frac{\nu^3}{\exp(\beta h\nu) - 1} \simeq \frac{8\pi V T}{c^3} \nu^2 \tag{6.59}$$

This relation is called "Rayleigh-Jeans law". At high frequencies, $\beta h\nu \gg 1$, one has

$$u(\nu, T) = \frac{8\pi h V}{c^3} \frac{\nu^3}{\exp(\beta h\nu) - 1} \simeq \frac{8\pi h V}{c^3} \exp(-\beta h\nu)\nu^2 \tag{6.60}$$

This limit is called "Wien law".

The above formulas for the low- and high-frequency limits explain satisfactorily the behavior of the curve shown in Fig. 6.3.

To determine the wave-length at which the energy density $u(\lambda, T)$ is maximum, it suffices to solve $du(\lambda, T)/d\lambda = 0$. One then finds the so-called "Wien displacement law" (see Problem 7):

$$\lambda_{max} T = \text{constante} \tag{6.61}$$

where λ_{max} is the wave-length at which the emitted energy is maximum. When T increases, λ_{max} diminishes. In other words, $\nu_{max} = c/\lambda_{max}$ should increase. This displacement of the maximum as a function of T is shown in Fig. 6.3.

The pressure of the thermal radiation in the cavity is calculated in Problem 8 below. One obtains

$$p = \frac{1}{3} \frac{\overline{E}}{V} \tag{6.62}$$

This equation is the equation of state of the photon gas.

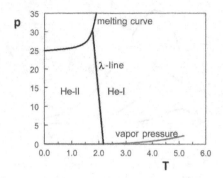

Fig. 6.4   Phase diagram in the $(p-T)$ space (p: pressure). p is in the unit of bar and T in Kelvin.

6.7.2   *Helium-4*

The nucleus of Helium-4 contains two protons and two neutrons. The resulting spin is zero, making Helium-4 a boson which obeys the Bose-Einstein statistics. In the normal conditions of temperature and pressure, atoms of Helium-4 form a boson gas which becomes a "normal" fluid (He-I phase) at 4.2 K with a density equal to 2.16×10^{22} atoms per cm^3. As the temperature decreases further, the liquid turns into a superfluid below $T_\lambda = 2.172$ K (He-II phase). The index λ was used to represent the shape of the peak of the specific heat around this temperature. The phase diagram in the $(p-T)$ space is shown in Fig. 6.4. Note that in the ground state, as bosons Helium-4 has zero impulsion, i. e. $\vec{p} = \hbar \vec{k} = 0$. At temperatures below T_λ their momenta are very small. The Heisenberg uncertainty principle $\Delta x \Delta p \simeq h$ implies that Δx is large. As a consequence, superfluid atoms are everywhere as if their viscosity is zero. If one uses the formula (6.21) for a free boson gas with the mass and the density 2.16×10^{22} atoms per cm^3 of Helium-4, one obtains $T_\lambda = 3.14$ K. The difference between this value and the observed temperature 2.172 K results from the fact that there is a weak interaction between Helium-4 atoms. In the He-II phase, if $T \neq 0$ then a very large number of atoms, N_0 in Eq. (6.31), are in the superfluid state with no thermal energy and no viscosity. The remaining N_e atoms are in the normal-fluid state. One of the most remarkable phenomena of the He-II is the so-called "fountain effect": superfluid atoms can climb up the wall of the container to get out of it by overflowing as on a fountain. This is due to a combination of several factors among which the zero viscosity enhances the motion by capillarity and by the gradient of the chemical potential. A treatment of the interaction between Helium-4 atoms by the second quantization is given in section 9.6 of chapter 9.

6.8 Conclusion

Fundamental properties of a free boson gas have been shown in this chapter. In particular, the behaviors of the gas at low- and high-temperature limits have been presented in details. The most remarkable phenomenon is no doubt the Bose-Einstein condensation which occurs below the Bose temperature T_B. This is one of rare quantum effects which manifest themselves at a macroscopic scale.

Some examples of bosonic systems have been shown to illustrate the application of the Bose-Einstein statistics. Thermal radiation, or black-body radiation, of photons in a cavity of an object was studied with some resulting laws which are very useful for example in the estimation of the temperature and emitted energy of faraway planets, from the observed spectrum. The case of a gas of Helium-4 was also shown to demonstrate the Bose-Einstein condensation. Systems of interacting bosons which need more sophisticated treatments will be shown in particular in chapters 9 and 12.

6.9 Problems

Problem 1. Gas of photons:
Show that the chemical potential of a gas of photons is zero.
Problem 2. Bose-Einstein condensation:
Using the expression of the energy of a free quantum particle in a box (2.34),

a) Calculate the energy of the first excited level of a Helium-4 atom
b) Calculate the chemical potential at $T = 1$ mK of a gas of Helium-4 having a density 10^{22} atoms per cm^3. Find the number of atoms occupying the first excited level using the result on the Bose-Einstein condensation shown in the chapter. If one uses the Maxwell-Boltzmann distribution, what is this number? Comment.

Problem 3. Pressure in a gas of bosons:

a) Show that at high temperatures, the expression (6.49) becomes the equation of state of a classical ideal gas.

b) Show that the pressure of a gas of bosons is always weaker than that of a classical ideal gas.

Problem 4. Two-dimensional gas of bosons:

Using the expression of N as a function of chemical potential μ in the thermodynamic limit for a two-dimensional gas of bosons, show that N can be written under the following form

$$N = AT \sum_{n=1}^{\infty} \exp(n\mu/k_B T)/n \qquad (6.63)$$

where A is a constant to be determined.

Show that there is no Bose-Einstein condensation in two dimensions.

Problem 5. Gas of bosons having internal degrees of freedom: Bose-Einstein condensation

Consider an ideal gas of N bosons, of volume V, maintained at the temperature T. One supposes that each particle, apart from its kinetic energy, possesses an internal degree of freedom yielding an internal energy equal to either $\epsilon_0 = 0$ or ϵ_1 ($\epsilon_1 > 0$).

a) Calculate the number de particles as a function of T and μ at high temperatures.

b) Find the Bose temperature T_B. Express T_B as a function of T_B^0 defined as the Bose temperature given by (6.21) without the internal degree of freedom ϵ_1.

c) One supposes that $\epsilon_1 \gg k_B T$. Find T_B.

Problem 6. Einstein's model for the vibration of atoms on a lattice:

Find the low- and high-temperature limits of the Einstein's model using the results of Problem 6 in chapter 3.

Problem 7. Wien displacement law:

Demonstrate the formula (6.61).

Problem 8. Equation of state of a gas of photons:

Demonstrate the formula (6.62).

Chapter 7

Systems of Interacting Particles: Method of Second Quantization

7.1 Introduction

To describe correctly a system of interacting particles, we have to include the interaction in the Schrödinger equation. In principle, the wave function contains all information on the system. However, a solution of the Schrödinger equation is often impossible because of the existence of a complicated potential and/or a large number of particles. The second quantization provides a technique to avoid the Schrödinger equation. Of course, we can show the equivalence of the two methods by going from the Schrödinger equation to the second quantization by just changing the way to describe the microstates of the system.

The second quantization method transforms the Hamiltonian in terms of creation and annihilation operators. It has the advantage that the Bose-Einstein or Fermi-Dirac statistics of the particles are implicitly incorporated in the Hamiltonian through the operators so that we do not need to use cumbersome products of individual wave functions to construct boson symmetric or fermion antisymmetric wave functions for the system as seen below.

We recall that the first quantization is the quantization of the particle energy by the boundary conditions applied to the solution of the Schrödinger equation. The wave function is not an operator. In the second quantization, the wave function is replaced by a field operator which is in fact a quantization of the wave function.

The method of second quantization is very useful in the study of systems of weakly interacting particles. In particular, the second quantization is an efficient tool to describe collective elementary excitations such as phonons and magnons.

7.2   First quantization: symmetric and antisymmetric wave functions

In this section one recalls some important points of the first quantization. One considers a system of N interacting particles. The general Hamiltonian is written as

$$\mathcal{H} = \sum_i H(\vec{r}_i) + \frac{1}{2} \sum_{i,j} V(\vec{r}_i, \vec{r}_j) \qquad (7.1)$$

where the first sum are terms each of which concerns one particle: $H(\vec{r}_i)$ can be a kinetic energy or an interaction of the particle at $\vec{r}_i$ with an external field etc. The second sum are interaction terms: $V(\vec{r}_i, \vec{r}_j)$ is the interaction between two particles at $\vec{r}_i$ and $\vec{r}_j$ (the case $i = j$ is excluded).

One considers the case of electrons where $V(\vec{r}_i, \vec{r}_j)$ is the Coulomb interaction:

$$V(\vec{r}_i, \vec{r}_j) = \frac{e^2}{|\vec{r}_i - \vec{r}_j|} \qquad (7.2)$$

where $-e$ is the electron charge. This interaction does not depend on the electron spin, but the wave function depends on the spin state, as will be seen below. Electrons are fermions, they obey the Pauli principle as seen in chapter 5. Each individual state is characterized by both its orbital and spin states. The wave function of electron i is written as

$$\psi_{\vec{k}\sigma}(\vec{r}_i, \zeta_i) = \phi_{\vec{k}}(\vec{r}_i) S_\sigma(\zeta_i) \qquad (7.3)$$

where $\phi_{\vec{k}}(\vec{r}_i)$ is the orbital state and $S_\sigma(\zeta_i)$ the spin state. One shall use the following notations:

$\sigma = +$ or $-$

$\zeta = \uparrow$ or $\downarrow$

and

$S_+(\uparrow) = 1, \quad S_+(\downarrow) = 0, \quad S_-(\uparrow) = 0, \quad S_-(\downarrow) = 1$

With these notations, it is clear that a particle of spin $\uparrow$ (respectively $\downarrow$) can occupy only the spin state $+$ (respectively $-$). Otherwise, the wave function $\psi_{\vec{k}\sigma}(\vec{r}_i, \zeta_i)$ is zero. One says $\psi_{\vec{k}\sigma}(\vec{r}_i, \zeta_i)$ is the wave function of the electron located at $\vec{r}_i$, having the spin ζ, occupying the state $(\vec{k}, \sigma)$.

For a system of N particles, the simplest wave function is a combination of products of individual wave functions $\psi_{\vec{k}\sigma}(\vec{r}_i, \zeta_i)$:

$$\psi_{f_1}(q_1)\psi_{f_2}(q_2)...\psi_{f_i}(q_i)\psi_{f_j}(q_j)...\psi_{f_N}(q_N) \qquad (7.4)$$

where, for simplicity, one has put $q_i = (\vec{r}_i, \zeta_i)$ and $f_i = (\vec{k}_i, \sigma_i)$. To satisfy the Pauli principle, the wave function of the system Ψ should be antisymmetric with respect to the exchange of two particles in their states: Ψ should change its sign. In addition, one has to take into account the indiscernibility of the particles, namely any particle can be in any individual state: the wave function should include all permutations of the particles $q_1, q_2, ..., q_N$ in the states $f_1, f_2, ..., f_N$. The form of Ψ which naturally satisfies these requirements is a determinant:

$$\Psi = A\bar{D} \tag{7.5}$$

where A denotes a constant determined by the normalization of Ψ and $\bar{D}$ is given by

$$\begin{vmatrix} \psi_{f_1}(q_1) & \psi_{f_1}(q_2) & \cdots & \psi_{f_1}(q_N) \\ \psi_{f_2}(q_1) & \psi_{f_2}(q_2) & \cdots & \psi_{f_2}(q_N) \\ \cdots & \cdots & \cdots & \cdots \\ \psi_{f_N}(q_1) & \psi_{f_N}(q_2) & \cdots & \psi_{f_N}(q_N) \end{vmatrix}$$

One sees that exchange of two particles ($q_i \leftrightarrow q_j$) is equivalent to exchange of two columns: the determinant changes its sign. In addition, if two particles have the same state, for example $f_i = f_j$, then two lines of the determinant are identical, so Ψ is zero. The Pauli principle is thus obeyed with a determinant wave function (7.5). The determinant $\bar{D}$ is called "Slater determinant".

One can rewrite Ψ as follows:

$$\Psi = A\sum_p (-1)^p \mathcal{P}\psi_{f_1}(q_1)\psi_{f_2}(q_2)...\psi_{f_N}(q_N) \tag{7.6}$$

where $\mathcal{P}$ is the two-particle permutation operator, and p its parity which can be determined by the following convention: to permute two particles q_i and q_l one counts the number of particles which lie on the left of q_i and q_l in the product of Eq. (7.6): let n_i and n_l be these numbers. If $|n_l - n_i|$ is even, then $p = 0$. Otherwise, $p = 1$. One can see this by considering the two-particle wave function Ψ: one has $\Psi = A[\psi_{f_1}(q_1)\psi_{f_2}(q_2) - \psi_{f_1}(q_2)\psi_{f_2}(q_1)]$.

Using the wave-function normalization, the constant A is calculated as follows:

$$1 = \int |\Psi|^2 dq_1 dq_2 ... dq_N \tag{7.7}$$

$$= A^2 \int \left[\sum_{p'} (-1)^{p'} \mathcal{P}' \psi_{f_1}^*(q_1) \psi_{f_2}^*(q_2) ... \psi_{f_N}^*(q_N) \right]$$

$$\times \left[\sum_{p} (-1)^{p} \mathcal{P} \psi_{f_1}(q_1) \psi_{f_2}(q_2) ... \psi_{f_N}(q_N) \right] dq_1 dq_2 ... dq_N \tag{7.8}$$

$$= A^2 \int [N! \text{ terms}][N! \text{ terms}] dq_1 dq_2 ... dq_N \tag{7.9}$$

$$\tag{7.10}$$

One supposes that the individual wave functions $\psi_{f_i}(q_i)$ are orthogonal with each other: the integral of the product between a p' term and a p term is not zero only if the each particle occupies the same state in ψ^* and in ψ. In other words, only direct products in the above expression give rise to non zero integrals, namely if $p = p'$. Each integral is equal to 1 due to the orthonormality of $\psi_{f_i}(q_i)$. One obtains

$$1 = A^2 N! \tag{7.11}$$

$$A = \sqrt{\frac{1}{N!}} \tag{7.12}$$

In the case of two particles, one can directly check the above relation:

$$1 = A^2 \int [\psi_{f_1}^*(q_1)\psi_{f_2}^*(q_2) - \psi_{f_1}^*(q_2)\psi_{f_2}^*(q_1)]$$

$$[\psi_{f_1}(q_1)\psi_{f_2}(q_2) - \psi_{f_1}(q_2)\psi_{f_2}(q_1)]dq_1 dq_2 \tag{7.13}$$

$$= A^2 [\int \psi_{f_1}^*(q_1)\psi_{f_1}(q_1)dq_1 \int \psi_{f_2}^*(q_2)\psi_{f_2}(q_2)dq_2$$

$$+ \int \psi_{f_1}^*(q_2)\psi_{f_1}(q_2)dq_2 \int \psi_{f_2}^*(q_1)\psi_{f_2}(q_1)dq_1$$

$$- \int \psi_{f_1}^*(q_1)\psi_{f_2}(q_1)dq_1 \int \psi_{f_2}^*(q_2)\psi_{f_1}(q_2)dq_2$$

$$- \int \psi_{f_1}^*(q_2)\psi_{f_2}(q_2)dq_2 \int \psi_{f_2}^*(q_1)\psi_{f_1}(q_1)dq_1] \tag{7.14}$$

$$= A^2[1 + 1 - 0 - 0] \tag{7.15}$$

$$A = \sqrt{\frac{1}{2}} \tag{7.16}$$

In the case of bosons, the total wave function is invariant with respect to permutation of particles: it has the form of (7.6) but $p = 0$ for all permutations.

It is important to note that the antisymmetric character of the fermion wave function gives rise to the exchange energy in a system of interacting fermions. In a first approximation, one obtains the Hartree-Fock equation

$$H(\vec{r}_1)\psi_{\vec{k}_i,\sigma_i}(\vec{r}_1) + \sum_{j\neq i}\left[\int d\vec{r}_2\psi^*_{\vec{k}_j,\sigma_j}(\vec{r}_2)V(\vec{r}_1,\vec{r}_2)\psi_{\vec{k}_j,\sigma_j}(\vec{r}_2)\right]\psi_{\vec{k}_i,\sigma_i}(\vec{r}_1)$$

$$-\sum_{j\neq i}\delta_{\sigma_i,\sigma_j}\left[\int d\vec{r}_2\psi^*_{\vec{k}_j,\sigma_j}(\vec{r}_2)V(\vec{r}_1,\vec{r}_2)\psi_{\vec{k}_i,\sigma_i}(\vec{r}_2)\right]\psi_{\vec{k}_j,\sigma_j}(\vec{r}_1)$$

$$= \lambda_i\psi_{\vec{k}_i,\sigma_i}(\vec{r}_1) \tag{7.17}$$

where λ_i is the energy of the state $\vec{k}_i$. One can write it under a more compact form

$$H(\vec{r}_1)\psi_{\vec{k}_i,\sigma_i}(\vec{r}_1) + \sum_{j\neq i}\langle\psi_{\vec{k}_j,\sigma_j}(\vec{r}_2)|V(\vec{r}_1,\vec{r}_2))|\psi_{\vec{k}_j,\sigma_j}(\vec{r}_2)\rangle\psi_{\vec{k}_i,\sigma_i}(\vec{r}_1)$$

$$-\sum_{j\neq i}\delta_{\sigma_i,\sigma_j}\langle\psi_{\vec{k}_j,\sigma_j}(\vec{r}_2)|V(\vec{r}_1,\vec{r}_2)|\psi_{\vec{k}_i,\sigma_i}(\vec{r}_2)\rangle\psi_{\vec{k}_j,\sigma_j}(\vec{r}_1)$$

$$= \lambda_i\psi_{\vec{k}_i,\sigma_i}(\vec{r}_1) \tag{7.18}$$

or, by multiplying on the left by $\psi^*_{\vec{k}_i,\sigma_i}$ and then integrating on $\vec{r}_1$,

$$\langle\psi_{\vec{k}_i,\sigma_i}|H(\vec{r}_1)|\psi_{\vec{k}_i,\sigma_i}(\vec{r}_1)\rangle + \sum_{j\neq i}[K_{ij} - J_{ij}] = \lambda_i \tag{7.19}$$

where K_{ij} and J_{ij} are defined by

$$K_{ij} = \int \psi^*_{f_i}(q_i)\psi^*_{f_j}(q_j)V(q_i,q_j)\psi_{f_i}(q_i)\psi_{f_j}(q_j)dq_idq_j \tag{7.20}$$

$$-J_{ij} = -\delta_{\sigma_i,\sigma_j}\int \psi^*_{f_i}(q_i)\psi^*_{f_j}(q_j)V(q_i,q_j)\psi_{f_i}(q_j)\psi_{f_j}(q_i)dq_idq_j \tag{7.21}$$

K_{ij} is called "direct interaction" or "Coulomb interaction" and J_{ij} "exchange interaction" which is not zero only if the two spins σ_i and σ_j are equal. One will return to this equation in section 7.7.

One can demonstrate the Hartree-Fock equation using the method of Lagrange multipliers to minimize the system average energy $\langle\Psi|\mathcal{H}|\Psi\rangle$ under the constraint of the normalization of the determinant wave function Ψ shown above (see for example Ref. [50]). One can also do it in a simpler manner using the second quantization method (see section 7.7). Application of the Hartree-Fock equation to a gas of interacting fermions is found in chapter 10.

7.3   Representation of microstates by occupation numbers

One considers a system of N identical, indiscernible particles. The Hamiltonian is written as

$$\mathcal{H} = \sum_i H_i + \frac{1}{2} \sum_{i,j} V(\vec{r}_i, \vec{r}_j) \tag{7.22}$$

where H_i is a single-particle term such as the kinetic energy of the particle i and $V(\vec{r}_i, \vec{r}_j)$ the interaction between two particles at $\vec{r}_i$ and $\vec{r}_j$. Since the particles are identical and indiscernible, one can imagine that they have the same "list" of individual states: each of them takes one state of the list. A state i is characterized by some physical parameters such as the wave vector and the spin state, $\vec{k}_i$ and σ_i $(i = 1, ..., N)$. This state of the list is occupied by n_i particles. One can define a microstate of the system by the numbers of particles $\{n_i\}$ in the individual states $(\vec{k}_i, \sigma_i)$, $i = 1, ..., N$. All possible different particle distributions $\{n_i\}$ constitute the ensemble of microstates of the system. One says that the system is defined by a "state vector" given by

$$|\Psi\rangle = |n_1, n_2, ..., n_k, ..., n_N\rangle \tag{7.23}$$

where n_k is the number of particles occupying the individual state k. This state vector replaces the wave function of the Schrödinger equation. As for wave functions, one imposes that the state vectors form a complete set of orthogonal states. One has

$$\langle \Psi'|\Psi\rangle = \langle n_1', ..., n_k', ...|n_1, ..., n_k, ...\rangle = \delta_{n_1',n_1}...\delta_{n_k',n_k}... \tag{7.24}$$

where δ_{n_k',n_k} is the Kronecker symbol.

In the first quantization, the postulate on the symmetry of the wave function allows us to distinguish bosons and fermions: for bosons the permutation of two particles in their states does not change the sign of the corresponding wave function, while for fermions the permutation does change its sign. One of the consequences of the symmetry postulate is that in the case of bosons an individual state can contain any number of particles while in the case of fermions, each individual state can have zero or one particle (see chapters 5 and 6). This is known as the Pauli principle. In the method of second quantization, it is the symmetry of the operators which allows one to distinguish bosons and fermions as seen in the following.

7.4 Second quantization: the case of bosons

7.4.1 *Hamiltonian in second quantization*

In the second quantization, we work with the Hamiltonian where the system symmetry is incorporated. The Hamiltonian can be obtained from the Schrödinger equation but there is no need to remember the demonstration. We can use as the starting point the general Hamiltonian written in the second quantization given by (7.38). However, for a curious reader, we shall give below the demonstration for the boson case. The fermion case can be obtained in a similar way using the antisymmetric wave function described in paragraph 7.2.

One considers a system of N bosons defined by (7.22). The system wave function $\Psi(\vec{r}_1, \vec{r}_2, \cdots \vec{r}_N)$ does not change its sign with particle permutation:

$$P_{ij}\Psi(\vec{r}_1, \cdots \vec{r}_i, \cdots \vec{r}_j, \cdots \vec{r}_N) = \Psi(\vec{r}_1, \cdots \vec{r}_j, \cdots \vec{r}_i, \cdots \vec{r}_N)$$

Each state of the system can be defined by the occupation numbers as seen in the previous section. One writes

$$\phi_{\{n_1 n_2 \cdots n_N\}}(\vec{r}_1, \vec{r}_2, \cdots \vec{r}_N) = A \sum_p P\varphi_{f_1}(\vec{r}_1)\varphi_{f_2}(\vec{r}_2) \cdots \varphi_{f_N}(\vec{r}_N) \quad (7.25)$$

where the permutation operator P acts only on two particles occupying two different individual states, and n_i represents the number of particles in the state f_i, A being the normalization constant. One notes that $\sum_i n_i = N$, and there are $\frac{N!}{\prod_i n_i!}$ combinations in (7.25). The constant A is calculated as follows:

$$1 = \int |\phi_{\{n\}}(\vec{r}_i \cdots \vec{r}_N)|^2 d\vec{r}_1 ... d\vec{r}_N \quad \{n\} = \{n_1 n_2 \cdots n_N\}$$

$$= A^2 \sum_p \sum_{p'} \int P\varphi_{f_1}^*(\vec{r}_1) \cdots \varphi_{f_N}^*(\vec{r}_N) P'\varphi_{f_1}(\vec{r}_1) \cdots \varphi_{f_N} d\vec{r}_1 \cdots d\vec{r}_N$$

If $P' \neq P$, the integral is zero because of the orthogonality of φ_{f_i}. If $P' = P$, the integral is equal to 1. Since there are $\frac{N!}{\prod_i n_i!}$ permutations, one has

$$A = \sqrt{\frac{\prod_i n_i!}{N!}}$$

Thus , each system state can be represented by an ensemble $\phi_{\{n\}}$. The total wave function $\Psi(\vec{r}_1, \cdots \vec{r}_N)$ is given by

$$\Psi(\vec{r}_1, \cdots \vec{r}_N) = \sum_{\{n\}} C(n_1 \cdots n_N) \phi_{\{n\}}(\vec{r}_1, \cdots \vec{r}_N) \qquad (7.26)$$

It follows that

$$\sum_{\{n\}} |C(n_1 \cdots n_N)|^2 = 1$$

$|C(n_1 \cdots n_N)|^2$ is the probability density of the individual state $\phi_{\{n\}}$. The system energy is given by $\mathcal{H}|\Psi\rangle = E|\Psi\rangle$.

To determine coefficients $C(n_1 \cdots n_N)$ in (7.26), one calculates $\langle \phi_0 | \mathcal{H} | \Psi \rangle = E \langle \phi_0 | \Psi \rangle$ where ϕ_0 is one element of the ensemble $\phi_{\{n\}}$, for example $\varphi_{f_1}(\vec{r}_1) \cdots \varphi_{f_N}(\vec{r}_N)$("standard arrangement"). One has

$$\langle \varphi_{f_1}(\vec{r}_1) \cdots \varphi_{f_N}(\vec{r}_N)| \left\{ \sum_i H(\vec{r}_i) + \frac{1}{2} \sum_{i,j} V(\vec{r}_i, \vec{r}_j) \right\} | \times$$

$$\times \sum_{\{n\}} C(n_1 \cdots n_N) \sqrt{\frac{\prod_i n_i!}{N!}} \sum_P P \varphi_{f_1}(\vec{r}_1) \cdots \varphi_{f_N}(\vec{r}_N) \rangle$$

$$= E \langle \varphi_{f_1}(\vec{r}_1) \cdots \varphi_{f_N}(\vec{r}_N)| \sum_{\{n\}} C(n_1 \cdots n_N) \sqrt{\frac{\prod_i n_i!}{N!}} \times$$

$$\times \sum_P P \varphi_{f_1}(\vec{r}_1) \cdots \varphi_{f_N}(\vec{r}_N) \rangle \qquad (7.27)$$

Let us calculate now the " one-particle terms" of the left-hand side. One considers the particle $\vec{r}_i$ in φ_{f_p} where there are n_p particles. The integral corresponding to the particle $\vec{r}_i$ is

$$\langle \varphi_{f_1}(\vec{r}_1) \cdots \underbrace{\varphi_{f_p}(\vec{r}_i) \varphi_{f_p}(\vec{r}_{i+1}) \varphi_{f_p}(\vec{r}_{i+2})}_{n_p \text{ particles in } \varphi_{f_p}} \cdots |H(\vec{r}_i)| \Psi(\vec{r}_1 \cdots \vec{r}_N) \rangle \quad (7.28)$$

Since $H(\vec{r}_i)$ acts only on particle $\vec{r}_i$ in the "ket", the other particles should stay in the same states as in the "bra", otherwise the integral is zero by the orthogonality of wave functions φ_{f_p}. As for particle $\vec{r}_i$, it can be in another state: one supposes that $\vec{r}_i$ is in φ_{f_r} in the "ket". The corresponding term is

$$C(\cdots n_p - 1 \cdots n_r + 1 \cdots) \sqrt{\frac{\cdots (n_p - 1)! \cdots (n_r + 1)! \cdots}{N!}} \times$$

$$\times \varphi_{f_1}(\vec{r}_1) \cdots \underbrace{\varphi_{f_p}(\vec{r}_{i+1}) \varphi_{f_p}(\vec{r}_{i+2})} \cdots$$

$$(n_p - 1) \text{ particles}$$

$$\cdots \underbrace{\varphi_{f_r}(\vec{r}_i) \varphi_{f_r}(\vec{r}_j) \varphi_{f_r}(\vec{r}_{j+1})}$$

$$(n_r + 1) \text{ particles} \tag{7.29}$$

The integral (7.28) becomes

$$\sum_{f_r} n_p \langle \varphi_{f_p} | H | \varphi_{f_r} \rangle \left[\frac{(n_p - 1)! \cdots (n_r + 1)! \cdots}{N!} \right] C(\cdots n_p - 1 \cdots n_r + 1 \cdots) \tag{7.30}$$

The sum on f_r in (7.30) indicates that φ_{f_r} can be any state of Ψ and the factor n_p in (7.30) results from the fact that n_p particles in φ_{f_p} of the "bra" are indiscernible. This means that for a given φ_{f_r}, there are n_p terms identical to (7.29) of the "ket" giving non zero integrals.

The expression (7.30) is obtained for particles in φ_{f_p}. For all particles, one has

$$\langle \phi_0 | \mathcal{H}(\vec{r}_i) | \Psi \rangle = \sum_{f_r} \sum_{f_p} \langle \varphi_{f_p} | H | \varphi_{f_r} \rangle \left[\frac{\cdots (n_p - 1)!(n_r + 1)! \cdots}{N!} \right]^{\frac{1}{2}}$$

$$\times n_p C(\cdots n_p - 1 \cdots n_r + 1 \cdots) \tag{7.31}$$

Let us consider now the "two-particle terms" $\langle \phi_0 | V(\vec{r}_i, \vec{r}_j) | \Psi \rangle$ of (7.27). One uses a similar procedure: one supposes that the two particles $\vec{r}_i$ and $\vec{r}_j$ are respectively in φ_{f_p} and φ_{f_q} of the "bra" and that φ_{f_p} and φ_{f_q} contain n_p and n_q particles. In order that $\langle \phi_0 | V | \Psi \rangle$ is not zero the other particles in the "ket" should be each in the same state as in the "bra". As for $\vec{r}_i$ and $\vec{r}_j$, they can be in the same or different states on the two sides. One supposes that in the "ket", $\vec{r}_i$ and $\vec{r}_j$ are respectively in φ_{f_r} and φ_{f_s}, one obtains for these particles in φ_{f_p} and φ_{f_q} in the "bra",

$$\langle \phi_0 | V(\vec{r}_i, \vec{r}_j) | \Psi \rangle = \sum_{f_r} \sum_{f_p} \langle \varphi_{f_p} \varphi_{f_q} | V | \varphi_{f_r} \varphi_{f_s} \rangle n_p (n_q - \delta_{pq}) \times$$

$$\times \left[\frac{(n_p - 1)!(n_q - 1)! \cdots (n_r + 1)!(n_s + 1)!}{N!} \right]^{\frac{1}{2}} \times$$

$$\times C(\cdots n_p - 1, n_q - 1, n_r + 1, n_s + 1 \cdots) \tag{7.32}$$

The sums on f_r and f_s mean that f_r and f_s can be any state. If $\varphi_{f_p} = \varphi_{f_q}$, one has to replace in the square root the factors $(n_p - 1)!(n_q - 1)!$ by $(n_p - 2)!$ and in the argument of C, $(n_p - 1, n_q - 1)$ by $(n_p - 2)$. Similarly, if $\varphi_{f_s} = \varphi_{f_s}$, one replaces $(n_r + 1)!(n_s + 1)!$ by $(n_r + 2)!$ and modifies the corresponding argument of C.

Finally, for all interactions $V(\vec{r}_i, \vec{r}_j)$, one obtains

$$\langle \phi_0 | \frac{1}{2} \sum_{i,j} V(\vec{r}_i, \vec{r}_j) | \Psi \rangle =$$

$$\frac{1}{2} \sum_{pqrs} \langle pq | V | rs \rangle n_p (n_q - \delta_{pq}) \left[\frac{(n_p - 1)!(n_q - 1)! \cdots (n_r + 1)!(n_s + 1)!}{N!} \right]^{\frac{1}{2}} \times$$

$$C(\cdots n_p - 1, n_q - 1, n_r + 1, n_s + 1 \cdots) \tag{7.33}$$

where $\langle pq | V | rs \rangle = \langle \varphi_{f_p} \varphi_{f_q} | V | \varphi_{f_r} \varphi_{f_s} \rangle$. The right-hand side of (7.27) is calculated and given by

$$E \langle \phi_0 | \Psi \rangle =$$

$$E \left[\frac{..n_p! .. n_q! .. n_r! .. n_s! ..}{N!} \right]^{\frac{1}{2}} C(\cdots n_p - 1, n_q - 1, n_r + 1, n_s \cdots) \tag{7.34}$$

To put the equation (7.27) under the form

$$\mathcal{H} C(...n_p n_q n_r n_s...) = E C(...n_p n_q n_r n_s...)$$

one should define the following operators a_p and a_p^+ by

$$a_p C(...n_p...) = \sqrt{n_p} C(...n_p - 1...) \tag{7.35}$$

$$a_p^+ C(...n_p...) = \sqrt{n_p + 1} C(...n_p + 1...) \tag{7.36}$$

a_p is called "annihilation operator" and a_p^+ "creation operator". The coefficients C in (7.31) and (7.33) are then expressed by

$$C(...n_p, n_r + 1...) = \frac{1}{\sqrt{n_p(n_r + 1)}} a_r^+ a_p C(...n_p...n_r...)$$

$$C(...n_p - 1, n_q - 1, n_r + 1, n_p + 1...) = \frac{1}{[n_p n_q (n_r + 1)(n_s + 1)]^{\frac{1}{2}}} \times$$

$$a_r^+ a_s^+ a_q a_p C(...n_p n_q n_r n_s...)$$

Equations (7.31), (7.33) and (7.34) yield

$$\hat{\mathcal{H}}C(...n_f...) = EC(...n_f...) \tag{7.37}$$

where operator $\hat{\mathcal{H}}$ is given by

$$\hat{\mathcal{H}} = \sum_{p,r}\langle r|H|p\rangle a_r^+ a_p + \frac{1}{2}\sum_{pqrs}\langle rs|V|pq\rangle a_r^+ a_s^+ a_p a_q \tag{7.38}$$

where one has replaced $\langle p|H|r\rangle$ and $\langle pq|V|rs\rangle$ by $\langle r|H|p\rangle$ and $\langle rs|V|pq\rangle$ because H and V are hermitian. One has

$$\langle r|H|p\rangle = \int d\vec{r}\,\phi_r^*(\vec{r})H(\vec{r})\phi_p(\vec{r}) \tag{7.39}$$

$$\langle rs|V|pq\rangle = \int\int d\vec{r}d\vec{r}'\,\phi_r^*(\vec{r})\phi_s^*(\vec{r}')V(\vec{r},\vec{r}')\phi_p(\vec{r})\phi_q(\vec{r}') \tag{7.40}$$

The wave function $\phi_i(\vec{r})$ describes the individual state i of the particle at $\vec{r}$. For example, in the case of a plane wave one has $\phi_i(\vec{r}) = \exp(i\vec{k}_i\cdot\vec{r})/\sqrt{\Omega}$ where $\vec{k}_i$ is the wave vector and Ω the system volume.

Hamiltonian (7.38) is the quantized version of the classical Hamiltonian (7.22). This quantization is called "second quantization" (the first quantization is the quantization of the energy by using the Schrödinger equation).

7.4.2 *Properties of boson operators*

Having demonstrated the equivalence of the second quantization with the Schrödinger equation, one uses from now the representation by the occupation numbers described in 7.3: one introduces the operators a_k and a_k^+ defined by the following relations

$$a_k^+|\Psi\rangle = \sqrt{n_k+1}|...,n_k+1,....\rangle \tag{7.41}$$

$$a_k|\Psi\rangle = \sqrt{n_k}|...,n_k-1,....\rangle \tag{7.42}$$

As seen in the kets, operator a_k^+ creates a particle while operator a_k destroys a particle in the state k when they operate on $|\Psi>$. For this reason, they are called creation and annihilation operators, respectively. By the above definitions, one sees that

$$a_k^+ a_k|\Psi\rangle = n_k|...,n_k,....\rangle \tag{7.43}$$

This relation shows that operator $a_k^+ a_k$ has the eigenvalue n_k which is the number of particles in the state k. One calls therefore $a_k^+ a_k$ operator "occupation number".

In addition, using (7.41) and (7.42), one gets

$$a_k a_k^+ |\Psi\rangle = (n_k + 1)|..., n_k, ...\rangle \qquad (7.44)$$

Comparing this relation to (7.43), one obtains

$$a_k a_k^+ - a_k^+ a_k = 1 \qquad (7.45)$$

Now, if $k \neq k'$, by using (7.24) one has

$$\langle \Psi | a_{k'} a_k^+ | \Psi \rangle = \sqrt{(n_k + 1)n_{k'}} \langle ..., n_k, n_{k'}, ...|..., n_k + 1, n_{k'} - 1, ...\rangle$$
$$= 0 \qquad (7.46)$$

$$\langle \Psi | a_k^+ a_{k'} | \Psi \rangle = \sqrt{(n_k + 1)n_{k'}} \langle ..., n_k, n_{k'}, ...|..., n_k + 1, n_{k'} - 1, ...\rangle$$
$$= 0 \qquad (7.47)$$

Combining with (7.45) one can write

$$[a_{k'}, a_k^+] \equiv a_{k'} a_k^+ - a_k^+ a_{k'} = \delta_{k,k'} \qquad (7.48)$$

One can show in the same manner that for arbitrary k and k', one has

$$[a_{k'}, a_k] = 0 \quad , \quad [a_{k'}^+, a_k^+] = 0 \qquad (7.49)$$

Relations (7.48) and (7.49) are called "commutation relations".

7.5   Second quantization: the case of fermions

In the case of fermions, one can demonstrate the general Hamiltonian starting from the Schrödinger equation as one has done to obtain (7.38) for bosons. However, this demonstration is lengthy to repeat here and it does not bring more physical insights. One therefore admits as the starting point in the study of systems of interacting fermions the Hamiltonian given in (7.60) below, using the creation and annihilation operators b_f^+ and b_f defined by

$$b_f |\Psi\rangle = b_f |...n_f...\rangle = (-1)^{[f]} n_f |...n_f - 1...\rangle \qquad (7.50)$$
$$b_f^+ |\Psi\rangle = b_f^+ |...n_f...\rangle = (-1)^{[f]} (1 - n_f)|...n_f + 1...\rangle \qquad (7.51)$$

where $[f]$ is, by convention, the number of particles occupying the states on the left of the state f in the ket. It is noted that in some books the coefficients in front of the ket of (7.50)-(7.51) are given by $\sqrt{n_f}$ and $\sqrt{1 - n_f}$ instead of n_f and $(1 - n_f)$. However, one can verify that they are equivalent because n_f is 0 or 1. One has

$$b_f b_g |...n_f...n_g...\rangle = (-1)^{[g]} n_g b_f |...n_f, ...n_g - 1, ...\rangle$$
$$= (-1)^{[g]+[f]} n_g n_f |...n_f - 1, ...n_g - 1, ...\rangle$$
$$b_g b_f |...n_f...n_g...\rangle = (-1)^{[f]} n_f b_g |...n_f - 1...n_g, ...\rangle$$
$$= (-1)^{[f]+[g]-1} n_f n_g |...n_f - 1, ...n_g - 1, ...\rangle$$

from which one has $b_f b_g = -b_g b_f$, or equivalently

$$[b_f, b_g]_+ \equiv b_f b_g + b_g b_f = 0 \tag{7.52}$$

In the same manner, one obtains for arbitrary f and g

$$\left[b_f^+, b_g^+ \right]_+ = 0 \tag{7.53}$$

and

$$\left[b_f^+, b_g \right]_+ = 0 \quad \text{if} \ f \neq g \tag{7.54}$$

Now if $f = g$, one has

$$b_f^+ b_f |...n_f...\rangle = (-1)^{[f]} n_f b_f^+ |...n_f - 1...\rangle$$
$$= (-1)^{2[f]} n_f (1 - n_f + 1) |...n_f...\rangle$$
$$= n_f (2 - n_f) |...n_f... \rangle = n_f |...n_f...\rangle \tag{7.55}$$

where in the last line, one has used $n_f(2 - n_f) = n_f$ because

$$n_f = \left\{ \begin{array}{l} 0 \Rightarrow n_f(2 - n_f) = 0 \\ 1 \Rightarrow n_f(2 - n_f) = 1 \end{array} \right\} \Rightarrow n_f(2 - n_f) = n_f \tag{7.56}$$

One calls $b_f^+ b_f$ operator "occupation number" because its eigenvalue when operating on the ket is n_f. Besides,

$$b_f b_f^+ |...n_f...\rangle = (-1)^{[-f]+[f]} (n_f + 1)(1 - n_f) |...n_f...\rangle$$
$$= (1 + n_f)(1 - n_f) |...n_f...\rangle \tag{7.57}$$

If $n_f = 0 \Rightarrow (1 + n_f)(1 - n_f) = 1$
If $n_f = 1 \Rightarrow (1 + n_f)(1 - n_f) = 0$ $\Big\} \Rightarrow (1 + n_f)(1 - n_f) = 1 - n_f$

from which

$$b_f b_f^+ |...n_f... \rangle = (1 - n_f) |..n_f... \rangle \tag{7.58}$$

Comparing (7.58) to (7.55), one obtains $b_f b_f^+ = 1 - b_f^+ b_f$, namely $\left[b_f, b_f^+ \right]_+ = 1$. Combining with (7.54), one can write

$$\left[b_g, b_f^+ \right]_+ = \delta_{g,f} \tag{7.59}$$

Relations (7.52), (7.53) and (7.59) are called "fermion anticommutation relations".

Hamiltonian (7.22) in the case of fermions is written in the second quantization as

$$\hat{\mathcal{H}} = \sum_{i,k} \langle k|H|i \rangle b_k^+ b_i - \frac{1}{2} \sum_{i,j,k,l} \langle kl|V|ij \rangle b_k^+ b_l^+ b_i b_j \tag{7.60}$$

where

$$\langle k|H|i \rangle = \int d\vec{r} \phi_k^*(\vec{r}) H(\vec{r}) \phi_i(\vec{r}) \tag{7.61}$$

$$\langle kl|V|ij \rangle = \int \int d\vec{r} d\vec{r}' \phi_k^*(\vec{r}) \phi_l^*(\vec{r}') V(\vec{r}, \vec{r}') \phi_i(\vec{r}) \phi_j(\vec{r}') \tag{7.62}$$

Due to the anticommutation of the operators, one should respect the order of the operators as well as that of the arguments $\vec{r}$ and $\vec{r}'$ of ϕ functions in $\langle kl|V|ij \rangle$. A permutation of the operators should obey the anticommutation relations (7.52), (7.53) and (7.59).

Hamiltonians (7.38) and (7.60) are very useful in the study of systems of weakly interacting particles [100].

Remarks:

(1) The state vector containing the occupation numbers of occupied individual states [see (7.23)] can be expressed by the operators b and b^+ as

$$|1_{k_1} 1_{k_2} ... 1_{k_N} \rangle \equiv b_{k_1}^+ b_{k_2}^+ ... b_{k_N}^+ |000... \rangle$$

where $|000... \rangle$ represents the vacuum state.

(2) One has the following relations

$$b_k |...0_k... \rangle = 0$$
$$b_k^+ |...0_k... \rangle = |...1_k... \rangle$$
$$b_k^+ |...1_k... \rangle = 0$$
$$b_k |...1_k... \rangle = |...0_k... \rangle$$

(3) If the individual state of a fermion is characterized by its wave vector $\vec{k}$ and its spin σ, then the anticommutation relations (7.52), (7.53) and (7.59) are explicitly written as

$$[b_{k\sigma}, b^+_{k'\sigma'}]_+ = \delta_{k,k'}\delta_{\sigma,\sigma'}$$

$$[b_{k\sigma}, b_{k'\sigma'}]_+ = [b^+_{k\sigma}, b^+_{k'\sigma'}]_+ = 0, \quad \forall(k\sigma, \ k'\sigma')$$

(4) The ensemble of state vectors $|n_{k_1} n_{k_2}...n_{k_N}\rangle$, with $n_{k_i}(i = 1, N) = 0$ or 1, constitute a complete set of orthonormal states :

$$\langle n'_{k_1} n'_{k_2}...n'_{k_i}...|n_{k_1} n_{k_2}...n_{k_i}...\rangle = \delta_{n'_{k_1}, n_{k_1}} \delta_{n'_{k_2}, n_{k_2}}...\delta_{n'_{k_i}, n_{k_i}}...$$

$$\sum_{n_{k_1}...} |n_{k_1} n_{k_2}...\rangle\langle n_{k_1} n_{k_2}...| = 1 \quad (n_{k_i} = 0 \ \text{ or } \ 1)$$

One has similar results for bosons. Note however that the sum over $n_{k_1}, n_{k_2}, ...$ is performed with all possible values of each n_{k_i}: for instance, in a system of N bosons, one has $n_{k_i} = 0, 1, ..., N$.

(5) Note that the state vector $|n_{k_1} n_{k_2}...n_{k_N}\rangle$ is not a wave function.

7.6 Field operators

Using $b^+_{k\sigma}$ and $b_{k\sigma}$, the destruction and the creation of a fermion in the state $(\vec{k}, \sigma)$ at $\vec{r}$ are given by the expressions

$$\frac{1}{\Omega^{1/2}} b_{k\sigma} e^{i\vec{k}.\vec{r}} \quad \text{and} \quad \frac{1}{\Omega^{1/2}} b^+_{k\sigma} e^{-i\vec{k}.\vec{r}}$$

Note that in these notations one has combined operators $b^+_{k\sigma}$ and $b_{k\sigma}$ with the corresponding wave functions, namely $\frac{1}{\Omega^{1/2}} e^{i\vec{k}.\vec{r}}$ where Ω is the system volume. One introduces now the "field operators" defined by

$$\hat{\Psi}_\sigma(\vec{r}) = \frac{1}{\Omega^{1/2}} \sum_{\vec{k}} b_{k\sigma} e^{i\vec{k}.\vec{r}} \tag{7.63}$$

$$\hat{\Psi}^+_\sigma(\vec{r}) = \frac{1}{\Omega^{1/2}} \sum_{\vec{k}} b^+_{k\sigma} e^{-i\vec{k}.\vec{r}} \tag{7.64}$$

$\hat{\Psi}_\sigma(\vec{r})$ and $\hat{\Psi}^+_\sigma(\vec{r})$ are called annihilation and creation field operators at $\vec{r}$. From these relations, one has

$$b_{k\sigma} = \frac{1}{\Omega^{1/2}} \int d\vec{r} \hat{\Psi}_\sigma(\vec{r}) e^{-i\vec{k}.\vec{r}} \tag{7.65}$$

$$b^+_{k\sigma} = \frac{1}{\Omega^{1/2}} \int d\vec{r} \hat{\Psi}^+_\sigma(\vec{r}) e^{i\vec{k}.\vec{r}} \tag{7.66}$$

Using the anticommutation relations of b^+ and b, (7.52)-(7.53) and (7.59), one obtains

$$[\hat{\Psi}_\sigma(\vec{r}_1), \hat{\Psi}_{\sigma'}^+(\vec{r}_2)]_+ = \delta_{\vec{r}_1, \vec{r}_2} \delta_{\sigma, \sigma'} \tag{7.67}$$

$$[\hat{\Psi}_\sigma(\vec{r}_1), \hat{\Psi}_{\sigma'}(\vec{r}_2)]_+ = [\hat{\Psi}_\sigma^+(\vec{r}_1), \hat{\Psi}_{\sigma'}^+(\vec{r}_2)]_+ = 0 \tag{7.68}$$

The definitions for bosons are similar to (7.63)-(7.64) with $b_{k\sigma}^+$ and $b_{k\sigma}$ replaced by a_k^+ and a_k (without spin). In addition, in the boson case, we have the commutation relations between $\hat{\Psi}(\vec{r})$ and $\hat{\Psi}^+(\vec{r})$ instead of (7.67)-(7.68).

In what follows, one limits oneself to the fermion case. The boson case can be obtained in the same manner. One can express various physical quantities in terms of field operators $\hat{\Psi}_\sigma(\vec{r})$ and $\hat{\Psi}_\sigma^+(\vec{r})$. One has seen above that the Hamiltonian of the first quantization is expressed in the second quantization by (7.60) with the help of operators b^+ and b. One can now express it in terms of $\hat{\Psi}$ and $\hat{\Psi}^+$ as follows:

$$\hat{\mathcal{H}} = \sum_\sigma \int d\vec{r} \hat{\Psi}_\sigma^+(\vec{r}) H(\vec{r}) \hat{\Psi}_\sigma(\vec{r})$$

$$- \frac{1}{2} \sum_{\sigma, \sigma'} \int \int d\vec{r}_1 d\vec{r}_2 \hat{\Psi}_\sigma^+(\vec{r}_1) \hat{\Psi}_{\sigma'}^+(\vec{r}_2) V(\vec{r}_1, \vec{r}_2) \hat{\Psi}_\sigma(\vec{r}_1) \hat{\Psi}_{\sigma'}(\vec{r}_2) \tag{7.69}$$

The "total occupation number" operator is given by

$$\hat{N} = \sum_{i,\sigma} b_{i\sigma}^+ b_{i\sigma} \tag{7.70}$$

Using (7.65)-(7.66), $\hat{N}$ can be rewritten as

$$\hat{N} = \sum_{i,\sigma} \int d\vec{r}_1 d\vec{r}_2 \hat{\Psi}_\sigma^+(\vec{r}_2) e^{i\vec{k}_i \cdot \vec{r}_2} e^{-i\vec{k}_i \cdot \vec{r}_1} \hat{\Psi}_\sigma(\vec{r}_1)$$

$$= \sum_{i,\sigma} \int d\vec{r}_1 d\vec{r}_2 \hat{\Psi}_\sigma^+(\vec{r}_2) e^{i\vec{k}_i \cdot (\vec{r}_2 - \vec{r}_1)} \hat{\Psi}_\sigma(\vec{r}_1)$$

$$= \sum_\sigma \int d\vec{r}_1 d\vec{r}_2 \hat{\Psi}_\sigma^+(\vec{r}_2) \delta_{\vec{r}_2, \vec{r}_1} \hat{\Psi}_\sigma(\vec{r}_1) = \sum_\sigma \int d\vec{r}_1 \hat{\Psi}_\sigma^+(\vec{r}_1) \hat{\Psi}_\sigma(\vec{r}_1)$$

$$= \sum_\sigma \int \hat{\rho}_\sigma(\vec{r}_1) d\vec{r}_1 \tag{7.71}$$

where

$$\hat{\rho}_\sigma(\vec{r}) \equiv \hat{\Psi}_\sigma^+(\vec{r})\hat{\Psi}_\sigma(\vec{r}) = \quad \text{operator density of particles} \qquad (7.72)$$

Operator density of states $\hat{\rho}(\vec{k})$ is given by the Fourier transform

$$\hat{\rho}(\vec{k}) = \sum_\sigma \int d\vec{r}\,\hat{\rho}_\sigma(\vec{r})e^{-i\vec{k}\cdot\vec{r}} = \sum_\sigma \sum_{\vec{k}_1,\vec{k}_2} \int d\vec{r}\,e^{i(\vec{k}_2-\vec{k}_1)\cdot\vec{r}}b_{\vec{k}_1\sigma}^+ b_{\vec{k}_2\sigma}e^{-i\vec{k}\cdot\vec{r}}$$

$$= \sum_\sigma \sum_{\vec{k}_1,\vec{k}_2} \int d\vec{r}\,e^{i(\vec{k}_2-\vec{k}_1-\vec{k})\cdot\vec{r}}b_{\vec{k}_1\sigma}^+ b_{\vec{k}_2\sigma}$$

$$= \sum_\sigma \sum_{\vec{k}_1,\vec{k}_2} \delta_{\vec{k}_2-\vec{k}_1-\vec{k},0}\,b_{\vec{k}_1\sigma}^+ b_{\vec{k}_2\sigma} = \sum_\sigma \sum_{\vec{k}_1} b_{\vec{k}_1\sigma}^+ b_{\vec{k}_1+\vec{k},\sigma} \qquad (7.73)$$

where one has used $\vec{k}_1$ instead of k_1 etc. for subscripts of b and b^+ operators to avoid confusion when performing the sums.

7.7 Hartree-Fock approximation

The equation of motion for $\hat{\Psi}_\sigma(\vec{r})$ reads

$$i\hbar\frac{d\hat{\Psi}_\sigma(\vec{r},t)}{dt} = -[\hat{\mathcal{H}}, \hat{\Psi}_\sigma(\vec{r},t)] \qquad (7.74)$$

where $\hat{\mathcal{H}}$ is the Hamiltonian in the second quantization of a system of fermions given by Eq. (7.69). One shows in the following that one can obtain the Hartree-Fock equation by taking a first-order approximation (linearization) of the right-hand side of the above equation of motion: to calculate $[\hat{\mathcal{H}}, \hat{\Psi}_\sigma(\vec{r},t)]$, one decomposes the chains of fermion operators in the commutators as follows:

$$[AB, C] = A\,[B, C]_+ - [A, C]_+\,B \qquad (7.75)$$

For example with the kinetic term of $\hat{\mathcal{H}}$, one has

$$\sum_{\sigma'} \int d\vec{r}_1 \left[\hat{\Psi}_{\sigma'}^+(\vec{r}_1) H(\vec{r}_1) \hat{\Psi}_{\sigma'}(\vec{r}_1), \hat{\Psi}_\sigma(\vec{r}) \right]$$

$$= \sum_{\sigma'} \int d\vec{r}_1 \{ \hat{\Psi}_{\sigma'}^+(\vec{r}_1) \left[H(\vec{r}_1)\hat{\Psi}_{\sigma'}(\vec{r}_1), \hat{\Psi}_\sigma(\vec{r}) \right]_+$$

$$- \left[\hat{\Psi}_{\sigma'}^+(\vec{r}_1), \hat{\Psi}_\sigma(\vec{r}) \right]_+ H(\vec{r}_1)\hat{\Psi}_{\sigma'}(\vec{r}_1) \}$$

$$= \sum_{\sigma'} \int d\vec{r}_1 \left\{ 0 - \delta_{\vec{r},\vec{r}_1} \delta_{\sigma,\sigma'} H(\vec{r}_1)\hat{\Psi}_{\sigma'}(\vec{r}_1) \right\}$$

$$= \frac{p^2}{2m} \hat{\Psi}_\sigma(\vec{r}) \tag{7.76}$$

where one has put

$$A = \hat{\Psi}_{\sigma'}^+(\vec{r}_1), \quad B = H(\vec{r}_1)\hat{\Psi}_{\sigma'}(\vec{r}_1), \quad C = \hat{\Psi}_\sigma(\vec{r})$$

For the interaction term of $\hat{\mathcal{H}}$, one has to decompose several times until one has only anticommutation relations between two field operators. One then gets

$$i\hbar \frac{d\hat{\Psi}_\sigma(\vec{r},t)}{dt} = -\frac{p^2}{2m}\hat{\Psi}_\sigma(\vec{r},t) + \sum_{\sigma'} \int d\vec{r}_1 \hat{\Psi}_{\sigma'}^+(\vec{r}_1,t) V(\vec{r},\vec{r}_1)\hat{\Psi}_{\sigma'}(\vec{r}_1,t)\hat{\Psi}_\sigma(\vec{r},t) \tag{7.77}$$

where $\hat{\Psi}_\sigma(\vec{r},t)$ and $\hat{\Psi}_\sigma^+(\vec{r},t)$ are interaction representations of (7.63) and (7.64). One has

$$\hat{\Psi}(\vec{r},t) = \sum_{\vec{k},\sigma} b_{\vec{k}\sigma} e^{-i\omega_{\vec{k}}t} \varphi_{\vec{k}}(\vec{r}) \tag{7.78}$$

$$\hat{\Psi}^+(\vec{r},t) = \sum_{\vec{k},\sigma} b_{\vec{k}\sigma}^+ e^{i\omega_{\vec{k}}t} \varphi_{\vec{k}}^+(\vec{r}) \tag{7.79}$$

Replacing these expressions in (7.77), one obtains

$$\sum_{\vec{k},\sigma} \hbar\omega_{\vec{k}} b_{\vec{k}\sigma} e^{-i\omega_{\vec{k}}t} \varphi_{\vec{k}}(\vec{r}) = \sum_{\vec{k},\sigma} \frac{\hbar^2 k^2}{2m} b_{\vec{k}\sigma} e^{-i\omega_{\vec{k}}t} \varphi_{\vec{k}}(\vec{r})$$

$$+ \sum_{\vec{k},\vec{k}',\vec{k}'',\sigma,\sigma',\sigma''} \int d\vec{r}_1 \varphi_{\vec{k}'}^+(\vec{r}_1) V(\vec{r},\vec{r}_1)\varphi_{\vec{k}''}(\vec{r}_1) e^{i(\omega_{\vec{k}'}-\omega_{\vec{k}''})t}$$

$$\times b_{\vec{k}'\sigma'}^+ b_{\vec{k}''\sigma''} b_{\vec{k}\sigma} e^{-i\omega_{\vec{k}}t} \varphi_{\vec{k}}(\vec{r}) \tag{7.80}$$

For a first-order approximation of the right-hand side of (7.80), one uses the following decoupling called "random-phase approximation" (RPA) for the fermion case

$$b^+_{\vec{k}'\sigma'} b_{\vec{k}''\sigma''} b_{\vec{k}\sigma} \simeq < b^+_{\vec{k}'\sigma'} b_{\vec{k}''\sigma''} > b_{\vec{k}\sigma} - < b^+_{\vec{k}'\sigma'} b_{\vec{k}\sigma} > b_{\vec{k}''\sigma''} \qquad (7.81)$$

where the negative sign of the right-hand side results from the permutation of $b_{\vec{k}''\sigma''}$ and $b_{\vec{k}\sigma}$, and $< ... >$ denotes the average. This decoupling supposes that only terms of the type $< b^+_{\vec{k}'\sigma'} b_{\vec{k}''\sigma''} >$ are not zero. In the ground state,

$$< b^+_{\vec{k}'\sigma'} b_{\vec{k}''\sigma''} >= n_{\vec{k}'\sigma'} \delta(\vec{k}', \vec{k}'') \delta(\sigma', \sigma'') \Theta(k_F - k').$$

Equation (7.80) becomes

$$\sum_{\vec{k},\sigma} \hbar \omega_{\vec{k}} \varphi_{\vec{k}}(\vec{r}) b_{\vec{k}\sigma} e^{-i\omega_{\vec{k}}t} = \sum_{\vec{k},\sigma} \frac{\hbar^2 k^2}{2m} b_{\vec{k}\sigma} e^{-i\omega_{\vec{k}}t} \varphi_{\vec{k}}(\vec{r})$$

$$+ \sum_{\vec{k},\sigma} \sum_{\vec{k}',\sigma'} \int d\vec{r}_1 \varphi^+_{\vec{k}'}(\vec{r}_1) V(\vec{r}, \vec{r}_1) \varphi_{\vec{k}'}(\vec{r}_1) n_{\vec{k}'\sigma'} \varphi_{\vec{k}}(\vec{r}) b_{\vec{k}\sigma} e^{-i\omega_{\vec{k}}t}$$

$$- \sum_{\vec{k}'',\sigma''} \sum_{\vec{k}',\sigma'} \delta(\sigma', \sigma) \int d\vec{r}_1 \varphi^+_{\vec{k}'}(\vec{r}_1) V(\vec{r}, \vec{r}_1) \varphi_{\vec{k}}(\vec{r}_1)$$

$$\times n_{\vec{k}'\sigma'} \varphi_{\vec{k}'}(\vec{r}) b_{\vec{k}''\sigma''} e^{-i\omega_{\vec{k}''}t}$$

Changing now the dummy variables $(\vec{k}'', \sigma'')$ into $(\vec{k}, \sigma)$ in the last term, removing the sums $\sum_{\vec{k},\sigma}$ then taking off the factor $b_{\vec{k}\sigma} e^{-i\omega_{\vec{k}}t}$ on both sides, one arrives at

$$\hbar \omega_{\vec{k}} \varphi_{\vec{k},\sigma}(\vec{r}) = \frac{\hbar^2 k^2}{2m} \varphi_{\vec{k},\sigma}(\vec{r})$$

$$+ \sum_{\vec{k}',\sigma'} \int d\vec{r}_1 \varphi^+_{\vec{k}',\sigma'}(\vec{r}_1) V(\vec{r}, \vec{r}_1) \varphi_{\vec{k}',\sigma'}(\vec{r}_1) \varphi_{\vec{k},\sigma}(\vec{r})$$

$$- \sum_{\vec{k}',\sigma'} \delta(\sigma', \sigma) \int d\vec{r}_1 \varphi^+_{\vec{k}',\sigma'}(\vec{r}_1) V(\vec{r}, \vec{r}_1) \varphi_{\vec{k},\sigma}(\vec{r}_1) \varphi_{\vec{k}',\sigma'}(\vec{r})$$

where one has replaced $n_{\vec{k}'\sigma'}$ by 1 (ground state) and one has transferred, for compactness, the spin indices σ, σ' of states $\vec{k}$ and $\vec{k}'$ on φ. This equation is the Hartree-Fock equation [see Eq. (7.17)]. The first term is the kinetic energy, the second term is the direct interaction which is the double average, on space and states, of the interaction $V(\vec{r}, \vec{r}_1)$, while the last term is the exchange interaction which results from the permutation of two particles in $\varphi_{\vec{k},\sigma}(\vec{r}_1)$ and $\varphi_{\vec{k}',\sigma'}(\vec{r})$ with respect to the direct term. This exchange is possible only if $\sigma = \sigma'$.

7.8   Conclusion

In this chapter, the second quantization method has been introduced using the representation of the microstates of a system by their occupation numbers. The general Hamiltonian including kinetic and interaction terms has been given in terms of creation and annihilation operators and in terms of field operators, for both boson and fermion cases. Using this Hamiltonian, one can compute physical quantities such as the energy of the system. This method is very useful for studying systems of weakly interacting quantum particles. In particular, it provides a relatively easy technique to calculate collective elementary excitation spectra such as magnons and phonons. Several examples are treated in chapters 9 and 12.

One has also introduced the Hartree-Fock equation which shows the effect of spin exchange in systems of interacting fermions. This exchange interaction is at the origin of magnetic properties of materials. Various spin interaction models stem from the exchange interaction (see chapter 12). These models serve to study not only magnetic properties of materials but also properties of various systems going from alloys to neural networks where physical quantities can be mapped into a spin language. The investigation of spin systems constitutes one of the central tasks of statistical physics. One has seen a few examples in Problems treated in chapters 2 and 3.

7.9   Problems

Problem 1. Exercise of commutation and anticommutation relations:
One defines the following operators

$$B_{\vec{k}\sigma}^{+} = b_{\vec{k}\sigma}^{+} b_{-\vec{k}-\sigma}^{+}$$
$$B_{\vec{k}\sigma} = b_{-\vec{k}-\sigma} b_{\vec{k}\sigma}$$

where $b_{\vec{k}\sigma}^{+}$ and $b_{\vec{k}\sigma}$ are fermion creation and annihilation operators of state $(\vec{k}, \sigma)$.

a) What are the meanings of $B_{\vec{k}\sigma}^{+}$ and $B_{\vec{k}\sigma}$?

b) Show the following commutation and anticommutation relations

$$\left[B_{\vec{k}\sigma}, B_{\vec{k}'\sigma}^{+}\right] = (1 - n_{\vec{k}\sigma} - n_{-\vec{k}-\sigma})\delta(\vec{k}, \vec{k}')$$

$$\left[B_{\vec{k}\sigma}, B_{\vec{k}'\sigma}\right] = 0$$

$$\left[B_{\vec{k}\sigma}, B_{\vec{k}'\sigma}\right]_{+} = 2B_{\vec{k}\sigma}B_{\vec{k}'\sigma}\left[1 - \delta(\vec{k}, \vec{k}')\right]$$

where $n_{\vec{k}\sigma} = b_{\vec{k}\sigma}^{+}b_{\vec{k}\sigma}$

Problem 2. Show that $[\hat{\mathcal{H}}, \hat{N}]=0$ where $\hat{N}$ is the field operator of occupation number defined in (7.70) and $\hat{\mathcal{H}}$ the Hamiltonian in the second quantization (7.69).

Problem 3. Show that $\hat{\Psi}(\vec{r})\hat{N} = (\hat{N}+1)\hat{\Psi}(\vec{r})$ for both boson and fermion cases.

Problem 4. Show that $\hat{\Psi}^{+}(\vec{r})|\text{vac}\rangle$ ("vac" stands for vacuum) is a state in which there is a particle localized at $\vec{r}$.

Problem 5. Boson Hamiltonian:

Show that the general Hamiltonian (7.38) can be rewritten as

$$\hat{\mathcal{H}} = \sum_{\vec{k}} \epsilon_{\vec{k}} a_{\vec{k}}^{+} a_{\vec{k}}$$

$$+\frac{1}{2} \sum_{\vec{k}_1,\vec{k}_2,\vec{k}_3,\vec{k}_4} \mathcal{V}(\vec{k}_1 - \vec{k}_3) a_{\vec{k}_1}^{+} a_{\vec{k}_2}^{+} a_{\vec{k}_3} a_{\vec{k}_4} \delta(\vec{k}_1 + \vec{k}_2 - \vec{k}_3 - \vec{k}_4)$$

$$(7.82)$$

where $\mathcal{V}(\vec{k}_1 - \vec{k}_3)$ is the Fourier transform of the interaction between bosons, $\epsilon_{\vec{k}}$ the kinetic energy of the state $\vec{k}$, a and a^{+} are annihilation and creation operators, respectively. This Hamiltonian is used to study phonons excited in a gas of Helium-4 in chapter 9.

Problem 6. Gas of interacting bosons:

Consider a homogeneous system of N particles of spin zero. The Hamiltonian is given by

$$\hat{\mathcal{H}} = \sum_{\vec{k}} \epsilon_{\vec{k}} a_{\vec{k}}^{+} a_{\vec{k}} + \frac{1}{2} \sum_{\vec{k}_1,\vec{k}_2,\vec{k}_3,\vec{k}_4} V_{\vec{k}_1\vec{k}_2,\vec{k}_3\vec{k}_4} a_{\vec{k}_1}^{+} a_{\vec{k}_2}^{+} a_{\vec{k}_3} a_{\vec{k}_4} \qquad (7.83)$$

where $a_{\vec{k}}^{+}$ and $a_{\vec{k}}$ are boson creation and annihilation operators, $\epsilon_{\vec{k}} = \frac{\hbar^2 k^2}{2m}$ and $V_{\vec{k}_1\vec{k}_2,\vec{k}_3\vec{k}_4} = \langle \vec{k}_1\vec{k}_2|V(\vec{r}_1, \vec{r}_2)|\vec{k}_3\vec{k}_4\rangle$, $V(\vec{r}_1, \vec{r}_2)$ being the interaction potential of two bosons at $\vec{r}_1$ and $\vec{r}_2$.

a) Show that the average value of $\hat{\mathcal{H}}$ in the ground state is given by

$$\frac{E_1}{N} = \frac{N-1}{2\Omega}\mathcal{V}(0) \qquad (7.84)$$

where

$$\mathcal{V}(\vec{q}) = \int V(\vec{r})e^{-i\vec{q}\cdot\vec{r}} \tag{7.85}$$

which is a Fourier component of $V(\vec{r})$.

b) Show that the next order of the energy is given by

$$\frac{E_2}{N} = -\frac{N-1}{2\Omega} \int \frac{d\vec{q}}{(2\pi)^3} \frac{|\mathcal{V}(\vec{q})|^2}{\hbar^2 q^2/m} \tag{7.86}$$

Problem 7. Method of diagonalization of Hamiltonian in second quantization:

Consider the following Hamiltonian of a system of bosons

$$\hat{\mathcal{H}} = \sum_k H_k$$

$$H_k = [A_k(a_k^+ a_k + a_{-k}^+ a_{-k}) + B_k(a_k a_{-k} + a_{-k}^+ a_k^+)] \tag{7.87}$$

where A_k and B_k are constants, a_k and a_k^+ are annihilation and creation operators which obey the commutation relations. In order to transform the above Hamiltonian into a diagonal form, namely

$$H_k = \lambda_k c_k^+ c_k \tag{7.88}$$

one uses the following transformation for H_k:

$$c_k = u_k a_k - v_k a_{-k}^+ \tag{7.89}$$

$$c_k^+ = u_k a_k^+ - v_k a_{-k} \tag{7.90}$$

where u_k and v_k are real and satisfy the relation $u_k^2 - v_k^2 = 1$.

a) Show that $[c_k, c_{k'}^+] = \delta_{k,k'}$.

b) Calculate $[c_k^+, H_k]$ using H_k given by (7.87). Calculate $[c_k^+, H_k]$ using (7.88). Show that the latter calculation gives $[c_k^+, H_k] = -\lambda_k c_k^+$. Compare these two calculations and find the following result for λ_k:

$$\lambda_k = \sqrt{A_k^2 - B_k^2} \tag{7.91}$$

PART 2
Application to Condensed Matter

Chapter 8

Symmetry in Crystalline Solids

In this chapter, one briefly introduces some elements on the crystalline symmetry which are necessary in the calculation of physical properties of crystals. The periodicity of positions of atoms in a solid allows one to simplify the calculation not only in the real space but also in the reciprocal space as seen for example in chapters 9 and 12.

8.1 Crystalline symmetry

A perfect crystal is formed by the repetition of elementary identical blocks which can be atoms or groups of atoms (molecules). One can describe its symmetry by replacing each block position by a point called "site". These sites form a "lattice". In three dimensions, one can describe a lattice by the use of three elementary translation vectors called "basis vectors" or "lattice vectors" $(\vec{a}_1, \vec{a}_2, \vec{a}_3)$ which are defined in such a way that one can go from a site at $\vec{R}$ to any site of the lattice by the following translation

$$\vec{R}' = \vec{R} + n_1\vec{a}_1 + n_2\vec{a}_2 + n_3\vec{a}_3 \tag{8.1}$$

where n_1, n_2 and n_3 are integers. These three vectors define the unit cell of the lattice.

When all the sites in a lattice have the same environment, the lattice is called a Bravais lattice.

Let us take a first example: a chain of atoms with a distance a between two nearest neighbors constitutes a one-dimensional lattice. The distance a is called the lattice constant. The unit cell is a segment of length a between two nearest neighbors and the basis vector is $\vec{a} = a\vec{e}_1$ where $\vec{e}_1$ is the unit vector on the axis of the chain. Figure 8.1 shows the five Bravais lattices in two dimensions with basis vectors indicated for each of them.

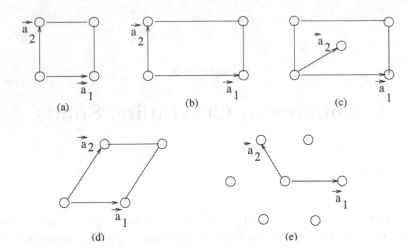

Fig. 8.1   In two dimensions, there are only five Bravais lattices: (a) square (b) rectangular (c) centered rectangular (d) oblique (e) triangular.

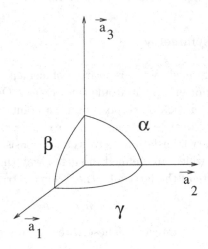

Fig. 8.2   Definition of the angles α, β and γ.

In three dimensions, the simplest Bravais lattice is the cubic lattice. The unit cell is the cube formed by three basis vectors $(\vec{a}_1 = a\vec{e}_1, \vec{a}_2 = a\vec{e}_2, \vec{a}_3 = a\vec{e}_3)$ where $(\vec{e}_1, \vec{e}_2, \vec{e}_3)$ are unit vectors on the Cartesian axes. There are 14 Bravais lattices in three dimensions. These lattices can be described using the angles α, β and γ defined in Fig. 8.2.

These 14 lattices are named according to their conventional cells, not always the elementary cells (primitive cells). Some are shown in Fig. 8.3.

1) *simple cubic*: $a_1 = a_2 = a_3 = a$, $\alpha = \beta = \gamma = \pi/2$.
2) *body-centered cubic*: this is a cubic cell with a site at the center. In the Cartesian system with $(\vec{e}_1, \vec{e}_2, \vec{e}_3)$ defined on three axes of the cube, the elementary cell is defined by

$$\vec{a}_1 = \frac{a}{2}(\vec{e}_1 + \vec{e}_2 - \vec{e}_3) \tag{8.2}$$

$$\vec{a}_2 = \frac{a}{2}(-\vec{e}_1 + \vec{e}_2 + \vec{e}_3) \tag{8.3}$$

$$\vec{a}_3 = \frac{a}{2}(\vec{e}_1 - \vec{e}_2 + \vec{e}_3) \tag{8.4}$$

3) *face-centered cubic*: the conventional cell is a cube with a site at the center of each of the 6 faces. The elementary cell is defined by

$$\vec{a}_1 = \frac{a}{2}(\vec{e}_1 + \vec{e}_2) \tag{8.5}$$

$$\vec{a}_2 = \frac{a}{2}(\vec{e}_2 + \vec{e}_3) \tag{8.6}$$

$$\vec{a}_3 = \frac{a}{2}(\vec{e}_3 + \vec{e}_1) \tag{8.7}$$

4) *tetragonal*: $a_1 = a_2 \neq a_3$, $\alpha = \beta = \gamma = \pi/2$. This lattice can be simple tetragonal or centered tetragonal.
5) *orthorhombic*: $a_1 \neq a_2 \neq a_3$, $\alpha = \beta = \gamma = \pi/2$. There are four varieties: simple, centered, base-centered and face-centered orthorhombic.
6) *monoclinic*: $a_1 \neq a_2 \neq a_3$, $\alpha = \gamma = \pi/2 \neq \beta$. There are two varieties: simple and base-centered monoclinic.
7) *triclinic*: $a_1 \neq a_2 \neq a_3$, $\alpha \neq \beta \neq \gamma$.
8) *trigonal*: $a_1 = a_2 = a_3$, $\alpha = \beta = \gamma < 2\pi/3$ and $\neq \pi/2$.
9) *stacked triangular* : $a_1 = a_2 \neq a_3$, $\alpha = \beta = \pi/2$, $\gamma = 2\pi/3$.

8.2 Reciprocal lattices

The reciprocal lattice of a real-space Bravais lattice is its counterpart in the Fourier space (or $\vec{k}$-space) constructed from the basis vectors $(\vec{a}_1, \vec{a}_2, \vec{a}_3)$ with the formulas given below. Note that calculations of physical properties are often carried out in the $\vec{k}$-space whose symmetry is given by that of the reciprocal lattice.

We consider a crystal of dimension $L_1 \times L_2 \times L_3$. The atomic distances, namely the lattice constants, in the three directions are a_1, a_2 and

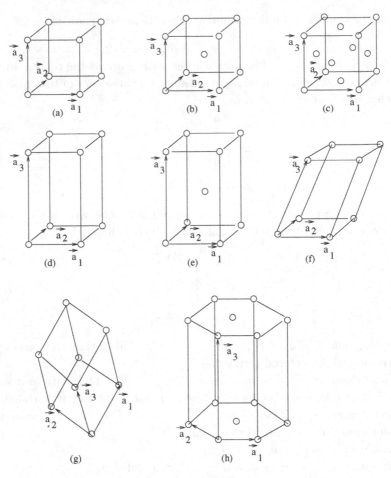

Fig. 8.3   Some Bravais lattices in three dimensions: (a) simple cubic (b) body-centered cubic (c) face-centered cubic (d) tetragonal (e) centered tetragonal (f) triclinic (g) trigonal (h) stacked triangular lattice.

a_3, respectively. Before showing the construction rules, let us give a physical meaning of these rules by taking an example. One knows that the microstate of a free particle is characterized by a wave vector $\vec{k}$ with the wave function $e^{i\vec{k}\cdot\vec{r}}$ (see subsection 2.4.2). The values of $\vec{k}$ are given by the periodic boundary conditions:

$$e^{i\vec{k}\cdot(\vec{r}+\vec{\Gamma})} = e^{i\vec{k}\cdot\vec{r}} \tag{8.8}$$

where $\vec{\Gamma} = L_1\vec{a}_1 + L_2\vec{a}_2 + L_3\vec{a}_3$ is the dimension of the system. One rewrites $\vec{\Gamma} = N_1 a_1 \vec{e}_1 + N_2 a_2 \vec{e}_2 + N_3 a_3 \vec{e}_3$, N_i $(i = 1, 2, 3)$ being the number of cells in the direction i with $\vec{e}_i$ as its unit vector. From the above expression, one has

$$e^{i\vec{k}\cdot\vec{r}} = 1 \tag{8.9}$$

from which

$$k_x = \frac{2\pi n_x}{L_1} \tag{8.10}$$

$$k_y = \frac{2\pi n_y}{L_2} \tag{8.11}$$

$$k_z = \frac{2\pi n_z}{L_3} \tag{8.12}$$

where n_i $(i = x, y, z) = 0, \pm 1, \pm 2, \pm 3, ...$ The phase space is thus determined by the above allowed values of $\vec{k}$. Let us take the one-dimensional lattice of constant a with $L_1 = L$. We have $L = Na$ where N is the number of atoms. We notice that among the allowed values of $k = \frac{2\pi n}{Na}$, there are particular values: each time when n is equal to a multiple of N $(n = mN$, m: integer), we have $k = \frac{2\pi mN}{Na} = \frac{2\pi m}{a}$, namely k is a multiple of $\frac{2\pi}{a}$. These points are noted by a capital letter

$$K_m = \frac{2\pi}{a}m \tag{8.13}$$

When m takes all values $(m=0,\pm 1, \pm 2, ...)$, K_m generates the reciprocal lattice with the lattice constant equal to $\frac{2\pi}{a}$ and the basis vector $\vec{b}_1 = \frac{2\pi}{a}\vec{e}_1$.

We can now generalize to get the rules for the reciprocal-lattice construction in three dimensions. The basis vectors of the reciprocal lattice are $\vec{b}_i$ $(i = 1, 2, 3)$ given by

$$\vec{b}_1 = \frac{2\pi}{\Omega}\vec{a}_2 \wedge \vec{a}_3 \tag{8.14}$$

$$\vec{b}_2 = \frac{2\pi}{\Omega}\vec{a}_3 \wedge \vec{a}_1 \tag{8.15}$$

$$\vec{b}_3 = \frac{2\pi}{\Omega}\vec{a}_1 \wedge \vec{a}_2 \tag{8.16}$$

where $\Omega = \vec{a}_1 \cdot (\vec{a}_2 \wedge \vec{a}_3)$ is the volume of the elementary lattice cell in the real space.

We can also use the following equivalent formulas:

$$\vec{b}_i \cdot \vec{a}_j = 2\pi \delta_{i,j} \tag{8.17}$$

where $(i, j = 1, 2, 3)$ and $\delta_{i,j}$ is the Kronecker symbol.

It is straightforward to find that

- the reciprocal lattice of a square lattice of constant a is a square lattice of constant $\frac{2\pi}{a}$.
- the reciprocal lattice of a simple cubic lattice of constant a is a simple cubic lattice of constant $\frac{2\pi}{a}$.
- the reciprocal lattice of a body-centered cubic lattice is a face-centered cubic lattice and vice-versa.

8.3   Wave-vector space - Brillouin zones

The wave vectors $\vec{k}$ take values allowed by the periodic boundary conditions. In one dimension, $k = \frac{2\pi n}{L}$. It is noted that the reciprocal lattice vectors K are special values in the k-space. Between two lattice sites in this space, the distance is $\frac{2\pi}{a}$ which contains N values of k: $k = \frac{2\pi n}{L}$ with $n \in [0, N[$ (open bound at one end which is attributed to the following segment).

For convenience, one continues to consider the one-dimensional case. One divides the values of k in zones called "Brillouin zones". The first zone is around the origin $k = 0$ and limited at $\pm\pi/a$, namely at the half distance, at each side, to the next lattice sites situated at $\pm 2\pi/a$. Here are some other examples of Brillouin zones:

- The first Brillouin zone of a reciprocal square lattice is a square centered at the origin $\vec{k} = 0$ and limited by $\pm\pi/a$ on each axis k_x and k_y.
- The first Brillouin zone of a reciprocal cubic lattice is a cube centered at the origin $\vec{k} = 0$ and limited by $\pm\pi/a$ on k_x, k_y and k_z.

One can construct the second, third, ... Brillouin zones. For example, the second Brillouin zone in one dimension contains values of k belonging to the two separated intervals: $[\pi/a, 2\pi/a[$ and $[-\pi/a, -2\pi/a[$. However, these zones are not useful for the purpose of this book.

Let us give the general rule to construct the first Brillouin zone in one or more dimensions. One considers a lattice site as the origin. For a given vector connecting the origin to one of its nearest neighboring site, one plots a plane perpendicular at its middle point (bisector plane). One plots bisector planes for all other vectors connecting the origin site to its nearest neighbors. The space around the origin, limited by these bisector planes constitutes the first Brillouin zone. For illustration, one applies this rule to the one-dimensional case: the two nearest neighbors of the origin are situated at $\pm\frac{2\pi}{a}$ so that the two bisector planes cut the nearest-neighbor vectors at $\pm\frac{\pi}{a}$. The first Brillouin zone in one dimension is thus the segment

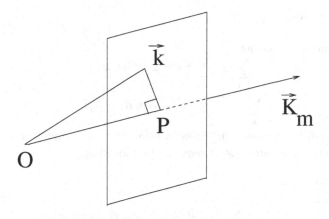

Fig. 8.4   A boundary plane of the first Brillouin zone.

$[-\frac{\pi}{a}, +\frac{\pi}{a}]$ around the origin as shown above. Applying now the construction rule to the simple cubic lattice, one sees that the 6 bisector planes, cutting the 6 nearest-neighbor vectors ($\pm a\vec{e}_1, \pm a\vec{e}_2, \pm a\vec{e}_3$) at their middle points, form a cube around the origin limited at ($\pm\frac{\pi}{a}, \pm\frac{\pi}{a}, \pm\frac{\pi}{a}$) in the three Cartesian directions.

One can verify that each Brillouin zone contains N values of k. This rule applies in any dimension where N is the total number of cells of the crystal.

One will see in chapter 11 that states lying on the boundaries of Brillouin zones play an important role in electronic properties. It is thus useful to give here the equation for zone boundaries. Let $\vec{K}_m$ be the vector connecting the origin of the reciprocal lattice to a neighboring site. The limit of the first Brillouin zone is found, by construction, on the bisector plane perpendicular to that vector. Let P be the intersection point of this plane with $\vec{K}_m$. A state on the bisector plane belongs thus to the boundary of the first Brillouin zone (see Fig. 8.4).

One has

$$\frac{1}{2}|\vec{K}_m| = \overline{OP} = |\vec{k}|\cos(\theta) = \frac{\vec{k}\cdot\vec{K}_m}{|\vec{K}_m|} \tag{8.18}$$

from which one obtains the following equation of the boundary:

$$2\vec{k}\cdot\vec{K}_m = \pm|\vec{K}_m|^2 \tag{8.19}$$

The two signs indicate two symmetric faces of the Brillouin zone boundary.

8.4   Sum rules

In one dimension, one has

$$\sum_i e^{-i(k-k')R_i} = N\delta(k - k') \tag{8.20}$$

$$\sum_k e^{-ik(R_i - R_j)} = N\delta(R_i - R_j) \tag{8.21}$$

One can generalize these sum rules in two and three dimensions. One can also write them as integrals as seen in the following.

- Sum on $\vec{k}$:
  One has

$$\sum_{\vec{k}} e^{i\vec{k}\cdot(\vec{R}-\vec{R}')} = N\delta(\vec{R} - \vec{R}') \tag{8.22}$$

where the sum is performed on allowed values of $\vec{k}$ in the first Brillouin zone, N is the total number of sites of the system, namely $N = N_1 \times N_2 \times N_3$, $\vec{R}$ and $\vec{R}'$ are positions of the lattice sites in the real space.

Demonstration:

One has

$$\vec{R} = m_1 a_1 \vec{e}_1 + m_2 a_2 \vec{e}_2 + m_3 a_3 \vec{e}_3,$$

$$\vec{R}' = m_1' a_1 \vec{e}_1 + m_2' a_2 \vec{e}_2 + m_3' a_3 \vec{e}_3, \quad (m_i, m_i' = \text{integers}),$$

$$\vec{k} = \frac{2\pi n_1}{N_1 a_1}\vec{e}_1 + \frac{2\pi n_2}{N_2 a_2}\vec{e}_2 + \frac{2\pi n_3}{N_3 a_3}\vec{e}_3, \quad [n_i(i = 1, 2, 3) = \text{integers}],$$

from which one has

$$\sum_{\vec{k}} e^{i\vec{k}\cdot(\vec{R}-\vec{R}')} = \sum_{n_1=0}^{N_1-1} \sum_{n_2=0}^{N_2-1} \sum_{n_3=0}^{N_3-1} e^{i2\pi[n_1(m_1-m_1')+n_2(m_2-m_2')+n_3(m_3-m_3')]}$$

$$= \sum_{n_1=0}^{N_1-1} e^{\frac{i2\pi n_1(m_1-m_1')}{N_1}} \sum_{n_2=0}^{N_2-1} e^{\frac{i2\pi n_2(m_2-m_2')}{N_2}} \sum_{n_3=0}^{N_3-1} e^{\frac{i2\pi n_3(m_3-m_3')}{N_3}}$$

Each sum in the last line is a geometric series of N_i terms $(i = 1, 2, 3)$, of ratio $e^{\frac{i2\pi(m_i-m_i')}{N_i}}$. One has thus

$$\sum_{n_i=0}^{N_i-1} e^{\frac{i2\pi n_i(m_i-m_i')}{N_i}} = \frac{1 - \left[e^{\frac{i2\pi(m_i-m_i')}{N_i}}\right]^{N_i}}{1 - e^{\frac{i2\pi(m_i-m_i')}{N_i}}}$$

$$= \frac{1 - e^{i2\pi q_i}}{1 - e^{\frac{i2\pi q_i}{N_i}}} \tag{8.23}$$

where $q = m_i - m_i'$ is an integer. One sees that if $q_i \neq 0$, namely $\vec{R} \neq \vec{R}'$, the numerator is equal to zero. However, when $q_i = 0$ the left-hand side of (8.23) gives immediately N_i. One then obtains (8.22).

- Sum on $\vec{R}$:

$$\sum_{\vec{R}} e^{i\vec{k}\cdot\vec{R}} = N_1 \delta_{k_x, l_x \frac{2\pi}{a_1}} N_2 \delta_{k_y, l_y \frac{2\pi}{a_2}} N_3 \delta_{k_z, l_z \frac{2\pi}{a_3}} \qquad (8.24)$$

where the sum runs over position vectors of the lattice sites in the real space. The numbers l_i $(i = x, y, z)$ are integers.

Demonstration:

As before, one writes

$$\vec{R} = m_1 a_1 \vec{e}_1 + m_2 a_2 \vec{e}_2 + m_3 a_3 \vec{e}_3.$$

The sum becomes

$$\sum_{\vec{R}} e^{i\vec{k}\cdot\vec{R}} = \sum_{m_1=0}^{N_1-1} e^{ik_x m_1 a_1} \sum_{m_2=0}^{N_2-1} e^{ik_y m_2 a_2} \sum_{m_3=0}^{N_3-1} e^{ik_z m_3 a_3}$$

$$= \sum_{m_1=0}^{N_1-1} e^{i\frac{2\pi n_x}{N_1 a_1} m_1 a_1} \sum_{m_2=0}^{N_2-1} e^{i\frac{2\pi n_y}{N_2 a_2} m_2 a_2} \sum_{m_3=0}^{N_3-1} e^{i\frac{2\pi n_z}{N_3 a_3} m_3 a_3}$$

$$= \sum_{m_1=0}^{N_1-1} e^{i\frac{2\pi n_x m_1}{N_1}} \sum_{m_2=0}^{N_2-1} e^{i\frac{2\pi n_y m_2}{N_2}} \sum_{m_3=0}^{N_3-1} e^{i\frac{2\pi n_z m_3}{N_3}}$$

$$= \frac{1 - e^{i2\pi n_x}}{1 - e^{\frac{i2\pi n_x}{N_1}}} \frac{1 - e^{i2\pi n_y}}{1 - e^{\frac{i2\pi n_y}{N_2}}} \frac{1 - e^{i2\pi n_z}}{1 - e^{\frac{i2\pi n_z}{N_3}}}$$

$$= N_1 \delta_{\frac{n_x}{N_1}, l_x} N_2 \delta_{\frac{n_y}{N_2}, l_y} N_3 \delta_{\frac{n_z}{N_3}, l_z} \qquad (8.25)$$

where the last line has been calculated in the same manner as for (8.23), with l_i $(i = x, y, z)$ being integers. Replacing $\frac{n_x}{N_1} = \frac{k_x a_1}{2\pi}$, $\frac{n_y}{N_2} = \frac{k_y a_2}{2\pi}$ and $\frac{n_z}{N_3} = \frac{k_z a_3}{2\pi}$ in (8.25), one obtains (8.24).

Note that the relation (8.24) shows that the sum is not zero only if the wave vector $\vec{k}$ is equal to a vector $\vec{K}$ of a site of the reciprocal lattice.

- Integral on $\vec{r}$:

In the case where the periodic boundary conditions apply, one has

$$\int_{\Omega} e^{i\vec{k}\cdot\vec{r}} d\vec{r} = \Omega \delta(\vec{k}) \qquad (8.26)$$

where the integral is performed in the volume $\Omega = L_1 \times L_2 \times L_3$ of the crystal.

Demonstration:

Integrating on x one has

$$\int_0^{L_1} dx e^{ik_x x} = \frac{1}{ik_x}(e^{ik_x L_1} - 1)$$

The periodic boundary condition implies $e^{ik_x L_1} = 1$ [see (8.9)], so that

$$\int_0^{L_1} dx e^{ik_x x} = \frac{1}{ik_x}(e^{ik_x L_1} - 1) = 0 \text{ if } k_x \neq 0$$
$$= L_1 \text{ if } k_x = 0$$

The last line is obtained by directly calculating $\int_0^{L_1} dx e^{ik_x x}$ with $k_x = 0$. Similar calculations are done in the two other directions. One has therefore (8.26).

Remarks:

1) One observes that

$$e^{i\vec{K}\cdot\vec{R}} = e^{i\left(\frac{l_x 2\pi}{a_1} m_1 a_1 + \frac{l_y 2\pi}{a_2} m_2 a_2 + \frac{l_z 2\pi}{a_3} m_3 a_3\right)} = 1 \qquad (8.27)$$

This is because the argument of the exponential is a multiple of 2π.

2) When the size of the system is very large (thermodynamic limit) one can replace (8.22) by an integral

$$\sum_{\vec{k}} \ldots = \frac{L_1 \times L_2 \times L_3}{(2\pi)^3} \int_{-\frac{\pi}{a_1}}^{\frac{\pi}{a_1}} dk_x \int_{-\frac{\pi}{a_2}}^{\frac{\pi}{a_2}} dk_y \int_{-\frac{\pi}{a_3}}^{\frac{\pi}{a_3}} dk_z \ldots \qquad (8.28)$$

The factor is from the fact that the number of microstates in the first Brillouin zone is equal to its volume divided by the volume of a microstate which is $\frac{2\pi}{L_1} \times \frac{2\pi}{L_2} \times \frac{2\pi}{L_3}$ (this volume is the space between two successive values of $\vec{k}$ in each of the three directions, see subsection 2.4.2).

8.5   Fourier analysis

One gives below some properties of a periodic function. These are useful when one deals with, for example, periodic potentials (see chapter 11).

Let $f(\vec{r})$ be a periodic function with the periodicity of the lattice, and $\vec{r}$ a real-space position. One has

$$f(\vec{r} + \vec{R}) = f(\vec{r}) \tag{8.29}$$

$\vec{R}$ being a vector connecting two lattice sites, namely

$$\vec{R} = m_1\vec{a}_1 + m_2\vec{a}_2 + m_3\vec{a}_3 \quad (m_1, m_2, m_3 = \text{ integers}).$$

One shows that

$$f(\vec{r}) = \frac{1}{\Omega^{1/2}} \sum_{\vec{K}} A_{\vec{K}} e^{i\vec{K}\cdot\vec{r}} \tag{8.30}$$

where the sum is taken over the vectors of the reciprocal lattice, $A_{\vec{K}}$ is the Fourier component of $f(\vec{r})$.

Demonstration:

The Fourier series of $f(\vec{r})$ in the $\vec{k}$-space is written as

$$f(\vec{r}) = \frac{1}{\Omega^{1/2}} \sum_{\vec{k}} A_{\vec{k}} e^{i\vec{k}\cdot\vec{r}}.$$

The periodicity of $f(\vec{r})$ [see (8.29)] implies that

$$e^{i\vec{k}\cdot\vec{R}} = 1.$$

Comparing to (8.27), one sees that $\vec{k}$ should be a vector of the reciprocal lattice $\vec{K}$. One has then the relation (8.30). The Fourier component $A_{\vec{K}}$ is given by

$$A_{\vec{K}} = \frac{1}{\Omega^{1/2}} \int_{\Omega} f(\vec{r}) e^{-i\vec{K}\cdot\vec{r}} d\vec{r} \tag{8.31}$$

where the integration is performed over the volume $\Omega = L_1 \times L_2 \times L_3$ of the crystal.

Demonstration:

Using (8.30) one has

$$\frac{1}{\Omega^{1/2}} \int_{\Omega} f(\vec{r}) e^{-i\vec{K}\cdot\vec{r}} d\vec{r} = \frac{1}{\Omega^{1/2}} \int_{\Omega} \frac{1}{\Omega^{1/2}} \sum_{\vec{K}'} A_{\vec{K}'} e^{i\vec{K}'\cdot\vec{r}} e^{-i\vec{K}\cdot\vec{r}} d\vec{r}$$

$$= \frac{1}{\Omega} \sum_{\vec{K}'} A_{\vec{K}'} \int_{\Omega} e^{-i(\vec{K}-\vec{K}')\cdot\vec{r}} d\vec{r}$$

$$= \frac{1}{\Omega} \sum_{\vec{K}'} A_{\vec{K}'} \Omega \delta_{\vec{K},\vec{K}'}$$

$$= A_{\vec{K}}$$

where one has used (8.26) before the last line. One finds thus the relation (8.31).

8.6   Representation in $\vec{k}$-space

Let us show a convenient way to show results, such as the energy $E = \frac{\hbar^2 k^2}{2m}$, in the $\vec{k}$-space. Instead of using all values of $\vec{k}$, one shall use values of k belonging to the first Brillouin zone. For values of k outside of the first zone, E is translated toward the first zone with a translation using $K_m = \frac{2\pi m}{a}$ where $m = \pm 1, \pm 2, \ldots$. For example, value of E for $k' \in [\pi/a, 2\pi/a[$ of the second zone is translated to the first zone at the value of k given by $k = k' + K_{-1}$ with $K_{-1} = -\frac{2\pi}{a}$. Of course, the value of E at k and k' is the same, only its wave-vector changes for a presentation convenience. For a given value of E, one should indicate where it comes from, namely its translation if any: one shall use an index ℓ to indicate its "origin": $\ell = 1$ for E of the first zone (no translation of E: $m = 0$), $\ell = 2$ for E translated from the second zone ($m = \pm 1$) etc. The presentation using only k of the first Brillouin zone is called "reduced-zone presentation". This is shown in Fig. 8.5.

Note that when one indicates a value of E one should precise which index ℓ it has: index ℓ indicates thus the "energy band" number: first band, second band, etc. As it has been said above, each Brillouin zone has N allowed values of k. Consequently, each energy band contains N microstates (not including the spin degeneracy).

8.7   Conclusion

One has shown in this chapter how to describe the symmetry of the crystalline structure and its counterpart in the $\vec{k}$ space. Since physical properties are often presented in the $\vec{k}$ space, it is very important to understand the vocabulary and its general fundamental properties such as Brillouin zones, the number of microstates in an energy band and the sum rules. One will see afterward that these are unavoidable means to describe physical properties of crystalline solids, such as properties of phonons, magnons and band structures.

8.8   Problems

Problem 1. Reciprocal lattice of a triangular lattice:
          Construct the reciprocal lattice of a triangular lattice of constant

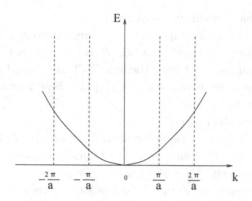

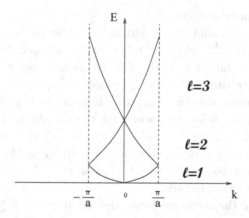

Fig. 8.5 Top: E versus k not limited in the first Brillouin zone. Bottom: E in the reduced zone presentation. Indices $\ell = 1, 2, 3$ indicate first, second and third energy bands, respectively.

    a. Define the first Brillouin zone.

Problem 2. Honeycomb lattice:

    Consider a honeycomb lattice (or hexagonal lattice) of constant a (distance between two nearest neighboring sites). Is it a Bravais lattice? How can one construct the reciprocal lattice of a honeycomb lattice?

Problem 3. Face-centered cubic lattice - Body-centered cubic lattice:

    Show that the reciprocal lattice of a face-centered cubic lattice is a body-centered cubic lattice. Show that the reciprocal lattice of a body-centered cubic lattice is a face-centered cubic lattice.

Problem 4. Chain of two types of atom:

Consider a chain composed of two types of atom A et B alternately. The distance between two neighboring atoms is a. Construct the reciprocal lattice. Define the first Brillouin zone. Give the number of microstates in this zone. Compare these results to those of a mono-atomic chain.

Problem 5. Periodic potential:

Let $V(\vec{r})$ be a periodic, real and symmetric potential, namely $V(\vec{r}) = V(-\vec{r})$. Show that the Fourier component $A_{\vec{K}}$ of (8.31) is real, namely $A^*_{\vec{K}} = A_{-\vec{K}}$ and $A_{\vec{K}} = A_{-\vec{K}}$.

Problem 6. Fourier transform of the Coulomb potential:

Calculate the Fourier component $A_{\vec{k}}$ of the potential $V(\vec{r}) = \frac{1}{r}$. Show that $A_{\vec{k}} = \frac{4\pi}{k^2}$.

Problem 7. Fourier analysis:

Consider a chain of N atoms, of constant a. Define $q_r = \frac{1}{N^{1/2}} \sum_k Q_k e^{ikr}$ where k takes values allowed by the periodic boundary condition in the first Brillouin zone, r is the position of a lattice site. Show that $Q_k = \frac{1}{N^{1/2}} \sum_r q_r e^{-ikr}$.

Problem 8. Fourier analysis of function in continuous real space:

Consider the following series with the variable x defined between $-L/2$ and $L/2$: $q(x) = \frac{1}{L^{1/2}} \sum_k Q_k e^{ikx}$ where k takes values allowed by the periodic boundary condition in the first Brillouin zone. Show that $Q_k = \frac{1}{L^{1/2}} \int_{-L/2}^{L/2} dx q(x) e^{-ikx}$.

Problem 9. Structure factors:

Calculate the structure factor defined by $\gamma_{\vec{k}} = \frac{1}{Z} \sum_{\vec{R}} e^{i\vec{k} \cdot \vec{R}}$ where $\vec{R}$ is a vector connecting a lattice site to one of its neighbors and the sum is taken on its Z neighbors. Give the result for the simple cubic lattice, face-centered cubic lattice, body-centered cubic lattice, triangular lattice.

Chapter 9

Interacting Atoms in Crystals: Phonons

Elementary excitations of interacting particles such as atoms or spins on a lattice obey the Bose-Einstein statistics developed in chapter 6. In this chapter, one shows in some details the case of phonons which are the quantization of atom vibrations in crystals. Some other popular subjects such as magnons (collective motions of interacting spins), plasmons (oscillations of charges in plasmas) and excitons are very important in condensed matter where statistical physics demonstrates its efficiency. The theory of magnons, or quantized spin waves, which is very important in magnetism, deserves a large space in chapter 12.

9.1 Introduction

In a crystalline solid, atoms interact with each other, giving rise to a collective vibration at low temperatures. At a high temperature, called melting temperature, the crystal melts into a liquid phase. In this chapter, we are interested in the collective motion of the atoms. When quantized, its energy quanta is called "phonon", by analogy with photons for quantized light waves.

In a crystal, atoms occupy the sites of a lattice. The symmetries of the lattice and its counterpart in the reciprocal space (wave-vector space) have been shown in chapter 8. The main properties of crystalline symmetry, in particular the sum rules, are widely used as seen below and in the following chapters.

In classical mechanics, interacting atoms in the lattice are considered as a system of masses interacting via a potential which provides a restoring force. Among potential models, the simplest one is no doubt the Lennard-Jones potential given by

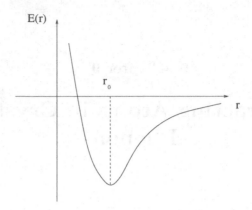

Fig. 9.1   Lennard-Jones potential.

$$U(r) = U_0(\frac{1}{r^6} - \frac{1}{r^{12}}) \tag{9.1}$$

where U_0 is a constant and r the relative distance between two atoms (see Fig. 9.1).

At $T = 0$, two atoms are at the distance r_0 where $U(r)$ is minimum. When one atom moves to r near r_0, a restoring force due to the other atom tries to move it back to r_0. If $(r - r_0)/r_0$ is small, one can replace the potential by a harmonic potential of the form $K(r - r_0)^2/2$ where K is a constant. The system is thus equivalent to a system of harmonic oscillators coupled with each other by springs of constant K. Each oscillator has an energy equal to $E = (\vec{p})^2/2m + K(r - r_0)^2/2$, dependent on 6 degrees of freedom in three dimensions (3 for the momentum and 3 for the position). The system energy is given by the theorem of energy equipartition shown in section 3.6.4: each degree of freedom gives a contribution of $k_B T/2$ to the total energy. Thus, for a system of N harmonic oscillators, the total energy is equal to $N \times 6 \times Nk_B T/2 = 3Nk_B T$. As a consequence, the heat capacity C_V of the atom vibrations of the crystal is equal to $3Nk_B$ independent of temperature T. However, experimentally C_V is found to be proportional to T^3 at low temperatures. This suggests that there is a need to incorporate quantum effects via the Bose-Einstein statistics in the calculation of the vibration energy.

9.2 Vibrations in one dimension

One shows in this section how to calculate the dispersion relation for the vibration of a one-dimensional chain of atoms. One considers a chain of N atoms, of lattice constant a. At equilibrium, atoms occupy the lattice sites X_i ($i = 1, N$). Let ζ_i be the instantaneous position of atom i. The distance to its equilibrium position is thus $x_i = \zeta_i - X_i$. Let $\phi(\zeta_i, \zeta_j)$ be the potential between atoms i and j. One supposes that the interaction between them depends only on their relative distance so that $\phi(\zeta_i, \zeta_j) = \phi(|\zeta_i - \zeta_j|)$. Expanding this potential around the equilibrium distance, namely around $X_i - X_j$, one has

$$\phi(\zeta_i - \zeta_j) = \phi(X_i - X_j) + \left[\frac{\partial \phi}{\partial(\zeta_i - \zeta_j)} \right]_{X_i - X_j} [X_i - X_j - (\zeta_i - \zeta_j)]$$

$$+ \frac{1}{2} \left[\frac{\partial^2 \phi}{\partial(\zeta_i - \zeta_j)^2} \right]_{X_i - X_j} [X_i - X_j - (\zeta_i - \zeta_j)]^2 + \ldots$$

$$= \phi(X_i - X_j) + \frac{1}{2} \left[\frac{\partial^2 \phi}{\partial(\zeta_i - \zeta_j)^2} \right]_{X_i - X_j} (x_i - x_j)^2 + \ldots \quad (9.2)$$

where the second term of the first line is zero because it is the force at the equilibrium position. The approximation where terms of order higher than the harmonic one $(x_i - x_j)^2$ are neglected, is called "harmonic approximation". This approximation is valid for small displacements around equilibrium positions.

9.2.1 *Equation of motion*

One defines the interaction coupling between two atoms i and j at ζ_i and ζ_j by

$$K_{ij} \equiv \left[\frac{\partial^2 \phi}{\partial(\zeta_i - \zeta_j)^2} \right]_{X_i - X_j}.$$

If interactions are limited between nearest neighbors, one puts $K_{ij} = K$ for nearest i and j, namely $|X_i - X_j| = a$, and $K_{ij} = 0$ otherwise. Omitting the constant $\sum_{i,j} \phi(X_i - X_j)$, the Hamiltonian in the harmonic approximation is written as

$$\mathcal{H} = \sum_i \frac{p_i^2}{2m} + \phi(\zeta_i - \zeta_j) \simeq \sum_i \frac{p_i^2}{2m} + \frac{K}{4} \sum_{i,j} (x_i - x_j)^2 \quad (9.3)$$

where one has introduced the factor $1/2$ to remove the double counting due to the double sum on i and j. Using the equation of motion $\dot{p}_i = -\frac{\partial \mathcal{H}}{\partial x_i}$ of the classical mechanics, one obtains

$$m\ddot{x}_i = -K(2x_i - x_{i+1} - x_{i-1}) \tag{9.4}$$

There are N equations for N atoms of the chain.

9.2.2   *Dispersion relation*

One admits a solution is of the form

$$x_i(t) = Ae^{i(kX_i - \omega t)} \tag{9.5}$$

where k is the wave vector, ω the pulsation and t the time. The constant A is the vibration amplitude. Replacing $X_{i\pm1}$ by $X_i \pm a$, one has

$$x_{i\pm1}(t) = x_i(t)e^{\pm ika} \tag{9.6}$$

Equation (9.4) becomes

$$m\omega^2 x_i(t) = -K(2 - e^{ika} - e^{-ika})x_i(t) \tag{9.7}$$

from which,

$$\omega^2 = \frac{2K}{m}[1 - \cos(ka)]$$
$$\omega^2 = \frac{4K}{m}\sin^2\left(\frac{ka}{2}\right)$$

or

$$\omega_k = 2\sqrt{\frac{K}{m}}\left|\sin\left(\frac{ka}{2}\right)\right| \tag{9.8}$$

Equation (9.8) is called "dispersion relation" of phonon in one dimension.

Using the periodic boundary conditions for (9.5), namely $x_{i+N}(t) = x_i(t)$, one has

$$k = \frac{2\pi n}{Na} \tag{9.9}$$

where $n = 0, \pm1, \pm2, ...$ There are thus N values of k in the first Brillouin zone (see chapter 8). Figure 9.2 shows ω_k as a function of k.

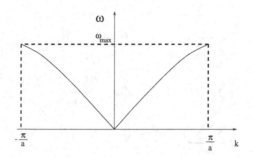

Fig. 9.2 Dispersion relation (9.8). $\omega_{max} = 2\sqrt{\frac{K}{m}}$.

It is noted that when $k \to 0$, one has a linear dependence of ω on small k:

$$\omega \to \sqrt{\frac{K}{m}}ka \tag{9.10}$$

The group velocity of a vibration wave is defined by

$$v_g = \frac{d\omega}{dk} = a\sqrt{\frac{K}{m}}\left|\cos\left(\frac{ka}{2}\right)\right| \tag{9.11}$$

The sound velocity v_s is the limit of v_g when $k \to 0$. One has

$$v_s = a\sqrt{\frac{K}{m}} \tag{9.12}$$

This velocity is that of the sound propagating in a solid. For this reason, one calls vibration modes of long wave length (small k) "acoustic modes".

9.3 Vibrations in two and three dimensions

In two and three dimensions the calculation is more complicated due to several components of atom positions. One can generalize the second term of (9.2) to two or three dimensions as

$$U = \frac{1}{4} \sum_{i,j} \sum_{\alpha,\beta} (u_i^\alpha - u_j^\alpha) \frac{\partial^2 \phi(\vec{R}_i - \vec{R}_j)}{\partial u_i^\alpha \partial u_j^\beta} (u_i^\beta - u_j^\beta)$$

$$= \frac{1}{2} \sum_{i,j} \sum_{\alpha,\beta} (u_i^\alpha u_i^\beta - u_i^\alpha u_j^\beta) \frac{\partial^2 \phi(\vec{R}_i - \vec{R}_j)}{\partial u_i^\alpha \partial u_j^\beta}$$

$$= \frac{1}{2} \sum_{i,j} \sum_{\alpha,\beta} (u_i^\alpha u_i^\beta - u_i^\alpha u_j^\beta) K_{\alpha\beta}(\vec{R}_i - \vec{R}_j) \qquad (9.13)$$

where $\vec{u}_i$ is the deviation from the equilibrium position $\vec{R}_i$ of the i-th atom and

$$K_{\alpha\beta}(\vec{R}_i - \vec{R}_j) = \frac{\partial^2 \phi}{\partial u_i^\alpha \partial u_j^\beta} \qquad (9.14)$$

with $\alpha, \beta = x, y, z$.

One considers the matrix M defined by

$$M_{\alpha\beta}(\vec{R}_i - \vec{R}_j) = \delta_{i,j} \sum_l K_{\alpha\beta}(\vec{R}_i - \vec{R}_l) - K_{\alpha\beta}(\vec{R}_i - \vec{R}_j) \qquad (9.15)$$

where the sum on l is performed over all lattice sites including i. Using this definition, one rewrites (9.13) as

$$U = \frac{1}{2} \sum_{i,j} \sum_{\alpha,\beta} M_{\alpha\beta}(\vec{R}_i - \vec{R}_j) u_i^\alpha u_j^\beta \qquad (9.16)$$

The form of (9.15) is rather complicated but it is helpful when one wishes to simplify the calculation using the symmetry of the crystal. For example, one has the following sum rule:

$$\sum_j M_{\alpha\beta}(\vec{R}_i - \vec{R}_j) = 0 \qquad (9.17)$$

Demonstration:

$$\sum_j M_{\alpha\beta}(\vec{R}_i - \vec{R}_j) = \sum_j \left[\delta_{i,j} \sum_l K_{\alpha\beta}(\vec{R}_i - \vec{R}_l) - K_{\alpha\beta}(\vec{R}_i - \vec{R}_j) \right]$$

$$= \sum_l K_{\alpha\beta}(\vec{R}_i - \vec{R}_l) - \sum_j K_{\alpha\beta}(\vec{R}_i - \vec{R}_j)$$

$$= 0 \qquad (9.18)$$

Using the equation of motion

$$\dot{p}_i^\alpha = -\frac{\partial \mathcal{H}}{\partial u_i^\alpha} = -\frac{\partial U}{\partial u_i^\alpha},$$

one obtains

$$m\ddot{u}_i^\alpha(t) = -\sum_j \sum_\beta M_{\alpha\beta}(\vec{R}_i - \vec{R}_j)u_j^\beta(t) \tag{9.19}$$

There are $3N$ equations for N atoms of the crystal. Physically, $M_{\alpha\beta}(\vec{R}_i - \vec{R}_j)$ is the force which acts on atom i in the direction α when atom j moves in the direction β. The α-component of the force by atom j is

$$F_{ij}^\alpha(t) = -\sum_\beta M_{\alpha\beta}(\vec{R}_i - \vec{R}_j)u_j^\beta(t)) \tag{9.20}$$

One bears in mind that in the harmonic approximation $|\vec{u}_i|$ are considered very small so that $\vec{u}_i - \vec{u}_j \simeq \vec{R}_i - \vec{R}_j$: the force between atoms i and j is thus a central force approximatively lying on the axis $\vec{R}_i - \vec{R}_j$. This simplification yields

$$\vec{F}_{ij} = -K(\vec{R}_i - \vec{R}_j)\left[(\vec{u}_i(t) - \vec{u}_j(t)) \cdot \frac{(\vec{R}_i - \vec{R}_j)}{|\vec{R}_i - \vec{R}_j|}\right]\frac{(\vec{R}_i - \vec{R}_j)}{|\vec{R}_i - \vec{R}_j|} \tag{9.21}$$

where $K(\vec{R}_i - \vec{R}_j)$ is the constant of the restoring force between atoms i and j. The potential energy U in (9.13) becomes

$$U = \frac{1}{2}\sum_{i,j} K(\vec{R}_i - \vec{R}_j)\left[(\vec{u}_i(t) - \vec{u}_j(t)) \cdot \frac{(\vec{R}_i - \vec{R}_j)}{|\vec{R}_i - \vec{R}_j|}\right]^2 \tag{9.22}$$

One uses the solution of the form

$$u_i^\alpha(t) = A_\alpha e^{i(\vec{k}\cdot\vec{R}_i - \omega t)} \tag{9.23}$$

where $\vec{k}$ is the wave vector, ω the pulsation and t the time. Vector $\vec{A} = (A_\alpha, \alpha = x, y, z)$ represents the amplitude of vibration. Putting $\vec{\rho}_{ij} = \vec{R}_j - \vec{R}_i$, one has from (9.19):

$$m\omega^2 A_\alpha = \sum_j \sum_\beta M_{\alpha\beta}(\vec{R}_i - \vec{R}_j)e^{i\vec{k}\cdot\vec{\rho}_{ij}}A_\beta$$

$$= \sum_\beta D_{\alpha\beta}(\vec{k})A_\beta \tag{9.24}$$

where $[D_{\alpha\beta}(\vec{k})]$ is called the "dynamic matrix" defined by

$$D_{\alpha\beta}(\vec{k}) = \sum_j M_{\alpha\beta}(\vec{R}_i - \vec{R}_j)e^{i\vec{k}\cdot\vec{\rho}_{ij}} \qquad (9.25)$$

This is a 3×3 matrix in the case of three dimensions. The sum on j runs over neighbors of i. However, including $j = i$ while summing does not change the problem since the corresponding contribution is zero. One writes thus

$$\begin{aligned} D_{\alpha\beta}(\vec{k}) &= M_{\alpha\beta}(0) + \sum_{j\neq i} M_{\alpha\beta}(\vec{R}_i - \vec{R}_j)e^{i\vec{k}\cdot\vec{\rho}_{ij}} \\ &= \sum_{j\neq i} M_{\alpha\beta}(\vec{R}_i - \vec{R}_j)\left[e^{i\vec{k}\cdot\vec{\rho}_{ij}} - 1\right] \end{aligned} \qquad (9.26)$$

where one has put

$$M_{\alpha\beta}(0) = -\sum_{j\neq i} M_{\alpha\beta}(\vec{R}_i - \vec{R}_j) \qquad (9.27)$$

Finally, (9.24) is rewritten as

$$\sum_\beta [D_{\alpha\beta}(\vec{k}) - m\omega^2\delta_{\alpha,\beta}]A_\beta = 0 \qquad (9.28)$$

where $\delta_{\alpha,\beta}$ is the Kronecker symbol.

There are three coupled equations for $\alpha = x, y, z$. This is a Kramer system: non trivial solutions imply that

$$\det[D(\vec{k}) - m\omega^2 I] = 0 \qquad (9.29)$$

I being a unitary 3×3 matrix. Equation (9.29) yields in general three eigenvalues of ω for each $\vec{k}$. The corresponding eigenvectors given by (9.28) describe the nature of atomic motion of each mode. The three modes include one longitudinal and two transversal modes. These modes, according to the system symmetry, can be degenerate.

9.4   Quantization of vibration: phonons

9.4.1   *Normal coordinates, vibration energy*

The vibration energy (9.3) can be expressed by normal coordinates which are historic terms indicating the Fourier transforms of momenta and positions, p_i and u_i, of atoms:

$$u_i(t) = \frac{1}{\sqrt{N}} \sum_k X_k(t) e^{ikR_i} \tag{9.30}$$

$$p_i(t) = \frac{1}{\sqrt{N}} \sum_k P_k(t) e^{ikR_i} \tag{9.31}$$

The sum k is taken over all values in the first Brillouin zone. The Fourier components, or normal coordinates, $X_k(t)$ and $P_k(t)$ are given by the inverse transformations

$$X_k(t) = \frac{1}{\sqrt{N}} \sum_i u_i(t) e^{-ikR_i} \tag{9.32}$$

$$P_k(t) = \frac{1}{\sqrt{N}} \sum_i p_i(t) e^{-ikR_i} \tag{9.33}$$

Replacing (9.30) and (9.31) in (9.3), and using the sum rules (see chapter 8)

$$\sum_i e^{-i(k-k')R_i} = N\delta(k - k') \tag{9.34}$$

$$\sum_k e^{-ik(R_i - R_j)} = N\delta(R_i - R_j) \tag{9.35}$$

one obtains

$$E = \sum_k \frac{P_k^+ P_k}{2m} + \sum_k \frac{m\omega(k)^2}{2} X_k^+ X_k$$

$$= \sum_k E_k \tag{9.36}$$

where

$$E_k = \frac{P_k^+ P_k}{2m} + \frac{m\omega(k)^2}{2} X_k^+ X_k \tag{9.37}$$

E_k is the energy of a harmonic oscillator in the Fourier space with a pulsation $\omega(k)$. One sees that the total vibration energy (9.36) is a sum of energies of independent harmonic oscillators, much simpler than the expression (9.3) where the harmonic oscillators are coupled in the real space.

9.4.2   *Quantization of vibration*

For simplicity, one continues with the one-dimensional case. To understand the origin of the first quantum model for phonons, one transforms the normal coordinates X_k, X_k^+, P_k and P_k^+ into quantum operators $\hat{X}_k$, $\hat{X}_k^+$, $\hat{P}_k$ and $\hat{P}_k^+$ by the use of the correspondence principle of the early days of quantum mechanics:

$$\hat{X}_k = \frac{1}{\sqrt{N}} \sum_i \hat{u}_i e^{-ikR_i} \tag{9.38}$$

$$\hat{P}_k = \frac{1}{\sqrt{N}} \sum_i \hat{p}_i e^{-ikR_i} \tag{9.39}$$

The operators $\hat{u}_i$ and $\hat{p}_i$ have real eigenvalues which impose

$$\hat{X}_{-k} = \hat{X}_k^+ \tag{9.40}$$
$$\hat{P}_{-k} = \hat{P}_k^+ \tag{9.41}$$

Note that $\hat{u}_i$ and $\hat{p}_i$ obey the commutation relations such as $[\hat{u}_i, \hat{p}_j] = i\hbar\delta_{i,j}$, therefore one has the following commutation relations for $\hat{X}_k$ and $\hat{P}_k$:

$$[\hat{X}_k, \hat{X}_{k'}] = 0 \tag{9.42}$$
$$[\hat{P}_k, \hat{P}_{k'}] = 0 \tag{9.43}$$
$$[\hat{X}_k, \hat{P}_{k'}^+] = i\hbar\delta_{k,k'} \tag{9.44}$$

The quantum version of (9.36) and (9.37) is

$$\hat{E} = \sum_k \frac{\hat{P}_k^+ \hat{P}_k}{2m} + \sum_k \frac{m\omega(k)^2}{2} \hat{X}_k^+ \hat{X}_k$$
$$= \sum_k \hat{E}_k \tag{9.45}$$

where

$$\hat{E}_k = \frac{\hat{P}_k^+ \hat{P}_k}{2m} + \frac{m\omega(k)^2}{2} \hat{X}_k^+ \hat{X}_k \tag{9.46}$$

The energy operator (9.46) contains two kinds of operators, $\hat{X}_k$ and $\hat{P}_k$. To make it diagonal, one has to introduce new operators a_k and a_k^+ defined by

$$a_k = A_k \hat{X}_k + iB_k \hat{P}_k \tag{9.47}$$
$$a_k^+ = A_k \hat{X}_k^+ - iB_k \hat{P}_k^+ \tag{9.48}$$

where A_k and B_k are coefficients (not operators) to be determined. Since $\hat{X}_k$ and $\hat{P}_k$ obey the commutation relations (9.43)-(9.44), one can show that a_k and a_k^+ obey the commutation relations:

$$[a_k, a_{k'}] = 0 \tag{9.49}$$

$$[a_k^+, a_{k'}^+] = 0 \tag{9.50}$$

$$[a_k, a_{k'}^+] = \delta_{k,k'} \tag{9.51}$$

if one chooses

$$A_k = \sqrt{\frac{m\omega(k)}{2\hbar}} \tag{9.52}$$

$$B_k = \frac{1}{\sqrt{2m\hbar\omega(k)}} \tag{9.53}$$

Operators a_k and a_k^+ are thus boson operators. They are called "annihilation" and "creation" operators, respectively, as in chapter 7.

To rewrite (9.45) in terms of a_k and a_k^+, one replaces $\hat{X}_k$ and $\hat{P}_k$ by the following relations obtained from (9.47) and (9.48):

$$\hat{X}_k = \frac{1}{2}A_k^{-1}[a_k + a_{-k}^+] \tag{9.54}$$

$$\hat{P}_k^+ = -\frac{i}{2}B_k^{-1}[a_k - a_{-k}^+] \tag{9.55}$$

One then has

$$\hat{E} = \sum_k \hat{E}_k \tag{9.56}$$

where

$$\hat{E}_k = \frac{1}{2}\hbar\omega(k)[a_k a_k^+ + a_{-k}^+ a_{-k}] \tag{9.57}$$

The sum in (9.56) is performed in the first Brillouin zone. Equation (9.56) is rewritten as, using (9.51),

$$\begin{aligned}
\hat{E} &= \sum_k \frac{1}{2}\hbar\omega(k)[a_k a_k^+ + a_{-k}^+ a_{-k}] \\
&= \sum_k \frac{1}{2}\hbar\omega(k)[a_k^+ a_k + 1 + a_k^+ a_k] \\
&= \sum_k \hbar\omega(k)\left[a_k^+ a_k + \frac{1}{2}\right]
\end{aligned} \tag{9.58}$$

where one has used $\sum_k a^+_{-k} a_{-k} = \sum_k a^+_k a_k$ because k takes symmetrical positive and negative in the first Brillouin zone.

One recognizes that (9.58) is the Hamiltonian of an ensemble of N independent quantum harmonic oscillators with different eigenfrequencies $\omega(k)$. This is the quantum version of (9.36).

The quantization in the case of two and three dimensions is straightforward.

9.5   Thermal properties of phonons

One considers the following Hamiltonian of a quantum harmonic oscillator in one dimension

$$\hat{H}_k = \hbar\omega_k \left(a^+_k a_k + \frac{1}{2} \right) \tag{9.59}$$

where $\hbar\omega_k$ is the energy of mode k, namely the phonon energy of wave vector k, a^+_k and a_k are creation and annihilation boson operators. Note that ω_k is given by (9.8). The energy of the oscillator is

$$E_k = \hbar\omega_k(n_k + \frac{1}{2}) \tag{9.60}$$

where n_k is a positive integer or zero. Using the result of Problem 6 of chapter 3 in the case of one dimension, one has the following partition function z for an oscillator

$$z = \frac{1}{2\sinh(\beta\hbar\omega_k/2)} \tag{9.61}$$

One puts $\epsilon_k = \hbar\omega_k$. The average number of phonons of energy ϵ_k is given by (3.52)

$$\overline{n}_k = -k_B T \frac{\partial \ln z}{\partial \epsilon_k}$$

$$= \frac{1}{\exp(\beta\hbar\omega_k) - 1} + \frac{1}{2} \tag{9.62}$$

Without the constant $\frac{1}{2}$, $\overline{n}_k$ is the Bose-Einstein distribution (6.1) with $\mu = 0$. This shows the boson character of phonons. In three dimensions, the constant $\frac{1}{2}$ in (9.59) is replaced by $\frac{3}{2}$.

To calculate the average energy $\overline{E}$, one writes

$$\overline{E} = \sum_k \hbar\omega(k)(\overline{n}_k + \frac{1}{2})$$

$$= \sum_k \hbar\omega(k)(\frac{1}{e^{\beta\hbar\omega(k)} - 1} + \frac{1}{2}) \qquad (9.63)$$

In principle, if one knows $\hbar\omega(k)$ one can calculate $\overline{E}$ and then the heat capacity C_V of phonons. One replaces the sum in (9.63) by an integral over ω for a large system:

$$\overline{E} = \sum_k \hbar\omega(k)(\frac{1}{e^{\beta\hbar\omega(k)} - 1} + \frac{1}{2})$$

$$= \frac{L}{2\pi} \int_{-\pi/a}^{\pi/a} \hbar\omega(k)dk(\frac{1}{e^{\beta\hbar\omega(k)} - 1} + \frac{1}{2})$$

$$= \int_0^{\omega_{max}} g(\omega)\hbar\omega(k)d\omega(\frac{1}{e^{\beta\hbar\omega(k)} - 1} + \frac{1}{2}) \qquad (9.64)$$

where L denotes the length of the crystal and $g(\omega)$ the "density of phonon modes" in analogy with $\rho(E)$, the density of states (see chapter 2).

9.5.1 *Density of modes*

To calculate $\overline{E}$, one should know $g(\omega)$. One calculates it in the same manner as for the electron density of states: one observes that the modes $\pm k$ have the same ω. One has the following identity

$$g(\omega)d\omega = 2G(k)dk \qquad (9.65)$$

where $G(k)dk$ is the number of allowed values of k between k and $k + dk$. The factor 2 takes into account the degeneracy of $\pm k$. Since the size of a microstate is $\frac{2\pi}{L}$, one has

$$G(k)dk = \frac{dk}{\frac{2\pi}{L}} \qquad (9.66)$$

from which,

$$g(\omega) = \frac{L}{\pi}\frac{dk}{d\omega} \qquad (9.67)$$

Using the dispersion relation (9.8), one gets

$$\frac{d\omega}{dk} = a\sqrt{\frac{K}{m}}\cos\left(\frac{ka}{2}\right)$$

$$= a\sqrt{\frac{K}{m}}\sqrt{1 - \sin^2\left(\frac{ka}{2}\right)} \tag{9.68}$$

One sees that $\frac{d\omega}{dk} = 0$ when $k = \pm\frac{\pi}{a}$ (borders of the first Brillouin zone). $g(\omega)$ diverges thus at these values of k, namely at the upper limit of the integral in (9.64). One cannot therefore calculate $\overline{E}$.

In three dimensions, one writes

$$g(\omega)d\omega = G(\vec{k})d\vec{k} \tag{9.69}$$

where

$$G(\vec{k})d\vec{k} = \frac{1}{\frac{(2\pi)^3}{L^3}}d\vec{k} \tag{9.70}$$

One has then

$$g(\omega) = \frac{L^3}{(2\pi)^3}\frac{1}{\left|\vec{\nabla}_{\vec{k}}\omega\right|} \tag{9.71}$$

The divergence of $g(\omega)$ takes place at various points in the Brillouin zone whenever $\vec{\nabla}_{\vec{k}}\omega = 0$. These points are called "Van Hove singularities". Because of these singularities, the calculation of $\overline{E}$ of (9.64) is impossible. Therefore, the "real" density of modes will be replaced by some approximations as seen in the following.

9.5.2   *Einstein model and Debye model*

For $\omega(k)$, one uses two models. The first one is the Einstein model which assumes that all harmonic oscillators have the same pulsation $\omega(k)$. With this hypothesis, all oscillators have the same partition function z. The partition function of the system is thus $Z = z^N$. The results given in (16.83) and (16.85) of Problem 6 in chapter 3 apply. According to these results, as the temperature tends to zero, the heat capacity C_V tends exponentially to zero, in contradiction with experiments where $C_V \propto T^3$. At high temperatures, the Eisntein model gives the result of the classical mechanics which is $C_V = 3Nk_B$.

The second model is the Debye model which is based on the following assumptions:

- Atoms on the lattice are quantum harmonic oscillators each of which has its own vibration frequency, unlike the Einstein model
- The pulsation ω is supposed to be proportional to the modulus of the wave vector $\vec{k}$, namely $\omega = v_s k$ (v_s: sound velocity)
- The density of modes is assumed to be $g_D(\omega)$ which is calculated using the previous hypothesis
- ω_{max} is replaced by the Debye pulsation ω_D to normalize the Debye density of modes as follows:

$$\int_0^{\omega_D} g_D(\omega) d\omega = 3N \tag{9.72}$$

where $3N$ is the total number of modes in three dimensions [12, 34]

One calculates first the Debye density of modes. One writes

$$g(\omega)d\omega = G(\vec{k})d\vec{k}$$
$$= 3\frac{1}{\frac{(2\pi)^3}{L^3}}d\vec{k}$$
$$= 3\frac{1}{\frac{(2\pi)^3}{L^3}}4\pi k^2 dk \tag{9.73}$$

The factor 3 in the right-hand side is introduced to take into account the fact that there are 3 modes associated with each wave vector $\vec{k}$, one longitudinal and two transverse, in three dimensions [12, 34]. One gets

$$g(\omega) = \frac{3}{\frac{(2\pi)^3}{L^3}}4\pi k^2 \frac{dk}{d\omega} \tag{9.74}$$

The Debye density of modes $g_D(\omega)$ is obtained by replacing $\frac{dk}{d\omega}$ and k^2 in the above equation using the second assumption of Debye $\omega = v_s k$. One obtains

$$g_D(\omega) = \frac{3L^3}{(2\pi)^3}4\pi(\frac{\omega}{v_s})^2\frac{1}{v_s}$$
$$= 3\frac{L^3}{2\pi^2 v_s^3}\omega^2 \tag{9.75}$$

Remark: In one dimension,

$$g_D(\omega) = \frac{L}{\pi}\frac{dk}{d\omega} = \frac{L}{\pi v_s}, \quad \text{independent of } \omega. \tag{9.76}$$

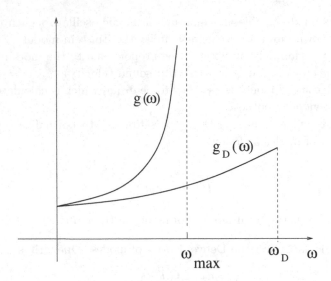

Fig. 9.3   Debye density of modes $g_D(\omega)$ is schematically compared to a real density of modes.

The last step is to determine the Debye pulsation ω_D which replaces ω_{max} in the normalization of $g_D(\omega)$. One has

$$
\begin{aligned}
3N &= \int_0^{\omega_D} g_D(\omega)d\omega \\
&= \frac{3L^3}{2\pi^2 v_s^3} \int_0^{\omega_D} \omega^2 d\omega \\
&= \frac{L^3}{2\pi^2 v_s^3}\omega_D^3
\end{aligned}
\tag{9.77}
$$

from which one has

$$
\omega_D^3 = 3N\frac{2\pi^2 v_s^3}{L^3}
\tag{9.78}
$$

An example of $g_D(\omega)$ is schematically shown in Fig. 9.3.

One is now ready for the calculation of the average energy using the Debye model. Equation (9.64) is written in three dimensions as

$$
\begin{aligned}
\overline{E} &= \int_0^{\omega_D} g_D(\omega)\hbar\omega d\omega \left(\frac{1}{e^{\beta\hbar\omega}-1} + \frac{3}{2} \right) \\
&= \frac{3L^3}{2\pi^2 v_s^3} \int_0^{\omega_D} \hbar\omega^3 d\omega \left(\frac{1}{e^{\beta\hbar\omega}-1} + \frac{3}{2} \right)
\end{aligned}
\tag{9.79}
$$

One omits in the following the constant $\frac{3}{2}$ in the integral which yields just a temperature-independent contribution, namely the zero-point energy $(T = 0)$. Putting $x = \beta\hbar\omega$, one obtains

$$
\begin{aligned}
\overline{E} &= \frac{3L^3}{2\pi^2 v_s^3} \int_0^{\omega_D} \frac{\hbar\omega^3 d\omega}{e^{\beta\hbar\omega} - 1} \\
&= \frac{3L^3(k_BT)^4}{2\pi^2\hbar^3 v_s^3} \int_0^{x_D} \frac{x^3 dx}{e^x - 1}
\end{aligned}
\tag{9.80}
$$

where $x_D = \frac{\hbar\omega_D}{k_BT}$.

At high temperatures, x is very small: the function in the integral of (9.80) is equivalent to x^2 so that

$$
\begin{aligned}
\overline{E} &= \frac{3L^3(k_BT)^4}{2\pi^2\hbar^3 v_s^3} \frac{x_D^3}{3} \\
&= \frac{3L^3(k_BT)^4}{2\pi^2\hbar^3 v_s^3} \frac{(\hbar\omega_D)^3}{3(k_BT)^3} \\
&= 3Nk_BT
\end{aligned}
\tag{9.81}
$$

where in the last equality ω_D^3 was replaced by (9.78). One recovers here the energy of a system of classical harmonic oscillators at high temperatures.

At low temperatures, $k_BT << \hbar\omega_D$, so that one can replace the upper limit of the integral (9.80) by ∞. Using the formula

$$
\int_0^\infty \frac{x^3 dx}{e^x - 1} = \frac{\pi^4}{15}
\tag{9.82}
$$

one obtains

$$
\overline{E} = \frac{3L^3(k_BT)^4}{2\pi^2\hbar^3 v_s^3} \frac{\pi^4}{15}
\tag{9.83}
$$

The heat capacity of the Debye model is thus

$$
C_V = \frac{2\pi^2 L^3 k_B^4 T^3}{5\hbar^3 v_s^3}
\tag{9.84}
$$

The above result is remarkable: it shows a T^3 dependence of C_V at low temperatures in agreement with experiments.

9.6   Phonons in a condensed gas of Helium-4

This section is devoted to a gas of Helium-4 at low temperatures. Helium-4 is a boson. Some properties of the Helium-4 liquid have been discussed in paragraph 6.7.2. Let us recall that superfluids exhibit many unusual properties. A superfluid is a mixture of a normal component and a superfluid component. The superfluid component has zero viscosity, zero entropy, and infinite thermal conductivity. L. Landau has introduced a phenological model which has two parts, phonons and rotons, in the He-4 spectrum shown below. This model explained with success the superfluid behaviors. One considers a gas of N weakly interacting atoms of Helium-4 at very low temperatures below the Bose temperature. By using the diagonalization method applied to the Hamiltonian in the second quantization, one shows in the following that phonons can be excited in this system.

The Hamiltonian of the gas is given by (7.82):

$$\mathcal{H} = \sum_{\vec{k}} \epsilon_{\vec{k}} a_{\vec{k}}^+ a_{\vec{k}}$$
$$+ \frac{1}{2} \sum_{\vec{k}_1, \vec{k}_2, \vec{k}_3, \vec{k}_4} V(\vec{k}_1 - \vec{k}_3) a_{\vec{k}_1}^+ a_{\vec{k}_2}^+ a_{\vec{k}_3} a_{\vec{k}_4} \delta(\vec{k}_1 + \vec{k}_2 - \vec{k}_3 - \vec{k}_4)$$

$$(9.85)$$

where $V(\vec{k}_1 - \vec{k}_3)$ is the Fourier transform of the interaction between bosons, $\epsilon_{\vec{k}}$ the kinetic energy of the state $\vec{k}$, a and a^+ are annihilation and creation operators, respectively.

At a very low temperature T below the Bose temperature, a small number n of particles is in excited states $\vec{k}$ ($|\vec{k}|$ different from zero) and the remaining large number of particles N_0 is in the ground state ($k = 0$). One has $N = n + N_0$ with $n << N$.

Let us examine the potential term H_p in (9.85):

$$H_p = \frac{1}{2} \sum_{\vec{k}_1,\vec{k}_2,\vec{k}_3,\vec{k}_4} V(\vec{k}_1 - \vec{k}_3) a^+_{\vec{k}_1} a^+_{\vec{k}_2} a_{\vec{k}_3} a_{\vec{k}_4} \delta(\vec{k}_1 + \vec{k}_2 - \vec{k}_3 - \vec{k}_4)$$

$$= \frac{1}{2} \sum_{\vec{k}_1,\vec{k}_2,\vec{k}_3,\vec{k}_4} V(\vec{k}_1 - \vec{k}_3) \delta(\vec{k}_1 + \vec{k}_2 - \vec{k}_3 - \vec{k}_4)$$

$$\times a^+_{\vec{k}_1} [\delta(\vec{k}_2 - \vec{k}_3) + a_{\vec{k}_3} a^+_{\vec{k}_2}] a_{\vec{k}_4}$$

$$= \frac{1}{2} \sum_{\vec{k}_1,\vec{k}_2,\vec{k}_3,\vec{k}_4} V(\vec{k}_1 - \vec{k}_3) a^+_{\vec{k}_1} a_{\vec{k}_4} \delta(\vec{k}_2 - \vec{k}_3) \delta(\vec{k}_1 + \vec{k}_2 - \vec{k}_3 - \vec{k}_4)$$

$$+ \frac{1}{2} \sum_{\vec{k}_1,\vec{k}_2,\vec{k}_3,\vec{k}_4} V(\vec{k}_1 - \vec{k}_3) a^+_{\vec{k}_1} a_{\vec{k}_3} a^+_{\vec{k}_2} a_{\vec{k}_4} \delta(\vec{k}_1 + \vec{k}_2 - \vec{k}_3 - \vec{k}_4)$$

$$(9.86)$$

In the ground state, all $k = 0$ so that the first term gives $\frac{1}{2} N_0^2 V(0)$ since

$$\text{1st term (GS)} = \frac{1}{2} \sum_{\vec{k}_1,\vec{k}_2,\vec{k}_3,\vec{k}_4} V(\vec{k}_1 - \vec{k}_3) a^+_{\vec{k}_1} a_{\vec{k}_4} \delta(\vec{k}_2 - \vec{k}_3)$$

$$\times \delta(\vec{k}_1 + \vec{k}_2 - \vec{k}_3 - \vec{k}_4)$$

$$= \frac{1}{2} V(0) \sum_{\vec{k}_1} n_{\vec{k}_1} \sum_{\vec{k}_2} 1$$

$$= \frac{1}{2} N_0^2 V(0) \qquad (9.87)$$

In excited states, the first term gives $(\vec{k}_1 = \vec{k}_4 \neq 0, \vec{k}_2 = \vec{k}_3 = 0)$:

$$\text{1st term (excited)} = 2 \frac{1}{2} N_0 \sum_{\vec{k}_1 \neq 0} V(\vec{k}_1) a^+_{\vec{k}_1} a_{\vec{k}_1} = N_0 \sum_{\vec{k} \neq 0} V(\vec{k}) a^+_{\vec{k}} a_{\vec{k}}$$

where the factor 2 takes into account the case $\vec{k}_1 = \vec{k}_4 = 0$, $\vec{k}_2 = \vec{k}_3 \neq 0$.

The second term of (9.86) with 4 operators $a^+_{\vec{k}_1} a_{\vec{k}_3} a^+_{\vec{k}_2} a_{\vec{k}_4}$ can be treated as follows:

* $\vec{k}_1 = \vec{k}_2 = 0$, $\vec{k}_3 = \vec{k}_4 \neq 0$:

$$\frac{1}{2} N_0 \sum_{\vec{k}_2 \neq 0} V(\vec{k}_3) a_{\vec{k}_3} a_{-\vec{k}_3}$$

* $\vec{k}_1 = \vec{k}_2 \neq 0$, $\vec{k}_3 = \vec{k}_4 = 0$:

$$\frac{1}{2} N_0 \sum_{\vec{k}_1 \neq 0} V(\vec{k}_1) a^+_{\vec{k}_1} a^+_{-\vec{k}_1}$$

* $\vec{k}_1 = \vec{k}_3 = 0$, $\vec{k}_2 = \vec{k}_4 \neq 0$:

$$\frac{1}{2} N_0 V(0) \sum_{\vec{k}_2 \neq 0} a^+_{\vec{k}_2} a_{\vec{k}_2}$$

* $\vec{k}_1 = \vec{k}_3 \neq 0$, $\vec{k}_2 = \vec{k}_4 = 0$:

$$\frac{1}{2} N_0 V(0) \sum_{\vec{k}_1 \neq 0} a^+_{\vec{k}_1} a_{\vec{k}_1}$$

In summary,

$$H_p = \frac{1}{2} N_0^2 V(0) + N_0 V(0) \sum_{\vec{k} \neq 0} a^+_{\vec{k}} a_{\vec{k}} + N_0 \sum_{\vec{k} \neq 0} V(\vec{k}) a^+_{\vec{k}} a_{\vec{k}}$$

$$+ \frac{1}{2} N_0 \sum_{\vec{k} \neq 0} V(\vec{k}) [a_{\vec{k}} a_{-\vec{k}} + a^+_{\vec{k}} a^+_{-\vec{k}}]$$

Therefore,

$$\mathcal{H} = \sum_{\vec{k}} \epsilon_{\vec{k}} a^+_{\vec{k}} a_{\vec{k}} + H_p$$

$$= \sum_{\vec{k}} \epsilon_{\vec{k}} a^+_{\vec{k}} a_{\vec{k}}$$

$$+ \frac{1}{2} N_0^2 V(0) + N_0 V(0) \sum_{\vec{k} \neq 0} a^+_{\vec{k}} a_{\vec{k}} + N_0 \sum_{\vec{k} \neq 0} V(\vec{k}) a^+_{\vec{k}} a_{\vec{k}}$$

$$+ \frac{1}{2} N_0 \sum_{\vec{k} \neq 0} V(\vec{k}) (a_{\vec{k}} a_{-\vec{k}} + a^+_{\vec{k}} a^+_{-\vec{k}}) + \text{terms of higher order}$$

$$(9.88)$$

where $V(\vec{k}) = V(-\vec{k})$ has been used.

Note that $N = N_0 + n$ with $n \ll N$ so that one can rewrite (9.88) in terms of N as

$$\mathcal{H} = \frac{1}{2} N^2 V(0) + \frac{1}{2} \sum_{\vec{k} \neq 0} H(\vec{k}) \qquad (9.89)$$

where

$$H(\vec{k}) = [\epsilon_{\vec{k}} + N V(\vec{k})] (a^+_{\vec{k}} a_{\vec{k}} + a^+_{-\vec{k}} a_{-\vec{k}})$$

$$+ N V(\vec{k}) (a_{\vec{k}} a_{-\vec{k}} + a^+_{\vec{k}} a^+_{-\vec{k}}) \qquad (9.90)$$

One can demonstrate this relation by replacing $N_0 = N - n$ in (9.88), neglecting terms of higher order in n.

The Hamiltonian (9.90) is not diagonal. In order to put it under a diagonal form, namely $H(\vec{k}) = \lambda_{\vec{k}} c_{\vec{k}}^+ c_{\vec{k}}$, one uses the following transformation for $H(\vec{k})$ (see method given in Problem 7 of chapter 7):

$$c_{\vec{k}} = u_{\vec{k}} a_{\vec{k}} - v_{\vec{k}} a_{-\vec{k}}^+ \tag{9.91}$$

$$c_{\vec{k}}^+ = u_{\vec{k}} a_{\vec{k}}^+ - v_{\vec{k}} a_{-\vec{k}} \tag{9.92}$$

where $u_{\vec{k}}$ and $v_{\vec{k}}$ are real and $u_{\vec{k}}^2 - v_{\vec{k}}^2 = 1$. One can show that operators $c_{\vec{k}}$ and $c_{\vec{k}}^+$ obey the commutation relations. Now, one determines $u_{\vec{k}}$ and $v_{\vec{k}}$ so that (9.90) is diagonal. From (9.91)-(9.92), one has

$$a_{\vec{k}}^+ = u_{\vec{k}} c_{\vec{k}}^+ + v_{\vec{k}} c_{-\vec{k}} \tag{9.93}$$

$$a_{\vec{k}} = u_{\vec{k}} c_{\vec{k}} + v_{\vec{k}} c_{-\vec{k}}^+ \tag{9.94}$$

where one puts $u_{\vec{k}}^2 - v_{\vec{k}}^2 = 1$. Replacing these in (9.90) and putting the coefficient of the term $(cc + c^+ c^+)$ equal to zero (see Problem 7 of chapter 7 and Problem 5 of chapter 12), one gets

$$\mathcal{H} = \sum_{\vec{k}} \lambda_{\vec{k}} c_{\vec{k}}^+ c_{\vec{k}} + \text{constant}$$

where the phonon spectrum $\lambda_{\vec{k}}$ is given by

$$\lambda_{\vec{k}}^2 = [\epsilon_{\vec{k}} + NV(\vec{k})]^2 - [NV(\vec{k})]^2$$

$$\lambda_{\vec{k}} = \sqrt{[\epsilon_{\vec{k}} + NV(\vec{k})]^2 - [NV(\vec{k})]^2} \tag{9.95}$$

This relation is known as the Bogoliubov spectrum [26]. When $V(\vec{k}) = 0$, one recovers the unperturbed spectrum $\epsilon_{\vec{k}}$. In the case of a condensate, using $\epsilon_{\vec{k}} = \hbar^2 k^2 / 2m$ one has

$$\lambda_{k \sim 0} = \hbar k \sqrt{\frac{NV(0)}{m}} = v_s \hbar k \tag{9.96}$$

where $v_s = \sqrt{\frac{NV(0)}{m}}$ is the phonon sound velocity.

This is a very important result: the small k part of dispersion of the superfluid is responsible for zero viscosity effect. Since the group velocity $d\lambda_{k \sim 0}/dk$ does not depend on k within a wide range of k, scattering processes which change k cannot change the propagation velocity of the superfluid.

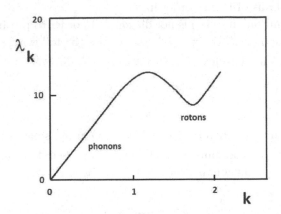

Fig. 9.4   Spectrum of Helium 4: energy $\lambda_{\vec{k}}$ (Kelvin) versus wave vector k (Å^{-1}).

One can now estimate the limit of k, say k_{lim}, up to which the Bogoliubov spectrum remains linear in k: to do this one neglects the V term in (9.95) and sets it equal to the linear part

$$v_s \hbar k_{lim} \simeq \epsilon_{\vec{k}_{lim}} = \frac{\hbar^2 k_{lim}^2}{2m}$$

$$\Rightarrow k_{lim} \simeq \frac{2}{\hbar} \sqrt{mNV(0)} \tag{9.97}$$

The spectrum of He-4 is shown in Fig. 9.4: the linear part at small k is due to phonons of the condensate, while the high-k part is due to "rotons", a kind of excitations suggested by Landau [109].

9.7   Conclusion

In this chapter, one has studied the vibration of atoms in crystalline solids using methods in statistical physics. In particular, using the harmonic approximation one has calculated the dispersion relation in one, two and three dimensions. One has introduced the quantization of atomic vibrations and used the Bose-Einstein statistics to study two particular models at finite temperatures, namely the Einstein and the Debye models. While both models recover the result of the classical mechanics at high temperatures, only the Debye model gives the correct temperature-dependence of the specific heat at low temperatures.

One has also presented the case of weakly interacting He-4 gas. This system is a boson gas which shows a Bose condensation at low temperatures (see chapter 6). One has calculated by the method of second quantization the phonon Bogoliubov spectrum which shows a linear dependence on k. This yields a k-independent sound velocity which characterizes the superfluid state observed in He-4 gas at low temperatures.

9.8 Problems

Problem 1. Prove (9.36).

Problem 2. Chain of two types of atoms:

Consider a chain made of two types of atom A and B alternately, of mass M and m, with $M > m$. The distance between two nearest neighbors is a. Let K be the constant of the restoring force between nearest neighbors in the harmonic approximation. One supposes the periodic boundary conditions.

a) Determine the reciprocal lattice and the first Brillouin zone (BZ).

b) Calculate the dispersion relation ω_k. Plot ω_k versus k.

c) Examine the modes at the center and at the border of the first BZ.

Problem 3. Interaction between next nearest neighbors in a chain:

Consider a chain of N identical atoms of mass m, of lattice constant a. One supposes the periodic boundary conditions. One uses the harmonic approximation taking into account interactions between nearest neighbors K_1 and between next nearest neighbors K_2.

a) Does the interaction between next nearest neighbors change the structure of the reciprocal lattice?

b) Calculate the dispersion relation ω_k. Plot ω_k versus k assuming $K_2/K_1 = 0.5$. Comment.

c) Calculate the sound velocity. Discuss the effect of K_2.

d) Discuss the limit $K_2/K_1 \to \infty$.

Problem 4. Chain of two types of distance:

Consider a chain of N identical atoms with periodic boundary conditions. The distance between an atom and its neighbor on the right is a_1 and its neighbor on the left is a_2.

a) Determine the reciprocal lattice and the first Brillouin zone (BZ).

b) Write the equations of motion for two neighboring atoms.

c) Calculate the dispersion relation ω_k. Plot ω_k versus k.

d) Describe the modes at the center and at the border of the first Brillouin zone.

Problem 5. Phonons in a rectangular lattice:

Study the vibration of atoms on a rectangular lattice taking into account interaction between nearest neighbors on each crystalline axis, within the harmonic approximation. One assumes the periodic boundary conditions and the same constant K on the two axes.

Problem 6. Phonons in a simple cubic lattice:

Calculate the dispersion relations of phonon modes in a simple cubic lattice, taking into account the interactions between nearest neighbors K_1 and between next nearest neighbors K_2.

Problem 7. Density of modes in a square lattice:

Calculate the density of modes for the square lattice by modifying the dispersion relation obtained in Problem 5 at small k.

Problem 8. Energy and specific heat at finite temperatures:

One assumes the following dispersion relation of a phonon branch:
$$\omega^2(\vec{k}) = 2C^2(3 - \cos k_x a - \cos k_y a - \cos k_z a) \qquad (9.98)$$
where $\vec{k}$ is the wave vector, a the lattice constant and C a constant. One uses the Debye approximation in this exercise.

a) Is the phonon branch acoustic or optical? If acoustic, calculate the sound velocity v_s. Show that it does not depend on the direction of $\vec{k}$.

b) Justify the Debye approximation at low temperatures.

c) Determine the density of modes in the Debye approximation, called $g_D(\omega)$.

d) Calculate the Debye pulsation ω_D assuming that there are N unit cells in the lattice with a volume Ω.

e) Calculate the phonon energy at the temperature T using the above phonon branch.

f) Find the contribution of this branch to the calorific capacity C_V. Comment.

Problem 9. Phonons in a chain with long-range interaction:

Consider a chain with a long-range interaction. The potential energy is assumed to be:

$$U = (1/2)\sum_n \sum_{m>0} K_m[u(na) - u([n+m]a)]^2.$$

a) Show that the dispersion relation is given by

$$\omega(k) = 2 \left(\sum_{m>0} (K_m/M) \sin^2(mka/2) \right)^{1/2}.$$

b) Determine the behavior at long wave length limit. Determine the sound velocity along the chain assuming that the series $\sum_m m^2 K_m$ converges.

c) Show that if $K_m \propto m^{-p}$ ($1 < p < 3$), at the limit of long wave length, one has $\omega \propto k^{(p-1)/2}$.

d) Show that for $p = 3$ one has $\omega \propto k|\ln k|^{1/2}$.

One supposes from here that the interaction is limited to nearest neighbors $K = K_1$):

e) Determine the density of modes $g(\omega)$ per length unit. Plot the dispersion relation and the density of modes. Analyze the singularities. Calculate the integral $\int_0^\infty g(\omega)d\omega$.

f) Find the contribution of phonons to the specific heat.

Problem 10. Phonons and melting:

a) Consider a harmonic oscillator in one dimension, of mass M, of pulsation ω in equilibrium with a heat reservoir at the temperature T. Show that the average value of the squared displacement is given by $<x^2> = <E>/M\omega^2$. What happens if $\omega \to 0$? What is the physical meaning of such a situation?

b) One uses the notation $|n\rangle$ to denote state vectors of the Hamiltonian of a harmonic oscillator and $< ... >$ to indicate the thermal average. Establish the following relations for the operators creation a^+ and annihilation a:

$$< a^+ a^+ > = < aa > = 0,$$

$$< a^+ a > = < aa^+ > -1 = \frac{1}{\exp(\beta\hbar\omega) - 1}$$

where $\beta = (k_b T)^{-1}$.

c) Generalize the above results to the case of a system composed of N independent harmonic oscillators characterized each by a pulsation ω_k.

d) Consider hereafter longitudinal vibrations of a mono-atomic chain of lattice constant d. Show that the operator of displacement of the m-th atom is written as

$$u_m = \sum_{k \in BZ} \left(\frac{\hbar}{2NM\omega_k} \right)^{1/2} \left[e^{ikmd} a_k + e^{-ikmd} a_k^+ \right].$$

Show that

$$< u_m^2 >= \sum_{k \in BZ} \frac{< E(k) >}{NM\omega_k^2}$$

where $< E(k) >$ is the average value of the energy of the phonon of wave vector k.

e) Express $< u_m^2 >$ by an integral using the density of modes $g(\omega)$. In one dimension, show that $< u_m^2 >$ diverges. What is the consequence of this situation?

f) One considers the case of two dimensions. What is the result on $< u_m^2 >$? Comment.

Chapter 10

Systems of Interacting Electrons — Fermi Liquids

10.1 Introduction

Systems of interacting electrons constitute one of the most important and most studied subjects in condensed matter. On the one hand, systems of interacting electrons provide an ideal testing ground for theories and approximations as seen in the intense theoretical development during the last five decades. On the other hand, one has witnessed in the same period an intensive development of devices based on properties of electrons in semiconductors, metals and superconductors under various geometries such as thin films, multilayers and superlattices. Today, thanks to technological advances it becomes possible to use for applications microscopic effects in transport properties of electrons not only by their charges but also by their spins.

This chapter is devoted to fundamental properties of gas of interacting electrons. The more one understands microscopic origins of various physical behaviors the better one conceives new innovative artificial devices. The main properties of Fermi liquids are also shown. One sees here once more that a combination of statistical physics and quantum mechanics is necessary to study such quantum systems with a large number of interacting particles.

10.2 Gas of interacting electrons

Let us study the gas of interacting electrons first by the Hartree-Fock approximation in this section and then by the second quantization in the next section. The advantage of the Hartree-Fock approximation is its simplicity which allows one to understand easily the main physical effects of the electron-electron interaction before using a more abstract but more efficient second-quantization method.

One uses the Hartree-Fock equation (7.18) with the following individual wave function

$$\psi_{\vec{k}_i,\sigma_i}(\vec{r},\zeta) = \frac{1}{\sqrt{\Omega}} e^{i\vec{k}_i \cdot \vec{r}} S_{\sigma_i}(\zeta) \tag{10.1}$$

Since ζ should be implicitly compatible with state σ_i, i. e. $S_{\sigma_i}(\zeta) = 1$, one can omit it in the following to simplify the writing. One just keeps the index σ_i of $\psi_{\vec{k}_i,\sigma_i}(\vec{r})$. Note that when $V(\vec{r}_i, \vec{r}_j)$ depends only on the relative distance between the two electrons as it is the case here, the plane wave (10.1) is always a solution of the Hartree-Fock equation. Replacing (10.1) in (7.18) one obtains

$$(H_0 + H_d + H_e)\psi_{\vec{k}_i,\sigma_i}(\vec{r}) = \epsilon_i \psi_{\vec{k}_i,\sigma_i}(\vec{r}) \tag{10.2}$$

where H_0, H_d and H_e are respectively kinetic energy, direct-interaction energy and exchange-interaction energy, of electron i. Explicitly,

$$H_0 = \frac{\hbar^2 k_i^2}{2m} \tag{10.3}$$

$$H_d = \sum_{\vec{k}_j} \sum_{\sigma_j=+,-} \frac{e^2}{\Omega} \int_\Omega \frac{d\vec{r}_1}{|\vec{r} - \vec{r}_1|} \tag{10.4}$$

where the sum over $\vec{k}_j$ is taken in the first Brillouin zone, and

$$\begin{aligned}
H_e\psi_{\vec{k}_i,\sigma_i}(\vec{r}) &= -\sum_{\vec{k}_j,\sigma_j} \delta_{\sigma_i,\sigma_j} \frac{e^2}{\Omega} \int_\Omega \frac{d\vec{r}_1 e^{-i\vec{k}_j \cdot \vec{r}_1 + i\vec{k}_i \cdot \vec{r}_1}}{|\vec{r} - \vec{r}_1|} \frac{1}{\sqrt{\Omega}} e^{i\vec{k}_j \cdot \vec{r}} \\
&= -\sum_{\vec{k}_j,\sigma_j} \delta_{\sigma_i,\sigma_j} \frac{e^2}{\Omega} \int_\Omega \frac{d\vec{r}_1 e^{i(\vec{k}_i - \vec{k}_j) \cdot (\vec{r}_1 - \vec{r})} e^{i(\vec{k}_i - \vec{k}_j) \cdot \vec{r}}}{|\vec{r} - \vec{r}_1|} \frac{1}{\sqrt{\Omega}} e^{i\vec{k}_i \cdot \vec{r}}
\end{aligned} \tag{10.5}$$

One notices that

$$\begin{aligned}
\int_\Omega d(\vec{r}_1 - \vec{r}) \frac{e^{i(\vec{k}_i - \vec{k}_j) \cdot (\vec{r}_1 - \vec{r})}}{|\vec{r}_1 - \vec{r}|} &= \int_\Omega \frac{d\vec{r}_2 e^{i(\vec{k}_i - \vec{k}_j) \cdot \vec{r}_2}}{|\vec{r}_2|} \quad (\vec{r}_2 \equiv \vec{r}_1 - \vec{r}) \\
&= v(\vec{k}_i - \vec{k}_j)
\end{aligned} \tag{10.6}$$

where $v(\vec{k}_i - \vec{k}_j)$ is the Fourier transform of $V(\vec{r}_2)$. Combining (10.3), (10.4) and (10.5), one has

$$\left\{ \frac{\hbar^2 k_i^2}{2m} + \frac{e^2}{\Omega} \sum_{\vec{k}_j} \sum_{\sigma_j=+,-} \left[v(0) - \delta_{\sigma_i,\sigma_j} v(\vec{k}_i - \vec{k}_j) \right] \right\} \psi_{\vec{k}_i,\sigma_i}(\vec{r}) = \epsilon_i \psi_{\vec{k}_i,\sigma_i}(\vec{r})$$

(10.7)

One sees here that the Hartree-Fock equation is transformed into a Schrödinger equation $\{...\} \psi_{\vec{k}_i,\sigma_i}(\vec{r}) = \epsilon_i \psi_{\vec{k}_i,\sigma_i}(\vec{r})$. The energy of electron i is thus

$$\epsilon_i = \frac{\hbar^2 k_i^2}{2m} + \frac{e^2}{\Omega} \sum_{\vec{k}_j} \sum_{\sigma_j=+,-} \left[v(0) - \delta_{\sigma_i,\sigma_j} v(\vec{k}_i - \vec{k}_j) \right] \qquad (10.8)$$

To complete our calculation, one has to calculate $v(0)$ and $v(\vec{k}_i - \vec{k}_j)$. If one supposes that the electron gas is superposed on a background of N positive charges of the ions as in the case of a metal model, and that these positive charges are uniformly distributed with a charge density

$$\rho^+ = \frac{Ne}{\Omega} \qquad (10.9)$$

then the interaction between electron i with the positive-charge background is given by

$$-e\rho^+ \int_\Omega \frac{d\vec{r}_i}{|\vec{r} - \vec{r}_i|} = -\frac{Ne^2}{\Omega} v(0) \qquad (10.10)$$

where the notation (10.6) has been used. One sees that the above term cancels the direct-interaction energy given in (10.8):

$$\frac{e^2}{\Omega} \sum_{\vec{k}_j} \sum_{\sigma_j=+,-} v(0) = \frac{e^2}{\Omega} v(0) \sum_{\vec{k}_j} \sum_{\sigma_j=+,-} 1 = \frac{e^2}{\Omega} v(0) N \qquad (10.11)$$

Therefore, the remaining energy of the electron in (10.8) is the kinetic energy and the exchange energy. One has

$$\epsilon_i = \frac{\hbar^2 k_i^2}{2m} - \frac{e^2}{\Omega} \sum_{\vec{k}_j} \sum_{\sigma_j=+,-} \delta_{\sigma_i,\sigma_j} v(\vec{k}_i - \vec{k}_j) \qquad (10.12)$$

One calculates now $v(\vec{k}_i - \vec{k}_j)$. Putting $\vec{k}' = \vec{k}_i - \vec{k}_j$, one has

$$v(\vec{k}') = \int_\Omega \frac{d\vec{r}' e^{i\vec{k}' \cdot \vec{r}'}}{r'}$$

$$= 2\pi \int_0^\infty r'^2 dr' \int_0^\pi \sin\theta d\theta \frac{e^{ik'r' \cos\theta}}{r'}$$

$$= 2\pi \lim_{\mu \to 0} \int_0^\infty e^{-\mu r'} r' dr' \int_1^{-1} d(-\cos\theta) e^{ik'r' \cos\theta}$$

$$= \frac{2\pi}{ik'} \lim_{\mu \to 0} \int_0^\infty dr' [e^{(-\mu + ik')r'} - e^{(-\mu - ik')r'}]$$

$$= \frac{2\pi}{ik'} \lim_{\mu \to 0} \left[\frac{-1}{-\mu + ik'} + \frac{-1}{\mu + ik'} \right]$$

$$= \lim_{\mu \to 0} \frac{4\pi}{k'^2 + \mu^2} \tag{10.13}$$

$$= \frac{4\pi}{k'^2} \tag{10.14}$$

In the above calculation, one has taken the infinite upper limit for the volume while introducing the factor $e^{-\mu r'}$ to avoid the otherwise divergence at $r' = \infty$. However, at the end of the calculation one has taken the $\lim_{\mu \to 0}$ to recover the original integrand. One will see below that this mathematical trick has a physical meaning: the factor $e^{-\mu r'}$ represents a screening effect due to the presence of other charges in the system. This screened Coulomb interaction $e^{-\mu r}/r$ makes the interaction damped faster with increasing distance r. The inverse of the screening length μ is shown to be proportional to $n^{2/3}$ (n: electron density) in the Thomas-Fermi approximation (see Problem 2). Equation (10.12) becomes

$$\epsilon_i = \frac{\hbar^2 k_i^2}{2m} - \frac{4\pi e^2}{\Omega} \sum_{\vec{k}_j} \sum_{\sigma_j = +,-} \delta_{\sigma_i, \sigma_j} \frac{1}{|\vec{k}_i - \vec{k}_j|^2} \tag{10.15}$$

Note that the sum on σ_j gives a factor 1 because of the condition $\sigma_j = \sigma_i$. The sum on $\vec{k}_j$, transformed into an integral, gives

$$\frac{4\pi e^2}{\Omega} \sum_{\vec{k}_j} \frac{1}{|\vec{k}_i - \vec{k}_j|^2} = \frac{4\pi e^2}{\Omega} \frac{\Omega}{(2\pi)^3} \int \frac{d\vec{k}_j}{(\vec{k}_i - \vec{k}_j)^2}$$

$$= \frac{e^2}{\pi} \int_0^{k_F} k_j^2 dk_j \int_1^{-1} \frac{d(-\cos\theta)}{k_i^2 + k_j^2 - 2k_i k_j \cos\theta}$$

$$= \frac{e^2}{\pi k_i} \int_0^{k_F} k_j dk_j \ln \left| \frac{k_j - k_i}{k_j + k_i} \right|$$

$$= \frac{e^2}{\pi k_i} \left[\frac{k_F^2 - k_i^2}{2} \ln \left| \frac{k_F + k_i}{k_F - k_i} \right| + k_i k_F \right] \tag{10.16}$$

where in the last line one has used an integration by parts by putting $u' = k_j dk_j$ and $v = \ln |...|$. This result is for $T = 0$ because one has limited the upper limit of the wave vector at the Fermi wave vector k_F.

Equation (10.12) becomes finally

$$\epsilon_i = \frac{\hbar^2 k_i^2}{2m} - \frac{e^2}{2\pi} \left[\frac{k_F^2 - k_i^2}{k_i} \ln \left| \frac{k_F + k_i}{k_F - k_i} \right| + 2k_F \right] \tag{10.17}$$

This is an important result: the exchange interaction between electrons shown by the second term has been neglected in the free-electron theory.

There is however a paradox: one can verify that the density of states $\rho(\epsilon_i)$ [see its definition (2.36)] calculated by using (10.17) is equal to zero at $k_i = k_F$ in contradiction with the fact that at $T = 0$, all energy levels up to the Fermi level are occupied by definition. This paradox can be solved by introducing the screening in the Coulomb interaction, as is seen in Problem 3. Therefore, this paradox is not an artefact of the Hartree-Fock approximation. It is a problem of the unscreened Coulomb potential (7.2).

10.2.1 *Kinetic and exchange energies*

One considers an interacting electron gas at $T = 0$.

10.2.1.1 *Total kinetic energy*

One has

$$E_c = \sum_{\vec{k}_i} \sum_{\sigma_i} \frac{\hbar^2 k_i^2}{2m} = 2 \sum_{\vec{k}_i} \frac{\hbar^2 k_i^2}{2m}$$

$$= 2 \frac{\Omega}{(2\pi)^3} \int d\vec{k}_i \frac{\hbar^2 k_i^2}{2m}$$

$$= 2 \frac{\Omega}{(2\pi)^3} \frac{\hbar^2}{2m} \int_0^{k_F} 4\pi k_i^4 dk_i = \frac{\Omega}{2\pi^2} \frac{k_F^5}{5}$$

$$= \frac{3}{5} N \frac{\hbar^2 k_F^2}{2m} = \frac{3}{5} N \epsilon_F \tag{10.18}$$

where one has used [see (5.24), (5.19) and (5.21)]

$$N = \sum_{\vec{k}_i} \sum_{\sigma_i} 1 = 2 \frac{\Omega}{(2\pi)^3} \int d\vec{k}_i = 2 \frac{\Omega}{(2\pi)^3} \int_0^{k_F} 4\pi k_i^2 dk_i$$

$$= 2 \frac{\Omega}{(2\pi)^3} \frac{4\pi k_F^3}{3} \tag{10.19}$$

For a later comparison, let us express E_c in the atomic unity Rydberg: from (10.19) one has

$$k_F^3 = 3\pi^2 \frac{N}{\Omega} = 3\pi^2 \frac{1}{\frac{4\pi r_0^3}{3}} = \frac{9\pi}{4}\frac{1}{r_0^3} \quad \Rightarrow \quad k_F = \frac{1}{ar_0} \tag{10.20}$$

where one has defined r_0 by $\frac{\Omega}{N}$ = volume of an electron = $\frac{4\pi r_0^3}{3}$ and $a = (\frac{4}{9\pi})^{1/3} \simeq 0.52$. The average kinetic energy per electron is thus

$$\bar{\epsilon}_c = E_c/N = \frac{3}{5}\frac{\hbar^2}{2m}[\frac{1}{ar_0}]^2 = \frac{3}{5}\frac{me^4}{2\hbar^2}\frac{1}{(ar_s)^2}$$

$$= \frac{3}{5}\frac{1}{(ar_s)^2} \quad \text{Rydberg} \tag{10.21}$$

where $r_s = \frac{r_0}{a_H}$ with $a_H = \frac{\hbar^2}{me^2}$ (Bohr radius) and 1 Rydberg=$\frac{me^4}{2\hbar^2}$. One has then

$$\bar{\epsilon}_c = \frac{2.21}{r_s^2} \quad \text{Rydberg} \tag{10.22}$$

10.2.1.2 *Total exchange energy*

In principle, one can integrate the exchange term of (10.17) with respect to $\vec{k}_i$ to find the total exchange energy E_{ex}. There is however an alternative way which is simpler as seen in the following. From (10.15) one has

$$E_{ex} = -\frac{1}{2}\frac{4\pi e^2}{\Omega}\sum_{\vec{k}_i,\vec{k}_j}\sum_{\sigma_i,\sigma_j}\delta_{\sigma_i,\sigma_j}\frac{1}{|\vec{k}_i - \vec{k}_j|^2}$$

$$= -2\frac{2\pi e^2}{\Omega}\frac{\Omega^2}{(2\pi)^6}\int d\vec{k}_i \int d\vec{k}_j \frac{1}{|\vec{k}_i - \vec{k}_j|^2} \tag{10.23}$$

where one has added a factor $1/2$ in the first line to remove the double counting, and a factor 2 in the second line for the sum on spin σ_i. One writes

$$\frac{1}{|\vec{k}_i - \vec{k}_j|^2} = \frac{1}{k_i^2 + k_j^2 - 2k_ik_j\cos\theta} = \frac{1}{k_i^2}\frac{1}{1 + (\frac{k_j}{k_i})^2 - 2\frac{k_j}{k_i}\cos\theta} \tag{10.24}$$

If one supposes $k_j < k_i$, then one recognizes that the last fraction is the square of the generating function of the Legendre polynomials $P_l(\cos\theta)$

$$\frac{1}{(1 - 2zx + x^2)^{1/2}} = \sum_{l=0}^{\infty} x^l P_l(z) \tag{10.25}$$

where $x < 1$, $-1 \le z \le 1$. One has

$$\frac{1}{|\vec{k}_i - \vec{k}_j|^2} = \frac{1}{k_i^2} \left[\sum_{l=0}^{\infty} \left(\frac{k_j}{k_i} \right)^l P_l(\cos\theta) \right]^2$$

$$= \frac{1}{k_i^2} \sum_{l=0}^{\infty} \sum_{l'=0}^{\infty} \left(\frac{k_j}{k_i} \right)^{l+l'} P_l(\cos\theta) P_{l'}(\cos\theta) \qquad (10.26)$$

Using this relation, one writes

$$I = \int d\vec{k}_i \int d\vec{k}_j \frac{1}{|\vec{k}_i - \vec{k}_j|^2} = 2 \int_{k_i \le k_F} d\vec{k}_i \int_{k_j < k_i} 2\pi (\frac{k_j}{k_i})^2 dk_j \int_1^{-1} d(-\cos\theta)$$

$$\times \sum_{l=0}^{\infty} \sum_{l'=0}^{\infty} \left(\frac{k_j}{k_i} \right)^{l+l'} P_l(\cos\theta) P_{l'}(\cos\theta) \qquad (10.27)$$

where factor 2 in the first line is for taking into account the symmetric case $k_j > k_i$. Using the following orthogonality of Legendre polynomials

$$\int_{-1}^{1} d(\cos\theta) P_l(\cos\theta) P_{l'}(\cos\theta) = \frac{2}{2l+1} \delta_{l,l'} \qquad (10.28)$$

one obtains

$$I = 32\pi^2 \int_0^{k_F} k_i^2 dk_i \int_0^{k_i} dk_j \sum_{l=0}^{\infty} \left(\frac{k_j}{k_i} \right)^{2l+2} \frac{1}{2l+1}$$

$$= 32\pi^2 \int_0^{k_F} k_i^2 dk_i k_i \sum_{l=0}^{\infty} \frac{1}{(2l+1)(2l+3)}$$

$$= 8\pi^2 k_F^4 \sum_{l=0}^{\infty} \frac{1}{(2l+1)(2l+3)}$$

$$= 8\pi^2 k_F^4 \frac{1}{2} \sum_{l=0}^{\infty} [\frac{1}{2l+1} - \frac{1}{2l+3}]$$

$$= 8\pi^2 k_F^4 \frac{1}{2} [1 - 1/3 + 1/3 - 1/5 + 1/5 - 1/7...]$$

$$= 4\pi^2 k_F^4 \qquad (10.29)$$

Equation (10.23) becomes

$$E_{ex} = -\frac{2e^2 \Omega}{(2\pi)^3} k_F^4 \qquad (10.30)$$

The exchange energy per electron is thus

$$\bar{\epsilon}_{ex} = E_{ex}/N = -\frac{2\pi e^2}{(2\pi)^3} \frac{\Omega}{N} k_F^4 = -\frac{2e^2}{(2\pi)^3} \frac{4\pi r_0^3}{3} k_F^4 \qquad (10.31)$$

Replacing k_F by (10.20) and r_0 by $r_s a_H$ one gets, in the same manner as for (10.22),

$$\bar{\epsilon}_{ex} = -\frac{0.916}{r_s} \quad \text{Rydberg} \tag{10.32}$$

The total energy per electron is

$$\bar{\epsilon} = \frac{2.21}{r_s^2} - \frac{0.916}{r_s} \quad \text{Rydberg} \tag{10.33}$$

One recalls that r_s, in unity of Bohr radius, is the radius of the average volume occupied by an electron. When r_s is large, namely when the electron density is small, the exchange term where spins are parallel dominates in the total energy, giving rise to a ferromagnetic state.

10.2.2   *Effective mass*

When an electron interacts weakly with its environment, one can consider it as a free electron but with a modified mass which takes into account the interaction. The modified mass is called "effective mass". It can be defined by the following tensor

$$\frac{1}{m^*_{\alpha\beta}} = \frac{1}{\hbar^2} \frac{\partial^2 \epsilon_{\vec{k}_i}}{\partial k_\alpha \partial k_\beta} \tag{10.34}$$

where $\alpha, \beta = x, y, z$. The effect of the interaction with the environment is contained in $\epsilon_{\vec{k}_i}$. It can be an interaction with other electrons of the gas as in what described above or an interaction with a periodic potential as in the theory of almost-free electrons, or an interaction with electrons of neighboring atoms in the tight-binding theory (see chapter 11). The energy of electron i can be written as

$$\epsilon_{\vec{k}_i} = \frac{\hbar^2 k_i^2}{2m^*}$$

In the case of an electron gas, using (10.17) one can calculate m^*. m^* contains the electron-electron interaction as seen in Problem 3.

10.3   Gas of interacting electrons by second quantization

One considers a gas of N electrons superposed on a background of uniformly distributed positive charges with density $\frac{Ne}{\Omega}$. This is a simple model for a metal. The Hamiltonian is written as

$$\hat{\mathcal{H}} = \mathcal{H}_b + \mathcal{H}_e + \mathcal{H}_{e-b} \tag{10.35}$$

where $\mathcal{H}_b$ is the interaction between the positive charges of the background, $\mathcal{H}_e$ that between electrons, and $\mathcal{H}_{e-b}$ the interaction between the background of positive charges and the electrons. One supposes that the interactions are screened by a screening parameter μ:

$$\hat{\mathcal{H}}_b = \frac{1}{2} \int d\vec{r}_1 \int d\vec{r}_2 n(\vec{r}_1) n(\vec{r}_2) \frac{e^{-\mu(|\vec{r}_1 - \vec{r}_2|)}}{|\vec{r}_1 - \vec{r}_2|}$$

$$= \frac{1}{2} e^2 \left(\frac{N}{\Omega}\right)^2 \int d\vec{r}_1 \int d\vec{r}_2 \frac{e^{-\mu(|\vec{r}_1 - \vec{r}_2|)}}{|\vec{r}_1 - \vec{r}_2|}$$

$$= \frac{1}{2} e^2 \frac{N^2}{\Omega} \frac{4\pi}{\mu^2} \tag{10.36}$$

where one has used the result of the Fourier transformation of the Coulomb potential. The factor $\frac{1}{2}$ is to remove the double counting due to the double integral. Similarly,

$$\hat{\mathcal{H}}_{e-b} = -e^2 \sum_i \frac{N}{\Omega} \int d\vec{r} \frac{e^{-\mu(|\vec{r} - \vec{r}_i|)}}{|\vec{r} - \vec{r}_i|} \qquad (\vec{r}_i:\ \text{position of the electron } i)$$

$$= -e^2 \frac{N}{\Omega} \sum_i \frac{4\pi}{\mu^2}$$

$$= -e^2 \frac{N^2}{\Omega} \frac{4\pi}{\mu^2} \tag{10.37}$$

Replacing (10.36) and (10.37) in (10.35), one obtains

$$\hat{\mathcal{H}} = -\frac{1}{2} e^2 \frac{N^2}{\Omega} \frac{4\pi}{\mu^2} + \mathcal{H}_e \tag{10.38}$$

The interaction term of $\mathcal{H}_e$ of (7.60) is written as

$$\hat{V} = \frac{e^2}{2} \sum_{\vec{k}_1,\vec{k}_2,\vec{k}_3,\vec{k}_4} \sum_{\sigma_1,\sigma_2,\sigma_3,\sigma_4} \langle \vec{k}_1\vec{k}_2|V|\vec{k}_3\vec{k}_4\rangle \times$$

$$b^+_{\vec{k}_1,\sigma_1} b^+_{\vec{k}_2,\sigma_2} b_{\vec{k}_4,\sigma_4} b_{\vec{k}_3,\sigma_3}$$

$$= \frac{e^2}{2\Omega^2} \sum_{\vec{k}_1,\vec{k}_2,\vec{k}_3,\vec{k}_4} \sum_{\sigma_1,\sigma_2,\sigma_3,\sigma_4} S_{\sigma_1} S_{\sigma_2} S_{\sigma_3} S_{\sigma_4} \times$$

$$\int d\vec{r}_2 \int d\vec{r}_1 e^{-i(\vec{k}_1\cdot\vec{r}_1+\vec{k}_2\cdot\vec{r}_2)} \frac{e^{-\mu|\vec{r}_1-\vec{r}_2|}}{|\vec{r}_1-\vec{r}_2|} \times$$

$$e^{i(\vec{k}_3\cdot\vec{r}_1+\vec{k}_4\cdot\vec{r}_2)} b^+_{\vec{k}_1,\sigma_1} b^+_{\vec{k}_2,\sigma_2} b_{\vec{k}_4,\sigma_4} b_{\vec{k}_3,\sigma_3}$$

$$= \frac{e^2}{2\Omega^2} \sum_{\vec{k}_1,\vec{k}_2,\vec{k}_3,\vec{k}_4} \sum_{\sigma_1,\sigma_2,\sigma_3,\sigma_4} \int d\vec{r}_2 e^{-i(\vec{k}_1+\vec{k}_2-\vec{k}_3\vec{k}_4)\cdot\vec{r}_2} \times$$

$$\int d\vec{r} e^{i(\vec{k}_3-\vec{k}_1)\cdot\vec{r}} \frac{e^{-\mu|\vec{r}|}}{|\vec{r}|} S_{\sigma_1} S_{\sigma_2} S_{\sigma_3} S_{\sigma_4} b^+_{\vec{k}_1,\sigma_1} b^+_{\vec{k}_2,\sigma_2} b_{\vec{k}_4,\sigma_4} b_{\vec{k}_3,\sigma_3}$$

$$= \frac{e^2}{2\Omega^2} \sum_{\vec{k}_1,\vec{k}_2,\vec{k}_3,\vec{k}_4} \sum_{\sigma_1,\sigma_2,\sigma_3,\sigma_4} \Omega \delta_{\vec{k}_1+\vec{k}_2,\vec{k}_3+\vec{k}_4} \delta_{\sigma_1,\sigma_3} \delta_{\sigma_2,\sigma_4} \times$$

$$\frac{4\pi}{(\vec{k}_1-\vec{k}_3)^2+\mu^2} S_{\sigma_1} S_{\sigma_2} S_{\sigma_3} S_{\sigma_4} b^+_{\vec{k}_1,\sigma_1} b^+_{\vec{k}_2,\sigma_2} b_{\vec{k}_4,\sigma_4} b_{\vec{k}_3,\sigma_3} \qquad (10.39)$$

where one has put $\vec{r} = \vec{r} - \vec{r}'$ and one has used the Fourier transform (10.14).

One puts for convenience $\vec{k}_3 = \vec{k}$, $\vec{k}_4 = \vec{p}$, $\vec{k}_1 = \vec{k} + \vec{q}$ and $\vec{k}_2 = \vec{p} - \vec{q}$. The condition $\vec{k}_1 + \vec{k}_2 = \vec{k}_3 + \vec{k}_4$ is thus satisfied. Taking into account the restrictions due to the Kronecker symbols, one has

$$\hat{V} = \frac{e^2}{2\Omega} \sum_{\vec{k},\vec{p},\vec{q}} \sum_{\sigma_1,\sigma_2} \frac{4\pi}{(\vec{k}_1-\vec{k}_3)^2+\mu^2} b^+_{\vec{k}+\vec{q},\sigma_1} b^+_{\vec{p}-\vec{q},\sigma_2} b_{\vec{p},\sigma_2} b_{\vec{k},\sigma_1} \qquad (10.40)$$

One distinguishes two cases: $\vec{q} = 0$ and $\vec{q} \neq 0$.

In the case where $\vec{q} = 0$, using the anticommutation relations one rewrites the chain of operators as

$$b^+_{\vec{k}+\vec{q},\sigma_1} b^+_{\vec{p}-\vec{q},\sigma_2} b_{\vec{p},\sigma_2} b_{\vec{k},\sigma_1} = b^+_{\vec{k},\sigma_1} b^+_{\vec{p},\sigma_2} b_{\vec{p},\sigma_2} b_{\vec{k},\sigma_1} = -b^+_{\vec{k},\sigma_1} b^+_{\vec{p},\sigma_2} b_{\vec{k},\sigma_1} b_{\vec{p},\sigma_2}$$

$$= -b^+_{\vec{k},\sigma_1} [\delta_{\sigma_1,\sigma_2} \delta_{\vec{p},\vec{k}} - b_{\vec{k},\sigma_1} b^+_{\vec{p},\sigma_2}] b_{\vec{p},\sigma_2}$$

$$= -b^+_{\vec{k},\sigma_1} b_{\vec{p},\sigma_2} \delta_{\sigma_1,\sigma_2} \delta_{\vec{p},\vec{k}}$$

$$+ b^+_{\vec{k},\sigma_1} b_{\vec{k},\sigma_1} b^+_{\vec{p},\sigma_2} b_{\vec{p},\sigma_2} \qquad (10.41)$$

from which one has

$$\frac{e^2}{2\Omega} \sum_{\vec{k},\vec{p},\vec{q}} \sum_{\sigma_1,\sigma_2} \frac{4\pi}{(\vec{k}_1 - \vec{k}_3)^2 + \mu^2} b^+_{\vec{k}+\vec{q},\sigma_1} b^+_{\vec{p}-\vec{q},\sigma_2} b_{\vec{p},\sigma_2} b_{\vec{k},\sigma_1}$$

$$= \frac{e^2}{2\Omega} \sum_{\vec{k},\vec{p}} \sum_{\sigma_1,\sigma_2} \frac{4\pi}{\mu^2} \left[b^+_{\vec{k},\sigma_1} b_{\vec{k},\sigma_1} b^+_{\vec{p},\sigma_2} b_{\vec{p},\sigma_2} - b^+_{\vec{k},\sigma_1} b_{\vec{p},\sigma_2} \delta_{\sigma_1,\sigma_2} \delta_{\vec{p},\vec{k}} \right]$$

$$= \frac{e^2}{2\Omega} \frac{4\pi}{\mu^2} \left[\hat{N}^2 - \hat{N} \right] \qquad (10.42)$$

where one has used (7.70). The first term of this equation cancels the first term of (10.38). The second term is proportional to $= \frac{N}{\Omega}$ (density of electrons), it is negligible before the other remaining terms which are proportional to a power of N larger than 1: one has seen that the total kinetic energy at $T = 0$ is $E_c = \frac{3}{5} N \epsilon_F = \frac{3}{5} N \frac{\hbar^2}{2m} (3\pi^2 N/\Omega)^{2/3} \propto N^{5/3}$ [see (10.18)] and the exchange energy $E_{ex} \propto k_F^4 \propto N^{4/3}$ [see (10.30)].

Therefore, there remains in $\hat{V}$ the sum on $\vec{q} \neq 0$ to be evaluated. The Hamiltonian (10.38) becomes

$$\hat{\mathcal{H}} = \sum_{\vec{k}_i} \sum_{\sigma_i} \frac{\hbar^2 k_i^2}{2m} b^+_{\vec{k}_j,\sigma_j} b_{\vec{k}_i,\sigma_i}$$

$$+ \frac{e^2}{2\Omega} \sum_{\vec{k},\vec{p},\vec{q}\neq 0} \sum_{\sigma_1,\sigma_2} \frac{4\pi}{q^2 + \mu^2} b^+_{\vec{k}+\vec{q},\sigma_1} b^+_{\vec{p}-\vec{q},\sigma_2} b_{\vec{p},\sigma_2} b_{\vec{k},\sigma_1} \qquad (10.43)$$

Expression (10.43) is the general Hamiltonian of an interacting electron gas.

Remark:

In an electron gas of high density, the kinetic term dominates due to its highest power in N. The interaction term $\hat{V}$ is then considered as a perturbation. If one calculates the mean value $< \hat{\mathcal{H}} >$ at $T = 0$ at the first-order perturbation in $\hat{V}$, one surely recovers the result obtained by solving the Hartree-Fock equation.

10.3.1 *Kinetic energy*

One has $\hat{\mathcal{H}} = \hat{\mathcal{H}}_0 + \hat{\mathcal{H}}_1$ where $\hat{\mathcal{H}}_0$ and $\hat{\mathcal{H}}_1$ are respectively the kinetic and interaction terms in (10.43). The non perturbed energy is the kinetic term. In the second quantization, this term is given by (7.60). One has

$$\hat{\mathcal{H}}_0 = \sum_{i,j} \langle i|H|j\rangle b_j^+ b_i = \sum_{i,j} \frac{1}{2m} \langle i|(-i\hbar\vec{\nabla})^2|j\rangle b_j^+ b_i$$

$$= \sum_{\vec{k}_i,\vec{k}_j} \sum_{\sigma_i,\sigma_j} S_{\sigma_i} S_{\sigma_j} \frac{1}{\Omega} \frac{1}{2m} \int d\vec{r} e^{-i\vec{k}_i\cdot\vec{r}}(-i\hbar\vec{\nabla})^2 e^{-i\vec{k}_j\cdot\vec{r}} b_{\vec{k}_j,\sigma_j}^+ b_{\vec{k}_i,\sigma_i}$$

$$= \frac{1}{\Omega} \sum_{\vec{k}_i,\vec{k}_j} \sum_{\sigma_i,\sigma_j} \frac{\hbar^2 k_j^2}{2m} \Omega \delta_{\sigma_i,\sigma_j} \delta_{\vec{k}_i,\vec{k}_j} b_{\vec{k}_j,\sigma_j}^+ b_{\vec{k}_i,\sigma_i}$$

$$= \sum_{\vec{k}_i} \sum_{\sigma_i} \frac{\hbar^2 k_i^2}{2m} b_{\vec{k}_i,\sigma_i}^+ b_{\vec{k}_i,\sigma_i} \qquad (10.44)$$

where one has used the orthogonality of individual wave functions. Let $|f\rangle$ be the state vector of the ground state. The kinetic energy at $T = 0$ is

$$E_0 = \langle f|\hat{\mathcal{H}}_0|f\rangle = \sum_{\vec{k}_i,\sigma_i} \frac{\hbar^2 k_i^2}{2m} \langle f|b_{\vec{k}_i,\sigma_i}^+ b_{\vec{k}_i,\sigma_i}|f\rangle$$

$$= \sum_{\vec{k}_i,\sigma_i} \frac{\hbar^2 k_i^2}{2m} \langle f|\hat{n}_{\vec{k}_i,\sigma_i}|f\rangle = \sum_{\vec{k}_i,\sigma_i} \frac{\hbar^2 k_i^2}{2m} n_{\vec{k}_i,\sigma_i} \langle f|f\rangle$$

$$= \sum_{\vec{k}_i(k_i\leq k_F)} \sum_{\sigma_i} \frac{\hbar^2 k_i^2}{2m} \qquad (10.45)$$

where one has replaced in the last line $n_{\vec{k}_i,\sigma_i}$ by 1 for $k_i \leq k_F$ and by 0 otherwise. Transforming the sum into an integral one obtains E_0 given by (5.24). In Rydberg, the average kinetic energy per electron at $T = 0$ is $2.21/r_s^2$ as shown by Eq. (10.22).

10.3.2   *Energy at first-order perturbation*

Using the second quantization method, one calculates the energy of an interacting electron gas. At the first-order perturbation, the energy is written as

$$E_1 = \langle f|\hat{\mathcal{H}}_1|f\rangle = \frac{e^2}{2\Omega} \sum_{\vec{k},\vec{p},\vec{q}\neq 0} \sum_{\sigma_1,\sigma_2} \frac{4\pi}{q^2}$$

$$\langle f|b_{\vec{k}+\vec{q},\sigma_1}^+ b_{\vec{p}-\vec{q},\sigma_2}^+ b_{\vec{p},\sigma_2} b_{\vec{k},\sigma_1}|f\rangle \qquad (10.46)$$

where one has used the unscreened Coulomb potential. For a non zero braket, the orthogonality of the state vectors imposes that the operation of $b^+_{\vec{k}+\vec{q},\sigma_1} b^+_{\vec{p}-\vec{q},\sigma_2} b_{\vec{p},\sigma_2} b_{\vec{k},\sigma_1}$ on the ket $|f\rangle$ should leave it unchanged. Since there are two creation operators and two annihilation operators, the condition imposed by the orthogonality is satisfied if the two particles created by the two creation operators are removed by the two annihilation operators. There are two possible ways:

(i) $(\vec{k}+\vec{q},\sigma_1) = (\vec{k},\sigma_1)$; $(\vec{p}-\vec{q},\sigma_2) = (\vec{p},\sigma_2)$

(ii) $(\vec{k}+\vec{q},\sigma_1) = (\vec{p},\sigma_2)$; $(\vec{p}-\vec{q},\sigma_2) = (\vec{k},\sigma_1)$

The first case is impossible because $\vec{q} \neq 0$. The second case gives

$$E_1 = \frac{e^2}{2\Omega} \sum_{\vec{k},\vec{p},\vec{q}\neq 0} \sum_{\sigma_1,\sigma_2} \frac{4\pi}{q^2}$$

$$\delta_{\vec{k}+\vec{q},\vec{p}} \delta_{\sigma_1,\sigma_2} \langle f | b^+_{\vec{k}+\vec{q},\sigma_1} b^+_{\vec{p}-\vec{q},\sigma_2} b_{\vec{k}+\vec{q},\sigma_1} b_{\vec{k},\sigma_1} | f \rangle$$

$$= -\frac{e^2}{2\Omega} \sum_{\vec{k},\vec{p},\vec{q}\neq 0} \sum_{\sigma_1,\sigma_2} \frac{4\pi}{q^2}$$

$$\delta_{\vec{k}+\vec{q},\vec{p}} \delta_{\sigma_1,\sigma_2} \langle f | b^+_{\vec{k}+\vec{q},\sigma_1} b_{\vec{k}+\vec{q},\sigma_1} b^+_{\vec{p}-\vec{q},\sigma_2} b_{\vec{k},\sigma_1} | f \rangle$$

$$= -\frac{e^2}{2\Omega} \sum_{\vec{k},\vec{p},\vec{q}\neq 0} \sum_{\sigma_1} \frac{4\pi}{q^2} n_{\vec{k}+\vec{q},\sigma_1} n_{\vec{k},\sigma_1} \langle f | f \rangle$$

$$= -\frac{e^2}{2\Omega} \sum_{\vec{k},\vec{q}\neq 0} \sum_{\sigma_1} \frac{4\pi}{q^2} \Theta(k_F - |\vec{k}+\vec{q}|)\Theta(k_F - |\vec{k}|) \qquad (10.47)$$

where one has used the Heavyside function Θ to indicate that $n_{\vec{k}+\vec{q},\sigma_1}$ and $n_{\vec{k},\sigma_1}$ are equal to 1 if their wave vectors are inside the sphere of radius k_F, because $T=0$. They are zero otherwise. Transforming now the sums on $\vec{k}$ and $\vec{q}$ into integrals in the domains where $k_F - |\vec{k}+\vec{q}| > 0$ and $k_F - |\vec{k}| > 0$, one obtains the results of the Hartree-Fock approximation (10.23), (10.30) and (10.32), namely $E_1 = -\frac{0.916}{r_s}$.

10.3.3 *Energy at second-order perturbation*

One calculates now the perturbation energy at the second order by using the Rayleigh-Schrödinger formula of quantum mechanics:

$$E_2 = \sum_{j \neq 0} \frac{|\langle f|\hat{\mathcal{H}}_1|j\rangle|^2}{E_0 - E_j}$$

$$= \sum_{j \neq 0} \frac{\langle f|\hat{\mathcal{H}}_1^*|j\rangle\langle j|\hat{\mathcal{H}}_1|f\rangle}{E_0 - E_j} \tag{10.48}$$

where $|j\rangle$ is the state vector of the excited state j of energy E_j of the electron gas. An excited state is a state lies outside of the Fermi sea $(k > k_F)$. One has $\langle f|\hat{\mathcal{H}}_1^*|j\rangle = (\langle j|\hat{\mathcal{H}}_1|f\rangle)^*$.

As before, in each of the brakets $< f|...|j >$ and $< j|...|f >$ there are two creation operators and two annihilation ones. The actions of these operators on the ket should result in a state identical to that in the bra, so that the considered braket is not zero. Consider

$$b^+_{\vec{k}+\vec{q},\sigma_1} b^+_{\vec{p}-\vec{q},\sigma_2} b_{\vec{p},\sigma_2} b_{\vec{k},\sigma_1}|f\rangle$$

There are two combinations for a non zero braket:

10.3.3.1 *Direct interaction:*

- Excited state: Operator $b_{\vec{k},\sigma_1}$ destroys the state $(\vec{k},\sigma_1)$ in the ground state, operator $b^+_{\vec{k}+\vec{q},\sigma_1}$ creates the excited state $(\vec{k}+\vec{q},\sigma_1)$ outside the Fermi sphere of radius k_F, operator $b_{\vec{p},\sigma_2}$ destroys the state $(\vec{p},\sigma_2)$ in the ground state and operator $b^+_{\vec{p}-\vec{q},\sigma_2}$ creates the excited state $(\vec{p}-\vec{q},\sigma_2)$ outside the Fermi sphere. The excited state of the system is thus $|j\rangle = b^+_{\vec{k}+\vec{q},\sigma_1} b^+_{\vec{p}-\vec{q},\sigma_2} b_{\vec{p},\sigma_2} b_{\vec{k},\sigma_1}|f\rangle$ with $|\vec{p}-\vec{q}|, |\vec{k}+\vec{q}| > k_F$, $|\vec{p}|, |\vec{k}| \leq k_F$. One then has

$$\langle j|b^+_{\vec{k}+\vec{q},\sigma_1} b^+_{\vec{p}-\vec{q},\sigma_2} b_{\vec{p},\sigma_2} b_{\vec{k},\sigma_1}|f\rangle \propto \langle j|j\rangle$$

- Back to the ground state: one considers the term $\langle f|...|j\rangle$ of (10.48). Explicitly, this term is written as

$$\langle f|\hat{\mathcal{H}}_1^*|j\rangle \propto < f|b^+_{\vec{k},\sigma_1} b^+_{\vec{p},\sigma_2} b_{\vec{p}-\vec{q},\sigma_2} b_{\vec{k}+\vec{q},\sigma_1}|j\rangle$$

The return to the ground state can be realized in the following manner: operators $b_{\vec{k}+\vec{q},\sigma_1}$ and $b_{\vec{p}-\vec{q},\sigma_2}$ destroy respectively the states $(\vec{k}+\vec{q},\sigma_1)$ and $(\vec{p}-\vec{q},\sigma_2)$ outside the Fermi sphere, operators $b^+_{\vec{k},\sigma_1}$ and $b^+_{\vec{p},\sigma_2}$ create the states $(\vec{k},\sigma_1)$ and $(\vec{p},\sigma_2)$ in the ground state. One finds thus the initial state of the system.

The above direct-interaction process corresponds to the schema displayed in Fig. 10.1.

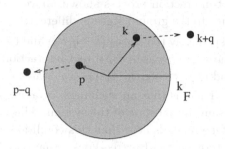

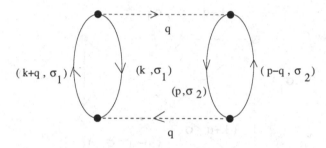

Fig. 10.1 Excited state (top) and direct-interaction process (bottom) shown with the following diagram convention: the transfer of the momentum $\vec{q}$ is indicated on the horizontal broken lines which represent the interaction $V(\vec{q})$. Note the conservation of the momentum at each vertex.

In the direct-interaction process, one has

$$E_0 - E_j = \frac{\hbar^2}{2m}\left[k^2 + p^2 - (\vec{k} + \vec{q})^2 - (\vec{p} - \vec{q})^2\right] = \frac{\hbar^2}{2m}\vec{q}\cdot(\vec{k} - \vec{p} + \vec{q}) \quad (10.49)$$

which gives

$$\begin{aligned}
E_2^d &= -\sum_{\vec{k},\vec{p},\vec{q}\neq 0}\sum_{\sigma_1,\sigma_2}\left(\frac{4\pi e^2}{\Omega q^2}\right)^2 \frac{n_{\vec{k},\sigma_1}(1 - n_{\vec{k}+\vec{q},\sigma_1})n_{\vec{p},\sigma_2}(1 - n_{\vec{p}-\vec{q},\sigma_2})}{\frac{\hbar^2}{2m}\vec{q}\cdot(\vec{k} - \vec{p} + \vec{q})} \\
&= -4\sum_{\vec{k},\vec{p},\vec{q}\neq 0}\left(\frac{4\pi e^2}{\Omega q^2}\right)^2 \frac{n_{\vec{k},\sigma_1}(1 - n_{\vec{k}+\vec{q},\sigma_1})n_{\vec{p},\sigma_2}(1 - n_{\vec{p}-\vec{q},\sigma_2})}{\frac{\hbar^2}{2m}\vec{q}\cdot(\vec{k} - \vec{p} + \vec{q})} \quad (10.50)
\end{aligned}$$

where n are the coefficients generated when applying the operators on the kets [see (7.50)-(7.50)] and the factor 4 results from the spin sums.

10.3.3.2 *Exchange interaction:*

In the exchange-interaction process, the excited state of the system is the same as the direct-interaction process shown above [see Fig. 10.1 (top)]. However, the return to the ground state is different: operators $b_{\vec{k}+\vec{q},\sigma_1}$ and $b_{\vec{p}-\vec{q},\sigma_2}$ destroy respectively the states $(\vec{k}+\vec{q},\sigma_1)$ and $(\vec{p}-\vec{q},\sigma_2)$ outside the Fermi sphere in the same way as in the direct-interaction process. However, operators $b^+_{\vec{k},\sigma_1}$ and $b^+_{\vec{p},\sigma_2}$ create the respective states $(\vec{p},\sigma_2)$ and $(\vec{k},\sigma_1)$ in the ground state. There is thus an exchange, or a crossing, even if at the end one finds the same initial state of the system. This exchange is possible only if $\sigma_1 = \sigma_2$. One reminds here that this condition is the same as that for the exchange term of the Hartree-Fock equation. One shows in Fig. 10.2 the diagram of the exchange-interaction process.

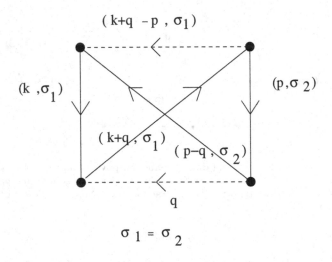

Fig. 10.2   Exchange-interaction process.

It is noted that $E_0 - E_j$ is given by the same expression as that for the direct interaction (10.49). However, the momentum transfer during the interaction which brings the system back to the ground state is $\vec{k} + \vec{q} - \vec{p}$ (see Fig. 10.2). As a consequence, the factor $v(q) = \frac{4\pi e^2}{q^2}$ becomes in the exchange process $\frac{4\pi e^2}{q^2}\frac{4\pi e^2}{(\vec{k}+\vec{q}-\vec{p})^2}$ [$v(q)$ is the Fourier transform of the Coulomb potential]. One finally has

$$E_2^x = + \sum_{\vec{k},\vec{p},\vec{q}\neq 0} \sum_{\sigma_1,\sigma_2} \delta_{\sigma_1,\sigma_2} \left(\frac{4\pi e^2}{\Omega q^2}\right) \left(\frac{4\pi e^2}{\Omega(\vec{k}+\vec{q}-\vec{p})^2}\right)$$

$$\times \frac{n_{\vec{k},\sigma_1}(1-n_{\vec{k}+\vec{q},\sigma_1})n_{\vec{p},\sigma_2}(1-n_{\vec{p}-\vec{q},\sigma_2})}{\frac{\hbar^2}{2m}\vec{q}\cdot(\vec{k}-\vec{p}+\vec{q})}$$

$$= +2 \sum_{\vec{k},\vec{p},\vec{q}\neq 0} \left(\frac{4\pi e^2}{q^2}\right) \left(\frac{4\pi e^2}{\Omega(\vec{k}+\vec{q}-\vec{p})^2}\right)$$

$$\times \frac{n_{\vec{k},\sigma_1}(1-n_{\vec{k}+\vec{q},\sigma_1})n_{\vec{p},\sigma_2}(1-n_{\vec{p}-\vec{q},\sigma_2})}{\frac{\hbar^2}{2m}\vec{q}\cdot(\vec{k}-\vec{p}+\vec{q})}$$

$$(10.51)$$

where the change of sign with respect to the direct term is due to the permutation of the operators in their action on the way back to the ground state. The factor 2 results from the sum on σ_1.

One notices that E_2^d given by Eq. (10.50) diverges at $q = 0$. To avoid this divergence, one has to introduce the screening effect into the Coulomb potential in the way one did to rectify the density of states at the Fermi level in the Hartree-Fock approximation. The theorem by Gell-Mann-Brueckner [61] indicates that the energy due to the perturbation of order n is given by

$$E_n \propto r_s^{n-2} \int \frac{dq}{q^{2n-3}}.$$

This shows that for $n = 2$, the divergence is logarithmic.

10.4 Fermi liquids

In the previous sections, one has examined effects of the electron-electron interaction on the electron energy. One has seen a very important role of the electron spin in the exchange dynamics. But all those calculations were restricted to $T = 0$. On the other hand, in chapter 5, one has considered the zero-, low- and high-temperature regimes but one did not take into account the electron-electron interaction.

To fill the gap, one shall discuss hereafter low-temperature properties of an interacting electron gas. Historically, this problem has been mainly investigated in the 1950's by Landau [106–108] in the weak-interaction limit.

The system was called "Fermi liquid" or "Landau's Fermi liquid" since then. The idea was to consider the interacting Fermi gas as a non interacting gas perturbed by the interaction. The wave function is the linear combination of unperturbed wave functions. The picture of a non interacting Fermi gas remains but the electron mass m is replaced by an effective mass m^* defined in (10.34). Electrons are called "quasiparticles" to remind that their wave functions and energies are not the same as in the free Fermi gas. The specific heat and susceptibility have similar forms as (5.30) and (5.44) found for the free Fermi gas. One has

$$C_V = \frac{1}{3}\pi^2 \rho(E_F) k_B^2 T = \gamma T \tag{10.52}$$

$$\chi = = \frac{2\mu_B^2}{1 + F_0}\rho(E_F) \tag{10.53}$$

where the density of states at the Fermi level is given by (2.39) with m replaced by m^*:

$$\rho(E_F) = \frac{\Omega}{4\pi^2}[\frac{2m^*}{\hbar^2}]^{3/2}E_F^{1/2} \tag{10.54}$$

and where F_0 is a Landau parameter related to the f functional introduced in the energy of an excited state outside the Fermi sphere:

$$\Delta E = \sum_{\vec{k},\sigma} \frac{\hbar^2 k_F \Delta k}{m^*}\delta n_{\vec{k}\sigma} + \frac{1}{2}\sum_{\vec{k},\sigma}\sum_{\vec{k}',\sigma'} f_{\vec{k}\sigma,\vec{k}'\sigma'}\delta n_{\vec{k}\sigma}\delta n'_{\vec{k}'\sigma'} \tag{10.55}$$

where $\Delta k = k - k_F$ and $\delta n_{\vec{k}\sigma}$ the number of excited quasiparticles. Note that the interaction between quasiparticle densities in the second term of (10.55) allows for collective modes of $\delta n_{\vec{k}\sigma}$ around the Fermi surface with f playing the role of a restoring force. The so-called zero sound stems from these oscillations. Let us examine a quasiparticle which is excited outside the Fermi sphere with a small energy $\Delta E = \epsilon$ measured from the Fermi level. The lifetime of this excited state can be estimated by the Fermi's golden rule:

$$\frac{1}{\tau_\epsilon} = \frac{2\pi}{\hbar}\sum_f |V_{if}|^2\delta(\epsilon - \epsilon_f) \tag{10.56}$$

where V_{if} is the scattering matrix between the initial state i of energy ϵ and the final state f which is initially empty, lying outside the Fermi sphere. The scattering leading to the final state can be described by the following steps (i) the quasiparticle lowers its energy from ϵ to $\epsilon' = \epsilon - \omega$ (with $\omega < \epsilon$ so that ϵ' is initially empty, outside the Fermi sphere), (ii) a quasiparticle inside the Fermi sea absorbs ω to be excited at ϵ'' outside the Fermi sphere (to be able to do this, this quasiparticle should initially be close enough to the Fermi level). The final state thus has the energy $\epsilon_f = \epsilon' + \omega - \epsilon''$. Supposing V_{if} be a constant, one has

$$\frac{1}{\tau_\epsilon} \simeq \frac{2\pi}{\hbar}|V|^2 \int_0^\epsilon \rho(E_F)d\omega \int_0^\omega \rho(E_F)d\epsilon' \int_{-\infty}^\infty \delta(\epsilon - \epsilon' - \omega + \epsilon'')\rho(E_F)d\epsilon''$$

$$= \frac{\pi}{\hbar}|V|^2[\rho(E_F)]^3\epsilon^2 \tag{10.57}$$

The excited state thus decays very slowly as ϵ^2 (ϵ is supposed to be very small). The quasiparticle is therefore very well defined with a long-enough lifetime. At low temperatures, one has $\epsilon \propto T^2$ [see (5.29) with m replaced by m^* in E_F]. Thus, the resistivity R which is proportional to $\frac{1}{\tau_\epsilon}$ behaves as T^2 at low temperatures.

The behavior of the Landau's Fermi liquid has been experimentally confirmed in the isotropic Helium-3 liquid [177]. However, in other systems such as electrons in periodic lattices, electrons in systems with disorder, in high-temperature superconductors, ... the Laudau's Fermi-liquid theory fails to account even for low-energy excited states. In the next section, one examines the main features of the Kondo effect due to the interaction between a magnetic impurity with conducting electrons at very low temperatures. For high-temperature superconductors, theoretical mechanisms to explain the superconductivity is still under debate. A discussion is given in chapter 14.

10.5 Kondo effect

The Kondo effect was discovered in 1964 [101]. For a simple description, let us consider a fixed impurity of spin one-half in a gas of non interacting conducting electrons. At high temperatures, the impurity spin is free to flip so that the susceptibility is proportional to $1/T$ (Curie law). However, as the temperature decreases conducting electrons scatter from the impurity spin. Note that the impurity prefers to align its spin antiparallel to that of a nearby conducting electron. Kondo has shown that the resistivity $R(T)$ due to the scattering by the impurity grows logarithmically as the temperature decreases. To understand the regime below the so-called Kondo temperature T_K, one has to recourse to the scaling analysis by the renormalization group [180]. It shows that as the temperature decreases, the antiferromagnetic coupling increases, making a logarithmic enhancement of $R(T)$. However, below the Kondo temperature, the impurity spin captures the spin of a nearby electron to form an inert singlet state: the susceptibility then saturates. Note that below T_K, the relative resistivity $R(T)/R(0)$ behaves as $R(T)/R(0) \sim 1 - aT^2$ ($a=$ a constant) and the specific heat falls to zero linearly with T. As seen in the previous section, these two properties are signatures of the Fermi liquid at very low temperatures. This is possible since the impurity is neutralized by the formation of an inert singlet at $T < T_K$.

10.6   Conclusion

In this chapter, one has seen that the electron-electron interaction gives rise to two energy contributions: the direct-interaction energy and the exchange-interaction energy. In a metal, the direct term is approximately canceled by the interaction between electrons and ion positive charges. So, the energy of a conducting electron is composed of its kinetic energy and the exchange energy. Note that the exchange energy is not zero only for pairs of electrons with parallel spins [see (10.12) and (10.33)]. Due to its negative value, the exchange interaction between electrons lowers the electron energy leading to the ferromagnetic state in a Fermi gas when the electron density becomes small (see Problem 4 below). In the high electron density limit, the distance between electrons is short so that neighboring electrons have their spins antiparallel to each other due to the Pauli exclusion principle (see the Problem 1).

The main features of the Landau's Fermi liquids as well as the Kondo effect at low temperatures have also been presented. These results have provided satisfactory explanations for a number of experiments. However, one needs new models with new mechanisms to understand behaviors of conducting electrons in more complicated systems such as high-temperature superconductors and metals near a quantum critical point.

10.7   Problems

Problem 1. System of two electrons - Fermi hole:

    a) Consider two electrons of a He atom. The Coulomb interaction between them at positions $\vec{r}$ and $\vec{r}'$ is of the form $V(\vec{r} - \vec{r}') = e^2/|\vec{r} - \vec{r}'|$ (e: electron charge). Write down the determinant wave function for this two-electron He atom in the ground state . Write down the integral allowing to calculate the total energy in terms of individual wave functions $\varphi_{1s}(\vec{r})$.

b) Consider now two electrons in a box of volume $\Omega = L^3$ where L is the linear dimension of the box. Using the plane wave functions for the electrons, show that the probability to find two electrons of opposite spins does not depend on their relative distance $\vec{r} - \vec{r}\,'$, but the probability to find two electrons of parallel spins depends on their distance. The small probability at short distance is called the "Fermi hole".

Problem 2. Screened Coulomb potential, Thomas-Fermi approximation: Consider an extra charge q (trial charge) placed in an electron gas. The density of the gas around the charge reacts in a way so as to screen the effect of q. This phenomenon is called "screening effect".

a) Write down the Poisson equation for the potential $\varphi(\vec{r})$ at $\vec{r}$ from the trial charge.

b) In the perturbed region around the trial charge, the chemical potential is given by $\mu = \epsilon_F(\vec{r}) - e\varphi(\vec{r})$ ($\epsilon_F(\vec{r})$: local Fermi energy, $-e$: electron charge). One supposes that the perturbation is small and spherical, show that

$$\varphi(r) = \frac{q \exp(-\lambda r)}{r} \tag{10.58}$$

where λ is given by

$$\lambda^2 = \frac{4n_0^{1/3}}{a_H} \tag{10.59}$$

with a_H (Bohr's radius)$=\hbar^2/me^2$ and n_0 the non perturbed electron density. Equation (10.58) is known as the screened Coulomb potential.

Problem 3. Hartree-Fock approximation:

a) Show that the density of states $\rho(\epsilon_i)$ of an electron calculated using the Hartree-Fock result (10.17) is equal to 0 at $k_i = k_F$. Calculate the electron effective mass m^*.

b) Introduce the screening factor $e^{-\mu r}$ to the Coulomb potential where μ is a screening constant [see Eq. (10.58)]. Using (10.13) without taking the limit $\mu \to 0$, calculate the electron energy using the method similar to that used to obtain (10.17). Show that the density of states does not diverge at $k_i = k_F$. Calculate the effective mass and the density of states at $k_i = k_F$.

Problem 4. Paramagnetic-ferromagnetic transition in an electron gas:
By comparing the total energy of a gas of interacting electrons in the ferromagnetic phase to that in the paramagnetic state, show that the ferromagnetic phase is more stable when the electron density is smaller than a critical value.

Problem 5. Gas of interacting fermions in second quantization:
Consider the following Hamiltonian of N interacting fermions

$$\hat{\mathcal{H}} = \sum_{\sigma} \int d\vec{r} \Psi_{\sigma}^{+}(\vec{r}) \frac{p^2}{2m} \Psi_{\sigma}(\vec{r})$$

$$+ \frac{1}{2} \sum_{\sigma,\sigma'} \int \int d\vec{r}_1 d\vec{r}_2 \Psi_{\sigma}^{+}(\vec{r}_1) \Psi_{\sigma'}^{+}(\vec{r}_2) V(\vec{r}_1 - \vec{r}_2)$$

$$\times \Psi_{\sigma'}^{+}(\vec{r}_2) \Psi_{\sigma}(\vec{r}_1) \tag{10.60}$$

where $\Psi_{\sigma}(\vec{r}_1)$ and $\Psi_{\sigma}^{+}(\vec{r}_1)$ are field operators with standard notations. The potential $V(\vec{r}_1 - \vec{r}_2)$ is given by $V(\vec{r}_1 - \vec{r}_2) = g\delta(\vec{r}_1 - \vec{r}_2)$ where g is a positive or negative constant.

a) Show that the second term of (10.60) is zero for parallel spins. Give the meaning of the result.

b) Write the equations of motiion for $\Psi_{\sigma}(\vec{r}, t)$ and $\Psi_{\sigma}^{+}(\vec{r}, t)$. Find the energy of a particle in the state $\vec{k}$ in the first-order approximation (equivalent to the Hartree-Fock approximation). One shall use the interaction representation

$$\Psi_{\sigma}(\vec{r}, t) = \sum_{\vec{k}} b_{\vec{k},\sigma} e^{-i\omega_k t} \varphi_{\vec{k}}(\vec{r})$$

where $b_{\vec{k},\sigma}$ is fermion annihilation operator and $\varphi_{\vec{k}}(\vec{r})$ the wave function of the state $(\vec{k}, \sigma)$.

c) Show that

$$\hat{\mathcal{H}} = \sum_{\vec{k},\sigma} \frac{\hbar^2 k^2}{2m} b_{\vec{k},\sigma}^{+} b_{\vec{k},\sigma}$$

$$+ \frac{g}{2\Omega} \sum_{\sigma' \neq \sigma} \sum_{\vec{k},\vec{k}',\vec{q}} b_{\vec{k}+\vec{q},\sigma}^{+} b_{\vec{k}'-\vec{q},\sigma'}^{+} b_{\vec{k},\sigma'} b_{\vec{k}',\sigma} \tag{10.61}$$

where Ω is the system volume and $\varphi_{\vec{k}}(\vec{r}) = e^{i\vec{k}\cdot\vec{r}}/\Omega^{1/2}$ has been used.

d) One supposes $g > 0$:
(1) Calculate the average energy of the ground state to the first order of g, as a function of N and ζ defined by the following:

$N = N^+ + N^-$ where $N^\pm$ are the numbers of up and down spins, ζ is given by $\zeta = (N^+ - N^-)/N$.

(2) Show that the system is partially magnetized, namely $\zeta \neq 0$, when

$$\frac{20}{9} < \frac{gN}{\Omega\epsilon_c} < \frac{5}{3}2^{2/3} \simeq 2.64$$

where ϵ_c is the average kinetic energy per particle in the paramagnetic (non magnetized) phase.

Problem 6. Fermion gas as a function of density:

Consider a system of N interacting particles of spin one-half, of volume Ω, defined by the following Hamiltonian

$$\hat{\mathcal{H}} = \sum_{\vec{k}\sigma,\vec{k}'\sigma'} \langle \vec{k}|T|\vec{k}'\rangle b^+_{\vec{k}\sigma} b_{\vec{k}'\sigma'}$$

$$+\frac{1}{2}\sum_{\sigma,\sigma'}\sum_{\vec{k}_1,\vec{k}_2,\vec{k}_3,\vec{k}_4} \langle \vec{k}_1\sigma\vec{k}_2\sigma'|V(\vec{r}_1 - \vec{r}_2)|\vec{k}_3\sigma\vec{k}_4\sigma'\rangle$$

$$\times b^+_{\vec{k}_1\sigma} b^+_{\vec{k}_2\sigma'} b_{\vec{k}_4\sigma'} b_{\vec{k}_3\sigma} \qquad (10.62)$$

where T is the kinetic term, $V(\vec{r}_1 - \vec{r}_2)$ the potential and $b^+_{\vec{k}\sigma}$ and $b_{\vec{k}\sigma}$ denote fermion operators.

a) Show that operator "occupation number" $\hat{N} = \sum_{\vec{k}\sigma} b^+_{\vec{k}\sigma} b_{\vec{k}\sigma}$ commutes with $\hat{\mathcal{H}}$.

b) Show that the system energy of the ground state is written as

$$\frac{E}{N} = \frac{\hbar^2}{2m}\frac{3}{5}\left(3\pi^2\right)^{2/3}\rho^{2/3} + g\frac{2\pi e^2}{\mu^2}\rho - g\frac{3e^2}{4\pi}\left(3\pi^2\right)^{1/3}\rho^{1/3}$$

where $\rho = N/\Omega$ (particle density), e the particle charge, $\mu > 0$ and $g \gtrless 0$ are constants given by

$$V(\vec{r}_1 - \vec{r}_2) = g\frac{e^2 e^{-\mu|\vec{r}_1 - \vec{r}_2|}}{|\vec{r}_1 - \vec{r}_2|}$$

c) If $g < 0$ (attractive interaction), show that the system energy becomes negative (bound state) beyond a critical value of ρ.

d) Consider $g = 1$ (electron gas). Define $N = N^+ + N^-$ where $N^\pm$ are the numbers of up and down spins, and ζ by $\zeta = (N^+ - N^-)/N$. Plot $\frac{E}{N}$ as a function of ζ for several values of $r_s = r_0/a_H$ where r_0 is defined by $\Omega/N = 4\pi r_0^3/3$ (volume occupied by a particle), $a_H = \hbar^2/(me^2)$ (Bohr radius). Show that when $r_s < 5.45$ the paramagnetic state ($\zeta = 0$) is stable, and when $r_s > 5.45$ the ferromagnetic state ($\zeta = 1$) is favorable as found in Problem 4.

Chapter 11

Electrons in Crystalline Solids: Energy Bands

So far, on the one hand, we have examined free electrons in metals in chapter 5 and effects of their interaction in chapter 10. On the other hand, we have examined how ions vibrate due to their own ion-ion interaction in chapter 9. However, we did not taken into account the interaction between electrons and the potential created by the ions in a crystal. This chapter fills the gap.

When interactions between electrons and their environment are taken into account, new remarkable results are found as seen in this chapter. Let us mention a few of them: i) when interaction between electrons and the periodic potential created by ions is introduced, the electron energy is broken into allowed and forbidden energy bands, ii) when electrons, tightly bound to their ions, weakly interact with electrons of neighboring atoms in the crystal, the tight-binding approximation shows that atomic energy levels are enlarged to become energy bands.

These subjects are more conveniently treated in books on condensed matter physics (see for example [12, 34]). However, our aim is to outline some important concepts and principal physical properties of electrons in crystalline solids to allow for a general understanding of electronic applications in semiconductors which will be shown in chapter 15.

In what follows we are interested in effects of the crystalline environment on properties of conducting electrons. We will study two limiting cases: the first case concerns free electrons weakly perturbed by the crystalline potential and the second case concerns electrons strongly bound to their mother atoms but perturbed by the presence of electrons on neighboring atoms in the solid. Electrons in the first case are termed "almost-free electrons", while in the second case the approximation is called "tight-binding approximation". In both cases, we shall use perturbation theories to calculate the energy correction to the unperturbed energy. We shall see that energy bands are originated from the interaction between electrons and crystal ions.

11.1　Wave function of an electron in a periodic potential: Bloch function

To describe properties of an electron in a periodic potential, one needs a wave function which takes into account the periodic symmetry of the crystal. Among such wave functions, the Bloch wave function is the simplest one which can however adequately describe effects of crystalline symmetries on the energy of electrons. The Bloch wave function has the following form

$$\psi_{\vec{k}}(\vec{r}) = u_{\vec{k}}(\vec{r})e^{i\vec{k}\cdot\vec{r}} \tag{11.1}$$

where $u_{\vec{k}}(\vec{r})$ is a function having the periodic symmetry of the crystal, namely

$$u_{\vec{k}}(\vec{r} + \vec{R}) = u_{\vec{k}}(\vec{r}) \tag{11.2}$$

where $\vec{R}$ is a vector connecting two lattice sites: $\vec{R} = n_1\vec{a}_1 + n_2\vec{a}_2 + n_3\vec{a}_3$ where n_i are integers and $\vec{a}_i$ denote the "basis vectors" or "lattice vectors" (see chapter 8).

One can verify that the Bloch function (11.1) is an eigenfunction of a Hamiltonian having a periodic potential. For simplicity, let us consider a one-dimensional crystal of N atoms with lattice constant a. The Hamiltonian is given by

$$\mathcal{H} = -\frac{\hbar^2\Delta}{2m} + V(x) \tag{11.3}$$

where $V(x)$ is a periodic potential, namely

$$V(x + a) = V(x) \tag{11.4}$$

One considers the translation operator Γ defined by

$$\Gamma\psi_k(x) = \psi_k(x + a) \tag{11.5}$$

One first proves that the Bloch function is an eigenfunction of Γ. One proves next that Γ commutes with $\mathcal{H}$ so that the Bloch function is also an eigenfunction of $\mathcal{H}$ (quantum mechanics).

One looks for the eigenvalue λ of Γ:

$$\Gamma\psi_k(x) = \lambda\psi_k(x) \tag{11.6}$$

To find λ, one operates N times Γ on $\psi_k(x)$ and one uses the periodic boundary condition

$$\Gamma^N\psi_k(x) = \psi_k(x + Na) = \psi_k(x) \tag{11.7}$$

Using (11.6), one has

$$\Gamma^N\psi_k(x) = \lambda^N\psi_k(x) \tag{11.8}$$

Comparing (11.7) and (11.8) one has

$$\lambda^N = 1 \tag{11.9}$$

from which one gets

$$\lambda = e^{\frac{2\pi n}{N}} \tag{11.10}$$

where $n = \pm 1, \pm 2, \dots$ This result shows that $\psi_k(x)$ is an eigenfunction of Γ with eigenvalue λ. One shows now the commutation relation $[\Gamma, \mathcal{H}] = 0$:

$$\begin{aligned}
\Gamma\mathcal{H}\psi_k(x) &= \Gamma\left[-\frac{\hbar^2\Delta}{2m} + V(x)\right]\psi_k(x) \\
&= \left[-\frac{\hbar^2}{2m}\frac{d^2}{d(x+a)^2} + V(x+a)\right]\psi_k(x+a) \\
&= \left[-\frac{\hbar^2}{2m}\frac{d^2}{dx^2} + V(x)\right]\Gamma\psi_k(x) \\
&= \mathcal{H}\Gamma\psi_k(x) \tag{11.11}
\end{aligned}$$

from which one sees that Γ commutes with $\mathcal{H}$. The Bloch function is thus an eigenfunction of $\mathcal{H}$ given by (11.3). One has the following property

$$\psi_k(x + a) = u_k(x + a)e^{ik(x+a)} = e^{ika}u_k(x)e^{ikx} = e^{ika}\psi_k(x) \tag{11.12}$$

In three dimensions, one writes

$$\psi_{\vec{k}}(\vec{r} + \vec{R}) = e^{i\vec{k}\cdot\vec{R}}\psi_{\vec{k}}(\vec{r}) \tag{11.13}$$

Physically, one can say that the Bloch function (11.1) is the wave function of a free electron $e^{i\vec{k}\cdot\vec{r}}$ modulated by the factor $u_{\vec{k}}(\vec{r})$ having the symmetry of the periodic potential of the lattice. Schematically, $\psi_k(x)$ is shown in Fig. 11.1.

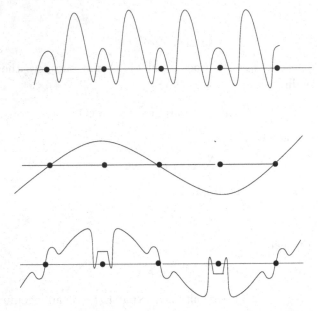

Fig. 11.1   Top: function $u_k(x)$ having the symmetry of the potential of the ions in the crystal, Middle: real part of a plane wave of a free electron, Bottom: Bloch function $u_k(x)\psi_k(x)$.

11.2   Theory of almost-free electrons

One supposes that the periodic potential due to ions on the lattice is weak so that one can consider it as a perturbation to the free-electron state.

11.2.1   *One-dimensional case*

Let $\psi_k(x)$ be the wave function of an electron in the state k in presence of a periodic potential $V(x)$. If $V(x)$ is weak, one can express $\psi_k(x)$ as a linear combination of unperturbed wave functions $\phi_j(x)$ of the electron

$$\psi_k(x) = \sum_j a_{kj}\phi_j(x) \tag{11.14}$$

where a_{kj} are coefficients to be determined and $\phi_j(x)$ are plane waves. The Schrödinger equation is

$$\mathcal{H}\psi_k(x) = [\mathcal{H}_0 + V(x)]\psi_k(x) = E_k\psi_k(x) \tag{11.15}$$

where $\mathcal{H}_0 = -\frac{\hbar^2 \Delta}{2m}$ is the unperturbed Hamiltonian (free electron), namely

$$\mathcal{H}_0\phi_k(x) = E_k^0\phi_k(x) \tag{11.16}$$

where $E_k^0 = \frac{\hbar^2 k^2}{2m}$ is the kinetic (unperturbed) energy. Replacing (11.14) in (11.15) one obtains

$$\sum_j a_{kj} \mathcal{H} \phi_j(x) = E_k \sum_j a_{kj} \phi_j(x) \qquad (11.17)$$

Multiplying $\phi_p^*(x)$ on the left of this equation and integrating on x one gets

$$\sum_j a_{kj} \mathcal{H}_{p,j} = E_k \sum_j a_{kj} \delta_{p,j} \qquad (11.18)$$

$$\sum_j a_{kj} [\mathcal{H}_{p,j} - E_k \delta_{p,j}] = 0 \qquad (11.19)$$

where the Kronecker symbol indicates the orthogonality of $\phi_j(x)$ and where $\mathcal{H}_{p,j}$ is the matrix element defined by

$$\mathcal{H}_{p,j} = \int_L \phi_p^*(x) \mathcal{H} \phi_j(x) dx \qquad (11.20)$$

$$= \int_L \phi_p^*(x) [\mathcal{H}_0 + V(x)] \phi_j(x) dx \qquad (11.21)$$

$$= E_j^0 \delta_{p,j} + \int_L \phi_p^*(x) V(x) \phi_j(x) dx \qquad (11.22)$$

$$= E_j^0 \delta_{p,j} + V_{p,j} \qquad (11.23)$$

A non trivial solution for a_{kj} implies

$$\det |\mathcal{H}_{p,j} - E_k \delta_{p,j}| = 0 \qquad (11.24)$$

This determinant is of dimension $N \times N$ where N is the number of states k in the first Brillouin zone. In principle (11.24) gives the values of E_k, but the large determinant dimension does not allow for an exact determination of E_k. One will take an approximation assuming that matrix elements $\mathcal{H}_{p,j}$ are non zero only between identical states or between degenerate states. This hypothesis is based on the fact that $V(x)$ is weak so that interaction between states can be limited to states of the same energy. In an approximation of the next order, one can take into account states with different but close energies. Now, in one dimension, the degenerate states are k and $-k$ corresponding to $\phi_k(x) = \frac{e^{ikx}}{L^{1/2}}$ and $\phi_{-k}(x) = \frac{e^{-ikx}}{L^{1/2}}$. Numbering the N states in the first Brillouin zone by $\pm k_i$ where $i = 1, ..., N/2$ and rewriting (11.24) by putting for clarity the states k_i and $-k_i$ one next to the other, one has

$$\begin{vmatrix} x & x & 0 & 0 & 0 & 0 & \cdots \\ x & x & 0 & 0 & 0 & 0 & \cdots \\ 0 & 0 & x & x & 0 & 0 & \cdots \\ 0 & 0 & x & x & 0 & 0 & \cdots \\ 0 & 0 & 0 & 0 & x & x & \cdots \\ 0 & 0 & 0 & 0 & x & x & \cdots \end{vmatrix} = 0$$

where x represents non zero elements. One can thus factorize the above determinant as a product of 2×2 determinants. The vanishing i-th determinant yields the i-th solution

$$\begin{vmatrix} \mathcal{H}_{i,i} - E_i & \mathcal{H}_{i,-i} \\ \mathcal{H}_{-i,i} & \mathcal{H}_{-i,-i} - E_{-i} \end{vmatrix} = 0$$

Explicitly, using (11.23) one has

$$\mathcal{H}_{i,i} = E_i^0 + V_{i,i} \tag{11.25}$$

$$\mathcal{H}_{i,-i} = V_{i,-i} \tag{11.26}$$

$$\mathcal{H}_{-i,-i} = E_{-i}^0 + V_{-i,-i} \tag{11.27}$$

$$\mathcal{H}_{-i,i} = V_{-i,i} \tag{11.28}$$

Since $E_i^0 = E_{-i}^0$ (degenerate states), one obtains

$$\begin{vmatrix} E_i^0 + V_{i,i} - E_i & V_{i,-i} \\ V_{-i,i} & E_i^0 + V_{-i,-i} - E_{-i} \end{vmatrix} = 0$$

Using $\phi_i(x) = \frac{1}{L^{1/2}} e^{ik_i x}$, one has

$$V_{i,i} = \int_L \phi_i^*(x) V(x) \phi_i(x) dx$$

$$= \frac{1}{L} \int_L V(x) dx$$

$$= V_0 \tag{11.29}$$

One can put V_0 of the last line equal to zero because $V(x)$ is periodic so that its spatial average is zero. One obtains similar result for $V_{-i,-i}$. One thus has

$$\begin{vmatrix} E_i^0 - E_i & V_{i,-i} \\ V_{-i,i} & E_{-i}^0 - E_{-i} \end{vmatrix} = 0$$

In addition, $E_i = E_{-i}$ so that one can decompose as follows

$$E_i = E_{-i} = E_i^0 + E_i^{(1)} \tag{11.30}$$

where $E_i^{(1)}$ is the energy correction due to the periodic potential. Since $V_{-i,i} = V_{i,-i}^*$, one has

$$(E_i^{(1)})^2 = |V_{i,-i}|^2 \tag{11.31}$$

from which

$$E_i^{(1)} = \pm|V_{i,-i}| \tag{11.32}$$

One calculates now $V_{i,-i}$.

11.2.2 Calculation of the energy correction

One has

$$\begin{aligned}
V_{i,-i} &= \int_L \phi_{k_i}^*(x)V(x)\phi_{-k_i}(x)dx \\
&= \frac{1}{L}\int_L e^{-ik_i x}V(x)e^{-ik_i x}dx \\
&= \frac{1}{L}\int_L e^{-i2k_i x}V(x)dx
\end{aligned} \tag{11.33}$$

As $V(x) = V(x + a)$ where a is the lattice constant, one replaces $\int_L \ldots$ by $\sum_{m=0}^{N-1}\int_{ma}^{(m+1)a} \ldots$, namely one integrates on an elementary cell and sum over all the cells:

$$\begin{aligned}
V_{i,-i} &= \frac{1}{L}\sum_{m=0}^{N-1}\int_{ma}^{(m+1)a} e^{-i2k_i x}V(x)dx \\
&= \frac{1}{L}\sum_{m=0}^{N-1}\int_0^a e^{-i2k_i(X+ma)}V(X+ma)dX \\
&= \frac{1}{L}\sum_{m=0}^{N-1}e^{-i2k_i ma}\int_0^a e^{-i2k_i X}V(X)dX
\end{aligned} \tag{11.34}$$

where one has put $X = x - ma$ and used $V(X + ma) = V(X)$. The sum on m in (11.34) is a geometric sum of N terms of ratio $e^{-i2k_i a}$:

$$S = \sum_{m=0}^{N-1} e^{-i2k_i ma} \tag{11.35}$$

$$= \frac{1 - e^{-2iNk_i a}}{1 - e^{-2ik_i a}}$$

$$= \frac{1 - e^{-4i\pi n}}{1 - e^{-\frac{4i\pi n}{N}}} \tag{11.36}$$

where one has replaced k_i by $\frac{2\pi n}{Na}$. The numerator of (11.36) is always zero. S is thus zero if the denominator is not zero, namely if $\frac{2n}{N}$ is not an integer. When

$$\frac{2n}{N} = l = \text{integer} \tag{11.37}$$

one directly calculates S using (11.35) and (11.37):

$$S = \sum_{m=0}^{N-1} e^{-i4\pi nm/N} = \sum_{m=0}^{N-1} e^{-i2\pi lm}$$

$$= \sum_{m=0}^{N-1} 1 = N \tag{11.38}$$

Equation (11.34) is rewritten as

$$V_{i,-i} = \frac{N}{L} \int_0^a e^{-i2k_i X} V(X) dX \quad \text{if } \frac{n}{N} = \frac{l}{2} = \text{half-integer}$$

$$= 0 \quad \text{otherwise.} \tag{11.39}$$

This result is very important: one sees that $V_{i,-i}$ is not zero only when the wave vector is on a border of a Brillouin zone (cf. chapter 8)

$$k_i = \frac{2\pi n}{Na} = l\frac{\pi}{a} \tag{11.40}$$

where l is an integer.

Equation (11.39) shows that the periodic potential has no effect on the electron energy except when the wave vector is at a zone border. In this case,

$$E_i = E_i^0 \pm |V_{i,-i}| \tag{11.41}$$

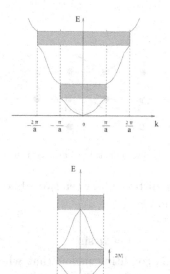

Fig. 11.2 Top: E versus k in extended-zone-scheme representation. Bottom: E in the reduced-zone scheme. The forbidden bands are in gray.

The unperturbed energy E_i^0 thus splits into two levels at $k_i = l\frac{\pi}{a}$. One concludes that the energy is discontinuous at these values of k_i: the discontinuity in energy creates "forbidden energy band" or "energy gap".

It is noted that if one takes into account the interaction between states near zone borders one should find the band gap is extended around $k_i = l\frac{\pi}{a}$ (see Problem 1 below). Figure 11.2 shows this result. The width of the forbidden band is $2|V_{i,-i}|$.

11.2.3 *Interpretation of the forbidden band gap*

The physical meaning of the forbidden band can be understood as follows: when an electron has a wave vector on a Brillouin zone border, it cannot propagate because of collisions with ions separated from each other by distances which correspond to these vectors in the k space. Electrons are reflected by ions. One shows below that wave vectors at Brillouin zone borders verify the Bragg condition for reflection by a crystal.

One considers a wave beam of wavelength λ arriving on atomic planes of a crystal as shown in Fig. 11.3. Let θ be the angle between the beam

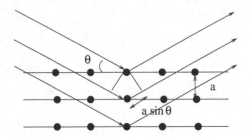

Fig. 11.3   Geometry of the Bragg reflection of an electron beam by a crystal.

and the atomic planes. The distance between two adjacent planes is a. The Bragg condition for reflection is

$$n\lambda = 2a \sin \theta \qquad (11.42)$$

where n is an integer. This condition means that when the difference in "optical path" between waves of the beam is equal to a multiple of λ, these waves are reflected by the crystal. For a one-dimensional lattice where $\theta = \pi/2$, this condition becomes $n\frac{2\pi}{k} = 2a$, namely $k = n\frac{\pi}{a}$. These wave vectors are on Brillouin zone borders. They correspond thus to reflected waves which cannot enter the crystal. In other words, electrons with energy corresponding to these wave vectors do not exist inside the crystal: this is the meaning of forbidden energy.

11.2.4   *Three-dimensional case*

The energy of an electron in three dimensions is $E = \frac{\hbar^2 |\vec{k}|^2}{2m}$. The degeneracy is equal to the number of states on the sphere of radius $|\vec{k}|$. As in one dimension, one supposes in the first approximation that only matrix elements $\mathcal{H}_{k,p}$ between degenerate states are not zero One has

$$V_{k,p} = \int_\Omega \phi^*_{\vec{k}}(\vec{r}) V(\vec{r}) \phi_{\vec{p}}(\vec{r}) d\vec{r}$$

$$= \frac{1}{\Omega} \sum_{m,n,j} e^{-i(\vec{k}-\vec{p})\cdot\vec{R}_{(m,n,j)}} \int_{\text{cell}} e^{-i(\vec{k}-\vec{p})\cdot\vec{r}'} V(\vec{r}') d\vec{r}' \qquad (11.43)$$

where one has put $\vec{r}' = \vec{r} + \vec{R}(m,n,j)$ and used $V(\vec{R}(m,n,j)+\vec{r}') = V(\vec{r}')$. Vector $\vec{R}(m,n,j)$ is the position of a lattice site:

$$\vec{R}(m,n,j) = m\vec{a}_1 + n\vec{a}_2 + j\vec{a}_3.$$

The sums on m, n and j in (11.43) can be written as a product of three geometric series for three Cartesian coordinates. For x component, writing $k_x = \frac{2\pi n_1}{N_1 a_1}$, and $p_x = \frac{2\pi j_1}{N_1 a_1}$, one obtains the sum

$$S = \sum_{m=1}^{N_1} e^{i\frac{2\pi(n_1-j_1)m}{N_1}} \tag{11.44}$$

$$= N_1 \delta(\frac{n_1 - j_1}{N_1} - l_1) \tag{11.45}$$

similarly to (11.36), where l_1 is an integer. Finally for three components, one obtains

$$V_{k,p} = N_1 \delta(\frac{n_1 - j_1}{N_1} - l_1) N_2 \delta(\frac{n_2 - j_2}{N_2} - l_2) N_3 \delta(\frac{n_3 - j_3}{N_3} - l_3)$$

$$\times \int_{\text{cell}} e^{-i(\vec{k}-\vec{p})\cdot\vec{r}'} V(\vec{r}')d\vec{r}' \tag{11.46}$$

One sees that $V_{k,p}$ is not zero only when

$$\frac{n_i - j_i}{N_i} = l_i = \text{integer} \tag{11.47}$$

This condition is equivalent to

$$\vec{k} - \vec{p} = \vec{K_l} \tag{11.48}$$

where $\vec{K_l}$ is a vector of the reciprocal lattice. One obtains thus

$$k^2 = (\vec{p} + \vec{K_l})^2 = p^2 + K_l^2 + 2\vec{p}\cdot\vec{K_l} \tag{11.49}$$

Since the states $\vec{k}$ and $\vec{p}$ are degenerate, namely $k^2 = p^2$, one obtains

$$2\vec{p}\cdot\vec{K_l} = \vec{K_l}^2 \tag{11.50}$$

This equation shows that $\vec{p}$ is on the boundary of a Brillouin zone [see (8.19) in chapter 8]. One finds here the result of the one-dimensional case: the electron energy becomes discontinuous at Brillouin zone boundaries.

11.3   Electrons in a periodic potential: the central equation

One shows here the general equation for an arbitrary periodic potential in one dimension for simplicity. The three-dimensional case is obtained in a straightforward manner.

One considers a periodic potential $V(x)$. One sees in chapter 8 that $V(x)$ can be expressed in the space of vectors K of the reciprocal lattice as

$$V(x) = \sum_K V_K e^{iKx} \tag{11.51}$$

Since $V(x) = V(-x)$, one has $V_K = V_{-K}$.

The Schrödinger equation is

$$\left[-\frac{\hbar^2 \Delta}{2m} + \sum_K V_K e^{iKx} \right] \psi(x) = E\psi(x) \tag{11.52}$$

The Fourier series of $\psi(x)$ is written as

$$\psi(x) = \sum_k C(k) e^{ikx} \tag{11.53}$$

where k is a wave vector allowed by the periodic boundary condition, namely $k = \frac{2\pi n}{Na}$, $C(k)$ are coefficients to be determined. Replacing (11.53) in (11.52) one has

$$\sum_k \frac{\hbar^2 k^2}{2m} C(k) e^{ikx} + \sum_k \sum_K C(k) V_K e^{i(k+K)x} = E \sum_k C(k) e^{ikx} \tag{11.54}$$

Putting $k' = k + K$, one obtains

$$\sum_k \frac{\hbar^2 k^2}{2m} C(k) e^{ikx} + \sum_{k'} \sum_K C(k' - K) V_K e^{ik'x} = E \sum_k C(k) e^{ikx}$$

$$\sum_k \left[\left(\frac{\hbar^2 k^2}{2m} - E \right) C(k) + \sum_K C(k - K) V_K \right] e^{ikx} = 0 \tag{11.55}$$

where one has replaced the dummy variable k' of the second sum in the first line by k. One arrives at the following general equation:

$$\left[\frac{\hbar^2 k^2}{2m} - E \right] C(k) + \sum_K C(k - K) V_K = 0 \tag{11.56}$$

This equation is sometimes called the "central equation" which allows one, in principle, to determine the coefficients $C(k)$. However, in reality, it is impossible to solve it because each coefficient is connected to all others: one has here an infinite system of equations. In practice, one takes some approximations. For example, if one takes into account only interaction between degenerate states one sees that in the first Brillouin zone, only two states on the zone border are degenerate: the point on the zone border $k = \frac{K_1}{2}$ where $K_1 = \frac{2\pi}{a}$ is degenerate with the point on the other border $k - K_1 = -\frac{K_1}{2}$. It follows that only coefficients $C(\frac{K_1}{2})$ and $C(-\frac{K_1}{2})$ are to be retained in Eq. (11.56). With $V_{K_1} = V_{-K_1}$, one has thus

$$\left[\frac{\hbar^2(\frac{K_1}{2})^2}{2m} - E\right] C(\frac{K_1}{2}) + C(-\frac{K_1}{2})V_{K_1} = 0 \tag{11.57}$$

$$\left[\frac{\hbar^2(-\frac{K_1}{2})^2}{2m} - E\right] C(-\frac{K_1}{2}) + C(\frac{K_1}{2})V_{K_1} = 0 \tag{11.58}$$

A non trivial solution requires that

$$\begin{vmatrix} \lambda - E & V_{K_1} \\ V_{K_1} & \lambda - E \end{vmatrix} = 0$$

where $\lambda = \frac{\hbar^2(K_1/2)^2}{2m} = \frac{\hbar^2(-K_1/2)^2}{2m}$. One then gets

$$E_\pm = \lambda \pm V_{K_1} \tag{11.59}$$

One recovers (11.41). Replacing these solutions in (11.57)-(11.58) one obtains

$$\frac{C(-\frac{K_1}{2})}{C(\frac{K_1}{2})} = \frac{\lambda - E}{V_K} = \pm 1 \tag{11.60}$$

The corresponding wave functions are the following stationary waves

$$\psi(x) = e^{i\frac{K_1 x}{2}} \pm e^{-i\frac{K_1 x}{2}} \tag{11.61}$$

Of course, (11.56) allows us to go to higher-order approximations. For example, when k is near the zone limit $\frac{K_1}{2}$, putting $\epsilon = k - \frac{K_1}{2}$ and supposing that ϵ is small, one finds (see Problem 1 below for demonstration):

$$E = E_\pm + \frac{\hbar^2 \epsilon^2}{2m}\left(1 \pm \frac{2\lambda}{V_{K_1}}\right) \tag{11.62}$$

This relation explains the parabolic curvature of E versus k near the Brillouin zone limits shown in Fig. 11.2.

11.3.1   *Band filling: classification of materials*

One has seen in chapter 8 that each Brillouin zone contains N values of $\vec{k}$ where N is the total number of lattice cells of the crystal. In the case of electrons, each state $\vec{k}$ can receive two electrons of opposite spins.

If there is one atom per lattice cell and each atom gives one electron to the crystal (monovalent atom) then in the ground state, N electrons occupy half of the first band (the lowest $N/2$ levels of $|\vec{k}|$, each with two electrons). There are thus $N/2$ levels unoccupied in the first band. When one applies an electric field of even very small amplitude, electrons receive the electric energy and they can move to unoccupied levels of higher energy, giving rise to an electric current. Such a crystal is a good metal.

On the other hand, if each cell has a two-atom basis, N cells of the crystal provide then $2N$ electrons which completely fill N levels of the first energy band (each level has two electrons). A weak applied electric field cannot change states of electrons because there is no unoccupied level to move in. There is thus no current for weak fields. This crystal corresponds to an insulator. If the field is so strong that electrons can jump across the gap to go to the second energy band, then a current may result. But in such a case the field can destroy the band structure. Of course, the temperature, if high enough, can excite electrons to the second band. One will see this situation in section 11.3.2.

In view of the above analysis based on the band filling, one can distinguish qualitatively solids by their conductivity. Insulators have energy bands completely filled (or completely empty). Metals have energy bands filled between 10% and 90%. Semiconductors and semimetals have the filling rate out of these two limits: bands almost empty (<10% filled) or almost full (>90% filled). Note that when a band is almost empty, there is a lot of unoccupied levels but there are very few electrons to contribute to the current. So, the current is weak. The same argument applies for almost filled bands: there are many electrons but a little place to move. Note also that in many materials, there are overlaps of different bands as shown in Fig. 11.4(d): in such a case, tunneling will occur between bands so that each band is partially filled. This yields currents of amplitude similar to those in semimetals and semiconductors.

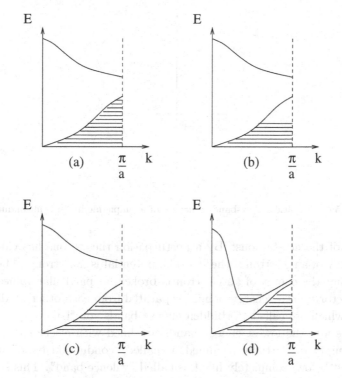

Fig. 11.4 Filling (hachured area) of energy bands in (a) an insulator, (b) a metal, (c) a semiconductor (d) a semimetal.

11.3.2 *Semiconductors*

One presents below the simplest model of a semiconductor and calculates some properties using the results obtained for a free Fermi gas (chapter 5) and the band structure given above.

In a crystal, if electrons are free to go anywhere inside the crystal, then they are conducting electrons and the crystal is a metal. On the contrary, if they are strongly localized on atoms, then the crystal is an insulator. Between these two limits, electrons are not free but not localized, the conduction can happen with a strong enough electric field. The crystal is a semimetal or a semiconductor. An analysis based on the band filling is given above to explain such a behavior of electron mobility.

In semiconductors, one cannot neglect interaction between electrons and ions of the lattice. This interaction can be described as the interaction between an electron and a periodic potential created by the periodic array of

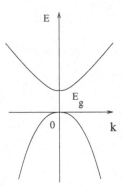

Fig. 11.5   Schematic energy-band structure of a simple model for semiconductors.

charges of the crystal ions. By a perturbation theory, one has calculated the energy of an electron if the periodic potential is not strong. The result shows that the energy of an electron is broken, at particular values of the wave vector $\vec{k}$, into "energy bands" separated from each other by discontinuities which are called "forbidden energy bands" or "band gaps". In the ground state, the lowest energy bands are filled with electrons. The last band, empty or partially occupied, is called "conduction band" and the band just below, completely filled, is called "valence band". This situation corresponds to semiconductors or insulators depending on the width of the forbidden band: the larger the forbidden band, the stronger the insulator. Note that good metals correspond to half-filled conduction bands as seen above.

One considers now the simplest model of a semiconductor in which the valence band is separated from the conduction band by an energy gap (forbidden energy band) of width E_g as shown in Fig. 11.5.

At $T = 0$, there are no electrons in the conduction band, while the valence band is completely filled. When the temperature is sufficiently high, a number of electrons crosses the forbidden band to go to the conduction band. For simplicity, one supposes here that the energy bands are parabolic in k:

$$E_{\vec{k}}^c = E_g + \frac{\hbar^2 k^2}{2m_c^*} \tag{11.63}$$

$$E_{\vec{k}}^v = -\frac{\hbar^2 k^2}{2m_v^*} \tag{11.64}$$

where m_c^* and m_v^* are the "effective masses" of an electron in the conduction and valence bands, respectively. The "effective mass" of an electron is defined in general as its mass at rest modified by its interactions with the surrounding environment. These interactions are supposed to be weak enough so that the energy of an electron still depends on k^2. In general, $m_c^* \neq m_v^*$. One calculates now the number of electrons n_c in the conduction band and the number of holes left behind in the valence band after the departure of electrons. One has

$$n_c = \int_{E_g}^{\infty} dE \rho(E) \frac{1}{e^{\beta(E-E_F)} + 1} \tag{11.65}$$

where the integral upper limit was replaced approximatively by ∞. This is justified by the fact that only microstates at the band bottom make important contributions to the integral. The density of states $\rho(E)$ given by (2.39) should be rewritten taking into account the new origin of the energy of the conduction band which is situated at $E = E_g$. One has

$$\rho(E) = B(E - E_g)^{1/2} = \frac{L^3}{2\pi^2} \left(\frac{2m_c^*}{\hbar^2} \right)^{3/2} (E - E_g)^{1/2} \tag{11.66}$$

where B is given by (5.20). Putting $x = E - E_g$, and asssuming that the semiconductor is at a high temperature, namely $\exp(\frac{E-E_F}{k_B T}) \gg 1$, one obtains

$$
\begin{aligned}
n_c &= B \int_0^{\infty} dx \frac{x^{1/2}}{e^{\beta(x+E_g-E_F)} + 1} \\
&\simeq B e^{-\beta(E_g - E_F)} \int_0^{\infty} dx \, x^{1/2} e^{-\beta x} \\
&= B e^{-\beta(E_g - E_F)} \frac{1}{\beta^{3/2}} \int_0^{\infty} dy \, y^{1/2} e^{-y} \\
&= B e^{-\beta(E_g - E_F)} \frac{1}{\beta^{3/2}} \Gamma\left(\frac{3}{2}\right) \\
&= B e^{-\beta(E_g - E_F)} \frac{1}{\beta^{3/2}} \frac{\sqrt{\pi}}{2} = N_c e^{-\beta(E_g - E_F)}
\end{aligned}
\tag{11.67}
$$

where N_c, called the "effective density of states" of conduction electrons, is defined by

$$N_c = 2L^3 \left(\frac{m_c^* k_B T}{2\pi \hbar^2} \right)^{3/2} \tag{11.68}$$

To calculate the number of holes n_v in the valence band, one replaces the distribution of particles f by the distribution of holes, namely $(1 - f)$. One has

$$f_p = 1 - f = 1 - \frac{1}{e^{\beta(E - E_F)} + 1} = \frac{1}{e^{-\beta(E - E_F)} + 1} \qquad (11.69)$$

In addition, $\rho(E)$ should be replaced by $\rho(-E)$ because the hole energy E is negative in the valence band as shown in Fig. 11.5. One replaces the lower limit of the valence band by $-\infty$ by the argument that only microstates at the top of the band make important contributions to the following integral:

$$\begin{aligned}
n_v &= B \int_{-\infty}^{0} dE (-E)^{1/2} \frac{1}{e^{-\beta(E - E_F)} + 1} \\
&\simeq B \int_{-\infty}^{0} dE (-E)^{1/2} e^{\beta(E - E_F)} \\
&= B e^{-\beta E_F} \int_{0}^{\infty} dE E^{1/2} e^{-\beta E} \qquad E \to -E \\
&= B e^{-\beta E_F} \frac{1}{\beta^{3/2}} \int_{0}^{\infty} dy y^{1/2} e^{-y} \\
&= B e^{-\beta E_F} \frac{1}{\beta^{3/2}} \frac{\sqrt{\pi}}{2} = N_v e^{-\beta E_F} \qquad (11.70)
\end{aligned}$$

where N_v, called the "effective density of states" of holes in the valence band, is defined by

$$N_v = 2L^3 \left(\frac{m_v^* k_B T}{2\pi \hbar^2}\right)^{3/2} \qquad (11.71)$$

Note that

$$n_c n_v = N_c N_v e^{-\beta E_g} = 4\left(\frac{k_B T}{2\pi \hbar^2}\right)^3 (m_c^* m_v^*)^{3/2} e^{-\beta E_g} \qquad (11.72)$$

This product is independent of E_F (one has used $L^3 = 1$, taken as volume unit). The relation (11.72) is called the "law of mass action".
Note that the following definition is often used in the literature

$$n_i^2 \equiv n_c n_v = N_c N_v e^{-\beta E_g} = 4\left(\frac{k_B T}{2\pi \hbar^2}\right)^3 (m_c^* m_v^*)^{3/2} e^{-\beta E_g} \qquad (11.73)$$

where n_i is called intrinsic concentration. It has an obvious meaning only in the case of an intrinsic semiconductor defined below. Otherwise, it is used under the definition $n_i^2 = n_c n_v$ even when the semiconductor is not intrinsic.

One considers now a particular case where the number of electrons in the conduction band is equal to the number of holes in the valence band, namely $n_c = n_v = n_i$. This case corresponds an intrinsic semiconductor. One has

$$n_c = n_v = \sqrt{n_c n_v} = \sqrt{N_c N_v} e^{-\beta E_g/2} \qquad (11.74)$$

Equalizing this equation to (11.67), one gets

$$e^{\beta E_F} = (\frac{N_v}{N_c})^{1/2} e^{\beta E_g/2}$$

hence $\qquad E_F = \frac{E_g}{2} + \frac{k_B T}{2} \ln \frac{N_v}{N_c} = \frac{E_g}{2} + \frac{3k_B T}{4} \ln \frac{m_v^*}{m_c^*} \quad (11.75)$

If $m_v^* = m_c^*$, the Fermi level lies at the middle of the band gap. One gives here some experimental values. For Ge, $E_g = 0.6$ eV, $m_c^* = m_v^* = 0.1m_0$ where m_0 is the mass at rest of an electron. At $T = 290$ K, $N_c = N_v = 75 \times 10^{22}$ m^{-3}, $n_c = n_v = 4.6 \times 10^{12}$ cm^{-3}. For Si, one has $E_g = 1$ eV, $m_c^* = m_v^* = 0.2m_0$. At $T = 290$ K, $N_c = N_v = 2.1 \times 10^{24}$ m^{-3}, $n_c = n_v = 4.3 \times 10^9$ cm^{-3}.

In an intrinsic semiconductor the conductivity takes place at high temperatures where the numbers of electrons and holes are sufficiently large, of the order of 10^{13}-10^{17} per cm^3. One recalls that the concentration of charge carriers (electrons and holes) of a metal is of the order of 10^{22}-10^{23} per cm^3), very large with respect to that in semiconductors. The numerical values given above show that at room temperature (290 K), Ge in its pure state (without impurities) is a semiconductor, but not Si.

11.4 Tight-binding approximation

In this section, one studies the effect of the periodic potential on electrons strongly bound to their "mother" ions. The interaction of an electron with electrons on neighboring ions is supposed to be weak enough to be considered as a perturbation to the atomic state of the electron. One uses below the "tight-binding approximation", the word which expresses the strong binding of electrons to their ions.

11.4.1 *One-dimensional case*

To illustrate the method, one first considers a one-dimensional lattice of N identical atoms, of lattice constant a.

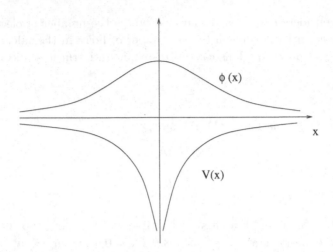

Fig. 11.6   Example of an s atomic wave function $\phi(x)$ and an atomic potential $V(x)$ of spherical symmetry.

If the atomic state of the electron under consideration is weakly perturbed by the presence of electrons of neighboring atoms, then one can use a perturbation theory. Let $\phi(x)$ be the atomic wave function of the electron and $V(x)$ the atomic potential acting on the electron from its mother ion. The Schrödinger equation is written as

$$\left[-\frac{\hbar^2 \Delta}{2m} + V(x) \right] \phi(x) = E^{(0)} \phi(x) \tag{11.76}$$

An example of $\phi(x)$ and $V(x)$ is shown in Fig. 11.6.

In the crystalline state where there is a periodic potential due to the lattice of ions, the atomic state of the electron under consideration is perturbed. One can describe the crystalline potential $U(x)$ at x as a linear combination of atomic potentials created by the ions of the lattice at x:

$$U(x) = \sum_n V(x - na) \tag{11.77}$$

where na is the position of an ion of the lattice and $V(x-na)$ is its potential acting at x. The Schrödinger equation of the electron in the crystalline state is

$$\mathcal{H}\Psi(x) = \left[-\frac{\hbar^2 \Delta}{2m} + U(x) \right] \Psi(x) = E\Psi(x) \tag{11.78}$$

where the wave function $\Psi(x)$ can be constructed as a linear combination of atomic orbitals $\phi(x)$:

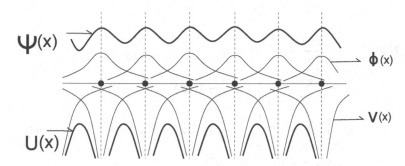

Fig. 11.7   Example of a crystalline potential $U(x)$ and a crystalline wave function $\Psi(x)$. $\phi(x)$ and $V(x)$ denote an atomic s wave function and an atomic potential. $U(x)$ and $\Psi(x)$ are given by (11.77) and (11.79).

$$\Psi(x) = \sum_n C_n \phi(x - na) \qquad (11.79)$$

where C_n are coefficients to be determined. The Bloch condition $\Psi(x+a) = e^{ika}\Psi(x)$ and the normalization of $\Psi(x)$ imply that

$$C_n = \frac{1}{\sqrt{N}} e^{ikna} \qquad (11.80)$$

where k is the wave vector verifying the periodic boundary condition. Figure 11.7 shows schematically the shapes of $\Psi(x)$ and $U(x)$.

   The energy of the electron is

$$E = \int_{\text{cell}} \Psi^*(x) \left[-\frac{\hbar^2 \Delta}{2m} + U(x) \right] \Psi(x) dx \qquad (11.81)$$

One first calculates $\mathcal{H}\Psi(x)$:

$$\mathcal{H}\Psi(x) = \left[-\frac{\hbar^2 \Delta}{2m} + U(x) \right] \Psi(x)$$

$$= \frac{1}{\sqrt{N}} \sum_n e^{ikna} \left[-\frac{\hbar^2 \Delta}{2m} + U(x) \right] \phi(x - na)$$

$$= \frac{1}{\sqrt{N}} \sum_n e^{ikna} \left[-\frac{\hbar^2 \Delta}{2m} + U(x) + V(x - na) - V(x - na) \right]$$
$$\times \phi(x - na)$$

$$= \frac{1}{\sqrt{N}} \sum_n e^{ikna} \left[-\frac{\hbar^2 \Delta}{2m} + V(x - na) \right] \phi(x - na)$$

$$+ \frac{1}{\sqrt{N}} \sum_n e^{ikna} \left[U(x) - V(x - na) \right] \phi(x - na)$$

$$= \frac{1}{\sqrt{N}} E^{(0)} \sum_n e^{ikna} \phi(x - na)$$

$$+ \frac{1}{\sqrt{N}} \sum_n e^{ikna} \left[U(x) - V(x - na) \right] \phi(x - na)$$

The first term corresponds to the atomic state of energy $E^{(0)}$, while the second term corresponds to the perturbation resulting from the difference of potentials between atomic and crystalline states. One has, from (11.81),

$$E = E^{(0)} + \frac{1}{N} \sum_{n,m} e^{ik(n-m)a} \int_{\text{cell}} \phi^*(x - ma) \left[U(x) - V(x - na) \right]$$
$$\times \phi(x - na) dx \tag{11.82}$$

Putting $X = x - na$, one writes

$$\int_{\text{cell}} \phi^*(x - ma) \left[U(x) - V(x - na) \right] \phi(x - na) dx$$

$$= \int_{\text{cell}} \phi^*[X + (m - n)a] \left[U(X + na) - V(X) \right] \phi(X) dX$$

$$= \int_{\text{cell}} \phi^*(X + la) \left[U(X) - V(X) \right] \phi(X) dX$$

where one has used $U(X + na) = U(X)$, $l = n - m$ being an integer. One obtains

$$E = E^{(0)} + \frac{1}{N} \sum_{n,l} e^{ikla} \int_{\text{cell}} \phi^*(X + la) \left[U(X) - V(X) \right] \phi(X) dX$$

$$= E^{(0)} + \sum_{l} e^{ikla} \int_{\text{cell}} \phi^*(X + la) \left[U(X) - V(X) \right] \phi(X) dX \quad (11.83)$$

where one has used $\sum_n 1 = N$. If one retains only interactions between electrons of nearest neighboring atoms, the values of l in the sum in (11.83) are $l = 0, \pm 1$. Note that in the tight-binding approximation interactions between electrons are weak, therefore interactions between atoms further than next nearest neighbors are often neglected. When only interaction between nearest neighbors are taken into account, one has

$$E = E^{(0)} + E_0^{(0)} + E_{+1} e^{ika} + E_{-1} e^{-ika} \quad (11.84)$$

where

$$E_0^{(0)} = \int_{\text{cell}} \phi^*(X) \left[U(X) - V(X) \right] \phi(X) dX \quad (11.85)$$

$$E_{+1} = \int_{\text{cell}} \phi^*(X + a) \left[U(X) - V(X) \right] \phi(X) dX \quad (11.86)$$

$$E_{-1} = \int_{\text{cell}} \phi^*(X - a) \left[U(X) - V(X) \right] \phi(X) dX \quad (11.87)$$

If $\phi(X)$ are s waves (spherical symmetry), then $E_{-1} = E_{+1}$. One has

$$E = E^{(0)} + E_0^{(0)} + 2E^{(1)} \cos(ka) \quad (11.88)$$

where one has put $E^{(1)} \equiv E_{-1} = E_{+1}$.

One sees that the atomic energy level $E^{(0)}$ is first displaced by $E_0^{(0)}$ then widened to become a band of width $4|E^{(1)}|$. This is shown in Fig. 11.8. $E_0^{(0)}$ and $E^{(1)}$, supposed to be negative in the figure, represent the cohesive energies which are at the origin of the formation of the crystal.

When k is small, Eq. (11.88) gives

$$E \simeq E^{(0)} + E_0^{(0)} + 2E^{(1)} - E^{(1)}(ka)^2 = C + \frac{\hbar^2 k^2}{2m^*} \quad (11.89)$$

where C is a constant and m^* the effective mass defined by

$$m^* = -\frac{\hbar^2}{2a^2 E^{(1)}} \quad (11.90)$$

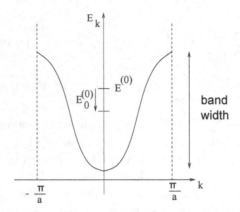

Fig. 11.8   Atomic level $E^{(0)}$ of an electron is widened by interaction with electrons on neighboring atoms in a crystal to become an energy band ($E_0^{(0)}$ and $E^{(1)}$ are assumed to be negative).

Under the form of (11.89) ($E^{(1)} < 0$), one sees that the energy of an electron in a crystalline state, to a constant, is the energy of a free electron with an effective mass which contains interactions with its neighbors. The general definition of the effective mass is given in paragraph 11.4.3.

When k is close to π/a, an expansion of $\cos(ka)$ in (11.88) gives

$$E \simeq C + \frac{\hbar^2 k^2}{2m^*} \tag{11.91}$$

with a negative effective mass.

11.4.2   *Three-dimensional case*

In three dimensions atomic state $\phi(\vec{r})$ is described by the following Schrödinger equation

$$\left[-\frac{\hbar^2 \Delta}{2m} + V(\vec{r}) \right] \phi(\vec{r}) = E^{(0)} \phi(\vec{r}) \tag{11.92}$$

and in the crystalline state, the wave function $\Psi(\vec{r})$ is given by

$$\left[-\frac{\hbar^2 \Delta}{2m} + U(\vec{r}) \right] \Psi(\vec{r}) = E \Psi(\vec{r}) \tag{11.93}$$

where the crystalline periodic potential is

$$U(\vec{r}) = \sum_j V(\vec{r} - \vec{R}_j) \tag{11.94}$$

$\vec{R}_j$ being the position vector of the site j.

$\Psi(\vec{r})$ is written as

$$\Psi(\vec{r}) = \sum_i C_i \phi(\vec{r} - \vec{R}_i) \tag{11.95}$$

The same calculation as in the one-dimensional case leads to

$$E_{\vec{k}} = E^{(0)} + \sum_l e^{i\vec{k}\cdot\vec{R}_l} \int_{\text{cell}} \phi^*(\vec{r} + \vec{R}_l) \left[U(\vec{r}) - V(\vec{r}) \right] \phi(\vec{r}) d\vec{r} \tag{11.96}$$

The sum over $\vec{R}_l$ can be limited to nearest and next nearest neighbors.

In the case of atomic spherical waves, the interaction depends only on the distance r. Equation (11.96) becomes simple. For example, in the case of a simple cubic lattice one has

$$E_{\vec{k}} = E^{(0)} + E_0^{(0)} + 2E^{(1)} \left[\cos(k_x a) + \cos(k_y a) \cos(k_z a) \right] \tag{11.97}$$

in analogy with (11.88), where

$$E_0^{(0)} = \int_{\text{cell}} \phi^*(\vec{r}) \left[U(\vec{r}) - V(\vec{r}) \right] \phi(\vec{r}) d\vec{r} \tag{11.98}$$

$$E^{(1)} = \int_{\text{cell}} \phi^*(\vec{r} + \vec{R}_l) \left[U(\vec{r}) - V(\vec{r}) \right] \phi(\vec{r}) d\vec{r} \tag{11.99}$$

$\vec{R}_l$ being a vector connecting one atom and a nearest neighbor.

When $\vec{k}$ is close to 0 or π/a, one can write, in the same manner as for (11.89),

$$E_{\vec{k}} \simeq C + \frac{\hbar^2 k^2}{2m^*}.$$

The equi-energetic surfaces are thus spherical near these points. The approximation which consists in assuming the form $\frac{\hbar^2 k^2}{2m^*}$ for the electron energy is called "effective-mass approximation".

11.4.3 *Velocity, acceleration, effective mass*

The velocity of an electron is defined by

$$\vec{v}_{\vec{k}} = \frac{1}{\hbar} \vec{\nabla}_{\vec{k}} E_{\vec{k}} \tag{11.100}$$

where $\vec{\nabla}_{\vec{k}}$ denotes the gradient with respect to $\vec{k}$, $E_{\vec{k}}$ is given by (11.97) for the simple cubic lattice, or by (11.96) in the general case.

The acceleration is therefore

$$\vec{a}_{\vec{k}} = \frac{d\vec{v}_{\vec{k}}}{dt} = \frac{1}{\hbar}\frac{d}{dt}\vec{\nabla}_{\vec{k}}E_{\vec{k}}$$

$$= \frac{1}{\hbar}\vec{\nabla}_{\vec{k}}\left[\frac{d\vec{k}}{dt}\cdot\vec{\nabla}_{\vec{k}}E_{\vec{k}}\right] = \frac{1}{\hbar^2}\vec{\nabla}_{\vec{k}}[\vec{F}\cdot\vec{\nabla}_{\vec{k}}E_{\vec{k}}] \qquad (11.101)$$

where $\vec{F}$ is the force. Writing

$$\vec{F}\cdot\vec{\nabla}_{\vec{k}}E_{\vec{k}} = \sum_{i=x,y,z}\frac{\partial E_{\vec{k}}}{\partial k_i}F_i \qquad (11.102)$$

one obtains the i-component $(i = x, y, z)$ of the acceleration as

$$a_i = \frac{1}{\hbar^2}\sum_{j=x,y,z}\frac{\partial^2 E_{\vec{k}}}{\partial k_i\partial k_j}F_j \qquad (11.103)$$

The effective mass is a tensor defined by

$$\frac{1}{m_{ij}^*} = \frac{1}{\hbar^2}\frac{\partial^2 E_{\vec{k}}}{\partial k_i\partial k_j} \qquad (11.104)$$

It is obvious that m_{ij}^* $(i, j = x, y, z)$ depends on the dispersion relation $E_{\vec{k}}$. In the case of the simple cubic lattice, using (11.97) one sees that $m_{ij}^* = m_i^*\delta_{i,j}$ where

$$\frac{1}{m_i^*} = -\frac{1}{\hbar^2}2a^2E^{(1)}\cos(k_ia),$$

in agreement with (11.90) at $k \sim 0$.

The velocity, the acceleration and the effective mass of an electron depend on the interaction energy with its environment, therefore on $\vec{k}$.

11.5   Conclusion

This chapter is very important: one has shown that the interaction between electrons and ions of the crystal is the origin of energy bands in crystalline solids.

The first part of the chapter treats the case of electrons moving in a weak periodic potential of the ions. One has used the theory of almost-free electrons to show that the electron energy becomes discontinuous at the boundaries of Brillouin zones. The interaction electron-lattice opens thus forbidden bands (or gaps) in the energy curve. There are many applications

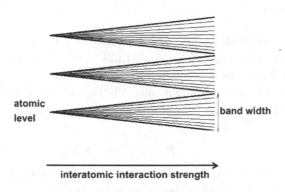

Fig. 11.9   Electron atomic levels (on the left) are widened by interaction with neighboring electrons in a crystal: energy bands are formed. Materials having narrow band widths (left part of the figure) are bad conductors while those having large band widths (right part of the figure) are good conductors.

in the electron transport using the energy gap in semiconductors. One will study some complementary aspects of semiconductors in chapter 15.

The second part of the chapter treats the case where electrons are strongly bound to their atoms in a crystalline solid, but their energy is perturbed by the interaction with electrons of neighboring atoms. One has shown that the electron atomic level is widened by this interaction to become an energy band. One has shown that near $k = 0$ and near zone boundaries the electron can be considered as a free electron with a mass modified by the interaction with the crystal. This modified mass is called effective mass. The effective-mass approximation is very useful in describing electronic properties, in particular in electron charge and spin transport phenomena. Note that the band width depends on the interaction strength: the stronger interaction with the crystal the larger the band width. This is schematically shown in Fig. 11.9. When band widths are large, gaps between bands become small, and bands can overlap. In such a situation, electrons can go from one band to another: the material is a good conductor. If the electron-lattice interaction is weak, electrons are strongly localized on atoms, the band width is narrow as in the case of d-orbital transition metals. In such a case, the material is a bad conductor.

11.6   Problems

Problem 1. Perturbation near the boundary of a Brillouin zone:
Demonstrate the expression (11.62).

Problem 2. Electrons in a square lattice:
Consider an electron in a square lattice of constant a, of surface S with the following Hamiltonian

$$H = \vec{p}^2/2m + V(\vec{r}).$$

The potential $V(\vec{r}) = V(x, y)$ perturbs weakly the kinetic term $H_0 = \vec{p}^2/2m$ and has the following properties:

$$V(x, y) = V(x + a, y) = V(x, y + a) =$$

$$= V(y, x) = V(-x, y) = V(x, -y).$$

One writes the eigenvectors and eigenvalues of H_0 as

$$H_0|\vec{k}\rangle_0 = \epsilon_0(\vec{k})|\vec{k}\rangle_0.$$

a) Write the general expression of vectors $\vec{K}$ of the reciprocal lattice and plot the first Brillouin zone.

b) Show that one can write the potential V as a Fourier series of the form:

$$V(\vec{r}) = \sum_{\vec{K}} W_{\vec{K}} e^{i\vec{K}\cdot\vec{r}}$$

and give the Fourier components $W_{\vec{K}}$ as a function of V. Show that $W_{\vec{K}}$ are real and $W_{-\vec{K}} = W_{\vec{K}}$.

c) Using the perturbation theory for non degenerate energy levels, determine the energy correction $\epsilon_1(\vec{k})$ at first order in V for energies $\epsilon_0(\vec{k})$ of the free electron. Verify that the correction $\epsilon_1(\vec{k})$ depends only on W_0.

d) Show that in the same conditions the correction at second order $\epsilon_2(\vec{k})$ is given by

$$\epsilon_2(\vec{k}) = \sum_{\vec{G} \neq 0} W_{\vec{G}}^2 / \left(\epsilon_0(\vec{k}) - \epsilon_0(\vec{k} + \vec{G}) \right).$$

e) Explain why the above equation is no more valid when $\vec{k}$ has its extremity on the boundary of the first Brillouin zone.

f) Let $\vec{K}_1$ be the vector in the reciprocal space of coordinates $(\pi/a, \pi/a)$. Show that $\epsilon_0(\vec{K}_1)$ is a four-fold degenerate energy:

$$\epsilon_0(\vec{K}_1) = \epsilon_0(\vec{K}_1 + \vec{G}_j)$$

for $j = 1, 2, 3$. Indicate what are the three vectors $\vec{G}_j$. Verify that two of them, called $\vec{G}_1$ and $\vec{G}_2$, are orthogonal. Using the symmetry properties of V, show that $W_{\vec{G}_1} = W_{\vec{G}_2}$ and $W_{\vec{G}_3} = W_{\vec{G}_2 - \vec{G}_1}$.

One shall put $\vec{G}_0 = 0$.

g) Using the perturbation theory for degenerate levels, determine the energy levels corresponding to $\epsilon_0(\vec{K}_1)$ at the first order in V.

Guide: To do this, write the perturbed state as a linear combination of unperturbed states:

$$\Psi_{\vec{K}_1} = \alpha_0 |\vec{K}_1\rangle_0 + \sum_{j=1}^{3} \alpha_j |\vec{G}_j\rangle_0$$

and diagonalize H in the subspace E of H_0 associated to $\epsilon_0(\vec{K}_1)$. For simplicity, one shall suppose that $W_0 = 0$ and one shall put $W = W_{\vec{G}_1} = W_{\vec{G}_2}$ and $W' = W_{\vec{G}_3}$.

Verify that the solutions are

$$\epsilon = \epsilon_0(\vec{K}_1) - W'$$

and

$$\epsilon = \epsilon_0(\vec{K}_1) + W' \pm 2W.$$

Does the potential V remove the degeneracy of the level $\epsilon_0(\vec{K}_1)$?

Problem 3. Electrons in a rectangular lattice:

One takes back the previous problem but with a rectangular lattice of constants a and b ($b > a$) in the directions x and y, respectively. The state $\vec{K}_1$ is defined by $(\pi/a, 0)$. Determine the degenerate states of $\vec{K}_1$ when $V = 0$. Does the potential V remove the degeneracy of the level $\epsilon_0(\vec{K}_1)$?

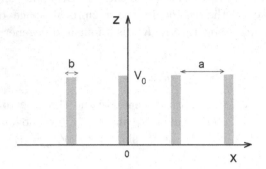

Fig. 11.10   Kronig-Penney model.

Problem 4. Energy band in the Kronig-Penney potential model:

Consider a one-dimensional lattice of N sites, of constant $c = a + b$ with $a >> b$ as shown in Fig. 11.10. The potential in this lattice is defined in the first cell by

$$V(x) = \begin{cases} 0 & \text{if } 0 \le x < a \\ V_0 & \text{if } a \le x < a + b \end{cases} \tag{11.105}$$

This potential is periodic, namely $V(x + c) = V(x)$. V_0 is finite but can go to infinity.

a) Write the Schrödinger equation for the wave function $\varphi(x)$ in the well ($0 \le x \le a$) and for $\phi(x)$ inside the barrier ($a \le x \le a + b$). Using the conservation of energy at a well-barrier boundary show that

$$k^2 = -q^2 + 2mV_0 \tag{11.106}$$

where k and q are wave vectors in the well and barrier, respectively, and m is the electron mass.

b) Show that

$$\begin{pmatrix} \varphi(a) \\ \frac{\partial \varphi(a)}{\partial x} \end{pmatrix} = \begin{pmatrix} \cos(ka) & \sin(ka)/k \\ k\sin(ka) & \cos(ka) \end{pmatrix} \begin{pmatrix} \varphi(0) \\ \frac{\partial \varphi(0)}{\partial x} \end{pmatrix}$$

where one has used for simplicity the notation $\partial \varphi(a)/\partial x \equiv [\partial \varphi(x)/\partial x]_{x=a}$ etc.

c) For x inside a barrier, show that

$$\begin{pmatrix} \phi(a+b) \\ \frac{\partial\phi(a+b)}{\partial x} \end{pmatrix} = \begin{pmatrix} \cosh(qb) & \sinh(qb)/q \\ q\sinh(qb) & \cosh(qb) \end{pmatrix} \begin{pmatrix} \phi(a) \\ \frac{\partial\phi(a)}{\partial x} \end{pmatrix}$$

Deduce that for the translation $0 \to a+b$, the wave function and its derivative are transformed by the transfer matrix $T = T_a T_b$ where T_a and T_b are the 2×2 matrices given by the above equations.

d) Using the periodic boundary condition, show that the eigenvalues of T are given by $\lambda = \exp(i2\pi n/N)$ where n is an integer.

e) By writing

$$\begin{pmatrix} \varphi(a+b) \\ \frac{\partial\varphi(a+b)}{\partial x} \end{pmatrix} = \begin{pmatrix} A & B \\ C & D \end{pmatrix} \begin{pmatrix} \varphi(0) \\ \frac{\partial\varphi(0)}{\partial x} \end{pmatrix}$$

calculate A, B, C and D. Calculate directly the eigenvalues of T given above by diagonalization. Show that the secular equation gives

$$2\cos(ka)\cosh(qb) + \left(\frac{k}{q} + \frac{q}{k}\right)\sin(ka)\sinh(qb) = 2\cos\left(\frac{2\pi n}{N}\right) \tag{11.107}$$

f) When $b \to 0$ and $V_0 \to \infty$ but qb remains finite, show that (11.107) is reduced to

$$\cos(ka) + mV_0 ab\frac{\sin(ka)}{ka} = \cos\left(\frac{2\pi n}{N}\right) \tag{11.108}$$

Plot the function of the left-hand side, Y_1, versus ka. Solutions of (11.108) require that $|Y_1| = |\cos\left(\frac{2\pi n}{N}\right)| \le 1$: identify the zones where there are no solutions (forbidden bands).

Problem 5. Tight-binding approximation:

Consider a lattice with one atom per elementary cell. The electron on an atom has the following atomic state defined by the orbital $\phi(\vec{r})$ which is an eigenfunction of the "atomic" Hamiltonian

$$H_a = \vec{p}^2/2m + V_a(\vec{r})$$

$$H_a\phi(\vec{r}) = \epsilon_a\phi(\vec{r}),$$

where $V_a(\vec{r})$ is the potential of the atom acting on its electron. The lattice is defined by the ensemble of translation vectors $\vec{t}_n$ with the lattice constant $a = 1$. The orbital centered at the site n is

$$\langle \vec{r}|n \rangle = \phi(\vec{r} - \vec{t}_n).$$

This orbital has an extension of the order of the interatomic distance so that orbitals at sites n and m do not overlap if $|n-m| > 1$, namely $\langle n|m \rangle = 0$. All atoms are supposed identical and their orbitals have a spherical symmetry.

a) Write the expression of the Hamiltonian H of an electron in the crystalline state. Give the form of the crystalline potential.

b) Writing H under the form:

$$H = \sum_n \epsilon_n |n\rangle\langle n| + \sum_{n \neq m} V_{n,m} |n\rangle\langle m|,$$

show that ϵ_n is independent of n and $V_{n,m} = V_{n-m}$.
In the following, one supposes that $V_{n-m} = V$ if n and m are nearest neighbors and $V_{n-m} = 0$ otherwise.

c) One looks for an eigenfunction $|\vec{k}\rangle$ of H: to do that, one supposes $|\vec{k}\rangle$ is a linear combination of atomic orbitals:

$$|\vec{k}\rangle = \sum_n c_n(\vec{k})|n\rangle.$$

Calculate $c_n(\vec{k})$ provided $|\vec{k}\rangle$ a Bloch function. Show that $|\vec{k}\rangle$ so obtained is an eigenfunction of H. Calculate the eigenvalue $\epsilon(\vec{k})$ of $|\vec{k}\rangle$.

Problem 6. Tight-binding approximation in a square lattice:
Consider a square lattice of constant a, of $N \times N$ atoms, with periodic boundary conditions.

a) Define the first Brillouin zone. Give the expression of the energy $\epsilon(\vec{k})$.

b) Calculate the density of states $\rho(\epsilon)$ when $k \to 0$.

Problem 7. Tight-binding approximation in a face-centered cubic lattice:
Consider a lattice of face-centered cubic structure. The elementary cell is defined by three basis vectors:

$$\vec{a}_1 = a(\vec{e}_2 + \vec{e}_3)/2$$
$$\vec{a}_2 = a(\vec{e}_3 + \vec{e}_1)/2$$
$$\vec{a}_3 = a(\vec{e}_1 + \vec{e}_2)/2$$

where $\vec{e}_1$, $\vec{e}_2$ and $\vec{e}_3$ are unit vectors along the Ox, Oy and Oz axes.

a) Construct the reciprocal lattice and define the first Brillouin zone. One shall use the cubic group symmetry to plot $1/8$ of the first Brillouin zone in the reciprocal space.

b) Modify the expression of $E_{\vec{k}}$ obtained in (17.81) (see solution of Problem 6) for the present face-centered cubic lattice.

c) Determine the points of the first Brillouin zone where the energy is extremal (maximum or minimum). Give the band width for $V < 0$ where

$$V \equiv E^{(1)} = \int_{\text{cell}} \phi^*(\vec{r} + \vec{R_l})[U(\vec{r}) - V(\vec{r})]\phi(\vec{r})d\vec{r},$$

$\vec{R_l}$ being vectors connecting a site to its first neighbors.

d) Find the dispersion relation near the point Γ ($\vec{k} = 0$) of the first Brillouin zone and give the corresponding tensor of the effective mass. Calculate the density of states near Γ.

Chapter 12

Systems of Interacting Spins: Magnons

This chapter is devoted to systems of interacting spins. On the one hand, spin systems describe magnetic materials which have been extensively studied both theoretically and experimentally for decades due to numerous applications. On the other hand, spin models can be also used to describe systems where parameters and symmetry can be mapped into a spin language. Such systems are not limited to physics, they are found in other domains such as biology, chemistry, and even social sciences. Spin systems constitute one of the central activities in statistical physics. As for other systems of interacting particles, systems of interacting spins can be exactly solved only in a very limited number of particular cases such as systems with short-range interaction in one and two dimensions. Otherwise, one has to use various approximations to obtain their properties.

In this chapter, one studies elementary excitations of a spin system. These collective motions are called spin waves, or magnons when they are quantized, in analogy with phonons which are energy quanta of vibrations of coupled atoms in crystals treated in chapter 9. Spin waves dominate magnetic properties of materials at low temperatures. Phase transitions in spin systems which occur at higher temperatures are studied in chapter 13.

12.1 Spin models

12.1.1 *Heisenberg model*

The Heisenberg model for the interaction between two spins $\vec{S}_i$ and $\vec{S}_j$ localized at $\vec{r}_i$ and $\vec{r}_j$ is given by

$$-2J_{ij}\vec{S}_i \cdot \vec{S}_j \tag{12.1}$$

where J_{ij} is the exchange interaction resulting from the Coulomb interaction between two electrons (or atoms) bearing $\vec{S}_i$ and $\vec{S}_j$. Though the model (12.1) is often used as the starting point for authors who want to apply it in their studies, its origin is demonstrated in Appendix C for completeness. In general, J_{ij} depends on the distance between the spins and on the orientation of $\vec{r}_j - \vec{r}_i$ with respect to the crystalline axes.

In the quantum model, $\vec{S}_i$ is a quantum spin whose components obey the spin commutation relations. For example, in the case of spin one-half, let $\vec{\sigma}$ be $2\vec{S}$, then the components of $\vec{\sigma}$ are the well-known Pauli matrices

$$\sigma_x = \begin{pmatrix} 0 & 1 \\ 1 & 0 \end{pmatrix} \tag{12.2}$$

$$\sigma_y = \begin{pmatrix} 0 & -i \\ i & 0 \end{pmatrix} \tag{12.3}$$

$$\sigma_z = \begin{pmatrix} 1 & 0 \\ 0 & -1 \end{pmatrix} \tag{12.4}$$

Note that σ_z is diagonal. The eigenvalues of S_z are thus $-1/2$ and $1/2$. As can be verified using the above matrices, the spin operators obey the following commutation relations

$$[\sigma^+, \sigma^-] = 2\sigma_z \tag{12.5}$$
$$[\sigma_z, \sigma^\pm] = \pm 2\sigma^\pm \tag{12.6}$$
$$[\sigma_x, \sigma_y] = i\sigma_z + \text{relations by circular permutations } of\ x, y, z \tag{12.7}$$

where $\sigma^\pm = \sigma_x \pm i\sigma_y$.

In the classical model, $\vec{S}_i$ is considered as a vector.

One considers hereafter the simplest case where the exchange interaction is limited between nearest neighbors and this interaction is identical and equal to J for all pairs of nearest neighbors regardless of their relative positions with respect to the crystalline axes. In this case, the Hamiltonian reads

$$\mathcal{H} = -2J \sum_{<i,j>} \vec{S}_i \cdot \vec{S}_j \tag{12.8}$$

where the sum is made over all pairs of nearest neighbors. One sees that if $J > 0$, $\mathcal{H}$ is minimum when all spins are parallel. This spin configuration corresponds to the ferromagnetic ground state.

In the case where $J < 0$, one has to distinguish classical and quantum spin models:

(i) In the classical model, the spins are vectors. Except for lattices composed of equilateral triangular faces in their elementary cell such as the two-dimensional triangular lattice, the face-centered cubic lattice and the hexagonal-close-packed lattice, $\mathcal{H}$ is minimum in the other lattices when all nearest neighbors are antiparallel throughout the crystal. This spin configuration is called "classical antiferromagnetic ground state" or "Néel state".

(ii) In the quantum model, the spins can be decomposed into spin operators [see (C.8)-(C.10)]: $(S_i^z, S_i^+ \equiv S_i^x + iS_i^y, S_i^- \equiv S_i^x - iS_i^y)$. These spin operators obey the spin commutation relations

$$[S_l^+, S_m^-] = 2S_l^z \delta_{l,m} \quad \text{and} \quad [S_l^z, S_m^\pm] = \pm S_l^\pm \delta_{l,m} \tag{12.9}$$

It is known that the Néel state is not the quantum ground state for antiferromagnets. As seen in Problem 7, at zero temperature the so-called zero-point quantum fluctuations make the spins contracted from its full length. The zero-point spin contraction is of the order of a few percents depending on the lattice structure. The real quantum ground state of antiferromagnets is not known, though one knows that it is not far from the classical Néel state, except in low dimensions where strong quantum fluctuations can destroy the Néel order. Note that in the case of ferromagnets, the perfect parallel spin configuration is the real ground state in both classical and quantum spin models.

When the geometry of the lattice does not allow a spin to "fully" satisfy all the interactions with its neighbors, the spin is frustrated. This happens when the elementary lattice cell is composed of equilateral triangles such as in the triangular lattice. There are many spectacular effects due to the frustration. Remarkable properties of frustrated spin systems have been found in the last two decades (see Ref. [49]).

12.1.2 *Ising, XY and Potts models*

Besides the Heisenberg model, there are three other very popular spin models in magnetism and statistical physics:

- Ising model: The Ising model is defined by

$$\mathcal{H} = -\sum_{(i,j)} J_{ij} \sigma_i \sigma_j \tag{12.10}$$

where σ_i (σ_j) is the spin at lattice site i (j) with two possible values $+1$ and -1. Such a spin is called "Ising spin". The Ising model can be used to study spin systems with a strong uniaxial anisotropy. It is also used as a simple model for phase transition investigation (see chapter 13). Any system of interacting entities where each entity has two individual states can be mapped into a system of Ising spins. For instance, in a binary alloy where each lattice site can be occupied by an A atom or a B atom, one can assign an up spin (+1) for an A atom and a down spin (-1) for a B atom. One can then study for example the structure of the alloy by studying the Ising model as a function of its composition.

- XY model: The XY model is defined by two-component spins. It is sometimes called "model of plane rotators". The Hamiltonian is given by

$$\mathcal{H} = -\sum_{(i,j)} J_{ij}(S_{ix}S_{jx} + S_{iy}S_{jy}) \tag{12.11}$$

The XY model is used to study spin systems with a strong planar anisotropy called "easy-plane anisotropy". It is used not only to describe some magnetic materials but also in statistical physics. A very interesting ferromagnetic XY model in two dimensions has been extensively studied because it gives rise to a very special phase transition known as the "Kosterlitz-Thouless" transition discovered in the 70's [102, 8, 34, 189].

- Potts model: The q-state Potts model is defined by

$$\mathcal{H} = -\sum_{(i,j)} J_{ij}\delta_{\sigma_i,\sigma_j} \tag{12.12}$$

where σ_i (σ_j) is a parameter taking q values, $\sigma_i = 1, 2, ..., q$ for example, and $\delta_{\sigma_i,\sigma_j}$ denotes the Kronecker symbol. The sum is often made over pairs of nearest neighbors. If interaction $J_{ij} = J > 0$ for nearest neighbors (i, j), then in the ground state there is only one value of q: it is ferromagnetic. Note that if $q = 2$, the model is equivalent to the Ising model. One defines the Potts order parameter Q by

$$Q = \frac{[q \ \max(Q_1, Q_2, ..., Q_q) - 1]}{q - 1} \tag{12.13}$$

where Q_n is the spatial average defined by

$$Q_n = \frac{1}{N}\sum_{j=1}^{N}\delta_{\sigma_j,n} \tag{12.14}$$

where $n = 1, ..., q$, the sum runs over all lattice sites, and N is the total site number. From this definition one sees that the ground state containing only one kind of spin has $Q = 1$, while in the disordered state q kinds of spin are equally present in the system, namely $Q_1 = Q_2 = ... = Q_q = 1/q$, so that $Q = 0$.

The q-state Potts model is used to study systems of interacting particles where each particle has q individual states. Exact methods to treat the Potts models in two dimensions are shown in a book by Baxter [20].

Two other models which deserve a description in details are the Hubbard model and the Kosterlitz-Thouless model. These are done in Appendices D and E.

12.2 Spin waves in ferromagnets

Elementary excitations of a system of interacting spins have a wave nature as in many other systems such as collective vibrations of coupled atoms in crystals (phonons), waves of charge density in plasmas, etc. In spin systems, the collective excitations are called spin waves or magnons when they are quantized. Spin waves propagate in magnetically ordered systems in the way phonons do in crystalline solids. At finite temperatures, as long as the magnetic order exists, i. e. $T < T_c$, spin waves are the only physical process that determines low-T magnetic properties of the system.

This section is devoted to spin waves in ferromagnets. Spin waves in antiferromagnets, ferrimagnets and helimagnets are given as problems listed at the end of this chapter. A classical treatment is first introduced to give a simple picture of spin waves. A quantum method is next presented to study in a more efficient manner finite-temperature properties of spin systems. One will see that spin waves obey the Bose-Einstein statistics which is used to calculate low-temperature properties.

12.2.1 *Classical treatment*

In ferromagnets, spins are parallel in the ground state. One supposes that the spins are aligned along the Oz axis. One considers each spin as a vector of modulus S with three components. This is the classical Heisenberg model. As the temperature increases, the spins rotate each around its z axis in a collective manner as shown in Fig. 12.1. The energy brought about by the temperature is shared by the whole system.

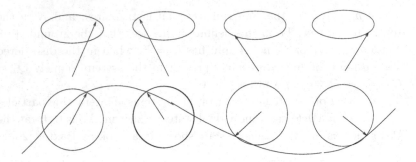

Fig. 12.1   A spin wave : side view (upper) and top view (lower).

Let us calculate the spin-wave energy. We consider the classical Heisenberg spin model on a lattice. The spins are supposed to interact with each other via a nearest-neighbor exchange interaction J. For the spin $\vec{S}_l$, the interaction with its nearest neighbors is written as

$$\mathcal{H}_l = -2J\vec{S}_l \cdot \sum_{\vec{R}} \vec{S}_{l+\vec{R}} \tag{12.15}$$

where $\vec{R}$ is a vector connecting the spin $\vec{S}_l$ with one of its neighbors. The sum is performed over all nearest neighbors. Equation (12.15) can be rewritten as

$$\mathcal{H}_l = -\vec{M}_l \cdot \vec{H}_{ex} \tag{12.16}$$

where

$$\vec{M}_l = g\mu_B \vec{S}_l \tag{12.17}$$

and

$$\vec{H}_{ex} = \frac{2J}{g\mu_B} \sum_{\vec{R}} \vec{S}_{l+\vec{R}} \tag{12.18}$$

g and μ_B are the Landé factor and Bohr magneton. $\vec{M}_l$ and $\vec{H}_{ex}$ are the magnetic moment of spin $\vec{S}_l$ and the field created by the nearest neighbors acting on $\vec{S}_l$.

The equation of motion of the kinetic moment $\hbar\vec{S}_l$ is written as

$$\hbar\frac{d\vec{S}_l}{dt} = \left[\vec{M}_l \wedge \vec{H}_{ex}\right] = 2J\left[\vec{S}_l \wedge \sum_{\vec{R}} \vec{S}_{l+\vec{R}}\right] \tag{12.19}$$

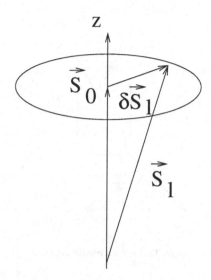

Fig. 12.2 Spin decomposition around the z axis.

If $\vec{S}_l$ is parallel to its neighbors then one has $\hbar\frac{d\vec{S}_l}{dt} = 0$, i.e. $\vec{S}_l$ is equal to a constant vector, thus there is no spin wave.

In an excited state due to a spin rotation around the z axis, one can decompose $\vec{S}_l$ into components as shown in Fig. 12.2:

$$\vec{S}_l = \vec{S}_0 + \delta\vec{S}_l \tag{12.20}$$

where $\delta\vec{S}_l$ represents the deviation and $\vec{S}_0$ the z spin component. For a homogenous system and for a given spin-wave mode, it is natural to suppose that $\vec{S}_0$ is time-independent. Equation (12.19) thus becomes

$$\hbar\frac{d(\delta\vec{S}_l)}{dt} = 2J \sum_{\vec{R}} \left[(\vec{S}_0 + \delta\vec{S}_l) \wedge (\vec{S}_0 + \delta\vec{S}_{l+\vec{R}}) \right]$$

$$\simeq 2J \sum_{\vec{R}} \left[(\delta\vec{S}_l - \delta\vec{S}_{l+\vec{R}}) \wedge \vec{S}_0 \right] \tag{12.21}$$

where $\delta\vec{S}_l$ and $\delta\vec{S}_{l+\vec{R}}$ are supposed to be small. This hypothesis is justified at low temperatures.

Let x and y be the two other Cartesian coordinates defining the plane perpendicular to the z axis. One has $\vec{S}_0 \simeq S\hat{k}$, $(\delta\vec{S}_l)^x = S_l^x$ and $(\delta\vec{S}_l)^y = S_l^y$, $\hat{k}$ being the unit vector on the z axis (see Fig. 12.3).

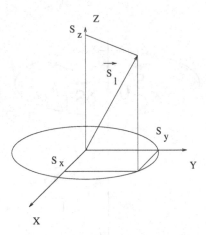

Fig. 12.3   Spin components.

Equation (12.21) reads

$$\hbar\frac{dS_l^x}{dt} = 2JS\sum_{\vec{R}}\left(S_l^y - S_{l+\vec{R}}^y\right) \tag{12.22}$$

$$\hbar\frac{dS_l^y}{dt} = -2JS\sum_{\vec{R}}\left(S_l^x - S_{l+\vec{R}}^x\right) \tag{12.23}$$

$$\hbar\frac{dS_l^z}{dt} = 0 \tag{12.24}$$

S_l^z is thus a constant of motion. One looks for the solutions of S_l^x and S_l^y of the form

$$S_l^x = Ue^{i(\vec{k}\cdot\vec{l}-\omega_{\vec{k}}t)} \tag{12.25}$$

$$S_l^y = Ve^{i(\vec{k}\cdot\vec{l}-\omega_{\vec{k}}t)} \tag{12.26}$$

where $\vec{k}$ and $\vec{l}$ are the wave vector and the position of $\vec{S}_l$, respectively. Replacing (12.25)-(12.26) in (12.22)-(12.23), one obtains

$$-i\hbar\omega_{\vec{k}}U = 2JSZ\left[1 - \frac{1}{Z}\sum_{\vec{R}}e^{i\vec{k}\cdot\vec{R}}\right]V \tag{12.27}$$

$$i\hbar\omega_{\vec{k}}V = 2JSZ\left[1 - \frac{1}{Z}\sum_{\vec{R}}e^{i\vec{k}\cdot\vec{R}}\right]U \tag{12.28}$$

where Z is the coordination number (number of nearest neighbors). The non trivial solutions of U and V verify

$$\begin{vmatrix} i\hbar\omega_{\vec{k}} & 2JSZ(1-\gamma_{\vec{k}}) \\ -2JSZ(1-\gamma_{\vec{k}}) & i\hbar\omega_{\vec{k}} \end{vmatrix} = 0$$

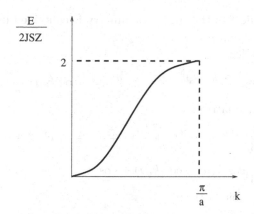

Fig. 12.4   Spin-wave dispersion relation of a ferromagnet in one dimension.

from which one has

$$\hbar\omega_{\vec{k}} = 2JSZ\left(1 - \gamma_{\vec{k}}\right) \tag{12.29}$$

where

$$\gamma_{\vec{k}} = \frac{1}{Z}\sum_{\vec{R}} e^{i\vec{k}\cdot\vec{R}} \tag{12.30}$$

With (12.30) and (12.29) one finds $V = -iU$. This relation indicates that the spin rotation has a circular precession around the z axis.

The relation (12.29) is called "dispersion relation" of spin waves in ferromagnets.

Example : In the case of a chain of spins one has

$$\gamma_k = \frac{1}{2}(e^{ika} + e^{-ika}) = \cos(ka) \tag{12.31}$$

where a is the lattice constant. Equation (12.29) becomes

$$\hbar\omega_k = 2JSZ\left[1 - \cos(ka)\right] \tag{12.32}$$

Figure 12.4 shows $\hbar\omega_k$ versus k in the first Brillouin zone. When $ka << 1$, with $Z = 2$ for one dimension, one has

$$\hbar\omega_k \simeq 2JS(ka)^2 \tag{12.33}$$

ω_k is thus proportional to k^2 for small k (long wave-length), in contrast to the case of phonons where ω_k is proportional to k. A consequence of this difference is that macroscopic physical properties which are averaged over these elementary excitations will show different temperature-dependent behaviors as seen below.

Remark: Effect of the crystal symmetry is contained in the factor $\gamma_{\vec{k}}$. Here are a few examples:

1) Square lattice:

$$\gamma_{\vec{k}} = \frac{1}{4}\left(e^{ik_x a} + e^{-ik_x a} + e^{ik_y a} + e^{-ik_y a}\right) = \frac{1}{2}[\cos(k_x a) + \cos(k_y a)] \quad (12.34)$$

2) Simple cubic lattice:

$$\gamma_{\vec{k}} = \frac{1}{6}(e^{ik_x a} + e^{-ik_x a} + e^{ik_y a} + e^{-ik_y a} + e^{ik_z a} + e^{-ik_z a})$$

$$= \frac{1}{3}[\cos(k_x a) + \cos(k_y a) + \cos(k_z a)] \quad (12.35)$$

3) Centered cubic lattice:

$$\gamma_{\vec{k}} = \cos(\frac{k_x a}{2})\cos(\frac{k_y a}{2})\cos(\frac{k_z a}{2}) \quad (12.36)$$

4) Face-centered cubic lattice:

$$\gamma_{\vec{k}} = \frac{1}{3}\left[\cos(\frac{k_x a}{2})\cos(\frac{k_y a}{2}) + \cos(\frac{k_y a}{2})\cos(\frac{k_z a}{2}) + \cos(\frac{k_z a}{2})\cos(\frac{k_x a}{2})\right]$$

$$(12.37)$$

12.2.2   *Quantum theory*

One considers now quantum spins. The spin $\vec{S}_l$ at the lattice site l can be decomposed into spin operators S_l^z and $S_l^{\pm} = S_l^x \pm iS_l^y$. The spin operators obey the following commutation relations:

$$[S_l^+, S_{l'}^-] = 2S_l^z \delta_{ll'} \quad (12.38)$$

$$[S_l^z, S_{l'}^{\pm}] = \pm S_l^{\pm} \delta_{ll'} \quad (12.39)$$

Let Θ_S^m be a spin function where S indicates the spin amplitude and m the spin component along the z axis. The quantum theory of angular momentum gives

$$S^+\Theta_S^m = \sqrt{(S-m)(S+m+1)}\Theta_S^{m+1} \quad (12.40)$$

$$S^-\Theta_S^m = \sqrt{(S+m)(S-m+1)}\Theta_S^{m-1} \quad (12.41)$$

$$S^z\Theta_S^m = m\Theta_S^m \quad (12.42)$$

In the particular case where $S = 1/2$, one has $\Theta_{1/2}^{1/2} = 1/2$, $\Theta_{1/2}^{-1/2} = -1/2$, $S^+\Theta_{1/2}^{1/2} = 0$, $S^-\Theta_{1/2}^{1/2} = -1/2$, $S^+\Theta_{1/2}^{-1/2} = 1/2$, $S^-\Theta_{1/2}^{-1/2} = 0$.

The eigenvalue of $S^+ = S^x + iS^y$ "increases" m by a unity while that of $S^- = S^x - iS^y$ decreases. As m varies from $-S$ to S, it has $(2S+1)$ values.

In general, one writes $S^z = S - n$ where n is called the excited spin-wave "number".

Now, instead of Θ_S^m, one can use the following corresponding functions in n

$$\Theta_S^m \leftrightarrow F_S(n) \tag{12.43}$$
$$\Theta_S^{m+1} \leftrightarrow F_S(n-1) \tag{12.44}$$
$$\Theta_S^{m-1} \leftrightarrow F_S(n+1) \tag{12.45}$$

A decrease of m corresponds to an increase of n and vice-versa. Equations (12.40)-(12.42) become

$$
\begin{aligned}
S^+ F_S(n) &= \sqrt{n(2S-n+1)}F_S(n-1) \\
&= \sqrt{2S}\sqrt{1-\frac{n-1}{2S}}\sqrt{n}F_S(n-1)
\end{aligned}
\tag{12.46}
$$

$$
\begin{aligned}
S^- F_S(n) &= \sqrt{(2S-n)(n+1)}F_S(n+1) \\
&= \sqrt{2S}\sqrt{n+1}\sqrt{1-\frac{n}{2S}}F_S(n+1)
\end{aligned}
\tag{12.47}
$$

$$S^z F_S(n) = (S-n)F(n) \tag{12.48}$$

The Holdstein-Primakoff method consists in introducing the operators a and a^+ as follows:

$$a F_S(n) = \sqrt{n}F_S(n-1) \tag{12.49}$$
$$a^+ F_S(n) = \sqrt{n+1}F_S(n+1) \tag{12.50}$$

These operators obey the commutation relations (see Problem 3). Using Eqs. (12.49) and (12.50), one has

$$a^+ a F_S(n) = n F_S(n) \tag{12.51}$$

This relation shows that n is an eigenvalue of $a^+ a$. Thus, $a^+ a$ is called "spin-wave number operator".

From (12.49)-(12.51) one has

$$S^z = S - a^+ a \tag{12.52}$$
$$S^+ = \sqrt{2S}f(S)a \tag{12.53}$$
$$S^- = \sqrt{2S}a^+ f(S) \tag{12.54}$$

where

$$f(S) = \sqrt{1-\frac{a^+ a}{2S}} \tag{12.55}$$

The transformations (12.52)-(12.55) are called "Holstein-Primakoff transformations".

Remark: One can apply S^z, S^+ and S^- given by (12.52)-(12.54) on the function $F_S(n)$ then make use of (12.49)-(12.51) to find again (12.46)-(12.48).

One considers now the following Heisenberg Hamiltonian

$$\mathcal{H} = -2J \sum_{(l,m)} \vec{S}_l \cdot \vec{S}_m - g\mu_B H \sum_l S_l^z \tag{12.56}$$

where interactions are limited to nearest neighbor pairs (l, m) with an exchange integral $J > 0$ (ferromagnetic). H is the amplitude of a magnetic field applied along the z direction, g and μ_B are respectively the Landé factor and Bohr magneton.

Using $S^\pm = S^x \pm iS^y$ for $\vec{S}_l$ and $\vec{S}_m$ one rewrites $\mathcal{H}$ as

$$\mathcal{H} = -2J \sum_{(l,m)} \left[S_l^z S_m^z + \frac{1}{2}(S_l^+ S_m^- + S_l^- S_m^+) \right] - g\mu_B H \sum_l S_l^z \tag{12.57}$$

Replacing $S^\pm$ and S^z by (12.52)-(12.54) while keeping position indices l and m of operators a and a^+ one obtains

$$\mathcal{H} = -2J \sum_{(l,m)} [S^2 + Sa_l^+ f_l(S) f_m(S) a_m + S f_l(S) a_l a_m^+ f_m(S) - Sa_l^+ a_l -$$

$$Sa_m^+ a_m + a_l^+ a_l a_m^+ a_m] - g\mu_B H \sum_l (S - a_l^+ a_l) \tag{12.58}$$

It is impossible to find a solution of this equation because of nonlinear terms such as $a_l^+ a_l a_m^+ a_m$, $f_l(S)$ and $f_m(S)$. In a first approximation , one can assume that the number of excited spin-waves n is small with respect to $2S$ (namely $a^+ a << 2S$) so that one can expand $f_l(S)$ and $f_m(S)$ as follows:

$$f_l(S) \simeq 1 - \frac{a_l^+ a_l}{4S} + \dots \tag{12.59}$$

$$f_m(S) \simeq 1 - \frac{a_m^+ a_m}{4S} + \dots \tag{12.60}$$

Equation (12.58) becomes, to the quadratic order in a and a^+,

$$\mathcal{H} \simeq -ZJNS^2 - g\mu_B HNS + 4JS \sum_{(l,m)} (a_l^+ a_l - a_l^+ a_m) + g\mu_B H \sum_l a_l^+ a_l \tag{12.61}$$

where one has used the following relation

$$\sum_{(l,m)} 1 = \frac{1}{2} \sum_{l} \sum_{\vec{R}} 1 = \frac{Z}{2} \sum_{l} 1 = \frac{Z}{2} N \qquad (12.62)$$

$\vec{R}$ being the vector connecting the spin at l to one of its nearest neighbors, Z the coordination number and N the total number of spins.

The first term of (12.61) is the energy of the ground state where all spins are parallel and the second is a constant which will be omitted in the following. One introduces next the following Fourier transformations

$$a_l^+ = \frac{1}{\sqrt{N}} \sum_{\vec{k}} e^{-i\vec{k}\cdot\vec{l}} a_{\vec{k}}^+ \qquad (12.63)$$

$$a_l = \frac{1}{\sqrt{N}} \sum_{\vec{k}} e^{i\vec{k}\cdot\vec{l}} a_{\vec{k}} \qquad (12.64)$$

One can show that $a_{\vec{k}}$ and $a_{\vec{k}}^+$ obey the boson commutation relations just as real-space operators a_l and a_l^+ (see Problem 3 below). Putting (12.64) in (12.61) one finds

$$\mathcal{H} = \sum_{\vec{k}} \left[2ZJS(1-\gamma_{\vec{k}}) + g\mu_B H \right] a_{\vec{k}}^+ a_{\vec{k}} = \sum_{\vec{k}} \epsilon_{\vec{k}} a_{\vec{k}}^+ a_{\vec{k}} \qquad (12.65)$$

where

$$\epsilon_{\vec{k}} = 2ZJS(1-\gamma_{\vec{k}}) + g\mu_B H \qquad (12.66)$$

In the case where $H = 0$ one finds again $\epsilon_{\vec{k}} = \hbar\omega_{\vec{k}}$, the magnon dispersion (12.29) obtained above by the classical treatment. The Holstein-Primakoff method allows however to go further by taking into account terms of order higher than quadratic in a^+ and a. By using expansions (12.59)-(12.60) one obtains terms of four operators, six operators, ... which represent interactions between spin waves. These terms play an important role when the temperature increases.

12.2.3 *Properties at low temperatures*

One studies here some low-temperature properties of spin waves using the dispersion relation (12.66).

12.2.3.1   *Magnetization*

One has seen above that a and a^+ are boson operators. The number of spin waves (or magnons) of $\vec{k}$ mode at temperature T is therefore given by the Bose-Einstein distribution

$$< n_{\vec{k}} >=< a_{\vec{k}}^+ a_{\vec{k}} >= \frac{1}{\exp[\beta \epsilon_{\vec{k}}] - 1} \qquad (12.67)$$

The magnetization is defined by

$$M = g\mu_B \sum_j < S_j^z >= g\mu_B \sum_{j=1}^N (S- < a_j^+ a_j >) \qquad (12.68)$$

where the sum is performed over all spins. One notes that the magnetization is defined as the magnetic moment per unit of volume. In the above expression, one takes thus the volume $\Omega = 1$.

With (12.64), Eq. (12.68) becomes

$$M = g\mu_B \sum_j < S_j^z >= g\mu_B N(S - \frac{1}{N} \sum_{\vec{k}} < n_{\vec{k}} >) \qquad (12.69)$$

where $< n_{\vec{k}} >$ is given by (12.67).

One shows here how to calculate M in the case of a simple cubic lattice and $H = 0$. The sum in (12.69) reads

$$\frac{1}{N} \sum_{\vec{k}} < n_{\vec{k}} >= \frac{\Omega}{(2\pi)^3} \frac{1}{N} \int \int \int \frac{dk_x dk_y dk_z}{\exp[\beta 2ZJS(1-\gamma_{\vec{k}})] - 1} \qquad (12.70)$$

Using $\gamma_{\vec{k}}$ of (12.35), one has

$$1 - \gamma_{\vec{k}} = 1 - \frac{1}{3}[\cos(k_x a) + \cos(k_y a) + \cos(k_z a)]$$

$$\simeq \frac{1}{6}[(k_x a)^2 + (k_y a)^2 + (k_z a)^2 - O(k^4)] \qquad (12.71)$$

where one used an expansion for small $\vec{k}$ because at low temperatures (large β) the main contribution to the integral (12.70) comes from the region of small $\vec{k}$. With $\Omega = 1$, $Z = 6$ (simple cubic lattice) and (12.71), Eq. (12.70) becomes

$$\frac{1}{N} \sum_{\vec{k}} < n_{\vec{k}} > \simeq \frac{1}{(2\pi)^3} \int_0^\infty \frac{4\pi k^2 dk}{e^{\beta 2JS(ka)^2} - 1}$$

$$= \frac{1}{2\pi^2} \int_0^\infty \sum_{l=1}^\infty \exp[-l\beta 2JS(ka)^2] k^2 dk \qquad (12.72)$$

where the upper limit of the integral which is the border of the first Brillouin zone has been replaced by ∞ as justified by the fact that important contributions are due to small k. Putting $x = l\beta 2JS(ka)^2$, one obtains

$$\frac{1}{N}\sum_{\vec{k}} < n_{\vec{k}} > \simeq \frac{1}{(2\pi)^2}\left(\frac{k_B T}{2JS}\right)^{3/2}\sum_{l=1}^{\infty}\frac{1}{l^{3/2}}\int_0^{\infty}e^{-x}x^{1/2}dx \quad (12.73)$$

One notes that the integral of the right-hand side is equal to $\frac{\sqrt{\pi}}{2}$ and that the sum on l is the Riemann's series $\zeta(3/2)$. Finally, one arrives at

$$\frac{M}{g\mu_B N} = S - \zeta(3/2)(\frac{k_B}{8\pi JS})^{3/2}T^{3/2} \quad (12.74)$$

The magnetization decreases with increasing T by a term proportional to $T^{3/2}$. This is called the Bloch's law.

As T increases further one has to take into account higher-order terms in (12.71). In doing so, one obtains

$$\frac{M}{g\mu_B N} = S - \zeta(3/2)t^{3/2} - \frac{3\pi}{4}\zeta(5/2)t^{5/2} - \frac{33\pi^2}{32}\zeta(7/2)t^{7/2} - \dots \quad (12.75)$$

where $t = \frac{k_B T}{8\pi JS}$.

This result, exact at low temperatures, has been confirmed by experiments at least up to $T^{5/2}$.

Note: $\zeta(3/2) = 2.612$, $\zeta(5/2) = 1.341$, $\zeta(7/2) = 1.127$.

12.2.3.2 *Energy and heat capacity*

The energy of a ferromagnet is calculated in the same manner. One obtains

$$E = -ZJNS^2 + \sum_{\vec{k}}\epsilon_{\vec{k}} < n_{\vec{k}} > \quad (12.76)$$

$$E \simeq -ZJNS^2 + 12N\pi JS\zeta(5/2)t^{5/2} + \dots \quad (12.77)$$

where the first term is the ground-state energy and the second term the energy of excited magnons to the quadratic order.

The magnetic heat capacity C_V^m is thus

$$C_V^m = \frac{dE}{dT} = \frac{dE}{dt}\frac{dt}{dT} \simeq \frac{15}{4}Nk_B\zeta(5/2)t^{3/2} + \dots \quad (12.78)$$

One notes that at low temperatures the power of T in C_V^m is different from the heat capacity of an electron gas where C_V^e is proportional to T. It is also different from that of phonons where $C_V^p \simeq T^3$. This dependence of C_V^m on T has been experimentally confirmed.

12.3   Other magnets

12.3.1   *Antiferromagnets*

Antiferromagnets are a class of magnetic materials in which neighboring spins are antiparallel. The magnon theory can be applied to antiferromagnets as shown in Problem 6 below. Two remarkable differences between ferromagnets and antiferromagnets are found at low temperatures:

(i) the magnon spectrum in ferromagnets at small k is proportional to k^2 [see (12.33)] while that in antiferromagnets is proportional to k as seen in Problem 6. This difference yields different thermodynamic behaviors in ferromagnets and antiferromagnets at low temperatures.

(ii) in the ground state, antiferromagnets do not have a full spin length as in ferromagnets. Spins in antiferromagnets suffer the so-called "zero-point spin contraction" at $T = 0$ due to quantum fluctuations, in analogy with the zero-temperature energy in phonons. That is the reason why the perfectly antiparallel spin configuration, namely the Néel state, is not the true ground state of antiferromagnets. The zero-point spin contraction in antiferromagnets is calculated in Problem 7.

12.3.2   *Ferrimagnets*

Ferrimagnets are antiferromagnetic materials in which the spin amplitudes of the two sublattices are different. The mean-field theory as well as the magnon theory can be used for ferrimagnets in the similar manner as in the case of antiferromagnets. A simple model is given in Problem 9 of chapter 13.

12.3.3   *Helimagnets*

Helimagnets are a family of materials in which the spins are not collinear in the low-T ordered phase, unlike systems considered above. Due to a competition between various kinds of interaction, the neighboring spins make an angle different from 0 and π. Helimagnets are thus frustrated systems which present many spectacular properties [48, 49].

One considers here a simplest example of helimagnet which is a chain with a ferromagnetic interaction $J_1(> 0)$ between nearest neighbors and an antiferromagnetic interaction $J_2(< 0)$ between next nearest neighbors. When $\varepsilon = |J_2|/J_1$ is larger than a critical value ε_c, the spin configuration of the ground state becomes non collinear. One shows that the helical configuration displayed in Fig. 12.5 is obtained by minimizing the following interaction energy between a spin with its neighbors:

$$E_i = -J_1 \mathbf{S}_i \cdot \mathbf{S}_{i+1} + |J_2| \mathbf{S}_i \cdot \mathbf{S}_{i+2}$$
$$= S^2 \left[-J_1 \cos\theta + |J_2| \cos(2\theta) \right] \tag{12.79}$$
$$\frac{\partial E_i}{\partial \theta} = S^2 \left[J_1 \sin\theta - 2|J_2| \sin(2\theta) \right]$$
$$= S^2 \left[J_1 \sin\theta - 4|J_2| \sin\theta \cos\theta \right] = 0, \tag{12.80}$$

where one has supposed that the angle between nearest neighbors is θ.

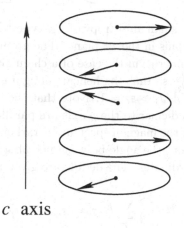

c **axis**

Fig. 12.5   Helical configuration when $\varepsilon = |J_2|/J_1 > \varepsilon_c = 1/4$ ($J_1 > 0$, $J_2 < 0$).

The solutions are

- Ferromagnetic and antiferromagnetic configurations:

$$\sin\theta = 0 \longrightarrow \theta = 0, \pi$$

- Helical configuration:

$$\cos\theta = \frac{J_1}{4|J_2|} \longrightarrow \theta = \pm \arccos\left(\frac{J_1}{4|J_2|} \right). \tag{12.81}$$

The last solution is possible if $-1 \leq \cos\theta \leq 1$, i.e. $J_1/|J_2| \leq 4$ or $|J_2|/J_1 \geq 1/4 \equiv \varepsilon_c$. There are two degenerate configurations corresponding to clockwise and counter-clockwise turning angles. Replacing the above solutions into Eq. (12.79), one sees that the antiferromagnetic solution ($\theta = \pi$) corresponds to the maximum of E. It is to be discarded. The ferromagnetic solution has an energy lower than that of the helical solution for $|J_2|/J_1 < 1/4$. It is therefore more stable in this range of parameters. For $|J_2|/J_1 > 1/4$, the reverse is true.

Let us consider a three-dimensional version of the helical chain considered above which has a body-centered cubic lattice with Heisenberg spins interacting with each other via i) a ferromagnetic interaction $J_1 > 0$ between nearest neighbors, ii) an antiferromagnetic interaction $J_2 < 0$ between next nearest neighbors along the y axis. The Hamiltonian is given by

$$\mathcal{H} = -J_1 \sum_{<i,j>} \vec{S}_i \cdot \vec{S}_j - J_2 \sum_{<i,l>} \vec{S}_i \cdot \vec{S}_l + D \sum_i (S_i^y)^2 \qquad (12.82)$$

where $D > 0$ is a very small anisotropy of the type "easy-plane anisotropy" which stabilizes the spins in the xz plane. The ground state can be calculated in the same manner as in the case of a chain given above. The result shows that $\cos\theta = \frac{J_1}{|J_2|}$ so that the helical configuration in the y direction is stable when $\varepsilon = |J_2|/J_1 > \varepsilon_c = 1$. Note that the spins belonging to the same xz plane perpendicular to the y axis are parallel.

The calculation of the magnon spectrum is rather complicated since one has to take into account the angle between neighboring spins. This can be done in the local coordinates. The details are given in Problem 8 with the result

$$\mathcal{H} = \frac{S}{2} \sum_{\vec{k}} \hbar\omega_{\vec{k}} [\alpha_{\vec{k}}^+ \alpha_{\vec{k}} + \alpha_{\vec{k}} \alpha_{\vec{k}}^+] \qquad (12.83)$$

where the energy of the magnon of mode $\vec{k}$ is

$$\hbar\omega_{\vec{k}} = \sqrt{A(\vec{k}, \vec{Q})^2 - B(\vec{k}, \vec{Q})^2} \qquad (12.84)$$

with $A(\vec{k}, \vec{Q})$ and $B(\vec{k}, \vec{Q})$ given by (17.147) and (17.148). Figure 12.6 shows the magnon spectrum for J_2/J_1 corresponding to $Q = \pi/3$. One observes that the magnon frequency is zero not only at $k = 0$ but also at $k_z = Q$.

12.3.4   *Frustrated magnets*

Frustrated spin systems are systems in which neighboring spins cannot find orientations to fully satisfy their interaction bonds. As a consequence, the spin configuration is often non collinear. The helimagnetic case given above belongs to the family of frustrated spin systems. One of the most studied systems in the last three decades is no doubt the phase transition in the triangular antiferromagnets with XY and Heisenberg spins (see reviews in Ref.

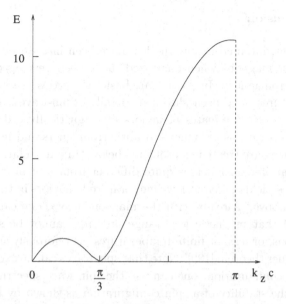

Fig. 12.6   Magnon spectrum $E = \hbar\omega_{\vec{k}}$, Eq. (17.153), versus $k_z c$ in a helimagnet defined by the Hamiltonian (12.82), with $Q = \pi/3$, $k_x = k_y = 0$.

[49]). In these models, the spin configuration is the so-called "120-degree configuration" shown as shown in Fig. 12.7 and in Problem 9. The Villain's model is also shown in this figure and in Problem 10. More discussion on frustrated spin systems is given in section 13.7.3.

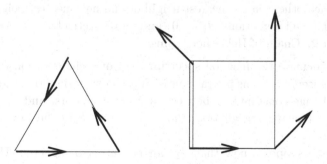

Fig. 12.7   Examples of frustrated spin systems. Left: antiferromagnetic triangular lattice with vector spins (XY or Heisenberg spins), Right: Villain's XY model, the single and double bonds are ferromagnetic and antiferromagnetic, respectively.

12.4 Conclusion

In this chapter, interacting spin models have been introduced. Spin-wave excitations, or magnons when quantized, have been developed to a large extent for ferromagnets. In particular, basic properties of magnons at finite temperatures have been shown in details. Spin-waves are the main phenomenon which dominates behaviors of magnetically ordered materials at low temperatures. Magnons in antiferromagnets and in some other kinds of magnets are treated as problems below. In particular, the magnon spectrum of antiferromagnets is quite different from that of ferromagnets. This difference yields different low-temperature behaviors in these magnets as discussed above. Already with the magnon theory, one can see (Problem 4 below) that magnetic long-range ordering cannot be stabilized in two dimensions, or less, at finite temperatures, as rigorously stated by the Mermin-Wagner theorem [129]. Note that spin waves can be excited only on a stable spin configuration: one can use the spin-wave spectrum to detect the limit of the stability of a spin configuration as shown by the example given in Problem 2 below. Understanding the role of the spin-wave excitations is the first step toward the comprehension of behaviors of magnetic materials which are intensively used in applications.

12.5 Problems

Problem 1. Ground-state spin configuration of a chain of Ising spins:
Determine the ground state a chain of Ising spins interacting with each other via the nearest-neighbor and next-nearest neighbor exchange interactions, J_1 (> 0) and J_2 respectively.

Problem 2. Chain of Heisenberg spins:

a) Calculate the magnon spectrum $\epsilon(k)$ of a chain of constant a, of Heisenberg spins interacting with each other via ferromagnetic exchange constants J_1 between nearest neighbors, and J_2 between next nearest neighbors. Plot $\epsilon(k)$ versus k in the first Brillouin zone.

b) One supposes now that J_2 can be antiferromagnetic. Using the spectrum $\epsilon(k)$, show that the ferromagnetic ordering becomes unstable when $|J_2|$ is larger than a critical value.

Problem 3. Commutation relations:
Show that the operators a^+ and a defined in the Holstein-Primakoff

approximation, Eqs. (12.49) and (12.50), respect rigorously the commutation relations between the spin operators.

Problem 4. Heisenberg model in two dimensions:

Consider the Heisenberg spin model on a two-dimensional lattice with a ferromagnetic interaction J between nearest neighbors.

a) Calculate the magnon spectrum $\epsilon(\vec{k})$. Verify that $\epsilon(\vec{k}) \propto k^2$ when $\vec{k} \to 0$.

b) Write down the formal expression of the magnetization M at temperature T. Show that M is not defined as soon as T is not zero. Comment.

Problem 5. Magnon-phonon interaction:

Consider the following Hamiltonian describing the interaction between magnon and phonon:

$$\mathcal{H} = \sum_{\vec{k}} \left[\omega_{\vec{k}}^m a_{\vec{k}}^+ a_{\vec{k}} + \omega_{\vec{k}}^p b_{\vec{k}}^+ b_{\vec{k}} + V_{\vec{k}} (a_{\vec{k}} b_{\vec{k}}^+ + a_{\vec{k}}^+ b_{\vec{k}}) \right] \qquad (12.85)$$

where $V_{\vec{k}}$ is the coupling constant, $\omega_{\vec{k}}^m$ and $\omega_{\vec{k}}^p$ are eigenfrequencies of magnon and phonon, respectively, a and a^+ denote annihilation and creation operators of magnon, while b and b^+ denote those of phonon.

a) Using the following transformation

$$a_{\vec{k}} = \cos \theta_{\vec{k}} c_{\vec{k}} + \sin \theta_{\vec{k}} d_{\vec{k}}$$
$$b_{\vec{k}} = \cos \theta_{\vec{k}} d_{\vec{k}} - \sin \theta_{\vec{k}} c_{\vec{k}}$$

where $\theta_{\vec{k}}$ is real, show that operators c, c^+ and d, d^+ obey the commutation relations.

b) Show that $\mathcal{H}$ is diagonal in $c_{\vec{k}}^+ c_{\vec{k}}$ and $d_{\vec{k}}^+ d_{\vec{k}}$ if

$$\tan 2\theta_{\vec{k}} = 2V_{\vec{k}} / (\omega_{\vec{k}}^p - \omega_{\vec{k}}^m) \qquad (12.86)$$

Show that when $\omega_{\vec{k}}^p = \omega_{\vec{k}}^m = \omega$ ("cross-over"), $\mathcal{H}$ is given by

$$\mathcal{H} = \sum_{\vec{k}} \left[(\omega - V_{\vec{k}}) c_{\vec{k}}^+ c_{\vec{k}} + (\omega + V_{\vec{k}}) d_{\vec{k}}^+ d_{\vec{k}} \right] \qquad (12.87)$$

Comment.

Problem 6. Magnons in antiferromagnets:

Consider the following Heisenberg Hamiltonian

$$
\mathcal{H} = J \sum_{(l,m)} \vec{S}_l \cdot \vec{S}_m - g\mu_B H \left(\sum_l S_l^z + \sum_m S_m^z \right)
$$

$$
= J \sum_{(l,m)} \left[S_l^z S_m^z + \frac{1}{2} \left(S_l^+ S_m^- + S_l^- S_m^+ \right) \right]
$$

$$
- g\mu_B H \left(\sum_l S_l^z + \sum_m S_m^z \right) \tag{12.88}
$$

where interactions are limited to pairs of nearest neighbors (l, m) with an exchange integral $J > 0$ (antiferromagnetic). H is the amplitude of a magnetic field applied in the z direction. l and m indicate the sites belonging respectively to ↑ and ↓ sublattices.

One uses the following Holstein-Primakoff method in the same manner as in the case of ferromagnets shown in (12.52)-(12.55) but with a distinction of up and down sublattices. For the up sublattice, one defines

$$
S_l^z = S - a_l^+ a_l \tag{12.89}
$$

$$
S_l^+ = \sqrt{2S} f_l(S) a_l \tag{12.90}
$$

$$
S_l^- = \sqrt{2S} a_l^+ f_l(S) \tag{12.91}
$$

where

$$
f_l(S) = \sqrt{1 - \frac{a_l^+ a_l}{2S}} \tag{12.92}
$$

For the down sublattice, one defines

$$
S_m^z = - S + b_m^+ b_m \tag{12.93}
$$

$$
S_m^+ = \sqrt{2S} b_m^+ f_m(S) \tag{12.94}
$$

$$
S_m^- = \sqrt{2S} f_m(S) b_m \tag{12.95}
$$

where

$$
f_l(S) = \sqrt{1 - \frac{b_m^+ b_m}{2S}} \tag{12.96}
$$

The operators a, a^+, b and b^+ obey the commutation relations (Problem 3 above).

Calculate the dispersion relation in antiferromagnets.

Problem 7. Properties at low temperatures of antiferromagnets:
Calculate the energy and the sublattice magnetization at low temperatures. Calculate the spin length at zero temperature.

Problem 8. Magnons in helimagnets:
Consider the helimagnet shown in Fig. 12.8. Let $(\vec{\xi}_i, \vec{\eta}_i, \vec{\zeta}_i)$ be the unit vectors making a direct trihedron at the site i, namely $\vec{\eta}_i$ is parallel to the y axis. One supposes in addition that the quantization axis of the spin $\vec{S}_i$ coincides with the local axis $\vec{\zeta}_i$. Calculate at the lowest order the magnon spectrum in a body-centered cubic lattice.

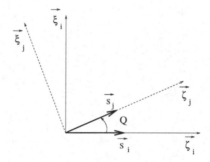

Fig. 12.8 Local coordinates defined for two spins $\vec{S}_i$ and $\vec{S}_j$. The axis $\vec{\eta}$ is common for the two spins.

Problem 9. Triangular antiferromagnet:
By minimizing the interaction energy, determine the ground-state spin configuration of a triangular lattice with XY spins interacting with each other via an antiferromagnetic exchange J between nearest neighbors. What can one say in the Heisenberg case?

Problem 10. Villain's model:
Consider the 2D Villain's model with XY spins defined on a square lattice as shown in Fig. 12.7: all horizontal bonds are ferromagnetic while one vertical line out of every two is antiferromagnetic (denoted by the double bond in that figure). Write the energy of the elementary plaquette. By minimizing this energy, determine the ground state as a function of the antiferromagnetic interaction $J_{AF} = -\eta J_F$ where η is a positive coefficient. Determine the angle between two neighboring spins as a function of η. Show that the critical value of η beyond which the spin configuration is not collinear is $\eta_c = 1/3$.

Chapter 13

Systems of Interacting Spins: Phase Transitions

13.1 Introduction

Phase transitions and critical phenomena constitute one of the most important domains of statistical physics. This domain includes not only activities in various areas of physics but also subjects in other sciences such as biology and chemistry.

To illustrate the theory of phase transition, interacting spin systems are used in this chapter. Many systems of different natures such as alloys, molecular and liquid crystals can be mapped into spin systems and solved using the spin language. In this chapter, one introduces methods to study phase transitions occurring in magnetic materials. There are many approximations with various degrees of precision. Among these approximations, the mean-field approximation is by far the simplest one which can give essential properties of the system under investigation, provided some precautions taken on its validity. One will see that the mean-field theory cannot determine with precision the nature of a phase transition. This nature is characterized by the so-called critical exponents (section 13.2.5) which are intimately related to microscopic interactions between particles and the system symmetry, therefore the more one knows them the better one understands the interactions inside a material which govern its properties. The renormalization group was introduced in the early 70's to provide a more accurate insight on the phase transition. The study of phase transitions has been spectacularly developed since then [8, 32, 178, 189] together with the use of Monte Carlo simulations. One has now a good understanding of second-order phase transitions in spite of the fact that there is still much to be done for more complex systems. The concept of the renormalization group has been formulated by K. G. Wilson in the early

70's [178, 179]. It has been applied with success in many fields of physics ranging from condensed matter to quantum field theory. Among the most remarkable results, the notion of universality is very interesting: transitions in different systems can belong to a same universality class. Other well-known successes in condensed matter include the Kondo problem [180] and the Kosterlitz-Thouless transition [102].

One first shows how the mean-field theory, sometimes called molecular-field approximation, is implemented. One will next briefly present the Landau-Ginzburg theory and the concept of the renormalization group. Other complementary methods are also introduced by using some simple systems for illustration.

13.2   Generalities

One presents hereafter some fundamental notions necessary to understand a phase transition. There exists a huge number of reviews and books specialized in this field [8, 32, 53, 189].

All systems do not have obligatorily a phase transition. In general, the existence of a phase transition depends on a few general parameters such as the space dimension, the nature of the interaction between particles and the system symmetry. For spin systems, one can give a brief summary here. In one dimension, in general there is no phase transition at a non zero temperature for systems of short-range interactions regardless of the spin model. The long-range ordering at $T = 0$ is destroyed as soon as $T \neq 0$. However, in two dimensions discrete spin models such as the Ising and Potts models have a phase transition at a finite temperature T_c, while continuous spin models such as the Heisenberg model do not have a transition at a finite temperature [129]. The XY spin model is a very particular case: in spite of the absence of a long-range order at finite temperatures, there is a phase transition of a special kind called the "Kosterlitz-Thouless" transition [102] (see Appendix E). In three dimensions, all known spin models have in general a phase transition at $T_c \neq 0$.

13.2.1   *Order parameter*

A transition from one phase to another may take place when an external parameter varies. Such a parameter can be the temperature or an external applied magnetic field. At the transition point, the system changes from one symmetry to another one. The most studied type of transition is no doubt the order-disorder transition. A popular transition of this kind is the loss of magnetic attraction of a permanent magnet with increasing temperature: this is a transition from a magnetically ordered phase to a magnetically disordered (or paramagnetic) phase.

In order to measure the degree of ordering, one defines an order parameter depending on the system symmetry. A good order parameter should be non zero in one phase and zero in the other phase, signaling thus the symmetry breaking when the system changes its phase. In some systems, the choice of the order parameter is natural. For example in a ferromagnet, the magnetization is the natural order parameter, namely the thermal average of the spin component on its quantization axis.

For a ferromagnetic system of N Ising spins, the order parameter is defined as

$$P = \frac{1}{N} \left| \sum_i \sigma_i \right| \tag{13.1}$$

where $\sigma_i = \pm 1$ is the spin at the site i. One sees that in the ground state where all spins are parallel, one has $P = 1$, and in the disordered state (paramagnetic state) where there is a random mixing of up and down spins at equal numbers $P = 0$.

For an antiferromagnetic system, the order parameter is the so-called staggered magnetization. For example, in one dimension with the Ising model, the staggered magnetization is given by

$$P = \frac{1}{N} \left| \sum_i (-1)^i \sigma_i \right| \tag{13.2}$$

where $(-1)^i$ is the parity of the site i. One can verify that in the antiferromagnetic ground state $P = 1$, and in the disordered phase $P = 0$.

For the ferromagnetic Potts model with q states: $\sigma_j = 1, ..., q$, the order parameter is defined as

$$P = \frac{q M_{\max} - 1}{q - 1} \tag{13.3}$$

where

$$M_{\max} = \frac{\max(M_1, M_2, ..., M_q)}{N} \tag{13.4}$$

with

$$M_i = \sum_j \delta_{\sigma_j,i} \quad (i = 1, 2, ..., q) \tag{13.5}$$

where the sum on j is performed over all sites of the system. One sees that in the ground state where there is only one kind of σ_j one has $P = 1$. In the disordered phase where all values of spin are equally present, namely $M_i = N/q$ for any $i = 1, ..., q$, one has $P = 0$.

For Heisenberg spins of amplitude 1 in a ferromagnet, the order parameter is the magnetization defined by

$$M = \frac{1}{N} \left| \sum_i \vec{S}_i \right| \tag{13.6}$$

One can verify that in the ground state, where all spins are parallel, M is 1 whatever the orientation of spins with respect to the crystal axes. In the disordered state, each spin has a random spatial orientation so that $\sum_i \vec{S}_i = 0$.

13.2.2   *Order of the phase transition*

A phase transition takes place when physical quantities of the system undergo an anomaly. In order to define properly a phase transition, one should examine various physical quantities at the transition point T_c. If physical quantities such as the specific heat C_V and the susceptibility χ which are second derivatives of the free energy F diverge (see chapter 3), one calls the corresponding phase transition a "phase transition of second order". In this case, the first derivatives of F such as the average energy $\overline{E}$ and the average magnetization $\overline{M}$ are continuous functions at T_c. On the other hand, in a first-order phase transition, these first-derivative quantities undergo a discontinuity at the transition point.

13.2.3   *Correlation function — Correlation length*

An important function in the study of phase transitions is the correlation function defined by

$$G(\vec{r}) = < \vec{S}(0) \cdot \vec{S}(\vec{r}) > \tag{13.7}$$

where $\vec{S}(0)$ is the spin at a site chosen as the origin and $\vec{S}(\vec{r})$ the spin at $\vec{r}$, $< ... >$ denotes the thermal average. In an isotropic system, $G(\vec{r})$ depends on r. In a phase where $\vec{S}(0)$ and $\vec{S}(\vec{r})$ are independent, namely their fluctuations are not correlated, $G(\vec{r})$ is zero. This is the case of the paramagnetic phase far from the transition temperature T_c at a large distance r.

When the temperature T_c is approached from the high temperature side, a correlation resulting from the interaction between spins sets in, $G(r)$ becomes non zero for spins at short distances. One can define the "correlation length" ξ as the distance beyond which $G(r)$ is no more significant. The fluctuations of two spins at a distance $r < \xi$ are said "correlated". The correlation length ξ is written as

$$G(r) = < \vec{S}(0) \cdot \vec{S}(r) > = A \frac{\exp(-r/\xi)}{r^{(d-1)/2}} \tag{13.8}$$

where d is the space dimension and A a constant. In a second-order transition, the correlation length diverges at the transition, namely all spins are correlated at the transition regardless of their distance. On the contrary, at a first-order transition, the correlation length is finite and there is a coexistence of the two phases at the transition point.

13.2.4 *Critical exponents*

When the transition is of second order, one can define in the vicinity of T_c the following critical exponents

$$C_V = A \left| \frac{T - T_c}{T_c} \right|^{-\alpha} \tag{13.9}$$

$$\overline{M} = B \left[\frac{T_c - T}{T_c} \right]^{\beta} \tag{13.10}$$

$$\chi = C \left| \frac{T - T_c}{T_c} \right|^{-\gamma} \tag{13.11}$$

$$\xi \propto \left[\frac{T - T_c}{T_c} \right]^{-\nu} \tag{13.12}$$

$$\overline{M} = H^{1/\delta} \tag{13.13}$$

Note that the same α is defined for $T > T_c$ and $T < T_c$ but the coefficient A is different for each side of T_c. This is also the case for γ. However, β is defined only for $T < T_c$ because $\overline{M} = 0$ for $T \geq T_c$. The definition of δ is valid only at $T = T_c$ when the system is under an applied magnetic field of amplitude H. Finally, at $T = T_c$ one defines exponent η of the correlation function by

$$G(r) \propto \frac{1}{r^{d-2+\eta}} \tag{13.14}$$

There is another exponent called "dynamic exponent" z defined via the relaxation time τ of the spin system for $T \geq T_c$:

$$\tau \propto \xi^z \propto \left[\frac{1}{T - T_c}\right]^{z\nu} \tag{13.15}$$

Only the six exponents α, β, γ, δ, ν and η are critical exponents. One sees below that there are four relations between them (see section 13.5). Therefore, there are only two of them which are to be determined in an independent manner.

It is obvious that one cannot define such exponents in a first-order transition because there is no divergence of physical quantities. For this reason, one says that first-order transitions are not critical. To be precise, one shall call "critical temperature" for second-order transitions and "transition temperature" for first-order transitions.

13.2.5   *Universality class*

Phase transitions of second order are distinguished by their "universality class". Phase transitions having the same values of critical exponents belong to the same university class. Renormalization group analysis shows that the universality class depends only on a few very general parameters such as the space dimension, the symmetry of the order parameter and the nature of the interaction. So, for example, Ising spin systems with short-range ferromagnetic interaction in two dimensions belong to the same universality class whatever the lattice structure is. Of course, the critical temperature T_c is not the same for square, hexagonal, rectangular, honeycomb, ...lattices, but T_c is not a universal quantity. It depends on the interaction value, the coordination number, ... but these quantities do not affect the values of the critical exponents.

The following table shows the critical exponents of some known universality classes.

Class	Symmetry	α	β	γ	ν	η
2d Ising	Z_2	0	1/8	7/4	1	1/4
2d Potts (q=3)	Z_3	1/3	1/9	13/9	5/6	4/15
2d Potts (q=4)	Z_4	2/3	1/12	7/6	2/3	1/4
3d Ising	Z_2	0.11	0.325	1.241	0.63	0.031
3d XY	$O(2)$	-0.007	0.345	1.316	0.669	0.033
3d Heisenberg	$O(3)$	0.115	0.3645	1.386	0.705	0.033
Mean-field		0	1/2	1	1/2	0

Note that when a system is invariant by the following local transformation ($J \to -J$, $\vec{S}_i \to -\vec{S}_i$), the universality class of the new system does not change. This is understood immediately if one looks at the partition function: such a local transformation does not change the argument of the exponential of the partition function. By consequence, physical properties do not change. However, the local transformation when operated on one spin out of every two in a square lattice for example (see Fig. 13.1), does change a ferromagnetic crystal into an antiferromagnetic one. One concludes that a ferromagnetic crystal and its antiferromagnetic counterpart have the same critical temperature and the same critical exponents.

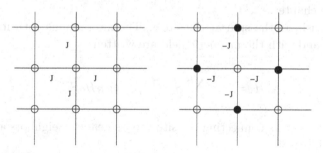

Fig. 13.1 Local transformation $J \to -J$, $\vec{S}_i \to -\vec{S}_i$ operated on one spin out of every two changes the square ferromagnetic lattice (left) into an antiferromagnetic lattice (right). White circles: ↑ spins, black circles: ↓ spins.

One notes that there exist many systems in which it is impossible to operate local transformation without changing the argument of the partition function. One of these systems is the triangular lattice: it is impossible to find a spin configuration to satisfy all interactions if one changes J into $-J$ everywhere. The energy of the system is not conserved, so the partition function changes. Such systems are called "frustrated systems" which are shown in section 13.7.3.

13.3   Ferromagnetism in mean-field theory

In this section, one presents the mean-field theory by using the Heisenberg model for ferromagnets. The extension of the method to other spin models and other magnets such as antiferromagnets and ferrimagnets can be done without difficulty.

One considers the following Hamiltonian

$$\mathcal{H} = -2 \sum_{(i,j)} J_{ij} \vec{S}_i \cdot \vec{S}_j - g\mu_B \sum_i \vec{H}_0 \cdot \vec{S}_i \qquad (13.16)$$

where $\vec{H}_0$ is a magnetic field applied in the z direction, g the Landé factor and μ_B the Bohr magneton. The first sum is performed over spin pairs $(\vec{S}_i, \vec{S}_j)$ occupying lattice sites i and j. For simplicity, one supposes in the following only interactions between nearest neighbors are not zero. Note that this hypothesis is not a hypothesis of the mean-field theory. It is used here for presentation commodity. The mean-field theory can be applied to systems with far-neighbor interactions as seen in Problem 2 given at the end of the chapter.

One considers the spin at the site i. The interactions with its nearest neighbors and with the magnetic field are written as

$$\mathcal{H}_i = -2J \sum_{\vec{\rho}} \vec{S}_i \cdot \vec{S}_{i+\vec{\rho}} - g\mu_B H_0 S_i^z \qquad (13.17)$$

where $\vec{\rho}$ are vectors connecting the site i to its nearest neighbors and J the exchange integral.

13.3.1   *Mean-field equation*

The first assumption of the mean-field theory is to replace all neighboring spins by an average value $< \vec{S}_{i+\vec{\rho}} >$ which is the same for all nearest neighbors. This value is to be computed in the following. The z axis is chosen as the spin quantization axis. The average values of the x and y spin components are then zero since the spin precesses circularly around the z axis:

$$< S_{i+\vec{\rho}}^x >=< S_{i+\vec{\rho}}^y >= 0.$$

For the z component, one has

$$< S_{i+\vec{\rho}}^z >=< S^z > + < \Delta S^z >$$

where $< S^z >$ is the average value in the absence of the magnetic field, and $< \Delta S^z >$ is the variation induced by the latter. One has

$$< \Delta S^z > \propto H_0$$

The spontaneous magnetization of the crystal is given by

$$M = Ng\mu_B < S^z > \tag{13.18}$$

where N is the total number of spins. $< S^z >$ depends on the temperature as seen below. Equation (13.17) becomes

$$\mathcal{H}_i \cong -2CJ\left[S_i^z \left(< S^z > + < \Delta S^z >\right)\right] - g\mu_B H_0 S_i^z \tag{13.19}$$

where C is the coordination number (number of nearest neighbors). One can express $\mathcal{H}_i$ as a function of the "molecular field" $\overline{H}$ by

$$\mathcal{H}_i = -g\mu_B \overline{H} S_i^z \tag{13.20}$$

where

$$\overline{H} = \frac{2CJ\left[< S^z > + < \Delta S^z >\right]}{g\mu_B} + H_0 \tag{13.21}$$

The average value $< S^z > + < \Delta S^z >$ is calculated by the canonical description (see chapter 3) as follows

$$< S^z > + < \Delta S^z >= \frac{\sum_{S_i^z=-S}^{S} S_i^z e^{-\beta \mathcal{H}_i}}{Z_i} \tag{13.22}$$

where $\beta = \frac{1}{k_B T}$ and Z_i the partition function defined by

$$
\begin{aligned}
Z_i &= \sum_{S_i^z=-S}^{S} \exp(\beta g\mu_B \overline{H} S_i^z) \\
&= \frac{\sinh\left[\beta g\mu_B \overline{H}(S + \frac{1}{2})\right]}{\sinh\left[\frac{1}{2}\beta g\mu_B \overline{H}\right]}
\end{aligned} \tag{13.23}
$$

where $S = |\vec{S}_i|$. One obtains

$$\sum_{S_i^z=-S}^{S} S_i^z e^{-\beta \mathcal{H}_i} = \frac{\partial}{\partial \alpha} \sum_{S_i^z=-S}^{S} e^{\alpha S_i^z} \quad (\alpha \equiv \beta g \mu_B \overline{H})$$

$$= \frac{\partial}{\partial \alpha} Z_i$$

$$= \frac{(S + \frac{1}{2}) \cosh(S + \frac{1}{2})\alpha \sinh \frac{\alpha}{2} - \frac{1}{2} \sinh(S + \frac{1}{2})\alpha \cosh \frac{\alpha}{2}}{\sinh^2 \frac{\alpha}{2}}$$

from which one gets

$$< S^z > + < \Delta S^z > = S B_S(x) \tag{13.24}$$

where $B_S(x)$ is the Brillouin function defined by

$$B_S(x) = \frac{2S+1}{2S} \coth \frac{(2S+1)x}{2S} - \frac{1}{2S} \coth \frac{x}{2S} \tag{13.25}$$

with

$$x = \beta g \mu_B S \overline{H} = \beta [2CJS(< S^z > + < \Delta S^z >) + g \mu_B S H_0] \tag{13.26}$$

Equation (13.24) is called "mean-field equation". Since the argument x of $B_S(x)$ contains $< S^z > + < \Delta S^z >$, (13.24) is an implicit equation of $[< S^z > + < \Delta S^z >]$ which depends on the temperature. In the case of spin one-half, $S = \frac{1}{2}$, the Brillouin function is

$$B_{\frac{1}{2}}(x) = \tanh x.$$

In the case where $S \to \infty$, one has

$$B_\infty(x) = \coth x - \frac{1}{x} \equiv \text{Langevin function.}$$

If H_0 is very weak, $< \Delta S^z >$ is then very small. In such a case, one can expand the Brillouin function near $x_0 = \beta 2CJS < S^z >$. By identifying the second terms of the two sides of (13.24), one obtains

$$< \Delta S^z > = S B_S'(x_0) \left[\frac{1}{k_b T} (g \mu_B S H_0 + 2CJS < \Delta S^z >) \right] \tag{13.27}$$

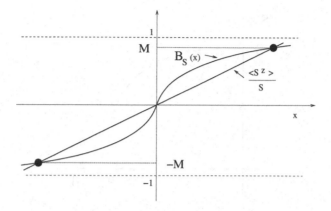

Fig. 13.2 Graphical solutions of Eq. (13.28).

where $B'_S(x_0)$ is the derivative of $B_S(x)$ with respect to x taken at x_0.

In zero applied field, (13.24) becomes

$$< S^z >= SB_S(x_0) \tag{13.28}$$

One can solve (13.28) by a graphical method: one looks for the intersection of the two curves $y_1 = \frac{<S^z>}{S} = \frac{k_B T}{2CJS^z}x$ and $y_2 = B_S(x)$ which represent the two sides of (13.28). The first curve, y_1, is a straight line with a slope proportional to the temperature. For a given value of T, there are two symmetric intersections at $\pm M$ as shown in Fig. 13.2. It is obvious that if the slope of y_1 is larger than the slope of y_2 at $x = 0$, there is no intersection other than the one at $x = 0$. The solution is then $< S^z >= 0$. The slope of y_2 at $x = 0$ thus determines the critical temperature T_c.

For each temperature, one finds thus two symmetrical solutions. The positive solutions of $< S^z >$ versus T are shown in Fig. 13.3.

13.3.2 *Critical temperature*

At high temperatures, $\beta < S^z >\ll 1$, one obtains from (13.25)

$$B_S(x) \simeq \frac{S+1}{3S}x - \frac{\left[S^2 + (S+1)^2\right](S+1)}{90S^3}x^3 + O(x^5) \tag{13.29}$$

Equation (13.28) becomes

$$< S^z > \left[\frac{2CJS(S+1)}{3k_BT} - 1\right] = \frac{S(S+1)\left[S^2 + (S+1)^2\right]}{90}\left(\frac{2CJ}{k_BT}\right)^3 <S^z>^3$$

$$+O(x^5) \tag{13.30}$$

This equation has a solution $< S^z >\neq 0$ only if

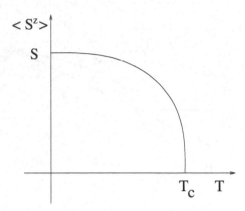

Fig. 13.3   Thermal average $< S^z >$ versus T.

$$\left[\frac{2}{3}CS(S+1) - \frac{k_B T}{J}\right] > 0$$

namely

$$T < \frac{2CJS(S+1)}{3k_B} \equiv T_c \qquad (13.31)$$

Once this condition is satisfied, $< S^z >$ is given by

$$< S^z >^2 \simeq \frac{10}{3} \frac{S^2(S+1)^2}{[S^2+(S+1)^2]} \left[\frac{T_c - T}{T_c}\right] \qquad (13.32)$$

T_c is called "critical temperature". When $T \geq T_c$, the solution of (13.30) is $< S^z >= 0$.

At low temperatures, $2CJS < S^z >$ is much larger than $k_B T$, the expansion of (13.25) gives

$$B_S(x) \simeq 1 - \frac{1}{S}e^{-x/S} + \cdots \qquad (13.33)$$

from which

$$< S^z >= SB_S(x) \simeq S - e^{-2JCS/k_B T} + \cdots \qquad (13.34)$$

If $T = 0$, one has $< S^z >= S$.

13.3.3 Specific heat

The average energy of a spin when $H_0 = 0$ is calculated by

$$\overline{E}_i = -\frac{\partial \ln Z_i}{\partial \beta} \simeq -2CJ < S^z >^2 \tag{13.35}$$

The total ferromagnetic energy of the crystal is

$$\overline{E} = \frac{1}{2} N \overline{E}_i \tag{13.36}$$

where the factor $\frac{1}{2}$ is added to remove the double counting. The specific heat is

$$C_V = \left(\frac{\partial \overline{E}}{\partial T}\right)_V = -2NCJ < S^z > \frac{\partial < S^z >}{\partial T} \tag{13.37}$$

At low temperatures, $< S^z > \cong S - e^{-2CJS/k_BT}$ [see (13.34)], one has

$$C_V(T \simeq 0) \simeq 4Nk_B \left[\frac{CJS}{k_BT}\right]^2 e^{-2CJS/k_BT} \tag{13.38}$$

When $T \to 0$, one has $C_V \simeq 0$. The third thermodynamic principle is thus satisfied.

For $T > T_c$, one has $\overline{E} = 0$, therefore $C_V = 0$. One calculates now C_V when $T \to T_c^-$. From (13.32) one has

$$\lim_{T \to T_c^-} < S^z > \frac{\partial < S^z >}{\partial T} = -\frac{5}{2} \frac{k_B}{JC} \frac{S(S+1)}{[S^2 + (S+1)^2]} \tag{13.39}$$

so that

$$C_V(T \to T_c^-) = 5Nk_B \frac{S(S+1)}{S^2 + (S+1)^2} \tag{13.40}$$

The discontinuity of C_V at T_c is thus

$$\text{For } S = \frac{1}{2} \Rightarrow \Delta C_V = \frac{3}{2} Nk_B$$

$$\text{For } S = \infty \Rightarrow \Delta C_V = \frac{5}{2} Nk_B$$

This discontinuity is a consequence of the mean-field theory resulting from the fact that critical fluctuations near T_c been neglected by replacing all spins by a uniform average. When fluctuations around the average values of spins are taken into account, C_V diverges at T_c when T_c is approached from both sides. Figure 13.4 shows C_V calculated by the mean-field theory as a function of T.

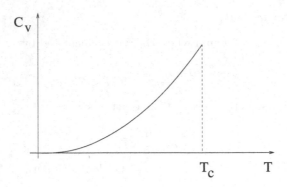

Fig. 13.4 C_V calculated by the mean-field theory versus T.

13.3.4   *Susceptibility*

By definition, the susceptibility is written as

$$\chi_\| = \left(\frac{\partial M}{\partial H_0}\right)_{H_0=0} = Ng\mu_B \left(\frac{\partial <S^z>}{\partial H_0}\right)_{H_0=0} \tag{13.41}$$

Equation (13.27) gives

$$<\Delta S^z> = \frac{SB'_S(x)\frac{g\mu_B S}{k_B T}H_0}{1 - SB'_S(x)\left(\frac{2CJS}{k_B T}\right)} \tag{13.42}$$

so that

$$\chi_\| = \frac{N(g\mu_B)^2 S^2 B'_S(x)}{k_B T - 2CJS^2 B'_S(x)} \tag{13.43}$$

where $x = \frac{2CJS<S^z>}{k_B T}$.

When $T \geq T_c$, one has $<S^z> = 0$ and $B'_x(0) = \frac{S+1}{3S}$. One gets

$$\chi_\|(T \geq T_c) = \frac{N(g\mu_B)^2 S(S+1)}{3k_B(T - T_c)} \qquad \text{(Curie-Weiss law)} \tag{13.44}$$

When $T \lesssim T_c$, one has $<S^z> \to 0$. Expanding $B'_S(x)$ with respect to $<S^z>$, one obtains

$$\chi_\|(T \lesssim T_c) = \frac{N(g\mu_B)^2 S(S+1)}{6k_B(T_c - T)} \tag{13.45}$$

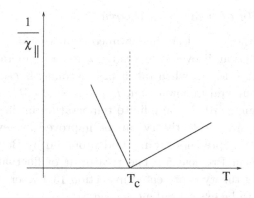

Fig. 13.5   Inverse of the susceptibility obtained by the mean-field theory versus T.

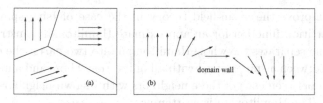

Fig. 13.6   (a) Ferromagnetic domains in an imperfect crystal (b) Example of a domain wall spin structure.

It is noted that the coefficient in this case is twice smaller than that in (13.44). When $T \to 0$, $\chi_\parallel \to 0$ because $M \to$ constant. The inverse of the susceptibility is schematically shown as a function of T in Fig. 13.5.

Note that $\chi_\parallel$ calculated above corresponds to a perfect ferromagnetic crystal where all spins are parallel to a crystalline direction. In reality, a ferromagnetic crystal can have several ferromagnetic domains with spins pointing in different directions. This is due to the presence of defects, dislocations, imperfections, ... during the formation of the crystal. The region between two magnetic domains is called "domain wall" in which the matching of two spin orientations is progressively realized. One shows schematically magnetic domains and a domain wall spin configuration in Fig. 13.6. The presence of domain walls makes it difficult to compare calculated and experimental susceptibilities.

13.3.5   *Validity of mean-field theory*

The mean-field theory neglects instantaneous fluctuations of spins. Due to this approximation, it overestimates the critical temperature T_c. Fluctuations favor disorder, so when taken into account, fluctuations cause a transition at a temperature lower than T_c given by (13.31), or even destroy magnetic long-range order at any finite temperature in dimensions $d = 1$ and $d = 2$. The mean-field theory can be improved by several methods. In the following one presents the method proposed by Bethe. A simpler version is shown in Problem 5. The treatment of fluctuations with the Landau-Ginzburg theory is presented in section 13.4 where the mean-field theory is shown to be exact for dimension $d > 4$.

13.3.6   *Improved mean-field theory: Bethe's approximation*

One can improve the mean-field theory in the case of Ising spins by using the partition function for an approximate Hamiltonian constructed as follows. One separates the whole Hamiltonian into two parts: the exact interaction between a spin, say σ_0, with all of its z neighbors and a mean-field interaction between each of these neighbors with its own neighbors outside of the cell. The Hamiltonian is written as

$$\mathcal{H} = -J\sum_{i=1}^{z} \sigma_0\sigma_i - \mu_B B\sigma_0 - \mu_B(B+H)\sum_{i=1}^{z}\sigma_i \tag{13.46}$$

where B is an applied magnetic field. The spin σ_0 and its z surrounding nearest neighbors σ_i form a cluster which is embedded in the crystal. H is the molecular field acting on σ_i by its $(z-1)$ neighbors outside of the cluster. H is thus given by $H = (z-1)J < \sigma > /\mu_B$. The mean-field equation will be obtained at the end by setting $< \sigma_0 >=< \sigma_i >\equiv< \sigma >$. One has

$$Z = \sum_{\sigma_0=\pm1}\sum_{\sigma_1=\pm1}\cdots\sum_{\sigma_z=\pm1} e^{\beta[J\sum_{i=1}^{z}\sigma_0\sigma_i+\mu_B B\sigma_0+\mu_B(B+H)\sum_{i=1}^{z}\sigma_i]}$$

$$= \sum_{\sigma_0=\pm1}\sum_{\sigma_1=\pm1}\cdots\sum_{\sigma_z=\pm1} e^{a\sigma_0\sum_{i=1}^{z}\sigma_i+b\sigma_0+(b+c)\sum_{i=1}^{z}\sigma_i} \tag{13.47}$$

where $a = J/k_B T$, $b = \mu_B B/k_B T$ and $c = \mu_B H/k_B T = (z-1)J < \sigma > /k_B T$. Summing on $\sigma_0 = \pm1$ and factorizing the other sums, one gets

$$Z = Z_+ + Z_- \quad \text{where}$$

$$Z_\pm = \sum_{\sigma_1=\pm1}\cdots\sum_{\sigma_z=\pm1} e^{\pm a\sum_{i=1}^{z}\sigma_i\pm b+(b+c)\sum_{i=1}^{z}\sigma_i}$$

$$= e^{\pm b}[2\cosh(\pm a + b + c)]^z \tag{13.48}$$

The averaged $< \sigma_0 >$ and $< \sigma_i > (i = 1, ..., z)$ are given by

$$< \sigma_0 > = \frac{Z_+ - Z_-}{Z} \tag{13.49}$$

$$< \sigma_i > = \frac{1}{z} \sum_{i=1}^{z} < \sigma_i > = \frac{1}{z} \frac{\partial Z/\partial c}{Z}$$

$$= [Z_+ \tanh(a + b + c) + Z_- \tanh(-a + b + c)]/Z \tag{13.50}$$

Setting $< \sigma_0 > = < \sigma_i > \equiv < \sigma >$, one has

$$Z_+ [1 - \tanh(a + b + c)] = Z_- [1 + \tanh(-a + b + c)] \tag{13.51}$$

Replacing $Z_\pm$ by (13.48), one obtains

$$\left[\frac{\cosh(a + b + c)}{\cosh(-a + b + c)} \right]^{z-1} = e^{2c} \tag{13.52}$$

This equation is used to determine self-consistently c, namely $< \sigma >$. In zero field ($B = 0$), one has $b = 0$ so that

$$\frac{\cosh(a + c)}{\cosh(-a + c)} = e^{2c/(z-1)}$$

$$\frac{c}{z - 1} = \frac{1}{2} \ln \frac{\cosh(a + c)}{\cosh(-a + c)} \tag{13.53}$$

One sees that $c = 0$ is a solution of the last equation. However there is a nonzero solution by making an expansion at small c (small $< \sigma >$) to the third order:

$$\frac{c}{z - 1} \simeq \frac{1}{2} \ln \frac{\cosh a + c \sinh a + (1/2)c^2 \cosh a + ...}{\cosh a - c \sinh a + (1/2)c^2 \cosh a + ...}$$

$$= [c - \frac{1}{3} \frac{c^3}{\cosh^2 a} + ...] \tanh a$$

$$\frac{1}{z - 1} = [1 - \frac{1}{3} \frac{c^2}{\cosh^2 a} + ...] \tanh a \tag{13.54}$$

from which one has

$$c^2 = 3 \frac{\cosh^3 a}{\sinh a} \left[\tanh a - \frac{1}{z - 1} + ... \right] \tag{13.55}$$

This equation admits a solution if $[\tanh a - \frac{1}{z-1}] > 0$ since the left-hand side is positive. This means that the nonzero solution exists if $T < T_c$ where T_c is given by

$$\tanh(J/k_B T_c) = \frac{1}{z - 1} \tag{13.56}$$

It is interesting to note that for $z = 1$ there is no solution for T_c and for $z = 2$ (one dimension) one has $T_c = 0$. This corresponds to results from exact solutions (see paragraphs 13.5.2 and 13.6, and Problems 10).

13.4   Landau-Ginzburg theory

One has shown above that the mean-field theory suffers from several serious problems due to the fact that instantaneous spin fluctuations in the critical region have been neglected in this theory. There exist several more efficient theories such as the high- and low-temperature series expansions [53], the Landau-Ginzburg theory and the renormalization group. In what follows, one presents the Landau-Ginzburg theory and the concepts of the renormalization group.

The Landau-Ginzburg theory is an extension of the mean-field theory which includes a great part of fluctuations so far neglected near the transition. The main idea is to start from an expansion of the free energy per spin f in the vicinity of T_c when the magnetization m is sufficiently small:

$$f = -\frac{k_B T}{N} \ln Z \tag{13.57}$$

$$f_{MF} = -\frac{k_B T}{N} \ln Z_{MF}$$

$$\simeq C + A(T - T_c^{MF})m^2 + Bm^4 + Dhm + ... \tag{13.58}$$

where C, A, B and D are constants, h is an applied magnetic field and MF denotes quantities coming from the mean-field theory. The form of this expansion and the sign of B reflect the system symmetry. In the case where $h = 0$ and $B > 0$, f_{MF} presents a minimum at $m = 0$ for $T > T_c^{MF}$ and two symmetric minima at $\pm m_0$ for $T < T_c^{MF}$ (see Fig. 13.7), indicating two degenerate ordered states. This degeneracy is removed when $h \neq 0$ as shown in Fig. 13.8.

When $B < 0$, a first-order transition is possible. At $T = T_c$, there are three equivalent minima of f_{MF} at $0, \pm m_0$ (see Fig. 13.9) contrary to the case $B > 0$. This means that at the transition the three phases $m = 0$ (paramagnetic state) and $\pm m_0$ (ordered states) coexist. The energy distribution at $T = T_c$ is thus bimodal, the peak at low energy corresponds to the energy of the ordered phase while that at high energy corresponds to the energy of the disordered state. The distance between the two peaks is the latent heat which is observed at a first-order transition.

When an m^3 term is present in the expansion (13.58), the transition is always of first order.

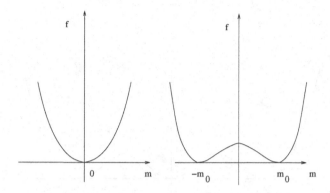

Fig. 13.7 Mean-field free energy at $T > T_c$ (left) and at $T < T_c$ (right).

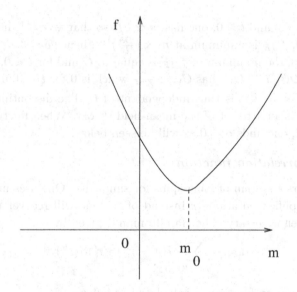

Fig. 13.8 Mean-field free energy at $T < T_c$ in an applied magnetic field.

13.4.1 *Mean-field critical exponents*

The critical exponents calculated by the mean-field theory are (see chapter 12) $\beta = 1/2$ [see (13.32)], $\gamma = 1$ [see (13.44)]. One can find them again here by using (13.58). Putting $t = (T - T_c^{MF})/T_c^{MF}$, one has

- When $t < 0$ and $h = 0$, f_{MF} is minimum at $m_0 \propto (-t)^{\frac{1}{2}}$, so that $\beta = 1/2$.

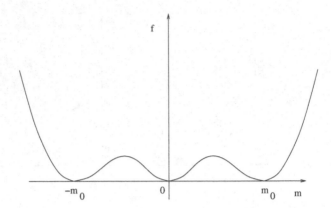

Fig. 13.9   Mean-field free energy at a first-order transition.

- When $h \neq 0$ and $t \geq 0$, one has $m \propto h/t$ so that $\chi \propto t^{-1}$, hence $\gamma = 1$.
- At $t = 0$, f_{MF} is minimum at $m \propto (\frac{Dh}{B})^{1/3}$, hence $\delta = 3$.
- For $t > 0$, the minimum of f_{MF} is equal to C and for $t < 0$, it is equal to $C + \mathcal{O}(\frac{A^2}{B}t^2)$. One has $C_V \propto \frac{\partial^2 f}{\partial T^2}$ which is 0 for $t > 0$ and equal to $\frac{A^2}{B}$ for $t < 0$. C_V is thus independent of t. The discontinuity of C_V at $t = 0$ is an artefact of the mean-field theory. When fluctuations are included, one finds $\alpha = 0$ as will be seen below.

13.4.2   *Correlation function*

One considers a system of Ising spins for simplicity. One uses in this paragraph a simplified notation: r instead of $\vec{r}$. One will recover the correct notation when necessary. The Hamiltonian is given by

$$\mathcal{H} = - \sum_{(r',r'')} J(r' - r'')S(r')S(r'') \tag{13.59}$$

The correlation function is calculated as follows

$$
\begin{aligned}
G(r) &= \; <S(0)S(r)> \\
&= \frac{\text{Tr}S(0)S(r)\exp(\frac{1}{2}\beta \sum_{r',r''} J(r' - r'')S(r')S(r''))}{\text{Tr}\exp(\frac{1}{2}\beta \sum_{r',r''} J(r' - r'')S(r')S(r''))} \\
&= \frac{\text{Tr}S(r)\exp(\frac{1}{2}\beta \sum_{r',r''} J(r' - r'')S(r')S(r''))}{\text{Tr}\exp(\frac{1}{2}\beta \sum_{r',r''} J(r' - r'')S(r')S(r''))} \\
&= \; <S(r)> = \tanh\left[\beta \sum_{r'} J(r - r')S(r')\right]
\end{aligned} \tag{13.60}
$$

where one has taken $S(0) = 1$. The trace was taken over all configurations with $S(0) = 1$. In the mean-field spirit, one replaces $S(r')$ on the right-hand side of (13.60) by its average value $< S(r') >$, then one makes an expansion around T_c when $< S(r') >$ is small, one obtains

$$m(r) \simeq \beta \sum_{r'} J(r - r')m(r') \tag{13.61}$$

using the notation $m(r) =< S(r) >$. The Fourier transform of this equation gives

$$m(k) = \beta J(k)m(k) + C \tag{13.62}$$

where C is a constant. For small k, one has

$$J(k) = \sum_{r=-\infty}^{\infty} J(r) \exp(ikr) \simeq \sum_{r=-\infty}^{\infty} J(r)[1 + ikr + (ikr)^2]$$

$$\simeq \sum_{r=-\infty}^{\infty} J(r)[1 - k^2 r^2] \simeq \sum_{r=-\infty}^{\infty} J(r) - k^2 \sum_{r=-\infty}^{\infty} r^2 J(r)$$

$$\simeq \sum_{r=-\infty}^{\infty} J(r) - k^2 \sum_{r'=-\infty}^{\infty} J(r') \sum_{r=-\infty}^{\infty} r^2 J(r)/ \sum_{r'=-\infty}^{\infty} J(r')$$

$$\simeq \sum_{r} \tilde{J}[1 - k^2 R^2] \tag{13.63}$$

where the sum on the term $J(r)r$ of the second equality is zero because $J(r)r$ is an odd function of r [$J(r)$ is an even function: $J(r) = J(-r)$]. $\tilde{J}$ is defined by

$$\tilde{J} = \sum_{r} J(r) \tag{13.64}$$

and R^2 is defined by

$$R^2 = \frac{\sum_{r} r^2 J(r)}{\sum_{r} J(r)} \tag{13.65}$$

R^2 is thus the order of the interaction range. One obtains from (13.62) and (13.63)

$$m(k) = \frac{C}{1 - \beta \tilde{J}(1 - R^2 k^2)} \tag{13.66}$$

One recalls here that in the mean-field theory $k_B T_c = \tilde{J}$. One writes thus

$$m(k) = \frac{C}{1 - \beta k_B T_c(1 - R^2 k^2)} = \frac{C}{1 - \frac{T_c}{T}(1 - R^2 k^2)}$$

$$= \frac{C}{t + \frac{T_c}{T} R^2 k^2} \simeq \frac{CR^{-2}}{tR^{-2} + k^2} \tag{13.67}$$

where one has taken $T \simeq T_c$. Putting

$$\xi^{-2} = tR^{-2} \tag{13.68}$$

and using $m(r) = G(r)$ of (13.60), one finally arrives at

$$G(k) = \frac{CR^{-2}}{\xi^{-2} + k^2} \tag{13.69}$$

The inverse Fourier transform of (13.69) gives (see also chapter 10)

$$G(r) \propto \frac{\exp(-r/\xi)}{r^{(d-1)/2}} \tag{13.70}$$

This form of $G(r)$ justifies the fact that ξ is called the "correlation length" in the mean-field theory. The expression (13.69) is called "Ornstein-Zernike correlation function".

From (13.68) one sees that $\xi \propto t^{-1/2}$, therefore $\nu = 1/2$ in this mean-field theory. In addition, at $t = 0$, ξ tends to ∞ (see the following paragraph), $G(k)$ of (13.69) is then proportional to $1/k^2$. The inverse Fourier transform of this function is $\frac{1}{r^{d-2}}$ for large r, indicating therefore that exponent η of the mean-field theory is equal to 0 [see definition (13.14)].

13.4.3 *Corrections to mean-field theory*

One decomposes the following term [see (12.20)]

$$J(r - r')\vec{S}_r \cdot \vec{S}_{r'} = J(r - r')[\vec{S}_r^z + \delta\vec{S}_r] \cdot [\vec{S}_{r'}^z + \delta\vec{S}_{r'}]$$

$$\simeq J(r - r') < \vec{S}_r^z > \cdot < \vec{S}_{r'}^z >$$

$$\simeq J(r - r')m(r)m(r') \tag{13.71}$$

where the average values of the linear terms in $\delta\vec{S}_r$ and $\delta\vec{S}_{r'}$ are zero by symmetry (rotating vectors in the xy plane). In the last equality, one has neglected, in the mean-field spirit, the following term

$$J(r - r')\delta\vec{S}_r \cdot \delta\vec{S}_{r'} \tag{13.72}$$

Using the correlation function

$$G(r - r') = < \vec{S}_r \cdot \vec{S}_{r'} > = \text{constant} + < \delta \vec{S}_r \cdot \delta \vec{S}_{r'} >$$

one writes

$$\sum_{r'} J(r - r') < \delta \vec{S}_r \cdot \delta \vec{S}_{r'} > = \sum_{r'} J(r - r') G(r - r') \tag{13.73}$$

One is interested in a long-distance behavior. Using the Fourier transform of (13.69) for small k, one writes

$$\sum_{r'} J(r - r') < \delta \vec{S}_r \cdot \delta \vec{S}_{r'} > \simeq \sum_{r'} J(r - r') \frac{C}{R^2} \int_{ZB} \frac{d^d k}{k^2 + \xi^{-2}}$$

$$= \tilde{J} \frac{C}{R^2} \int_{ZB} \frac{d^d k}{k^2 + \xi^{-2}} \tag{13.74}$$

where the integral is performed in the first Brillouin zone. One decomposes this integral as follows

$$\int_{ZB} \frac{d^d k}{k^2 + \xi^{-2}} = \int_{ZB} \frac{d^d k}{k^2} - \xi^{-2} \int_{ZB} \frac{d^d k}{k^2(k^2 + \xi^{-2})} \tag{13.75}$$

The first integral does not depend on ξ, namely independent of T. It contributes to shift the value of T_c calculated by the mean-field theory. The second integral depends on ξ thus on T: it converges if $k = \pm\pi/a \to \infty$, namely $a \to 0$ (continuum limit), and if d (space dimension) < 4. By a simple dimension analysis, one sees that this integral is proportional to ξ^{2-d} at the limit $\xi \to \infty$. The second term is thus proportional to $\frac{\tilde{J}}{R^2}\xi^{2-d}$. This term has been neglected before in the mean-field theory because it was wrongly considered as always small with respect to the term coming from the mean field. Let us examine when it can be neglected, namely when

$$\frac{\tilde{J}}{R^2}\xi^{2-d} \ll \tilde{J}t \tag{13.76}$$

where $t = \frac{T - T_c}{T}$. One knows that $\xi = Rt^{-1/2}$ [see (13.68)], expression (13.76) can be thus rewritten as

$$\xi^{4-d} \ll R^4 \tag{13.77}$$

This condition for neglecting the second term of (13.75) is called "Ginzburg's criterion". One sees that when $d < 4$ this criterion is not satisfied near T_c where $\xi \to \infty$. By consequence, the mean-field theory which neglects fluctuations characterized by ξ^{4-d} is not valid in the critical region for $d < 4$. The dimension $d = 4$ is called "upper critical dimension" for the Ising model with short-range interaction.

13.5   Renormalization group

13.5.1   *Transformation of renormalization group — Fixed point*

The central idea of the renormalization group is to replace the set of parameters which define the system by another set of parameters which are simpler to deal with but essential physical ingredients, in particular the system symmetry, are conserved. In the study of phase transitions, this approach consists in dividing the system into blocks of spins and replacing each block by a single spin. This "new" spin interacts with the others by the renormalized interactions calculated while decimating the block. The distances between the new spins are measured with a new lattice constant. This procedure is repeated with the new system of spins to obtain the next generation of spins which is used to generate the following generation, and so on. At each iteration, also called decimation or transformation, one has a relation between the new interaction $K' = \beta J'$ and the previous one $K = \beta J$:

$$K' = f(K) \tag{13.78}$$

The new correlation length ξ which is a function of K' is equal to the previous correlation length divided by the factor called "dilatation" b defined as the ratio between the new and old lattice constants. One has

$$\xi(K') = \frac{\xi(K)}{b} \tag{13.79}$$

At the phase transition, ξ becomes infinite, the measuring distance unit is no more important. The interaction constants K' and K become identical $K' = K = K^*$. This point is called the "fixed point" in the renormalization group language. The fixed point is thus determined by

$$K^* = f(K^*) \tag{13.80}$$

The one-dimensional case is simple to proceed as seen in paragraph 13.5.2 below. However, in the case of a general dimension $d > 1$, relation (13.78) is rather complicated. It is often impossible to find a solution of (13.80). One then has to take into account physical considerations to find appropriate approximations.

One considers a point K in the proximity of the fixed point K^*. If the iteration process takes K away from the fixed point, K^* is an unstable fixed point. This is a "run away" case. On the other hand, in the case where K tends to K^* by iteration, K^* is a stable fixed point (see Problem 13.5.2 below). The map of these trajectories near a fixed point with indicated moving directions is called a "flow diagram" . An example is shown in Fig. 13.10. This figure corresponds to the case of a square lattice of Ising spins with interactions $K_1 = \frac{J_1}{k_B T}$ in the x direction and $K_2 = \frac{J_2}{k_B T}$ in the y direction. P is the fixed point. For a given ratio K_2/K_1, namely one follows the discontinued line: intersection P_1 with the line separating regions of different flows is the critical point corresponding to that ratio K_2/K_1. The line of flow separation shown by the heavy solid line is the critical line. One sees that P_1 runs toward P on the critical line, therefore P_1 and P belong to the same universality class: the universality class does not thus depend on the ratio K_2/K_1.

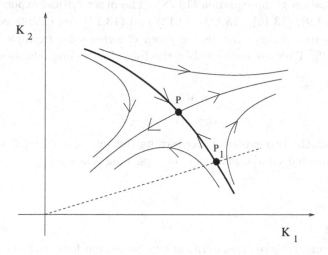

Fig. 13.10   Flow diagram for a square lattice of Ising spins with nearest-neighbor interactions K_1 along the x axis and K_2 along the y axis. See text for comments.

One can calculate the critical exponents using the renormalization group if one knows how K' depends on K in the vicinity of a fixed point K^*. One makes then an expansion of their relation (13.78) around K^*:

$$K' \simeq f(K^*) + (K - K^*)f'(K^*)$$
$$\simeq K^* + b^y(K - K^*) \tag{13.81}$$

where one has replaced $f(K^*)$ by K^* using (13.80), and one has used the following notation

$$y = \frac{\ln f'(K^*)}{\ln b}.$$

Now, near K^* one knows that $\xi \propto (K - K^*)^{-\nu}$ (definition of ν). Therefore, by using (13.79) and (13.81) one obtains

$$(K - K^*)^{-\nu} = b(K' - K^*)^{-\nu} = b[b^y(K - K^*)]^{-\nu} \qquad (13.82)$$

By identifying the two sides of the above equation, one gets

$$\nu = 1/y \qquad (13.83)$$

This example shows that the critical exponent ν is a derivative of the renormalization group equation (13.78). The other critical exponents, defined in (13.9), (13.10), (13.11), (13.13) and (13.14), are calculated by the use of the free energy. Details are given elsewhere, for example in Refs. [8, 32, 189]. They are connected by the following scaling relations

$$\alpha + 2\beta + \gamma = 2 \qquad (13.84)$$
$$\alpha + \beta(1 + \delta) = 2 \qquad (13.85)$$

In addition, the two exponents concerning the spin-spin correlation, ν and η, are connected via the following "hyperscaling relations"

$$\alpha = 2 - d\nu \qquad (13.86)$$
$$\gamma = \nu(2 - \eta) \qquad (13.87)$$

In summary, there are six critical exponents and four relations between them. It suffices to determine two among six exponents. The other four can be then calculated using the above four relations.

To close this section, one emphasizes that, in addition to the results shown above, another important result of the renormalization group is the relations connecting the system size to the critical exponents: physical quantities calculated at a finite system size are shown to depend on powers of the linear system size. These powers are simple functions of critical exponents. They are very useful for the determination of critical exponents by Monte Carlo simulation (see for example Refs. [24, 50, 105]).

13.5.2 *Renormalization group applied to an Ising chain*

One applies in the following the renormalization group to the case of a chain of Ising spins with a ferromagnetic interaction between nearest neighbors. One will show that there is no phase transition at finite temperature.

One considers the following Hamiltonian

$$\mathcal{H} = -K \sum_n \sigma_n \sigma_{n+1} \qquad (13.88)$$

where $K = J/k_B T$, J being a ferromagnetic interaction between nearest neighbors. The partition function is given by

$$Z = Tr \exp(-\mathcal{H}) \qquad (13.89)$$

To study this spin chain, one uses the decimation method. One divides the system into three-spin blocks as shown in Fig. 13.11.

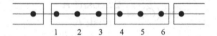

Fig. 13.11   Blocks of three spins used for the decimation.

One writes for the two blocks in the figure the corresponding factors in Z

$$\exp(K\sigma_2\sigma_3) \exp(K\sigma_3\sigma_4) \exp(K\sigma_4\sigma_5) \qquad (13.90)$$

Using the following equality for the case $\sigma_n = \pm 1$

$$\exp(K\sigma_2\sigma_3) = \cosh K (1 + x\sigma_2\sigma_3) \qquad (13.91)$$

where $x = \tanh K$, one rewrites (13.90) as

$$\cosh^3 K (1 + x\sigma_2\sigma_3)(1 + x\sigma_3\sigma_4)(1 + x\sigma_4\sigma_5) \qquad (13.92)$$

Writing explicitly each term of the product and making the sum on $\sigma_3 = \pm 1$ and $\sigma_4 = \pm 1$ (decimation of spins at block borders), one sees that the odd terms of these variables give zero contributions. There remains

$$2^2 \cosh^3 K (1 + x^3 \sigma_2 \sigma_5)$$

One can rewrite

$$2^2 \cosh^3 K(1 + x^3\sigma_2\sigma_5) = \exp[K'\sigma_2\sigma_5 + C] \qquad (13.93)$$

where C is a constant. If one forgets C, the right-hand side is similar to the initial Hamiltonian with a new interaction K' between the remaining spins σ_2 and σ_5. To calculate K' one writes

$$\exp[K'\sigma_2\sigma_5 + C] = \exp(C)\exp(K'\sigma_2\sigma_5)$$

$$= \exp(C)\cosh K'(1 + x'\sigma_2\sigma_5)$$

By identifying this with the left-hand side of (13.93), one obtains

$$x' = \tanh K' = x^3 \quad \text{where} \quad K' = \tanh^{-1}[\tanh^3 K]$$

and

$$\exp(C)\cosh K' = 2^2 \cosh^3 K$$

$$\exp(C) = \frac{2^2 \cosh^3 K}{\cosh K'}$$

$$C = -\ln[\frac{\cosh^3 K}{\cosh K'}] - 2\ln 2 \qquad (13.94)$$

One renumbers the spins after the first decimation as follows: $\sigma_1' = \sigma_2$, $\sigma_2' = \sigma_5$, ... (every three old spins). The new Hamiltonian is thus

$$\mathcal{H}' = -K'\sum_n \sigma_n'\sigma_{n+1}' - C\frac{N}{3} \qquad (13.95)$$

where $N/3$ is the number of three-spin blocks (N: initial number of spins).

The equation of the renormalization group is thus

$$K' = \tanh^{-1}[\tanh^3 K] \qquad (13.96)$$

One sees that $K' = K$ if $K = 0$ and $K = \infty$.

At high T, $K \to 0^+$, $\tanh^3 K < 1$, hence $K' \to 0$ after successive decimations. The fixed point at $T = \infty$ ($K = 0$) is thus stable.

At low T, $K \to \infty$, $\tanh^3 K \to 1^-$, hence $K' \to 0$ after successive decimations, namely a "run away" flow. The fixed point at $T = 0$ ($K = \infty$) is thus unstable.

The flow diagram is shown in Fig. 13.12.

Any point between $K = 0$ and $K = \infty$ moves to $K = 0$ after successive decimations. The nature of any point between these limits is therefore the same as that of $K = 0$ ($T = \infty$), namely it belongs to the paramagnetic phase.

Fig. 13.12   Flow diagram of a chain of Ising spins.

13.5.3   *Migdal-Kadanoff decimation method and Migdal-Kadanoff bond-moving approximation*

In the same spirit as the renormalization concept presented above, the Migdal-Kanadoff decimation method consists of decimating one spin out of every two on a chain to reduce the number of spins, namely the scale factor $b = 2$ is used in the real space. Let us describe how it works. One writes the partition function for an Ising spin chain in an applied magnetic field h:

$$Z = \sum_{\{\sigma_i = \pm 1\}} \exp \left\{ \sum_i [K\sigma_i \sigma_{i+1} + h\sigma_i] \right\}.$$

where the first sum runs over all spin configurations. Now, instead of dividing the chain into blocks as above, one decimates one spin out of every two as follows. One considers three consecutive spins i, k and j and one writes the corresponding part of the Hamiltonian with the sum on the middle spin σ_k:

$$-\beta H(i,j) = \frac{h}{2} \sum_{\sigma_k = \pm 1} (\sigma_i + \sigma_k + \sigma_j)$$

$$+ \ln \sum_{\{\sigma_k = \pm 1\}} \exp \left[K(\sigma_i \sigma_k + \sigma_k \sigma_j) + h\sigma_k \right]$$

$$= \frac{h}{2}(\sigma_i + \sigma_j) + \ln 2 + \ln \cosh \left[K(\sigma_i + \sigma_j) + h \right]$$

$$\simeq \frac{h}{2}(\sigma_i + \sigma_j) + \ln 2 + \ln \cosh \left[K(\sigma_i + \sigma_j) \right] + h \tanh \left[K(\sigma_i + \sigma_j) \right]$$

$$(13.97)$$

where in the last line one has assumed small h. Since $\sigma_i^2 = 1$, one has the following expressions

$$(\sigma_i + \sigma_j)^{2n} = 2^{2n-1}(1 + \sigma_i \sigma_j), \quad (\sigma_i + \sigma_j)^{2n+1} = 2^{2n}(\sigma_i + \sigma_j).$$

With these, one can write for any even and odd functions of $K(\sigma_i + \sigma_j)$:

$$f_e[K(\sigma_i + \sigma_j)] = \sum_{n=0} \frac{a_n K^{2n}(\sigma_i + \sigma_j)^{2n}}{n!}$$

$$= \sum_{n=0} \frac{a_n K^{2n} 2^{2n-1}(1 + \sigma_i \sigma_j)}{n!}$$

$$= \frac{1}{2} f_e(2K)(1 + \sigma_i \sigma_j) \tag{13.98}$$

$$f_o[K(\sigma_i + \sigma_j)] = \sum_{n=0} \frac{b_n K^{2n+1}(\sigma_i + \sigma_j)^{2n+1}}{n!}$$

$$= \sum_{n=0} \frac{b_n K^{2n+1} 2^{2n}(\sigma_i + \sigma_j)}{n!}$$

$$= \frac{1}{2} f_o(2K)(\sigma_i + \sigma_j) \tag{13.99}$$

Using these expressions, one gets from (13.97)

$$-\beta H(i,j) = \frac{1}{2} \ln \cosh(2K)\sigma_i \sigma_j + \frac{1}{2} h \left[1 + \tanh(2K)\right](\sigma_i + \sigma_j)$$

$$+ \ln 2 + \frac{1}{2} \ln \cosh(2K)$$

$$= K'\sigma_i \sigma_j + h'(\sigma_i + \sigma_j) \tag{13.100}$$

where

$$K' = \frac{1}{2} \ln \cosh(2K) \tag{13.101}$$

$$h' = h \left[1 + \tanh(2K)\right] + O(h^2) \tag{13.102}$$

Note that all quantities independent of spins have been omitted since they do not affect expectation values of physical quantities. Equation (13.100) has the same form as if the spins σ_i and σ_j are neighbors but with the parameters K' and h' instead of the original K and h in the initial Hamiltonian. If one looks for a fixed point one has to solve (13.101) when $K' = K = K^*$ with $h = 0$. But the equation $f(x) = \frac{1}{2} \ln \cosh(2x) = x$ has no solution for $0 < x < \infty$. There is a solution at $x = 0$ corresponding to the stable fixed point at $T = \infty$ as found in the previous paragraph. There is an unstable fixed point at $K = \infty$ as seen by the following expansion

$$K' = \frac{1}{2} \ln[\exp(2K) + \exp(-2K)] - \frac{\ln 2}{2} = K + \frac{1}{2} \ln[1 + \exp(-4K)] - \frac{\ln 2}{2}$$

$$\simeq K - \frac{\ln 2}{2} + O[\exp(-4K)] \tag{13.103}$$

One sees that K varies very slowly starting with the first iteration where $K = K_0 = J/T \gg 1$ ($T \sim 0$): between two iterations, K diminishes by $dK = -\frac{\ln 2}{2}$. On the other hand, the scaling parameter after n iterations is $b = 2^n$ so that $\ln b = n \ln 2$. For one step, one has $d(\ln b) = \ln 2$. Therefore,

$$dK = -\frac{d \ln b}{2} \Rightarrow \frac{dK}{d \ln b} = -\frac{1}{2}.$$

This yields an exponential law as seen below, instead of the power law in the case where there is a fixed point at a finite K. Integrating the above equation, one has

$$K(b) = K_0 - \frac{1}{2} \ln b \qquad (13.104)$$

The correlation length is measured in unit of b: one has the following scaling relation

$$\xi(K) = b\xi[K(b)]$$

where, under renormalization, $K(b)$ tends to zero. Since $\xi(0) \sim 1$ (paramagnetic state), one has $\xi(K) \simeq b$. Equation (13.104) becomes at the limit $K(b) \sim 0$

$$\ln b = 2K_0 = \frac{2J}{T} \Rightarrow b \simeq \exp(2K_0) = \exp(2J/T) \Rightarrow \xi(K) \simeq \exp(2J/T).$$

The Migdal-Kananoff decimation shown above is exact for one dimension. In two dimensions, one uses the so-called bond-moving approximation which consists of moving one horizontal bond line and one vertical bond line every two lines as shown in Fig. 13.13. In doing so, the number of lattice cells is reduced to a half and the spins left behind are free spins, they do not participate in the collective properties of the system. Each remaining bond has a new strength $2J$. Note that the spins on the new bonds (not at the crossings) have each two neighbors: one can decimate these spins by the Migdal-Kadanoff shown above with the result

$$K' = \frac{1}{2} \ln \cosh(4K) \qquad (13.105)$$

$$h' = h\left[1 + \tanh(4K)\right] + O(h^2) \qquad (13.106)$$

The fact that K becomes $2K$ changes a lot of things: the equation for fixed points ($h = 0$) is

$$K^* = \frac{1}{2} \ln \cosh(4K^*) \qquad (13.107)$$

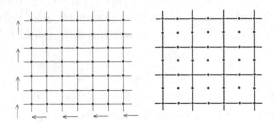

Fig. 13.13 Moving one horizontal line and one vertical line out of every two lines as indicated by arrows: the left lattice becomes the right lattice. Black circles denote the spins.

which admits a solution at a finite value of K. One sees this by examining the two limits:

$$\frac{1}{2}\ln\cosh(4K) \simeq \begin{cases} 4K^2 << K & \text{if } K \ll 1 \\ 2K > K & \text{if } K \gg 1 \end{cases} \qquad (13.108)$$

From this one sees that the function $K - \frac{1}{2}\ln\cosh(4K)$ should change its sign somewhere between 0 and ∞.

The phase transition occurs thus at a finite temperature. In the case of the square lattice considered here, one has $K^* \simeq 0.30469$, or $T_c/J \simeq 3.282$ which is below the mean-field value 4, but above the exact value 2.2692. Our conclusion is that the bond moving improves the mean-field theory but the moving procedure is not justified at all. In spite of this, the bond-moving approximation gives not very bad values of critical exponents.

To compute the critical exponents, one expands the recursion relation (13.107) around its fixed point. Putting $K = K^* + t$ where t is the reduced temperature, one has $t' = 1.6786t = b^{\lambda_t}t$ where one has used $b = 2$ and

$$\lambda_t = \frac{\ln 1.678}{\ln 2} \simeq 0.74674.$$

This value is higher than the mean-field value $\nu = 0.5$ but smaller than the exact value $\nu = 1$. Note that the bond moving approximation becomes exact for some hierarchical lattices which correspond to fractal spatial dimensions [72].

13.6 Transfer matrix method applied to an Ising chain

The transfer matrix method is very useful when the system can be divided into subsystems, each of which interacts only with its adjacent nearest neighboring subsystems. For example, the simple cubic lattice can be considered as composed of planes each of which interacts only with its neighboring planes. In the case of periodic boundary conditions, the partition function can be written as a product of partition functions of its N subsystems:

$$Z = \text{trace} \left[\prod_{i=1}^{N} W_i \right]$$

where W_i is the "transfer matrix" of dimension $n \times n$ representing the interaction connection between two adjacent blocks (subsystems). The trace of Z is the sum of the eigenvalues of Z. If the system is homogeneous, then all W_i are identical: each eigenvalue of the product of identical matrices W_i is equal the product of the corresponding eigenvalue of W_i (properties of trace). Let z_1, z_2, ...,z_n be the eigenvalues of W_i, one writes $Z = z_1^N + z_2^N + ... + z_n^N$. If $N \to \infty$, then $Z = z_{\max}^N$ where $z_{\max}$ is the largest eigenvalue among $z_1, z_2, ..., z_n$.

One applies in the following the transfer matrix method to the case of a chain of Ising spins using the periodic boundary condition.

Let N be the total number of spins. The Hamiltonian is given by

$$\mathcal{H}_0 = -J \sum_{n=1}^{N-1} \sigma_n \sigma_{n+1} - J \sigma_N \sigma_1 \tag{13.109}$$

where the last term expresses the periodic boundary condition. One has $\sigma_i = \pm 1$. One can define new variables $\alpha_n = \sigma_n \sigma_{n+1}$. α_n takes the values ± 1 as σ_i. One can rewrite $\mathcal{H}$ as

$$\mathcal{H}_0 = -J \sum_{n=1}^{N-1} \alpha_n - J \alpha_N \tag{13.110}$$

The partition function is then

$$Z = Tr \exp \left[\beta J \sum_{n=1}^{N} \alpha_n \right] = Tr \prod_{n=1}^{N} \exp(\beta J \alpha_n)$$

$$= \prod_{n=1}^{N} [\exp(\beta J) + \exp(-\beta J)] = [2 \cosh \beta J]^N \tag{13.111}$$

This result is the same as that obtained by the exact method shown in Problem 10. The average energy is calculated by $\overline{E} = -\partial \ln Z / \partial \beta$ [see (3.17)]:

$$\overline{E} = -\frac{\partial \ln Z}{\partial \beta} = -NJ \tanh(\beta J) \tag{13.112}$$

One obtains the following heat capacity

$$C_V = dE/dT = Nk_B \left[\frac{k_B T}{J} \cosh \frac{J}{k_B T} \right]^{-2} \tag{13.113}$$

In an applied magnetic field H, one proceeds as follows:

$$\mathcal{H} = \mathcal{H}_0 - H \sum_{n=1}^{N} \sigma_n \tag{13.114}$$

where $\mathcal{H}_0$ is given by (13.109). The partition function is $Z = \prod_{n=1}^{N} V_n$ where

$$V_n = \exp[\beta(J\sigma_n \sigma_{n+1} + H\sigma_n)] \tag{13.115}$$
$$V_N = \exp[\beta(J\sigma_N \sigma_1 + H\sigma_N)] \tag{13.116}$$

The matrix elements V_n, of dimension 2x2, depend on σ_n and σ_{n+1}. One has

$$V_n(1,1) = \exp[\beta(J+H)] \quad (\sigma_n = 1, \sigma_{n+1} = 1)$$
$$V_n(1,2) = \exp[\beta(-J+H)] \quad (\sigma_n = 1, \sigma_{n+1} = -1)$$
$$V_n(2,1) = \exp[\beta(-J-H)] \quad (\sigma_n = -1, \sigma_{n+1} = 1)$$
$$V_n(2,2) = \exp[\beta(J-H)] \quad (\sigma_n = -1, \sigma_{n+1} = -1)$$

The matrix V_n is called "transfer matrix". Note that all V_n $(n = 1, N)$ have the same elements, say V. One thus has $Z = Tr V^N$. Let z_1 and z_2 be the eigenvalues of V obtained by diagonalizing V:

$$z_1 = \exp(\beta J) \cosh(\beta H) + \sqrt{\exp(2\beta J) \cosh^2(\beta H) - 2 \sinh(2\beta J)}$$

$$z_2 = \exp(\beta J) \cosh(\beta H) - \sqrt{\exp(2\beta J) \cosh^2(\beta H) - 2 \sinh(2\beta J)}$$

One obtains then

$$Z = z_1^N + z_2^N = z_1^N (1 + \exp[-N \ln(z_1/z_2)]) \tag{13.117}$$

where z_1 denotes the larger eigenvalue. When $N \to \infty$, one has $Z = z_1^N$.

The susceptibility is calculated by $\chi = (dM/dH)_{H \to 0}$ where $M = -\partial F/\partial H$ with $F = -k_B T \ln Z$. One obtains

$$\chi \simeq \frac{1}{T} \exp[2J/k_B T] \tag{13.118}$$

This result shows that there is no phase transition in one dimension (absence of anomaly of χ with varying T as seen in Fig. 13.14).

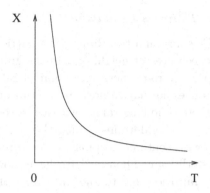

Fig. 13.14 χ versus T [Eq. (13.118)].

13.7 Phase transition in some particular systems

One has seen so far various methods used to study phase transitions and critical phenomena. In general, the nature of a phase transition depends on the symmetry of the order parameter, the spatial dimension and the nature of the interaction (short or long range). Standard methods presented above can be used to determine it with satisfactory precision. However, in some particular systems one needs special methods. Some of these remarkable systems are presented hereafter.

13.7.1 *Exactly solved spin systems*

There are several families of systems in one or two dimensions with short-range non-crossing interactions which can be exactly solved. The spin models in those solvable systems are often Ising and Potts models. One needs exact solutions in simple systems to test approximations conceived to be used in more complicated systems or systems in three dimensions. Methods for searching exact solutions are lengthy to present here. The reader is referred to the book by R. J. Baxter [20] for general methods and exactly solved models. For some exactly solved frustrated spin systems, the reader is referred to the review by Diep and Giacomini [51]. In a word, to find a solution, the most frequently used method is to transform the studied system into a vertex model where solutions for critical surfaces are known. Among the most popular models, one can mention the 8-, 16- and 32-vertex models [20].

13.7.2   *Kosterlitz-Thouless transition*

One considers the XY spins on a two-dimensional lattice with a ferromagnetic interaction between nearest neighbors. In the ground state, the spin configuration is a perfect ferromagnetic state, namely all spins are parallel. However, this system does not have a normal order-disorder transition at a finite temperature: there is no long-range ordering as soon as the temperature is not zero, following the Mermin-Wagner theorem [129] valid for two-dimensional systems with continuous spins (see the discussion in Problem 4 of chapter 12). Kosterlitz and Thouless [102] have shown that the system has a special phase transition due to the unbinding of vortex-antivortex pairs at a finite temperature below (above) which the correlation function decays as a power law (exponential law) with increasing distance. This transition, called Kosterlitz-Thouless (KT) or Kosterlitz-Thouless-Berezinskii transition, is of infinite order. For the reader interested in this special transition, Appendix E gives the main points explaining the mechanism lying behind the KT transition.

13.7.3   *Frustrated spin systems*

A system is said "frustrated" when the interaction bonds between a spin with its neighbors cannot be fully satisfied. An example is the triangular antiferromagnet: i) in the case of Ising spin model, the three spins on a triangle cannot find orientations to satisfy the three bonds, ii) in the case of XY or Heisenberg spin models, the spins make a "compromise" to form a non collinear configuration in order to partially satisfy each bond as shown in Fig. 12.7. The ground-state spin configuration of a few cases have been given in section 12.3.3 and in Problems 9 and 10 of chapter 12.

Effects due to the frustration are numerous and spectacular. One can mention a few of them: (i) high ground-state degeneracy, (ii) non collinear spin configuration, (iii) multiple phase transitions, (iv) reentrance phenomenon, (v) disorder lines, (vi) partial ordering at equilibrium, (vii) difficulty in determining the nature of phase transitions in several systems, etc. Some of these spectacular effects (iii)-(vi) have been observed in exactly solved two-dimensional systems [13, 38, 39, 47, 51]. It is believed that these effects persist in three-dimensional systems and in other more complicated unsolved models. For advanced reviews on frustrated systems, the reader is referred to Ref. [50].

13.8   Conclusion

In this chapter, one has introduced basic notions as well as some fundamental methods which are widely used in the field of phase transitions. The mean-field theory has been largely developed and commented. This theory provides a first approach which paves the way for other improving methods such as the Bethe's approximation and the Ginzburg's criterion. The renormalization group has been shown with simple examples to illustrate its concepts. In particular, the notion of universality class and the relations between the critical exponents have been discussed. An example of the transfer matrix method has been treated, and some complementary methods such as canonical and micro-canonical methods are also introduced as problems which are given below. More advanced methods, such as quantum phase transitions of low-dimensional systems [111, 132, 151] and the Hubbard model [56] are not included here to keep the contents of the book suitable for lectures in a graduate course.

13.9   Problems

Problem 1. Ising spin model in the mean-field approximation:
    Find the mean-field equation for the Ising spins with a nearest-neighbor ferromagnetic interaction.

Problem 2. Interaction between next-nearest neighbors in a centered cubic lattice:
    Consider a centered cubic lattice of Ising spins ± 1 with the following Hamiltonian:

$$\mathcal{H} = -J_1 \sum_{(i,j)} \sigma_i \sigma_j - J_2 \sum_{(i,k)} \sigma_i \sigma_k \qquad (13.119)$$

where σ_i is the spin at the site i, J_1 (> 0) ferromagnetic interaction between nearest neighbors (NN) and J_2 (> 0) that between next-nearest neighbors (NNN). The first and second sums run over NN and NNN spin pairs, respectively.

a) Find the magnetic order at zero temperature.

b) Give briefly the main hypotheses of the mean-field approximation.

c) Show by a qualitative argument (without calculation) that the interaction J_2 increases the transition temperature.

d) Using the mean-field approximation, calculate the partition function of a spin at temperature T. Find the equation allowing for calculating the average value $< \sigma >$ at T.

e) Determine the transition temperature T_c as a function of J_1 and J_2.

f) In the case where J_2 is negative (antiferromagnetic), the ferromagnetic ground state is valid up to a critical value of $|J_2|$. Determine this value called J_2^c. Beyond J_2^c, what is the ground state?

Problem 3. Interaction between next-nearest neighbors in a square lattice:

Answer the same questions as those in the preceding problem for the case of a square lattice.

Problem 4. System of two spins:

Consider two spins with the following Hamiltonian

$$\mathcal{H} = -2J\vec{S}_1 \cdot \vec{S}_2 - D[(S_1^z)^2 + (S_2^z)^2] - B(S_1^z + S_2^z) \quad (13.120)$$

where the exchange interaction J and the anisotropy D are assumed to be positive, B is the amplitude of an applied magnetic field parallel to the Oz axis.

Find the eigenvalues and eigenvectors of $\mathcal{H}$ for spins $1/2$.

Problem 5. Improvement of the mean-field approximation:

Consider the Heisenberg spin model with the following Hamiltonian:

$$\mathcal{H} = -2J\sum_{(i,j)} \vec{S}_i \cdot \vec{S}_j.$$

In order to improve the mean-field approximation, one can consider a cluster embedded in the mean field due to the surrounding neighbors. One starts with a cluster of two spins. The state of this cluster can be exactly calculated as shown in the preceding Problem. The mean-field approximation is then used for its surrounding neighbors. One writes for the cluster composed of spins $\vec{S}_i$ and $\vec{S}_j$:

$$\mathcal{H}_{ij} = -2J\vec{S}_i \cdot \vec{S}_j - 2(Z-1)J < S^z > (S_i^z + S_j^z) \quad (13.121)$$

where Z is the coordination number ($Z-1$ is thus the number of neighbors of a cluster spin), and $< S^z >$ the thermal average of the z-component of a neighboring spin.

Show that the critical temperature T_c for spin $1/2$ is given by

$$e^{-2J/k_B T_c} + 3 - 2(Z-1)J/k_B T = 0. \quad (13.122)$$

Comment.

Problem 6. Chain of Ising spins by micro-canonical method:

Consider a chain of N Ising spins interacting with each other via a nearest-neighbor coupling $J > 0$. The system is isolated with the Hamiltonian

$$\mathcal{H} = -J \sum_{i=1}^{N} \sigma_i \sigma_{i+1} \tag{13.123}$$

One supposes that the periodic boundary condition $\sigma_{N+1} = \sigma_1$ applies and N is even.

a) Calculate the energy of the ground state and its degeneracy.

b) Find the energy of the lowest excited state where there are two unsatisfied bonds, namely one reversed spin or two antiparallel spin pairs. Find its degeneracy. Deduce the energy $E(2n)$ of a state in which there are $2n$ unsatisfied bonds. Indicate the maximum energy of the system and its degeneracy.

c) For a given $E(2n)$ with $n \gg 1$, calculate the entropy and the micro-canonical temperature. Find the percentage x of unsatisfied bonds with respect to the total number of bonds. Find its low- and high-temperature limits.

Problem 7. Chain of Ising spins by canonical method:

Consider again the system in the preceding problem but put it now in the canonical situation at temperature T.

a) Calculate the partition function of the system using the Newton binomial relations: $(1 + u)^N = \sum_{n=0}^{N} C_N^n u^n$ and $(1 - u)^N = \sum_{n=0}^{N} (-1)^n C_N^n u^n$.

b) Calculate the system energy.

c) Calculate the average percentage $\bar{x}$ of unsatisfied bonds with respect to the total number of bonds. Find its low- and high-temperature limits. Compare these results to those of Problem 6.

Problem 8. Mean-field approximation for antiferromagnets:

Consider a system of Heisenberg spins interacting with each other via the Hamiltonian

$$\mathcal{H} = \sum_{(i,j)} J_{ij} \vec{S}_i \cdot \vec{S}_j - g\mu_B \sum_i \vec{H}_0 \cdot \vec{S}_i \tag{13.124}$$

where g and μ_B are the Landé factor and the Bohr magneton, respectively. $\vec{H}_0$ is a very small magnetic field applied along the z axis. One supposes that the exchange interaction J_{ij} is limited to the nearest neighbors with $J_{ij} = J$:

$$\mathcal{H} = J \sum_{(i,j)} \vec{S}_i \cdot \vec{S}_j - g\mu_B \sum_i \vec{H}_0 \cdot \vec{S}_i \qquad (13.125)$$

Note that the Hamiltonian is written with a positive sign so that the antiferromagnetic interaction corresponds to $J > 0$.

a) Describe the ground-state spin configuration. Illustrate it on a lattice with a figure.

b) Define two sublattices: one contains the up spins and the other the down spins. Write the mean-field equations for an up spin and a down spin.

c) Show that these two equations are reduced to a single one which is that obtained for ferromagnets. Comments.

d) Calculate the susceptibility. Show the difference compared to the ferromagnetic case.

Problem 9. Ferrimagnets by mean-field theory:

Ferrimagnetic materials have complicated crystalline structures. Ferrimagnets have very rich and complicated properties which are used in numerous applications in particular in recording industries thanks to their very high critical temperatures of the order of 500-800 Celcius degrees.

One introduces hereafter a very simple model to illustrate some remarkable properties of ferrimagnets. Consider a system of Heisenberg spins which is composed of two sublattices, sublattice A containing ↑ spins of amplitude S_A and sublattice B containing ↓ spins of amplitude S_B. The Hamiltonian is written as

$$\mathcal{H} = J_1 \sum_{(l,m)} \vec{S}_l \cdot \vec{S}_m \qquad (13.126)$$

where (l, l') and (m, m') indicate the sites of A and B, respectively. The interaction J_1 is between inter-sublattice nearest neighbors. One supposes $J_1 > 0$ (antiferromagnetic).

Using the mean-field approximation, calculate the sublattice magnetizations as functions of temperature.

Problem 10. Chain of Ising spins by exact method:

Consider a chain of N Ising spins with a ferromagnetic interaction between nearest neighbors, maintained at temperature T, with the following Hamiltonian

$$\mathcal{H} = -J \sum_{i=1}^{N} \sigma_i \sigma_{i+1} \qquad (13.127)$$

One supposes the periodic boundary condition $\sigma_{N+1} = \sigma_1$.

Calculate exactly the partition function of the system. Find the free energy, the average energy and the heat capacity, as functions of T. Show that there is no phase transition at finite temperature.

Chapter 14

Superconductivity

14.1 Introduction

In metals, conducting electrons scatter from various objects such as phonons, magnons, impurities and defects. As a consequence, the total resistivity has several contributions each of which depends on a different kind of diffusion processes. Let us summarize the most important contributions to the total resistivity $R_t(T)$ at low temperatures in the following expression

$$R_t(T) = R_0 + A_1 T^2 + A_2 T^5 + A_3 \ln \frac{\mu}{T} \qquad (14.1)$$

where A_1, A_2 and A_3 are constants. The first term is T-independent, the second term proportional to T^2 represents the scattering of itinerant spins at low T by lattice spin-waves. Note that the resistivity caused by a Fermi liquid is also proportional to T^2 (see section 10.4). The T^5 term corresponds to a low-T resistivity in metals which is due to the scattering of itinerant electrons by phonons. Note that at high T, metals show a linear-T dependence. The logarithm term is the resistivity due to the quantum Kondo effect caused by a magnetic impurity at very low T (see section 10.5). The temperature-independent term R_0 is due to scattering by impurities.

In superconductors, when the temperature is lowered to a critical temperature T_c the resistivity falls suddenly to zero: $R(T) = 0$ for $T \leq T_c$. The superconductivity has been discovered for the first time by Kamerlingh Onnes in 1911 in his experiment on mercury cooled below $T = 4.1$ K. Since then, many other materials, pure elements or compounds, have been found to be superconducting at low temperatures. In 1986, a new family of high-temperature superconductors, namely cuprate-perovskite ceramic materials, has been discovered with critical temperatures above 90 K [21]. While the low-temperature superconductivity can be explained by the Bardeen-Cooper-Schrieffer (BCS) theory, mechanisms lying behind the high-temperature superconductivity are still under debate.

This chapter is devoted to a presentation of two principal theories of superconductivity: the Ginzburg-Landau theory [68] and the BCS theory [16, 17]. A discussion on several possible mechanisms for the high-temperature superconductivity is given.

14.2   Properties of conventional superconductors

The resistivity in superconductors falls to zero below a critical temperature T_c. Under an applied magnetic field below T_c, the superconducting state remains as long as the magnetic field is lower than a critical value H_c. For fields lower than H_c, the magnetic field is pushed out of the material in the superconducting state. This field ejection is called the "Meissner effect" shown in Fig. 14.1. The transition from the superconducting state to the normal state occurs when the applied field is equal to H_c. For $H > H_c$ the material is in a normal state. Superconductors with such a single critical field is called "superconductors of type I". There exist superconductors of type II where there are two critical values of the applied field: H_{c1} and H_{c2}. For $H < H_{c1}$ one has the superconducting state with no magnetic field inside the superconductor. For $H_{c1} < H < H_{c2}$ there is a partial penetration of the field inside the superconductor. This intermediate phase, called "mixed phase" or "vortex phase", is a phase with a penetration of an amount of magnetic quantized flux called "fluxons". The superconducting state remains in the mixed phase as long as the applied field does not exceed H_{c2}. For $H > H_{c2}$ the superconductor transits to the normal state.

The nature of the phase transition between the normal and superconducting states depends on the material and the condition under which the phase transition takes place (with or without field, superconductors of type I or type II, ...).

Historically, the theory of superconductivity has started in the 1930's with, among others, F. and H. London [117, 118] and Gorter and Casimir [70]. A remarkable break-through has been realized with a phenomenological theory proposed by Ginzburg and Landau [68] in the 1950's. But one had to wait until 1957 where a microscopic theory was proposed by Bardeen, Cooper and Schrieffer (BCS) [16, 17]. The BCS theory remains a very complete theory to this day for conventional superconductors. The Ginzburg-Landau (GL) theory has been further developed by many people.

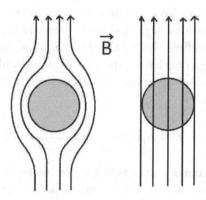

Fig. 14.1 Meissner effect: applied magnetic field $\vec{H}$ is expelled from a superconductor for $H < H_c$ (left), H_c being the critical field. For fields stronger than H_c, the superconducting state is destroyed, the field then penetrates inside the material (right).

In particular, Abrikosov has shown that the GL theory can predict the division of superconductors into two categories, namely type I and type II. He has shown that in the vortex phase of superconductors of type II, magnetic flux crossing a vortex is quantized [1]. On the other hand, Bogoliubov [27] has demonstrated that the BCS wave function, which had originally been derived from a variational argument, could be obtained using a canonical transformation of the electronic Hamiltonian (see below). In addition, Gor'kov [69] has microscopically derived the GL equations from the BCS theory near the phase transition.

In the following, a review is given on the main steps of these theories to understand the mechanism lying behind the conventional superconductivity.

14.3 Ginzburg-Landau theory of superconductivity

Let us start by recalling that the mean-field theory of second-order phase transition was developed by L. D. Landau before the GL theory was formulated for the superconductivity. This mean-field theory was shown in section 13.4 where mean-field critical exponents are found (see 13.4.1). In that section, one has also presented the correlation function which takes into account fluctuations neglected in the mean-field theory.

From their mean-field theory, Ginzburg and Landau have proposed the following free energy density for the case of superconductors where transition from the normal state to the superconducting state is characterized by an order parameter $|\Psi|$:

$$f_s = f_n + \alpha|\Psi|^2 + \frac{\beta}{2}|\Psi|^4 + \frac{1}{2m*}\left|\frac{\hbar}{i}\vec{\nabla}\Psi\right|^2 \qquad (14.2)$$

where one has assumed a zero field for the moment, α is a temperature-dependent coefficient and β a constant, as seen in (13.58). The indices s and n stand for superconducting and normal phases. One sees that the second and third terms describe the second-order phase transition, while the last term is new: it is the quantum-mechanical expression of the kinetic energy.

Under an applied magnetic field, one has to include in the above free energy the Hamiltonian of a charged particle in a magnetic field. The electric field and the magnetic field are given by

$$\vec{E} = -\vec{\nabla}\phi - \frac{\partial\vec{A}}{\partial t} \qquad (14.3)$$

$$\vec{B} = \vec{\nabla}\wedge\vec{A} \qquad (14.4)$$

where $\vec{A}$ is the vector potential and ϕ the scalar one. For a charged particle, its velocity is

$$m\vec{v}(t) = m\vec{v}(0) + q\int_0^t \vec{E}dt$$

$$= m\vec{v}(0) - q\vec{A} \qquad (14.5)$$

Note that $m\vec{v}(0) = m\vec{v}(t) + q\vec{A}$ is conserved after the application of the magnetic field so that one can define the particle momentum by $\vec{p} = m\vec{v}(t) + q\vec{A}$. The kinetic energy is thus

$$\frac{1}{2}m\vec{v}(t)^2 = \frac{1}{2m}\left(\vec{p} - q\vec{A}\right)^2 \qquad (14.6)$$

This term modifies the last term of (14.2).

In a gauge transformation $\vec{A} \rightarrow \vec{A}' = \vec{A} + \vec{\nabla}\chi$, $\phi \rightarrow \phi' = \phi - \frac{\partial\chi}{\partial t}$, where χ is an arbitrary function of $\vec{r}$, the equations (14.3) and (14.4) do not change. However, the particle wave function takes a phase factor, namely $\Psi' = \Psi \exp(ie\chi/\hbar)$. One has

$$\left(\vec{p} - e\vec{A}'\right)\Psi' = \left(\vec{p} - e\vec{A}'\right)\Psi\exp(ie\chi/\hbar)$$

$$= \exp(ie\chi/\hbar)\left[\left(\vec{p} - e\vec{A}'\right)\Psi + \vec{\nabla}\chi\Psi\right]$$

$$= \exp(ie\chi/\hbar)\left(\vec{p} - e\vec{A}\right)\Psi \qquad (14.7)$$

$$\left\langle\Psi'\left|\left(\vec{p} - e\vec{A}'\right)\right|\Psi'\right\rangle = \left\langle\Psi\left|\exp(-ie\chi/\hbar)\exp(ie\chi/\hbar)\left(\vec{p} - e\vec{A}\right)\right|\Psi\right\rangle$$

$$= \left\langle\Psi\left|\left(\vec{p} - e\vec{A}\right)\right|\Psi\right\rangle \qquad (14.8)$$

Since by definition $\vec{J} = \frac{1}{m^*}Re\left\langle\Psi\left|\left(\vec{p} - e\vec{A}\right)\right|\Psi\right\rangle$, (14.8) shows that $\vec{J}' = \vec{J}$ (gauge invariant). One also sees that the density is gauge-independent: $\rho' = |<\vec{r}|\Psi'>|^2 = |<\vec{r}|\Psi>|^2 = \rho$. One can also calculate $\left\langle\Psi'\left|\left(\vec{p} - q\vec{A}'\right)^2\right|\Psi'\right\rangle$ and show that $\mathcal{H}'\Psi' = \mathcal{H}\Psi$ where $\mathcal{H}' = \frac{1}{2m}\left(\vec{p} - q\vec{A}'\right)^2 + U$ (U: interaction term). The physics does not therefore change with the gauge transformation.

Let us give the meaning of $|\Psi|^2$ before showing the GL equations. One writes the kinetic energy density as

$$\frac{1}{2m*}\left|\left(\vec{p} - e^*\vec{A}\right)\Psi\right|^2 = \frac{1}{2m*}\left|\left(\frac{\hbar}{i}\vec{\nabla}|\Psi| + \hbar\vec{\nabla}|\Psi|\vec{\nabla}\varphi - e^*A|\Psi|\right)e^{i\varphi}\right|^2$$

$$= \frac{1}{2m*}\left[\hbar^2(\vec{\nabla}|\Psi|)^2 + (\hbar\vec{\nabla}\varphi - e^*A)^2|\Psi|^2\right] \qquad (14.9)$$

where one has used $\Psi = |\Psi|e^{i\varphi}$ and effective charge e^* and effective mass m^*. Examining the above equation, one sees that the first term is zero except where there is a gradient of particle density such as at the boundary between the normal phase and the superconducting phase. One will return to this point later. Now, the second term of (14.9) is the kinetic energy density of particles in the supercurrent. If the phase φ is a constant, then the kinetic energy density of the supercurrent is

$$\frac{e^{*2}A^2|\Psi|^2}{2m^*} \qquad (14.10)$$

One compares this expression to the particle density of the supercurrent n_s defined in the kinetic energy density ϵ_s in the superconducting state $\epsilon_s = n_s\frac{1}{2}m^*v_s^2$ where $\vec{v}_s = \frac{1}{m^*}(\vec{p}_s - e^*\vec{A})$. Assuming $\vec{\nabla}\varphi = 0$, one has $v_s = e^*A/m^*$. Therefore, $\epsilon_s = n_s\frac{e^{*2}A^2}{2m^*}$. Comparing this to (14.10), one has

$$n_s = |\Psi|^2 \qquad (14.11)$$

This is the meaning of $|\Psi|^2$ in the GL theory presented hereafter.

The full free energy density is

$$f_s = f_n + \alpha|\Psi|^2 + \frac{\beta}{2}|\Psi|^4 + \frac{1}{2m^*}\left|\left(\vec{p} - e^*\vec{A}\right)\Psi\right|^2 + \frac{\vec{B}^2}{2\mu_0} - \mu_0\vec{M}\cdot\vec{H} \quad (14.12)$$

The B-term is the magnetic field energy in the superconducting phase and the last term is that in the normal phase. M is the magnetization.

The two GL equations can be obtained by a variational method: one looks for the solution of $\partial \int_V f_s dV = 0$ (the total free energy is given by integrating f_s over the volume V). There are two variational parameters Ψ and $\vec{A}$ (or $\vec{B}$). One obtains two GL equations:

$$\alpha\Psi + \beta|\Psi|^2\Psi + \frac{1}{2m^*}\left(\frac{\hbar}{i}\vec{\nabla} - e^*\vec{A}\right)^2\Psi = 0, \quad (14.13)$$

$$\vec{J} = \frac{e^*\hbar}{2m^*i}\left(\Psi^*\vec{\nabla}\Psi - \Psi\vec{\nabla}\Psi^*\right) - \frac{e^{*2}\vec{A}}{m^*}|\Psi|^2 \quad (14.14)$$

The demonstration of the above equations, which is rather tedious, is given in Problem 1. Note that the second equation is often used with the following equivalent form

$$\vec{J} = \frac{e^*}{m^*}\left(\hbar\vec{\nabla}\varphi - e^*\vec{A}\right)|\Psi|^2 \quad (14.15)$$

Let us examine a few particular points:

- Deep inside the superconductor, there is no magnetic field (Meissner effect shown below) so that the second equation (14.14) gives $\vec{J} = 0$ because $\Psi^*\vec{\nabla}\Psi - \Psi\vec{\nabla}\Psi^* \propto \vec{\nabla}\varphi = 0$. The first equation (14.13) gives

$$\alpha\Psi + \beta|\Psi|^2\Psi - \frac{\hbar^2}{2m^*}\vec{\nabla}^2\Psi = 0 \quad (14.16)$$

The solution in the ground state of the superconducting phase [the kinetic term (last term)=0] is $\Psi = |\Psi|e^{i\varphi}$ with $|\Psi| = -\alpha/\beta \equiv |\Psi|_\infty$. Putting $f = |\Psi|/|\Psi|_\infty$ ($\alpha < 0$ in the superconducting state), Eq. (14.16) is rewritten, in one dimension for simplicity, as

$$-\frac{\hbar^2}{2m^*}\frac{d^2f}{dx^2} + \alpha f + \beta|\Psi|_0^2 f^3 = 0 \quad (14.17)$$

$$-\frac{\hbar^2}{2m^*|\alpha|}\frac{d^2f}{dx^2} + f - f^3 = 0 \quad (14.18)$$

One can define the Ginzburg-Landau coherence length ξ by

$$\xi^2 \equiv \frac{\hbar^2}{2m^*|\alpha|} \quad (14.19)$$

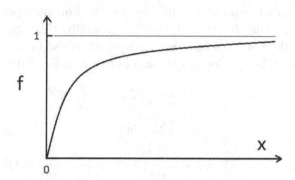

Fig. 14.2 The normalized order parameter $f = |\Psi|/|\Psi|_\infty$, representing the supercon-ducting electron density amplitude, versus x, distance from the normal-superconducting phase boundary defined at $x = 0$. f increases with increasing x as $\tanh(\frac{x}{\sqrt{2}\xi})$ where ξ is the Ginzburg-Landau coherence length.

Equation (14.18) can be exactly solved: the solution is of the form [one can verify this by substitution into (14.18)]

$$f = \tanh u = \frac{e^u - e^{-u}}{e^u + e^{-u}} \qquad (14.20)$$

where $u = \frac{x}{\sqrt{2}\xi}$. When u is large, $f \simeq 1 - 2e^{-\sqrt{2}x/\xi}$, and when u is small $f \simeq u = \frac{x}{\sqrt{2}\xi}$. Deep in the bulk, $f \to 1$. Figure 14.2 shows f as a function of x with a coherence length ξ.

- At the normal-superconducting boundary $x = 0$ ($x > 0$ for su-perconducting state): one uses the magnetic induction for $x > 0$ $B_z(x) = B_z(0)e^{-x/\lambda}$ where λ is the London's penetration length to be determined. From the equation $\vec{B} = \vec{\nabla} \wedge \vec{A}$, one has $B_z(x) = -\frac{\partial A_y}{\partial x}$. One chooses then the gauge $A_y(x) = \lambda B_z(x)$. Replacing this into the kinetic energy density (14.10), one has

$$\frac{e^{*2}A^2|\Psi|^2}{2m^*} = \frac{e^{*2}\lambda^2 B_z^2|\Psi|^2}{2m^*} \qquad (14.21)$$

This should be equal to the field energy density $\frac{B^2}{2\mu_0}$ in (14.12) (theorem of energy equipartition). Equating them, one obtains

$$\lambda^2 = \frac{m^*}{e^{*2}\mu_0|\Psi|^2} = \frac{m^*}{e^{*2}\mu_0 n_s} \qquad (14.22)$$

One sees that for $x \gg \lambda$, well inside the superconducting phase, the magnetic field is zero. This phenomenon is known as the "Meissner effect".

- London's equation: one has defined the London's penetration length λ by putting $B_z(x) = B_z(0)e^{-x/\lambda}$. Initially, this length was obtained from the London's equation. This equation can be recovered using the second GL equation (14.15) as follows when $\vec{\nabla}\varphi = 0$ for $x \gg \xi$:

$$\vec{J} = \frac{e^*}{m^*}\left(\hbar\vec{\nabla}\varphi - e^*\vec{A}\right)|\Psi|^2$$

$$= -\frac{e^{*2}\vec{A}}{m^*}|\Psi|^2$$

$$\vec{\nabla}\wedge\vec{J} = -\frac{e^{*2}}{m^*}|\Psi|^2\vec{\nabla}\wedge\vec{A} = -\frac{e^{*2}}{m^*}|\Psi|^2\vec{B}$$

$$= -\frac{1}{\mu_0\lambda^2}\vec{B} \tag{14.23}$$

where one has used (14.22) in the last line. Note that the Ampere's law in electromagnetism gives $\vec{J} = \frac{1}{\mu_0}\vec{\nabla}\wedge\vec{B}$ and $\vec{\nabla}\cdot\vec{B} = 0$. One rewrites the former as

$$\vec{\nabla}\wedge\vec{J} = \frac{1}{\mu_0}\vec{\nabla}\wedge\vec{\nabla}\wedge\vec{B} = \frac{1}{\mu_0}\left[\vec{\nabla}(\vec{\nabla}\cdot\vec{B}) - \vec{\nabla}^2\vec{B}\right]$$

$$= -\frac{1}{\mu_0}\vec{\nabla}^2\vec{B} \tag{14.24}$$

Comparing this to (14.23) one gets

$$\vec{\nabla}^2\vec{B} = \frac{1}{\lambda^2}\vec{B} \tag{14.25}$$

This is the London's equation with the solution $B_z(x) = B_0e^{-x/\lambda} = -\frac{A_0}{\lambda}e^{-x/\lambda}$ as one has anticipated earlier. Using (14.23), one obtains the following current in the y direction for $x \gg \xi$:

$$J_y(x) = \frac{A_0}{\mu_0\lambda^2}e^{-x/\lambda} \tag{14.26}$$

Figure 14.3 shows the geometry of the coordinates system for illustration: the superconducting phase is limited by the yz plane and defined for $x > 0$. The applied magnetic field $\vec{B}$ is on the z axis. The resulting current $\vec{J}$ is on the y axis. One sees that both $\vec{B}$ and $\vec{J}$ decay with increasing x. The decay length is the London penetration depth λ.

One imagines the superconductor is a cylinder with a large enough radius. The magnetic field is applied along the z direction parallel

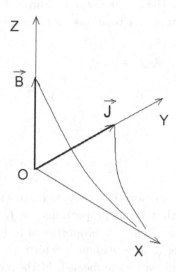

Fig. 14.3 $\vec{B}$ and $\vec{J}$ decay exponentially from the normal-superconducting phase boundary $x = 0$.

to the cylinder axis. The current of electrons circulating in the xy plane within a distance $\sim \lambda$ from the boundary of the cylinder creates an induced magnetic field which cancels the applied field inside the cylinder. That is the origin of the Meissner effect.

- Thermodynamic critical field: Deep inside the bulk superconducting state, there is no magnetic field (Meissner effect) one has $\vec{A} = 0$. The minimum of f_s of Eq. (14.12) is given by

$$\frac{\partial f_s}{\partial \Psi_\infty} = 0 \quad \Rightarrow \quad |\Psi_\infty|^2 = -\frac{\alpha}{\beta} \tag{14.27}$$

α being negative in the superconducting phase. One then has

$$f_s - f_n = -\frac{\alpha^2}{2\beta} \tag{14.28}$$

In order to destroy the superconducting state corresponding to the above free energy, one has to apply a field strong enough to have the field energy $-\frac{\mu_0 H_c^2}{2}$ to be equal to $f_s - f_n$ given above, namely

$$\frac{\mu_0 H_c^2}{2} = -(f_s - f_n) = \frac{\alpha^2}{2\beta} \tag{14.29}$$

H_c is called "thermodynamic critical field". From this, one has

$$H_c^2 = \frac{\alpha^2}{\mu_0 \beta} \tag{14.30}$$

Using (14.22) and the result obtained above, let us summarize in the following some important relations between the parameters in the GL theory:

$$|\Psi_\infty|^2 = -\frac{\alpha}{\beta}$$

$$\mu_0 H_c^2 = \frac{\alpha^2}{\beta} = -\alpha|\Psi_\infty|^2 = -\alpha n_s$$

$$\alpha = -\frac{\mu_0 H_c^2}{n_s} = -\frac{\mu_0^2 e^{*2}\lambda^2 H_c^2}{m^*}$$

$$\beta = -\frac{\alpha}{n_s} = -\frac{\mu_0^3 e^{*4}\lambda^4 H_c^2}{m^{*2}}$$

- Near the phase transition temperature T_c, one sees from the definition of H_c in (14.29) that H_c^2 is proportional to $f_s - f_n$. One knows that f_n is the free energy which is proportional to $(T - T_c)^2$ near T_c. This is seen by looking at the leading m^2 term in Eq. (13.58): m^2 near the transition is given in the mean-field theory by Eq. (13.32) which is proportional to $(T_c - T)$. The m^2 term of of the free energy f_n, namely $A(T - T_c^{MF})m^2$ of (13.58), is thus proportional to $(T - T_c)^2$. . However, one does not have such a simple expression for f_s. The result from a very early theory of Gorter and Casimir in 1934 [70] gives the temperature dependence of $H_c(T)$ by

$$H_c(T) = H_c(0)(1 - t^2)$$

where $t \equiv T/T_c$. This theory is reproduced in Problem 3 below. One expresses other parameters as follows:

$$\alpha = -\sqrt{\mu_0 \beta} H_c = -\sqrt{\mu_0 \beta} H_c(0)(1 - t^2)$$
$$= -\sqrt{\mu_0 \beta} H_c(0)(1 - t)(1 + t) \simeq 2\sqrt{\mu_0 \beta} H_c(0)(t - 1)$$

$$|\Psi_\infty|^2 = n_s = -\frac{\alpha}{\beta} = \frac{\alpha_0(1 - t)}{\beta} \propto (1 - t)$$

where $\alpha_0 = 2\sqrt{\mu_0 \beta} H_c(0)$. The above relations show that α changes its sign at the transition and n_s varies linearly with $T - T_c$ near the transition.

14.4   Superconductors of type II

One has seen above that there are two characteristic lengths λ and ξ. They have been obtained from different independent considerations. The coherence length ξ describes the length scale of variations in the magnitude of or the density of superconducting electrons $n_s = |\Psi|^2$, while λ describes the penetration depth of magnetic fields into a superconductor. In the above London's solution, the variation of n_s near the normal-superconducting boundary is not taken into account. Therefore, the London's solution is valid if ξ is small.

Let us define a new parameter κ by

$$\kappa = \frac{\lambda}{\xi} \tag{14.31}$$

Using (14.22) and (14.19) one has

$$\kappa = \sqrt{\frac{2m^{*2}\beta}{\mu_0 \hbar^2 e^{*2}}} \tag{14.32}$$

This expression shows that κ does not depend on the temperature. It depends on the material via β. Abrikosov [1] has shown that superconductors of type I correspond to $\kappa < \frac{1}{\sqrt{2}}$ and those of type II to $\kappa > \frac{1}{\sqrt{2}}$.

Let us discuss how to distinguish type I and type II. In type I superconductors, an applied magnetic field weaker than H_c is expelled from the superconducting state (Meissner effect). The transition from the superconducting state to the normal state occurs at H_c and is of first order. In type II superconductors, the magnetic field starts to penetrate into the superconducting space at a "critical field" H_{c1} to create a mixed phase of normal and superconducting states. As the field increases, this intermediate state will disappear at a second critical field H_{c2}. The material then undergoes a transition to the normal state. The mixed state between H_{c1} and H_{c2} is very interesting: Abrikosov [1] has shown that the vortices perpendicular to the applied field are created with a quantization of magnetic flux. These vortices, when numerous enough, forms a triangular lattice. Let us show the flux quantization and calculate the critical fields H_{c1} and H_{c2} in the following.

- Flux quantization:
  In the superconducting state, the density of superconducting electrons is constant, namely there is no current flow. Therefore from the second GL equation (14.14) one has

$$\oint \vec{J}.\vec{dl} = \frac{e^*}{m^*}|\Psi|^2 \left[\hbar \oint \vec{\nabla}\varphi.\vec{dl} - e^* \oint \vec{A}.\vec{dl} \right] = 0 \tag{14.33}$$

The first integral in the brackets is the variation of the phase φ on a closed loop: it is equal to $2\pi n$ where n is an integer. The second integral can be transformed as follows

$$\oint \vec{A}.\vec{dl} = \int_S \vec{\nabla} \wedge \vec{A}.\vec{dS} = \int_S \vec{B}.\vec{dS} = \phi \qquad (14.34)$$

where ϕ is the magnetic flux crossing the surface S. Equation (14.33) yields

$$\phi = 2\frac{\hbar}{e^*}\pi n = n\frac{h}{e^*} = n\phi_0 \qquad (14.35)$$

This equation shows the quantization of the magnetic flux in type II superconductors: the flux quantum is $\phi_0 = \frac{h}{e^*}$.

- Effective charge and mass:
  In the GL theory shown above one has used the effective charge and mass, e^* and m^*. As will be shown later, the GL theory will give the same results as the Bardeen-Cooper-Schrieffer theory if one identifies $e^* = 2e$ and $m^* = 2m$, where e and m are charge and mass of an electron. This equivalence shows that the quasiparticle in the GL theory is in fact a pair of electrons which will be called below "Cooper pair". The flux quantum is therefore

$$\phi_0 = 2\frac{\hbar}{e^*}\pi n = \frac{h}{2e} \qquad (14.36)$$

- Critical fields:
  Let H_{c1} be the lower critical field. For fields weaker than this, the Meissner effect takes place. For stronger fields, the field penetrates in the superconductor and creates vortices. One just considers one vortex. Let ϵ_1 be the energy of the vortex. One has (see the calculation given in Problem 4 below):

$$\epsilon_1 \simeq \frac{\phi_0^2}{4\pi\mu_0\lambda^2}\ln\frac{\lambda}{\xi} \qquad (14.37)$$

Supposing that λ/ξ is very large so the extent of $\vec{B}$ can be taken to infinity. The difference in Gibb's free energy between a superconductor containing no vortices and a superconductor containing one vortex, for a field parallel to a thick slab in the large λ/ξ limit, is then given by, at $T = 0$,

$$\Delta g_{GL} = \epsilon_1 - \int_0^\infty 2\pi r dr \vec{B}.\vec{H} = \epsilon_1 - \phi_0 H \qquad (14.38)$$

The phase separation occurs at $\Delta g_{GL} = 0$, namely at

$$H_{c1} = \frac{\epsilon_1}{\phi_0} \simeq \frac{\phi_0}{4\pi\mu_0\lambda^2}\ln\frac{\lambda}{\xi} \qquad (14.39)$$

where one has used (14.37). As the magnetic field increases above H_{c1}, the number of vortices increases. They come into interaction with each other when their distance is less than λ. A compact packing in a hexagonal lattice occurs, vortices then overlap to destroy the superconducting phase at the upper critical field H_{c2}. Let us calculate H_{c2}. We can first simplify the GL Hamiltonian in this situation before calculating. First, the order parameter Ψ is so small that one can neglect non collinear terms. Second, one can take the vector potential equal to that of the external field, namely $\vec{A} = (0, \mu_0 H x, 0)$ (or any others satisfying the gauge invariance $\vec{\nabla}.\vec{A} = 0$). The first GL equation (14.13) becomes

$$-\frac{\hbar^2}{2m^*}\frac{\partial^2\Psi}{\partial x^2} + \frac{1}{2m^*}(-i\hbar\frac{\partial}{\partial y} - e^*\mu_0 H x)^2\Psi - \frac{\hbar^2}{2m^*}\frac{\partial^2\Psi}{\partial z^2} = -\alpha\Psi \quad (14.40)$$

This is a Schrödinger equation which admits the following eigen-energy (see solution demonstrated in Problem 5 below):

$$|\alpha| = (n + \frac{1}{2})\omega_c + \frac{\hbar^2 k_z^2}{2m^*} \quad (14.41)$$

where n is an integer and ω_c denotes the cyclotron frequency

$$\omega_c = \frac{e^*\mu_0 H}{m^*} \quad (14.42)$$

Equation (14.41) gives the so-called Landau's levels in the case of a strong magnetic field applied on an electron gas (see Problem 5 below). Now, one has to choose for H_{c2} the strongest field among the magnetic fields satisfying (14.41): this corresponds obviously to the case where $n = 0$ and $k_z = 0$. One has

$$H_{c2} = \frac{2m^*|\alpha|}{\hbar\mu_0 e^*} = \frac{2m^*}{\hbar\mu_0 e^*}\frac{\hbar^2}{2m^*\xi^2} = \frac{\phi_0}{2\pi\mu_0\xi^2} \quad (14.43)$$

where one has used (14.19) and (14.36), with $e^* = 2e$ (see item "effective charge" above). Note that

$$\frac{H_{c2}}{H_{c1}} = 2(\frac{\lambda}{\xi})^2\frac{1}{\ln(\lambda/\xi)} \quad (14.44)$$

One sees that even for moderate values of λ/ξ, this ratio is far away from 1, i. e. the two critical fields are well separated.

14.5   Bardeen-Cooper-Schrieffer theory

The Bardeen-Cooper-Schrieffer (BCS) theory was published soon after the GL theory in 1957 [16, 17]. In this theory, the superconductivity is found to be originated from the motion of pairs of electrons bound via an electron-phonon interaction. Bound electron pairs, called Cooper pairs, are quasiparticles with effective charge $2e$ and effective mass $2m$. Two bound electrons have opposite spins so they are bosons of spin zero. One has seen in chapter 6 that bosons undergo the so-called Bose condensation below the Bose temperature and that superfluidity in He-4 can be explained by the condensation phenomenon. The case of the superconductivity bears a resemblance but details are quite different since in this case the superconductivity is caused by electron charges. The effect of magnetic field on moving charges is very important and plays thus an essential role in the mechanism of superconductivity as seen in the GL theory above.

One examines below the quantum microscopic BCS theory.

14.5.1   *Electron-phonon interaction*

It seems at a first sight that an attractive interaction between two electrons is impossible. However, Cooper [36] has shown that this coupling is possible via an intermediate phonon-electron interaction. To view this, let us imagine an electron is moving in a crystal constituted by an arrangement of ions of positive charges (the jellium model, where the ion positive charges are supposed to be uniformly distributed, treated in chapter 10 is not suitable to explain the superconductivity). When the electron passes through a crystal cell, it attracts ions toward its position. But by the time these ions are polarized, the electron is already far away because, due to its light mass with respect to that of ions, its velocity is much higher. The "retarded ion polarization" creates a surplus of positive charge at a space region which in turns attracts a nearby electron. The second electron will thus follow the path of the first one: there is therefore a "bound state" between them. This picture can be transformed into a formalism as seen below. At this stage, we are not immediately interested in writing down the phonon-electron interaction expression and trying to relate it to the electron-electron interaction. Rather, we will suppose an effective attractive interaction V between a pair of electrons and try to prove that the superconductivity comes from it. Let us briefly use the following picture: we consider a gas of N electrons with a filled Fermi sphere (ground state). We introduce two extra electrons into

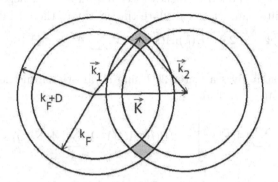

Fig. 14.4 Both $\vec{k}_1$ and $\vec{k}_2$ should lie in a shell of width $D = \hbar\omega_{\vec{q}}$ (phonon energy) above the Fermi sea represented by a sphere of radius k_F. The condition $\vec{K} = \vec{k}_1 + \vec{k}_2$=constant implies that $\vec{k}_1$ and $\vec{k}_2$ should lie in the gray areas.

the system, $[\vec{k}_1, E(\vec{k}_1)]$ and $[\vec{k}_2, E(\vec{k}_2)]$ where $k_1, k_2 > k_F$. The electron of wave vector $\vec{k}_1$ absorbs a phonon of wave vector $\vec{q}$, its energy is $E(\vec{k}_1 + \vec{q})$. This occurs only if $|E(\vec{k}_1 + \vec{q}) - E(\vec{k}_1)| \le \hbar\omega_{\vec{q}}$ where $\hbar\omega_{\vec{q}}$ is the phonon energy.

One considers first the wave function in the ground state $|GS\rangle$ where the Fermi sphere is filled. The creation of different pairs $(\vec{k}_1, \vec{k}_2)$ where $k_1, k_2 > k_F$ yields the wave function

$$\Psi_{N+2} = \sum_{\vec{k}_1, \vec{k}_2, \sigma_1, \sigma_2} A_{\sigma_1, \sigma_2}(\vec{k}_1, \vec{k}_2) c^+_{\vec{k}_1, \sigma_1} c^+_{\vec{k}_2, \sigma_2} |GS\rangle \qquad (14.45)$$

where $\sigma_i (i = 1, 2)$ are spins. To create a state with a predefined momentum, one should respect the conditions $\vec{K} = \vec{k}_1 + \vec{k}_2$=constant and $E(\vec{k}_1) - \epsilon_F \le \hbar\omega_{\vec{q}}$, while summing. These conditions are satisfied by the gray areas in Fig. 14.4.

The energy of the added electron pair is equal to their individual energies plus their interaction energy ΔE that we wish to calculate. We see that ΔE is largest if $\vec{K} = 0$: this case corresponds to $\vec{k}_2 = -\vec{k}_1 \equiv \vec{k}$. The wave-function becomes

$$\Psi_{N+2} = \sum_{\vec{k}} A(\vec{k}) c^+_{\vec{k}, \uparrow} c^+_{-\vec{k}, \downarrow} |GS\rangle \qquad (14.46)$$

The Hamiltonian of the above $N + 2$ electron gas is

$$\mathcal{H} = \sum_{\vec{k}, \sigma} E(\vec{k}) c^+_{\vec{k}, \sigma} c_{\vec{k}, \sigma} - \frac{V}{2} \sum_{\vec{k}, \vec{q}, \sigma} c^+_{\vec{k}+\vec{q}, \sigma} c^+_{-\vec{k}-\vec{q}, -\sigma} c_{-\vec{k}, -\sigma} c_{\vec{k}, \sigma} \qquad (14.47)$$

where V is not zero only for $|E(\vec{k}+\vec{q})-E(\vec{k})| \leq \hbar\omega_{\vec{q}}$. The value of $\omega_{\vec{q}}$ can be chosen as the Debye phonon frequency (see chapter 9). V is negative. Using the above Hamiltonian, let us calculate the energy of the added electrons:

$$E =< \Psi|\mathcal{H}|\Psi >= 2\sum_{\vec{k}} E(\vec{k})|A(\vec{k})|^2 - V\sum_{\vec{k},\vec{q}} A^*(\vec{k}+\vec{q})A(\vec{k}) \qquad (14.48)$$

where the spin sum gives a factor 2. Using a variational method with the constraint $\sum_{\vec{k}}|A(\vec{k})|^2 = 1$, one has

$$\frac{\partial}{\partial A^*(\vec{k}')}\left[E - \lambda\sum_{\vec{k}''}|A(\vec{k}'')|^2\right] = 2E(\vec{k}')A(\vec{k}')-V\sum_{\vec{q}} A(\vec{k}'-\vec{q})-\lambda A(\vec{k}') = 0$$
$$(14.49)$$

where λ is the Lagrange multiplier. Putting $\vec{k}' - \vec{q} = \vec{k}''$, one has

$$[2E(\vec{k}') - \lambda]A(\vec{k}') = V\sum_{\vec{k}''} A(\vec{k}'') \qquad (14.50)$$

Putting $C \equiv V\sum_{\vec{k}''} A(\vec{k}'')$, one has

$$A(\vec{k}) = \frac{VC}{2E(\vec{k}) - \lambda} \qquad (14.51)$$

Summing both sides over all states between ϵ_F and $\epsilon_F + \hbar\omega_{\vec{q}}$, one has

$$\sum_{\vec{k}} A(\vec{k}) = \sum_{\vec{k}} \frac{VC}{2E(\vec{k}) - \lambda}$$

$$C = \sum_{\vec{k}} \frac{VC}{2E(\vec{k}) - \lambda}$$

$$1 = \sum_{\vec{k}} \frac{V}{2E(\vec{k}) - \lambda}$$

Changing the sum into an integral, and noticing that λ is the energy variable E [see Eq. (7.18) for example], one writes

$$1 = \sum_{\vec{k}} \frac{V}{2E(\vec{k}) - E} = V\int_{\epsilon_F}^{\epsilon_F+\hbar\omega_{\vec{q}}} \frac{\rho(x)dx}{2x - E} \qquad (14.52)$$

with $x = E(\vec{k})$. Since the integration domain is very narrow, one replaces the density of states $\rho(x)$ by $\rho(\epsilon_F)$. One arrives at

$$E = 2\epsilon_F - \frac{2\hbar\omega_{\vec{q}}\exp[-2/\rho(\epsilon_F)V]}{1 - \exp[-2/\rho(\epsilon_F)V]}$$
$$\simeq 2\epsilon_F - 2\hbar\omega_{\vec{q}}\exp[-2/\rho(\epsilon_F)V] \qquad (14.53)$$

The last equality is for the case of weak V. The energy of the two added electrons is $2\epsilon_F$ if V is absent. When the interaction is turned on, the energy is diminished by the second term of (14.53). We will see below that the treatment of the Cooper pairs by the BCS theory gives the same result [see Eq. (14.82)].

14.5.2 *Cooper electron pairs — BCS Hamiltonian*

One defines the creation and annihilation operators of a bound Cooper electron pair by

$$B_{\vec{k}'}^+ = c_{\vec{k}',\uparrow}^+ c_{-\vec{k}',\downarrow}^+ \tag{14.54}$$

$$B_{\vec{k}'} = c_{-\vec{k},\downarrow} c_{\vec{k},\uparrow} \tag{14.55}$$

where $c_{\vec{k}',\uparrow}^+$ and $c_{\vec{k}',\uparrow}$ are creation and annihilation operators of the single-electron states $(\vec{k}',\uparrow)$, respectively.

One considers a gas of N electrons with the reduced Hamiltonian in the superconducting regime:

$$\mathcal{H} = \sum_{\vec{k}} \epsilon_{\vec{k}}(c_{\vec{k}}^+ c_{\vec{k}} + c_{-\vec{k}}^+ c_{-\vec{k}}) - V \sum_{\vec{k},\vec{k}'} c_{\vec{k}'}^+ c_{-\vec{k}'}^+ c_{-\vec{k}} c_{\vec{k}} \tag{14.56}$$

where $\epsilon_{\vec{k}}$ is the kinetic energy of the state $\vec{k}$ or $-\vec{k}$, and the second term represents excitations of Cooper pairs where one supposes that V is an attractive interaction between electrons of the pair in states near the Fermi level. This attractive interaction is due to a coupling with phonons discussed above. To simplify the writing of (14.56), the spins are implicitly supposed as follows: state $\vec{k}$ has a spin $\uparrow$ and state $-\vec{k}$ has a spin $\downarrow$.

Before showing the BCS theory, let us rapidly show a main result of the energy gap by an approximation: one writes the equation of motion for $c_{\vec{k}}$

$$i\hbar \frac{dc_{\vec{k}}}{dt} = [c_{\vec{k}}, \mathcal{H}] \tag{14.57}$$

and similarly that for $c_{-\vec{k}}^+$. Using the decomposition of operator chains shown in Problem 2 of chapter 7, one arrives at the two following equations of motion:

$$i\hbar \dot{c}_{\vec{k}} = \epsilon_{\vec{k}} c_{\vec{k}} - c_{-\vec{k}}^+ V \sum_{\vec{k}'} c_{-\vec{k}'} c_{\vec{k}'} \tag{14.58}$$

$$i\hbar \dot{c}_{-\vec{k}}^+ = -\epsilon_{\vec{k}} c_{-\vec{k}}^+ - c_{\vec{k}} V \sum_{\vec{k}'} c_{\vec{k}'}^+ c_{-\vec{k}'}^+ \tag{14.59}$$

To solve these equations, one linearizes them by replacing the products of operators by their average values, namely

$$\Delta_{\vec{k}} = V \sum_{\vec{k}'} \langle \phi_N^0 | c_{-\vec{k}'} c_{\vec{k}'} | \phi_{N+2}^0 \rangle \tag{14.60}$$

$$\Delta_{\vec{k}}^* = V \sum_{\vec{k}'} \langle \phi_N^0 | c_{\vec{k}'}^+ c_{-\vec{k}'}^+ | \phi_{N-2}^0 \rangle \tag{14.61}$$

where $|\phi_{N\pm2}^0\rangle$ is the ground state having $N \pm 2$ particles. This procedure is somewhat equivalent to the Hartree-Fock approximation or random-phase-approximation (RPA). One now replaces the sums in (14.58)-(14.59) by $\Delta_{\vec{k}}$ and $\Delta_{\vec{k}}^*$, one obtains two coupled equations. A solution of the form $c_{\vec{k}} \propto \exp(-i\omega_{\vec{k}}t)$ leads to

$$-i\hbar\omega_{\vec{k}} c_{\vec{k}} = \epsilon_{\vec{k}} c_{\vec{k}} - c_{-\vec{k}}^+ \Delta_{\vec{k}} \tag{14.62}$$

$$i\hbar\omega_{\vec{k}} c_{-\vec{k}}^+ = -\epsilon_{\vec{k}} c_{-\vec{k}}^+ - c_{\vec{k}} \Delta_{\vec{k}}^* \tag{14.63}$$

A non trivial solution imposes

$$\omega_{\vec{k}} = (\epsilon_{\vec{k}}^2 + \Delta^2)^{1/2} \tag{14.64}$$

where $\Delta^2 = \Delta_{\vec{k}} \Delta_{\vec{k}}^*$. Equation (14.64) is the energy of a quasiparticle (electron pair). Δ is the gap in the excitation spectrum. Note that this result can be exactly obtained (without the RPA approximation) by the Bogoliubov transformation defined by [27]:

$$a_{\vec{k}} = u_{\vec{k}} c_{\vec{k}} - v_{\vec{k}} c_{-\vec{k}}^+ \tag{14.65}$$

$$a_{\vec{k}}^+ = u_{\vec{k}} c_{\vec{k}}^+ - v_{\vec{k}} c_{-\vec{k}} \tag{14.66}$$

where $u_{\vec{k}} = u_{-\vec{k}}$ and $v_{\vec{k}} = -v_{-\vec{k}}$ (real) and $u_{\vec{k}}^2 + v_{\vec{k}}^2 = 1$. The operators $a_{\vec{k}}$ and $a_{\vec{k}}^+$ obey the anticommutation relations as shown here:

$$\begin{aligned}
[a_{\vec{k}}, a_{\vec{k}'}^+]_+ &= [u_{\vec{k}} c_{\vec{k}} - v_{\vec{k}} c_{-\vec{k}}^+, u_{\vec{k}'} c_{\vec{k}'}^+ - v_{\vec{k}'} c_{-\vec{k}'}]_+ \\
&= u_{\vec{k}} u_{\vec{k}'} [c_{\vec{k}}, c_{\vec{k}'}^+]_+ + v_{\vec{k}} v_{\vec{k}'} [c_{-\vec{k}}^+, c_{-\vec{k}'}]_+ \\
&= u_{\vec{k}}^2 \delta(\vec{k}, \vec{k}') + v_{\vec{k}}^2 \delta(\vec{k}, \vec{k}') \\
&= (u_{\vec{k}}^2 + v_{\vec{k}}^2) \delta(\vec{k}, \vec{k}') = \delta(\vec{k}, \vec{k}')
\end{aligned}$$

The other anticommutation relations are obtained in the same manner.

From (14.65)-(14.66), one has

$$c_{\vec{k}} = u_{\vec{k}} a_{\vec{k}} + v_{\vec{k}} a^+_{-\vec{k}} \tag{14.67}$$

$$c^+_{\vec{k}} = u_{\vec{k}} a^+_{\vec{k}} + v_{\vec{k}} a_{-\vec{k}} \tag{14.68}$$

$$c_{-\vec{k}} = u_{\vec{k}} a_{-\vec{k}} - v_{\vec{k}} a^+_{\vec{k}} \tag{14.69}$$

$$c^+_{-\vec{k}} = u_{\vec{k}} a^+_{-\vec{k}} + -v_{\vec{k}} a_{\vec{k}} \tag{14.70}$$

Replacing these relations in (14.56), one obtains a Hamiltonian with diagonal $(a^+_{\vec{k}} a_{\vec{k}} + a^+_{-\vec{k}} a_{-\vec{k}})$ term, non diagonal term and higher-order quadratic terms. This resulting Hamiltonian is diagonal if the coefficient of the non diagonal term is zero. This method of diagonalization is given in details in Problem 7 in chapter 7. Neglecting higher-order terms because one is in the ground state $(T = 0)$, one has

$$\mathcal{H} = 2 \sum_{\vec{k}} \epsilon_{\vec{k}} v^2_{\vec{k}} - V \sum_{\vec{k},\vec{k}'} u_{\vec{k}} v_{\vec{k}} u_{\vec{k}'} v_{\vec{k}'}$$

$$+ \sum_{\vec{k}} \left[\epsilon_{\vec{k}} (u^2_{\vec{k}} - v^2_{\vec{k}}) + 2 V u_{\vec{k}} v_{\vec{k}} \sum_{\vec{k}'} u_{\vec{k}'} v_{\vec{k}'} \right] (a^+_{\vec{k}} a_{\vec{k}} + a^+_{-\vec{k}} a_{-\vec{k}})$$

$$+ \sum_{\vec{k}} \left[2 u_{\vec{k}} v_{\vec{k}} \epsilon_{\vec{k}} - (u^2_{\vec{k}} - v^2_{\vec{k}}) V \sum_{\vec{k}'} u_{\vec{k}'} v_{\vec{k}'} \right] (a^+_{\vec{k}} a^+_{-\vec{k}} + a_{-\vec{k}} a_{\vec{k}})$$

$$\tag{14.71}$$

The first term of the first line contains the contribution of the free electron gas (when $k < k_F$). The second line represents the pair states in a diagonal form. Now, the whole Hamiltonian is diagonal if the coefficient in the brackets of the third line is zero, namely

$$\Delta (u^2_{\vec{k}} - v^2_{\vec{k}}) = 2 \epsilon_{\vec{k}} u_{\vec{k}} v_{\vec{k}} \tag{14.72}$$

where

$$\Delta = V \sum_{\vec{k}'} u_{\vec{k}'} v_{\vec{k}'} \tag{14.73}$$

Putting $u_{\vec{k}} = \cos(\theta_{\vec{k}}/2)$ and $v_{\vec{k}} = \sin(\theta_{\vec{k}}/2)$, one has

$$u^2_{\vec{k}} = \frac{1}{2}(1 + \cos\theta_{\vec{k}}) \quad v^2_{\vec{k}} = \frac{1}{2}(1 - \cos\theta_{\vec{k}})$$

$$u_{\vec{k}} v_{\vec{k}} = \cos(\theta_{\vec{k}}/2)\sin(\theta_{\vec{k}}/2) = \frac{1}{2}\sin\theta_{\vec{k}}$$

Replacing these into (14.72), one has

$$\Delta \cos \theta_{\vec{k}} = \epsilon_{\vec{k}} \sin \theta_{\vec{k}} \tag{14.74}$$

Hence

$$\tan \theta_{\vec{k}} = \Delta / \epsilon_{\vec{k}} \tag{14.75}$$

For a later use, let us give here some useful expressions derived from Eq. (14.74): putting it into square and adding $\Delta \sin^2 \theta_{\vec{k}}$ to both sides, one has

$$\Delta^2 [\cos^2 \theta_{\vec{k}} + \sin^2 \theta_{\vec{k}}] = [\epsilon_{\vec{k}}^2 + \Delta^2] \sin^2 \theta_{\vec{k}}$$

$$\Delta^2 = [\epsilon_{\vec{k}}^2 + \Delta^2] \sin^2 \theta_{\vec{k}}$$

$$\sin^2 \theta_{\vec{k}} = \frac{\Delta^2}{\epsilon_{\vec{k}}^2 + \Delta^2} \tag{14.76}$$

from which one has

$$\cos^2 \theta_{\vec{k}} = 1 - \sin^2 \theta_{\vec{k}} = 1 - \frac{\Delta^2}{\epsilon_{\vec{k}}^2 + \Delta^2} = \frac{\epsilon_{\vec{k}}^2}{\epsilon_{\vec{k}}^2 + \Delta^2} \tag{14.77}$$

Let us outline some important points in the following:

- Energy of an electron pair:
  The coefficient in front of $(a_{\vec{k}}^+ a_{\vec{k}} + a_{-\vec{k}}^+ a_{-\vec{k}})$ in (14.71) is by definition the energy of the electron pair $\omega_{\vec{k}}$. Using (14.76) and (14.77), one has

$$\omega_{\vec{k}} = \epsilon_{\vec{k}}(u_{\vec{k}}^2 - v_{\vec{k}}^2) + 2V u_{\vec{k}} v_{\vec{k}} \sum_{\vec{k}'} u_{\vec{k}'} v_{\vec{k}'}$$

$$= \epsilon_{\vec{k}}(u_{\vec{k}}^2 - v_{\vec{k}}^2) + 2u_{\vec{k}} v_{\vec{k}} \Delta$$

$$= \epsilon_{\vec{k}} \cos \theta_{\vec{k}} + \Delta \sin \theta_{\vec{k}} = \epsilon_{\vec{k}} \frac{\epsilon_{\vec{k}}}{(\epsilon_{\vec{k}}^2 + \Delta^2)^{1/2}} + \Delta \frac{\Delta}{(\epsilon_{\vec{k}}^2 + \Delta^2)^{1/2}}$$

$$= (\epsilon_{\vec{k}}^2 + \Delta^2)^{1/2} \tag{14.78}$$

This equation is the same as (14.64) obtained by an RPA approximation.

- Hamiltonian in the ground state:
  In the ground state no electron pairs are excited so the second line of (14.71) is zero because the "number" of electron pairs, i. e. $a_{\vec{k}}^+ a_{\vec{k}}$, is zero One rewrites thus the Hamiltonian as:

$$\mathcal{H} = 2 \sum_{\vec{k}} \epsilon_{\vec{k}} v_{\vec{k}}^2 - V \sum_{\vec{k}, \vec{k}'} u_{\vec{k}} v_{\vec{k}} u_{\vec{k}'} v_{\vec{k}'} \tag{14.79}$$

- Gap:

One calculates now the gap. Using (14.76) one has

$$\Delta = V \sum_{\vec{k}} u_{\vec{k}} v_{\vec{k}} = \frac{V}{2} \sum_{\vec{k}} \sin \theta_{\vec{k}} \qquad (14.80)$$

$$= \frac{V}{2} \sum_{\vec{k}} \frac{\Delta}{\sqrt{\epsilon_{\vec{k}}^2 + \Delta^2}} \qquad (14.81)$$

from which, by transforming the sum into an integral, one has

$$1 = \frac{V}{4} \int_{-\hbar\omega_{\vec{q}}}^{+\hbar\omega_{\vec{q}}} \frac{\rho(\epsilon) d\epsilon}{\sqrt{\epsilon^2 + \Delta^2}}$$

$$\simeq \frac{V}{4} \rho(\epsilon_F) \int_{-\hbar\omega_{\vec{q}}}^{+\hbar\omega_{\vec{q}}} \frac{d\epsilon}{\sqrt{\epsilon^2 + \Delta^2}}$$

where $\rho(\epsilon)$ is the density of states for one kind of spin, ρ of Eq. (2.39) divided by 2, taken approximately at the Fermi level $\rho(\epsilon_F)$. One has used the integral limits where $V = 0$ ($\epsilon_F \pm \hbar\omega_{\vec{q}}$). Changing the integral variable and integrating one obtains a logarithmic function which yields

$$\Delta = 2\hbar\omega_{\vec{q}} \exp[-2/\rho(\epsilon_F)V] \qquad (14.82)$$

This relation is similar to the binding energy of the Cooper pair calculated before Eq. (14.53).

14.5.3 *Ground-state wave function*

- One shows now that the ground state of the superconducting regime is given by the wave function:

$$|\phi^0\rangle = \prod_{\vec{k}} (u_{\vec{k}} + v_{\vec{k}} c_{\vec{k}}^+ c_{-\vec{k}}^+)|\text{vac}\rangle \qquad (14.83)$$

where only Cooper pairs ($\vec{k} \uparrow, -\vec{k} \downarrow$) appear. Other pairs with non opposite pairs of wave vectors may occur in the excited states. Cooper pairs lie near the Fermi level within a shell of width $\epsilon_F \pm \hbar\omega_{\vec{q}}$ where $\omega_{\vec{q}}$ is the phonon frequency interacting with electrons as discussed in paragraph 14.5.2. If one considers only the subspace of these states, one sees that ϕ^0 describes well these states thanks to operators $c_{\vec{k}}^+$ and $c_{-\vec{k}}^+$ which act on $|\text{vac}\rangle$. The quantities u and v are for normalization purpose. However, their choice should respect the fermion character of operators $c_{\vec{k}}^+$ and $c_{-\vec{k}}^+$ as one will see below.

- Note that that ϕ^0 is normalized:

$$\langle\phi^0|\phi^0\rangle = \prod_{\vec{k}}\langle\text{vac}|(u_{\vec{k}} + v_{\vec{k}}c_{-\vec{k}}c_{\vec{k}})(u_{\vec{k}} + v_{\vec{k}}c_{\vec{k}}^+c_{-\vec{k}}^+)|\text{vac}\rangle$$

$$= \prod_{\vec{k}}(u_{\vec{k}}^2 + v_{\vec{k}}^2)\langle\text{vac}|\text{vac}\rangle = 1 \qquad (14.84)$$

because

$$\langle\text{vac}|c_{-\vec{k}}c_{\vec{k}}c_{\vec{k}}^+c_{-\vec{k}}^+|\text{vac}\rangle = (1 - n_{\vec{k}})(1 - n_{-\vec{k}})\langle\text{vac}|\text{vac}\rangle$$

$$= \langle\text{vac}|\text{vac}\rangle = 1$$

where one has used $n_{\vec{k}} = n_{-\vec{k}} = 0$ in the vacuum state, and

$$\langle\text{vac}|c_{\vec{k}}c_{-\vec{k}}|\text{vac}\rangle = 0, \quad \text{and} \quad \langle\text{vac}|c_{\vec{k}}^+c_{-\vec{k}}^+|\text{vac}\rangle = 0$$

- Energy of the ground state in the superconducting regime:
One calculates $\langle\phi^0|c_{\vec{k}'}^+c_{\vec{k}'}|\phi^0\rangle$ and $\langle\phi^0|c_{\vec{k}'}^+c_{-\vec{k}'}^+c_{-\vec{k}''}c_{\vec{k}''}|\phi^0\rangle$ to deduce the energy of the ground state as follows:

$$\langle\phi^0|c_{\vec{k}'}^+c_{\vec{k}'}|\phi^0\rangle = \langle\text{vac}|(u_{\vec{k}'} + v_{\vec{k}'}c_{-\vec{k}'}c_{\vec{k}'})c_{\vec{k}'}^+c_{\vec{k}'}(u_{\vec{k}'}$$

$$+ v_{\vec{k}'}c_{\vec{k}'}^+c_{-\vec{k}'}^+)|\text{vac}\rangle\prod_{\vec{k}\neq\vec{k}'}\langle\phi^0|\phi^0\rangle$$

$$= \langle\text{vac}|[u_{\vec{k}'}^2 c_{\vec{k}'}^+c_{\vec{k}'} + v_{\vec{k}'}^2 c_{-\vec{k}'}c_{\vec{k}'}c_{\vec{k}'}^+c_{\vec{k}'}c_{\vec{k}'}^+c_{-\vec{k}'}^+$$

$$+ u_{\vec{k}'}v_{\vec{k}'}c_{\vec{k}'}^+c_{\vec{k}'}c_{\vec{k}'}^+c_{-\vec{k}'}^+ + ...]|\text{vac}\rangle$$

$$= [v_{\vec{k}'}^2] \qquad (14.85)$$

The other terms in [...] are 0. In the same manner, one obtains

$$\langle\phi^0|c_{\vec{k}'}^+c_{-\vec{k}'}^+c_{-\vec{k}''}c_{\vec{k}''}|\phi^0\rangle = u_{\vec{k}'}v_{\vec{k}'}u_{\vec{k}''}v_{\vec{k}''} \qquad (14.86)$$

The energy of the ground state is thus

$$E_g = \langle\phi^0|\mathcal{H}|\phi^0\rangle = 2\sum_{\vec{k}}\epsilon_{\vec{k}}v_{\vec{k}'}^2 - V\sum_{\vec{k}\neq\vec{k}'}u_{\vec{k}}v_{\vec{k}}u_{\vec{k}'}v_{\vec{k}'}$$

$$= \sum_{\vec{k}}\epsilon_{\vec{k}}(1 - \cos\theta_{\vec{k}}) - \frac{V}{4}\sum_{\vec{k}\neq\vec{k}'}\sin\theta_{\vec{k}}\sin\theta_{\vec{k}'} \qquad (14.87)$$

where the factor 2 in the first equality comes from the spin sum, and where one has used the definitions of $u_{\vec{k}}$ and $v_{\vec{k}}$. One obtains

$$E_g = -\sum_{\vec{k}} \epsilon_{\vec{k}} \cos \theta_{\vec{k}} - (\Delta^2/V) \tag{14.88}$$

where one has used Eq. (14.80) and the fact that $\sum_{\vec{k}} \epsilon_{\vec{k}} = 0$ because $\epsilon_{\vec{k}}$ is symmetric with respect to the Fermi level.

14.6 High-temperature superconductivity

The BCS theory presented above applies for conventional superconductors in which the electron-phonon interaction plays an important role in the formation of pairs of bound electrons. The temperature cannot break the electron bound energy at low temperatures so that the thermal dissipation cannot be transferred to electrons. The gap ensures thus the zero resistance. This mechanism however does not apply, at least with the form presented above, for high-T_c or high-temperature superconductors (HTS) where the critical temperature T_c ranges from 77 K, which is the temperature of boiling nitrogen liquid, to about 133 K (note that many authors call HTS for materials with T_c even less than 77 K). HTS have been discovered for the first time in 1986 by Bednorz and Müller [21]: they observed in a barium-doped compound of lanthanum and copper oxide the resistance dropped down to zero at a temperature around 35 K. Since then, many other HTS families have been found: in particular one can mention the cuprates family including YBaCuO, Bi-, Tl- and Hg-based high-Tc superconductors and the family of iron-based compounds which appears after 2008. The highest critical temperature is 133 K observed in mercury barium calcium copper oxide ($HgBa_2Ca_2Cu_3O_8$) [64, 155]. For the iron-based HTS, the highest critical temperatures are found in thin films of FeSe [66, 79, 116, 175] where a critical temperature in excess of 100 K has recently been reported. For applications, one needs materials not only with a high T_c but also with a high critical magnetic field. The BCS theory is limited by the strength of the electron-phonon interaction as seen above, it cannot explain T_c higher than about 23 K. For HTS, in spite of almost 30 years of efforts, there are no widely admitted theories. Certainly, the superconductivity in HTS is also due to bound pairs of electrons but mechanisms of electron pairing are not rigorously proved yet. However, there are some very interesting ideas such as the resonating valence bond theory [9] and the d-wave pairing [22, 73, 88, 104] which aimed at cuprate HTS.

Some common aspects in many HTS are (i) the existence of a long-range antiferromagnetic ordering in the proximity of the superconducting phase, this magnetic phase disappears when one enters the superconducting state as seen in the phase diagram (T_c, x) of YBaCuO (x: percentage of oxygen doping) (ii) the layered structure of alternating CuO_2 planes in compounds of perovskite structures. One of the properties of oxide superconductors is the superconductivity taking place between these CuO_2 layers. The more layers of CuO_2 the higher T_c. The first point has motivated researches in the d-wave pairing, focussing mainly on antiferromagnetic spin fluctuations [22, 73, 88, 104], while the second point has stimulated researches on the effects of the interlayer coupling which may enhance the conventional BCS mechanism. The debate is still under way.

14.7   Conclusion

In this chapter, one has presented the Ginzburg-Landau and Bardeen-Cooper-Schrieffer theories of superconductivity. The Ginzburg-Landau theory was based on the mean-field theory of the second-order phase transition. This theory shows the existence of two length scales, namely the London penetration depth λ and the coherence length ξ. By minimization of the phenomenological free energy, the Meissner effect was demonstrated and the spatial expansion of quasiparticles is shown by ξ. It was later improved by many people including Bogoliubov, Abrikosov and Gorkov who have shown in their respective works that the Ginzburg-Landau theory is a macroscopic version of the Bardeen-Cooper-Schrieffer theory. In particular, Abrikosov has demonstrated the existence of type II superconductors. The Bardeen-Cooper-Schrieffer theory was based on a microscopic mechanism, namely the electron-phonon interaction, which creates bound pairs of electrons near the Fermi surface. These bound pairs cause a gap in the electron spectrum which is at the origin of the superconductivity: this gap ensures the stability of the superconductivity and its zero resistance. To break this stability, one would need an appreciable amount of energy to break electron bound states. The Bardeen-Cooper-Schrieffer theory allows for a determination of the critical temperature and the gap energy.

14.8 Problems

Problem 1. Demonstrate the Ginzburg-Landau equations (14.13) and
(14.14).

Problem 2. Current density $\vec{J}$: gauge-invariance

a) Using (14.14), find (14.15).

b) Writing $\vec{J} = n_s e^* \vec{v}_s$ where n_s and $\vec{v}_s$ are the density and velocity
of superconducting electrons, calculate $\vec{v}_s$ and show that

$$m^* \vec{v}_s = \left(\hbar \vec{\nabla} \varphi - e^* \vec{A} \right)$$

c) Show that $\vec{J}$ is gauge-invariant.

Problem 3. Theory of Gorter-Casimir :

Prior to the Ginzburg-Landau theory, Gorter and Casimir [70] have
proposed in 1934 to describe the superconducting state as a state
with two components: the normal fluid with a concentration x and
the superfluid with a concentration $(1 - x)$. This picture is what
one has seen in the boson gas below the Bose temperature (chapter
6).

a) Writing the free energy as

$$F(x, T) = \sqrt{x} f_n(T) + (1 - x) f_s(T)$$

where $f_n(T) = -\gamma T^2/2$ (γ=constant) and $f_s(T) = -\beta$=constant,
calculate the value of x which minimizes $F(x, T)$.

b) Using Eq. (14.29), show that $H_c(T) = H_c(0)(1 - t^2)$ where $t = T/T_c$. Note that in spite of a phenomenological character in its
demonstration, this relation is experimentally verified.

Problem 4. Energy of a vortex:

a) Using the London's equation (14.25), calculate the energy of a
single vortex in type II superconductors assuming the condition
$\lambda \gg \xi$ (λ: London penetration length, ξ: Ginzburg-Landau coherence length).

b) Vortices interact with each other when their relative distance d
satisfies $d \lesssim \lambda$. The calculation of the interaction energy is rather
complicated. One admits the following interaction energy between
two vortices (see Refs. [103, 159] for demonstration):

$$E_{int} = A[K_0(d/\lambda) - K_0(\sqrt{2}d/\xi)]$$

where A is a positive constant. Show that the first term corresponds to a vortex-repulsive energy while the second to an attractive one.

c) Show that the first term dominates when

$$\kappa \equiv \frac{\lambda}{\xi} > \frac{1}{\sqrt{2}},$$

namely vortices stay separately, giving rise to type II superconductors.

Show that in the case where the second term dominates, i. e.

$$\kappa < \frac{1}{\sqrt{2}},$$

vortices attract each other, so that they disappear by merging into each other. This is the case of type I superconductors.

Problem 5. Electron gas in a strong magnetic field: Landau's levels, Landau diamagnetism

One shows below that a system of electrons of effective mass m^* under a strong magnetic field gives rise to a negative susceptibility. This phenomenon is called "Landau diamagnetism" which has the same mechanism as the vortices leading to the Meissner effect. When the applied magnetic field is strong, one cannot use the perturbation theory. One has to incorporate the action of the field in the Hamiltonian via the vector potential $\vec{A}(\vec{r})$. One supposes that the field $\vec{B}$ is applied along the z axis.

a) Solve Eq (14.40) to obtain the Landau's levels given in (14.41).

b) Give the degeneracy of each Landau's level using the conditions (i) the center x_0 of the circular trajectory should lie inside the sample (ii) the quantization of the in-plane wave-vector component k_y.

c) Show that the electron orbit is quantized.

d) Calculate the average magnetic moment of the system. Show that the susceptibility is negative, namely diamagnetic.

Chapter 15

Transport in Metals and Semiconductors

15.1 Introduction

This chapter is devoted to transport phenomena in metals and semiconductors which are at the heart of electronic applications. While the chapters presented in the first part of this book treat systems at equilibrium, the present chapter introduces systems out of equilibrium. For these systems, the most popular method is no doubt the use of the Boltzmann's equation [33]. Monte Carlo simulations [24] are more and more employed in the transport problem [124] using the standard Metropolis algorithm with some necessary modifications.

In what follows, one presents the formulation of the Boltzmann's equation and shows how to linearize its different terms for the case of systems not far from equilibrium. The relaxation-time approximation is next introduced to handle collision terms. Applications are given for metals and semiconductors which are the most used materials for electronic charge and spin transports. These subjects constitute one of the most important domains of research and development in view of its numerous industrial applications.

15.2 Boltzmann's equation

Moving particles carry with them physical quantities such as charges and spins which give rise to electric and spin currents. They also create heat currents resulting from a transfer of thermal energies from high- to low-temperature regions. It is obvious that a transport phenomenon takes place only when the system is out of equilibrium. To illustrate this, one considers an electric current in a metal. The current density is given by

$$\vec{j} = -e \int \vec{v} n(\vec{v}) d^3 v \tag{15.1}$$

where $n(\vec{v})$ is the density of electrons moving at the velocity $\vec{v}$ and $-e$ the electron charge. The integral is taken over all velocities. Let n_0 be the electron density at equilibrium. We have

$$n(\vec{v}) = n_0 + \Delta n \tag{15.2}$$

where Δn is the contribution induced by an external cause, for example an applied electric field. It is obvious that in the absence of Δn the integral (15.1) is zero because the integrand is an odd function of $\vec{v}$ which points in all directions with modulus from 0 to infinity. One thus has

$$\vec{j} = -e \int \vec{v} \Delta n d^3 v \tag{15.3}$$

To obtain $\vec{j}$, one has to know Δn. The objective for any transport theory is to estimate this quantity. One studies here this question with the Boltzmann's equation. One considers in the following only systems not far from equilibrium. Systems far from equilibrium which are very difficult to deal with [33], are not treated here.

15.2.1   *Classical formulation*

Let $n(x, y, z, p_x, p_y, p_z) dx dy dz dp_x dp_y dp_z$ be the number of particles in the volume $d\vec{r} d\vec{p}$ around the point $(\vec{r}, \vec{p})$ in the classical phase space. The variation of the density of particles in the volume results from collisions between particles and interaction between particles with an external field. One writes

$$\left(\frac{\partial n}{\partial t}\right)_C + \left(\frac{\partial n}{\partial t}\right)_F = \frac{dn}{dt} \tag{15.4}$$

where the indices C and F indicate "collision" and "field". In the regime out of equilibrium but stationary one has $\frac{dn}{dt} = 0$ which leads to

$$\left(\frac{\partial n}{\partial t}\right)_C + \left(\frac{\partial n}{\partial t}\right)_F = 0 \tag{15.5}$$

One gives below the explicit expressions of these terms:

(i) Expression of $\left(\frac{\partial n}{\partial t}\right)_F$:

The particles at (x, y, z, p_x, p_y, p_z) at the time $t + \Delta t$ were at $(x - \dot{x}\Delta t, y - \dot{y}\Delta t, z - \dot{z}\Delta t, p_x - \dot{p}_x\Delta t, p_y - \dot{p}_y\Delta t, p_z - \dot{p}_z\Delta t)$ at the time t. So, one has

$$
\begin{aligned}
\left(\frac{\partial n}{\partial t}\right)_F &= \lim_{\Delta t \to 0} \frac{n(x - \dot{x}\Delta t, ..., p_x - \dot{p}_x\Delta t, ...) - n(x, ..., p_x, ...)}{\Delta t} \\
&= -\left(\frac{\partial n}{\partial x}\right)\dot{x} - ... - \left(\frac{\partial n}{\partial p_x}\right)\dot{p}_x - ... \\
&= -\left[\dot{\vec{r}} \cdot \vec{\nabla}_{\vec{r}}n + \dot{\vec{p}} \cdot \vec{\nabla}_{\vec{p}}n\right] \\
&= -\left[\vec{v} \cdot \vec{\nabla}_{\vec{r}}n + \vec{F} \cdot \vec{\nabla}_{\vec{p}}n\right]
\end{aligned} \tag{15.6}
$$

where $\vec{F}$ is the force acting on the particles. The first term of the right-hand side is called "term of diffusion" due to the presence of $\vec{v}$.

(ii) Expression of $\left(\frac{\partial n}{\partial t}\right)_C$:

Let $P_{\vec{p}',\vec{p}}$ be the probability of the transition $\vec{p} \to \vec{p}'$ during a collision. $\left(\frac{\partial n}{\partial t}\right)_C$ is equal to the difference between the number of particles entering the volume $d\vec{r}d\vec{p}$ and the number of out-going particles. One writes

$$
\left(\frac{\partial n}{\partial t}\right)_C = \int [n(\vec{r}, \vec{p}')P_{\vec{p},\vec{p}'} - n(\vec{r}, \vec{p})P_{\vec{p}',\vec{p}}]\, d\vec{p}' \tag{15.7}
$$

The Boltzmann's equation (15.5) becomes

$$
\vec{v} \cdot \vec{\nabla}_{\vec{r}}n + \vec{F} \cdot \vec{\nabla}_{\vec{p}}n = \int [n(\vec{r}, \vec{p}')P_{\vec{p},\vec{p}'} - n(\vec{r}, \vec{p})P_{\vec{p}',\vec{p}}]\, d\vec{p}' \tag{15.8}
$$

15.2.2 Quantum formulation

For illustration, one considers the case of electrons. One replaces in (15.8) n by a distribution function $f(\vec{r}, \vec{k})$ which depends on the position $\vec{r}$ and on the wave vector $\vec{k} = \vec{p}/\hbar$. This function is the Fermi-Dirac distribution f_0 when the system is at equilibrium (see chapter 5). One also replaces

$$
n(\vec{r}, \vec{p}')P_{\vec{p},\vec{p}'} \quad \text{by} \quad f(\vec{r}, \vec{k}')[1 - f(\vec{r}, \vec{k})]P_{\vec{k},\vec{k}'}
$$
$$
n(\vec{r}, \vec{p})P_{\vec{p}',\vec{p}} \quad \text{by} \quad f(\vec{r}, \vec{k})[1 - f(\vec{r}, \vec{k}')]P_{\vec{k}',\vec{k}}
$$

The factor $[1 - f(\vec{r}, \vec{k})]$ is introduced to make sure that the state $f(\vec{r}, \vec{k})$ is empty, namely $f(\vec{r}, \vec{k}) = 0$, so that the transition $\vec{k}' \to \vec{k}$ is possible. The same thing is for the inverse transition. The Boltzmann's equation in the quantum case is thus written as

$$\frac{\hbar \vec{k}}{m^*} \cdot \vec{\nabla}_{\vec{r}} f(\vec{k}) + \frac{\vec{F}}{\hbar} \cdot \vec{\nabla}_{\vec{k}} f(\vec{k}) = \frac{\Omega}{(2\pi)^3} \int [f(\vec{k}')(1 - f(\vec{k})) P_{\vec{k},\vec{k}'}$$
$$- f(\vec{k})(1 - f(\vec{k}')) P_{\vec{k}',\vec{k}}] d\vec{k}' \qquad (15.9)$$

where m^* is the electron effective mass and Ω the system volume. For simplicity, $f(\vec{r}, \vec{k})$ was noted by $f(\vec{k})$ and for a large system the sum over all $\vec{k}'$ was replaced by an integral.

At equilibrium, the right-hand side of (15.9), $(\frac{\partial n}{\partial t})_C$, is zero, one has $f(\vec{k}) = f_0(\vec{k})$:

$$f_0(\vec{k}) = \frac{1}{e^{\beta(E_{\vec{k}} - \mu)} + 1}$$

The fluxes of in-going and out-going particles are equal at equilibrium, one has

$$f_0(\vec{k}')(1 - f_0(\vec{k})) P_{\vec{k},\vec{k}'} = f_0(\vec{k})(1 - f_0(\vec{k}')) P_{\vec{k}',\vec{k}}$$

$$P_{\vec{k},\vec{k}'} \left(\frac{1}{e^{\beta(E_{\vec{k}'} - \mu)} + 1} \right) \left(1 - \frac{1}{e^{\beta(E_{\vec{k}} - \mu)} + 1} \right) = P_{\vec{k}',\vec{k}} \left(\frac{1}{e^{\beta(E_{\vec{k}} - \mu)} + 1} \right)$$
$$\times \left(1 - \frac{1}{e^{\beta(E_{\vec{k}'} - \mu)} + 1} \right)$$

$$P_{\vec{k},\vec{k}'} e^{-\beta E_{\vec{k}'}} = P_{\vec{k}',\vec{k}} e^{-\beta E_{\vec{k}}} \qquad (15.10)$$

For an elastic collision, namely $E_{\vec{k}'} = E_{\vec{k}}$, one has $P_{\vec{k},\vec{k}'} = P_{\vec{k}',\vec{k}}$. The right-hand side of (15.9) is reduced to

$$\left(\frac{\partial f}{\partial t} \right)_C = \frac{\Omega}{(2\pi)^3} \int P_{\vec{k}',\vec{k}} \left[f(\vec{k}')(1 - f(\vec{k})) - f(\vec{k})(1 - f(\vec{k}')) \right] d\vec{k}'$$
$$= \frac{\Omega}{(2\pi)^3} \int P_{\vec{k}',\vec{k}} \left[f(\vec{k}') - f(\vec{k}) \right] d\vec{k}'$$
$$= \frac{\Omega}{(2\pi)^3} \int P_{\vec{k}',\vec{k}} \left[\varphi(\vec{k}') - \varphi(\vec{k}) \right] d\vec{k}' \qquad (15.11)$$

where

$$f(\vec{k}) \equiv f_0(\vec{k}) + \varphi(\vec{k}) \qquad (15.12)$$

$\varphi(\vec{k})$ represents the out-of-equilibrium contribution.

The Boltzmann's equation in its general form (15.9) is not possible to solve. When the system is not far from equilibrium, one can linearize and solve it without difficulty as seen below.

15.3 Linearized Boltzmann's equation

15.3.1 *Explicit linearized terms*

One linearizes the Boltzmann's equation (15.9) term by term as follows:

1. If the external force $\vec{F}$ is weak, f is not far from f_0, then

$$\vec{\nabla}_{\vec{k}} f \simeq \vec{\nabla}_{\vec{k}} f_0 = \frac{\partial f_0}{\partial E} \vec{\nabla}_{\vec{k}} E_{\vec{k}}$$

$$= \frac{\partial f_0}{\partial E} \frac{\hbar^2 \vec{k}}{m^*} \tag{15.13}$$

2. The term $\vec{\nabla}_{\vec{r}} f$ is linearized as

$$\vec{\nabla}_{\vec{r}} f \simeq \vec{\nabla}_{\vec{r}} f_0 = \frac{\partial f_0}{\partial T} \vec{\nabla}_{\vec{r}} T \qquad , \tag{15.14}$$

One has

$$\frac{\partial f_0}{\partial T} = \frac{\partial f_0}{\partial \beta} \frac{\partial \beta}{\partial T} + \frac{\partial f_0}{\partial \mu} \frac{\partial \mu}{\partial T}$$

$$= \frac{1}{k_B T^2} (E_{\vec{k}} - \mu) \frac{e^{\beta(E_{\vec{k}} - \mu)}}{(e^{\beta(E_{\vec{k}} - \mu)} + 1)^2} - \frac{\partial f_0}{\partial E_{\vec{k}}} \frac{\partial \mu}{\partial T}$$

$$= -\frac{(E_{\vec{k}} - \mu)}{T} [-\beta f_0 (1 - f_0)] - \frac{\partial f_0}{\partial E_{\vec{k}}} \frac{\partial \mu}{\partial T}$$

$$= -\left[\frac{E_{\vec{k}} - \mu}{T} + \frac{\partial \mu}{\partial T} \right] \frac{\partial f_0}{\partial E_{\vec{k}}} \tag{15.15}$$

where one has used the identity $\frac{\partial f_0}{\partial \mu} = -\frac{\partial f_0}{\partial E_{\vec{k}}}$. Finally,

$$\vec{\nabla}_{\vec{r}} f \simeq -\frac{\partial f_0}{\partial E_{\vec{k}}} \left[\frac{E_{\vec{k}} - \mu}{T} + \frac{\partial \mu}{\partial T} \right] \vec{\nabla}_{\vec{r}} T \tag{15.16}$$

The Boltzmann's equation (15.9) becomes

$$\left(\frac{\partial f}{\partial t} \right)_C = -\left(\frac{\partial f}{\partial t} \right)_F \simeq \frac{\hbar \vec{k}}{m^*} \cdot \frac{\partial f_0}{\partial E_{\vec{k}}} \vec{A} \tag{15.17}$$

where

$$\vec{A} = -e\vec{\varepsilon} - \left[\frac{E_{\vec{k}} - \mu}{T} + \frac{\partial \mu}{\partial T} \right] \vec{\nabla}_{\vec{r}} T \tag{15.18}$$

Note that one has used $\vec{F} = -e\vec{\varepsilon}$, $\vec{\varepsilon}$ being the applied electric field. One can also write $\vec{A}$ as

$$\vec{A} = -e\vec{\varepsilon} - \left(\frac{E_{\vec{k}} - \mu}{T}\right)\vec{\nabla}_{\vec{r}}T - \vec{\nabla}_{\vec{r}}\mu \tag{15.19}$$

Equation (15.17) is the linearized Bolzmann's equation. One will take another approximation before solving it: one introduces the relaxation-time approximation below for the left-hand side $(\frac{\partial f}{\partial t})_C$ of (15.17).

15.3.2 *Relaxation-time approximation*

If the external perturbation is weak, one can replace $(\frac{\partial f}{\partial t})_C$ by

$$\left(\frac{\partial f}{\partial t}\right)_C \simeq -\frac{f - f_0}{\tau} = -\frac{\varphi}{\tau} \tag{15.20}$$

where τ is the relaxation time between two collisions and (15.12) has been used. To be valid, τ should be independent of the external cause. One have to verify this condition when using this approximation. Writing

$$\left(\frac{\partial f}{\partial t}\right)_C = -\frac{\varphi}{\tau} \equiv +\frac{\phi}{\tau}\frac{\partial f_0}{\partial E} \tag{15.21}$$

one obtains the linearized Boltzmann's equation in the relaxation-time approximation:

$$\frac{\hbar\vec{k}}{m^*} \cdot \frac{\partial f_0}{\partial E}\vec{A} = \frac{\phi}{\tau}\frac{\partial f_0}{\partial E} \tag{15.22}$$

where $\vec{A}$ is given by (15.18). One has thus

$$\phi = \tau\vec{A} \cdot \frac{\hbar\vec{k}}{m^*} = \tau\vec{A} \cdot \vec{v} \tag{15.23}$$

Note that when one applies a magnetic field $\vec{B}$, the force becomes $\vec{F} = -e[\vec{\varepsilon} + \vec{v} \wedge \vec{B}]$. One cannot therefore replace f by f_0 in $\frac{\vec{F}}{\hbar} \cdot \vec{\nabla}_{\vec{k}}f$ because the field $\vec{B}$ will not appear in the final equation. To see the effect of $\vec{B}$, one has to go to the second order: one have to replace f by $f_0 + \varphi$, not by f_0. The solution (15.23) becomes (see demonstration in Problem 1)

$$\phi = \tau\vec{A} \cdot \vec{v} + \tau\frac{e}{\hbar^2}\vec{B} \cdot \left[\vec{\nabla}_{\vec{k}}\phi \wedge \vec{\nabla}_{\vec{k}}E\right] \tag{15.24}$$

This equation is a differential equation of ϕ because of the effect of $\vec{B}$.

15.4 Applications in general transport problems

From section 15.5, one focuses on the electric conductivity resulting from the motion of electrons using the Boltzmann's equation. However, this equation can be applied to general transport problems. One defines in this section some transport phenomena frequently encountered in applications.

15.4.1 *Heat current*

In the case where $\vec{\nabla}_{\vec{r}}T \neq 0$ in Eq. (15.19), one defines the heat current $\vec{Q}$ (or thermal current) as the surplus of heat with respect to the Fermi level:

$$\begin{aligned} \vec{Q} &= \frac{1}{4\pi^3}\int d\vec{k}(E - E_F)\varphi\vec{v} \\ &= \frac{1}{4\pi^3}\int d\vec{k}E\varphi\vec{v} - E_F\frac{1}{4\pi^3}\int d\vec{k}\varphi\vec{v} \\ &= \vec{W} + \frac{E_F}{e}\vec{j} \end{aligned} \qquad (15.25)$$

where

$$\vec{W} = \frac{1}{4\pi^3}\int d\vec{k}E\varphi\vec{v} \qquad (15.26)$$

Note that φ and ϕ are defined in (15.20)-(15.21).

15.4.2 *Thermo-electric current*

In the case where $\vec{\nabla}_{\vec{r}}T \neq 0$ and $\vec{\nabla}_{\vec{r}}\mu \neq 0$ in (15.19), the solution (15.23) gives, in the presence of an electric field,

$$\begin{aligned} \vec{j}_z &= -e\int \varphi v_z d^3v = e\int \phi\frac{\partial f_0}{\partial E}v_z d^3v \\ &= -e\left[K_0(-e\varepsilon_z - \frac{dE_F}{dz}) - (K_1 - E_F K_0)\frac{1}{T}\frac{dT}{dz}\right] \end{aligned} \qquad (15.27)$$

where

(i) one has supposed that the gradients of the temperature and the chemical potential are both in the z direction only, and $\vec{\varepsilon} \parallel \vec{Oz}$,

(ii) $\mu \simeq E_F$ which is valid at low temperatures,

(iii) one has used the following notation:

$$K_m \equiv -\frac{1}{4\pi^3}\int \frac{\partial f_0}{\partial E}E^m\tau v_z^2 d\vec{k} \qquad (15.28)$$

Replacing (15.27) in (15.25) one obtains

$$Q_z = (K_1 - E_F K_0)\left(-e\varepsilon_z - \frac{dE_F}{dz}\right) - (K_2 - 2E_F K_1 + E_F^2 K_0)\frac{1}{T}\frac{dT}{dz}$$

(15.29)

From (15.27), one gets ε_z:

$$\varepsilon_z = \frac{j_z}{e^2 K_0} - \frac{K_1 - E_F K_0}{eK_0 T}\frac{dT}{dz} - \frac{1}{e}\frac{dE_F}{dz}$$

(15.30)

Replacing ε_z in (15.29) one arrives at

$$Q_z = -\frac{K_1 - E_F K_0}{eK_0}j_z + \frac{K_1^2 - K_0 K_2}{K_0 T}\frac{dT}{dz}$$

(15.31)

The two above general equations relate four following effects: electric conductivity (Ohm's law), thermal conductivity, Seebeck effect and Peltier effect. One has separately

- Electric conductivity when $\frac{dT}{dz} = 0$, $\frac{dE_F}{dz} = 0$:

$$j_z = \sigma\varepsilon_z \quad \text{with } \sigma = e^2 K_0$$

(15.32)

- Thermal conductivity when $j_z = 0$, $\frac{dE_F}{dz} = 0$:

$$Q_z = -\lambda\frac{dT}{dz} \quad \text{with } \lambda = \frac{-K_1^2 + K_0 K_2}{K_0 T}$$

(15.33)

- Seebeck effect when $j_z = 0$, $\frac{dE_F}{dz} = 0$:

$$\varepsilon_z = S\frac{dT}{dz} \quad \text{with } \quad S = -\frac{K_1 - E_F K_0}{eK_0 T}$$

(15.34)

- Peltier effect when $\frac{dT}{dz} = 0$:

$$Q_z = Pj_z \quad \text{with } \quad P = -\frac{K_1 - E_F K_0}{eK_0}$$

(15.35)

The quantities σ, λ, S and P are expressed as functions of integrals K_m defined by (15.28). One calculates below these integrals.

One considers a surface $k=$constant in the $\vec{k}$ space. A variation of k gives rise to an energy variation dE given by

$$dE = \vec{\nabla}_{\vec{k}}E \cdot d\vec{k} = \left|\vec{\nabla}_{\vec{k}}E\right|dk_\perp$$

One writes

$$d\vec{k} \Rightarrow dk_{\perp}d^2s = \frac{dE_k}{\left|\vec{\nabla}_{\vec{k}}E\right|}d^2s$$

where d^2s is an elementary surface on the surface $k = $ constant. K_m is rewritten as

$$K_m = -\int_0^\infty dE \frac{\partial f_0}{\partial E} E^m G(E) \tag{15.36}$$

where

$$G(E) = \frac{1}{4\pi^3} \int_{E=\text{constant}} \tau v_z^2 \frac{d^2s}{\left|\vec{\nabla}_{\vec{k}}E\right|} \tag{15.37}$$

One has taken above the limit ∞ for the upper bound of (15.36). This approximation does not change the physical final result because at low temperatures $\frac{\partial f_0}{\partial E}$ has significant contribution only around the Fermi level (see chapter 5). It remains now to calculate $G(E)$. Integrating by parts (15.36) yields

$$K_m = -\left\{ f_0 E^m G(E)\big|_0^\infty - \int_0^\infty dE f_0 \left[(m-1)E^{m-1}G(E) + E^m G'(E)\right] \right\}$$

$$= \int_0^\infty dE f_0 \left[(m-1)E^{m-1}G(E) + E^m G'(E)\right] \tag{15.38}$$

At low temperatures, one can use the Sommerfeld expansion (5.26) for these integrals. One obtains

$$K_m = E_F^m G(E_F) +$$
$$\frac{\pi^2}{6}(k_B T)^2 \left[E^m \frac{d^2 G}{dE^2} + 2mE^{m-1}\frac{dG}{dE} + m(m-1)E^{m-2}G \right]_{E=E_F} \tag{15.39}$$

If one takes $E_F \simeq E_F^0 (T=0)$, one has at the lowest order

$$K_m \simeq (E_F^0)^m G(E_F^0) \tag{15.40}$$

To calculate $G(E)$, one supposes a spherical equi-energetic surface, namely $E = \frac{\hbar^2 k^2}{2m^*}$ (effective-mass approximation). One then has

$$\left|\vec{\nabla}_{\vec{k}}E\right| = \frac{\hbar^2 k}{m^*}$$
$$d^2s = 2\pi k^2 \sin\theta d\theta$$

One rewrites (15.37) as

$$G(E) = \tau \frac{k^3}{4\pi^3 m^*} \int_0^\pi \cos^2\theta \sin\theta d\theta$$

$$= \frac{1}{3\pi^2 m^*} \left(\frac{2m^*}{\hbar^2}\right)^{3/2} E^{3/2} \tau(E) \qquad (15.41)$$

where one has supposed that the relaxation time τ depends on E. The explicit form of $\tau(E)$ depends on the nature of collisions which take place in the system. One gives some concrete cases in the solution of Problem 2 below. Using (5.19), one has

$$G(E_F^0) = \frac{n}{m^*}\tau(E_F^0) \qquad (15.42)$$

where n is the electron density. Using (15.40) and (15.42), one obtains the coefficients defined in (15.32)-(15.35). One has explicitly

15.4.3   *Electric conductivity*

For the electric conductivity, one has

$$\sigma = e^2 G(E_F^0) = \frac{e^2 n \tau(E_F^0)}{m^*} \qquad (15.43)$$

This result will be obtained again in (15.55) by a direct calculation shown below.

15.4.4   *Thermal conductivity*

One sees that if one uses only the first term of (15.39) the numerator of λ defined in (15.33) is zero. One must therefore use the T^2 term of (15.39). One has then

$$\lambda = \frac{\pi^2}{3} k_B^2 T G(E_F^0) = \frac{\pi^2}{3} \frac{n}{m^*} k_B^2 T \tau(E_F^0) \qquad (15.44)$$

The dependence of λ on T is contained also in $\tau(E_F^0)$. Combining (15.43) and (15.44) one obtains the Wiedemann-Franz law:

$$\frac{\lambda}{\sigma} = \frac{\pi^2}{3}\left(\frac{k_B}{e}\right)^2 T \qquad (15.45)$$

The ratio $\frac{\lambda}{\sigma}$ is thus independent of the material. The coefficient $\frac{\pi^2}{3}\left(\frac{k_B}{e}\right)^2$ is called the "Lorenz number".

15.4.5 *Seebeck effect*

To calculate S defined in (15.33), one must go to the order of T^2 in (15.39). One obtains

$$S = -\frac{\pi^2}{3}\frac{k_B^2}{e}T\left[\frac{1}{G}\frac{dG}{dE}\right]_{E=E_F^0} \tag{15.46}$$

Note that the thermal power is defined by

$$U(T) = \int_0^T S dT \tag{15.47}$$

and the difference of voltage at two points A and B of the material in the z direction is written as

$$U_{AB} = -\int_A^B \varepsilon_z dz \simeq -\int_A^B S dT \tag{15.48}$$

where one has used (15.30) with $j_z = 0$, neglecting the variation of E_F between A and B.

15.4.6 *Peltier effect*

This effect describes the thermal current induced by an electric current in the absence of a temperature gradient. However, when an electric current crosses a junction between two metals A and B, it produces a thermal barrier $Q_{AB} = (P_A - P_B)j$ because of the fact $P_A \neq P_B$ (different metals). In order not to create a temperature gradient between the two sides of the junction (Peltier effect), one has two heat or to cool one of the two sides.

15.5 Resistivity

The study of the resistivity is one of the fundamental tasks in materials science because the transport properties are massively used in electronic devices. The resistivity has been studied since the discovery of the electron a century ago by the simple Drude theory using the classical free particle model with collisions due to atoms in the crystal. The following relation is established between the conductivity σ and the electronic parameters e (charge) and m (mass):

$$\sigma = \frac{ne^2\tau}{m} \tag{15.49}$$

where τ is the electron relaxation time, namely the average time between two successive collisions. In more sophisticated treatments of the resistivity where various interactions are taken into account, this relation is still valid provided two modifications (i) the electron mass is replaced by its effective mass which includes various effects due to the interactions with its environment (ii) the relaxation time τ is not a constant but dependent on collision mechanisms. The first modification is very important, the electron can have a "heavy" or "light" effective mass which modifies its mobility in crystals. The second modification has a strong impact on the temperature dependence of the resistivity: τ depends on some power of the electron energy, this power depends on diffusion mechanisms such as collisions with charged impurities, neutral impurities, magnetic impurities, phonons, magnons, etc. As a consequence, the relaxation time averaged over all energies $< \tau >$ depends differently on T according to the nature of the collision source. The properties of the total resistivity stem thus from different kinds of diffusion processes. Each contribution has in general a different temperature dependence. The most important contributions to the total resistivity $R_t(T)$ at low temperatures have been given in Eq. (14.1).

15.6   Spin-independent transport in metals — Ohm's law

In metals, conducting electrons are considered as completely free or free with an effective mass which includes weak interactions with their environment such as weak periodic potential of the ion lattice. The description of these electrons by the Fermi gas is valid.

To calculate the resistivity in metals, one considers the effect of an electric field: one assumes in (15.19) that $\vec{B} = 0$, $\vec{\nabla}_{\vec{r}} T = 0$ (no temperature gradient) and $\vec{\nabla}_{\vec{r}} \mu = 0$ (no gradient of chemical potential). From (15.24), one has

$$\phi = -\tau e \vec{\varepsilon} \cdot \vec{v} \qquad (15.50)$$

One obtains

$$\begin{aligned} f &= f_0 + \varphi = f_0 - \phi \frac{\partial f_0}{\partial E} \\ &= f_0 - \tau e \vec{\varepsilon} \cdot \vec{v} \frac{\partial f_0}{\partial E} \\ &= f_0 - \tau e \varepsilon_z v_z \frac{\partial f_0}{\partial E} \end{aligned} \qquad (15.51)$$

where one supposed $\vec{\varepsilon} \parallel \vec{Oz}$. The current density ($\Omega = 1$, volume unit) in the z direction is

$$j_z = -e \int \varphi v_z d^3 v = e \int \phi \frac{\partial f_0}{\partial E} v_z d^3 v$$

$$= -e^2 \varepsilon_z \int d^3 v v_z \tau \frac{\partial f_0}{\partial E}$$

or, in quantum version, $\int d^3 v ... = \frac{2\Omega}{(2\pi)^3} \int d\vec{k}...$ (factor 2 is spin degeneracy),

$$j_z = -e^2 \varepsilon_z \frac{2\Omega}{(2\pi)^3} \int \tau \frac{\hbar}{m^*} d\vec{k} \hbar k_z m^* \frac{\partial f_0}{\partial E}$$

$$= -\frac{e^2 \varepsilon_z \hbar^2}{(m^*)^2} \frac{\Omega}{4\pi^3} \int \tau 2\pi k^2 dk \frac{\partial f_0}{\partial E} \int_0^\pi k^2 \cos^2 \theta \sin \theta d\theta$$

$$= -\frac{e^2 \varepsilon_z \hbar^2}{(m^*)^2} \frac{\Omega}{4\pi^3} \int \tau 2\pi k^4 dk \frac{\partial f_0}{\partial E} \left[-\frac{\cos^3 \theta}{3} \right]_0^\pi$$

$$= -\frac{e^2 \varepsilon_z \hbar^2}{(m^*)^2} \frac{\Omega}{4\pi^3} \frac{2}{3} \int \tau 2\pi k^4 dk \frac{\partial f_0}{\partial E} \tag{15.52}$$

Transforming this integral into an integral on E, $\frac{\Omega}{4\pi^3} \int 4\pi k^2 dk... = \int dE \rho(E)...$, one gets

$$j_z = -\frac{e^2 \varepsilon_z \hbar^2}{(m^*)^2} \frac{1}{3} \int \tau \rho(E) \left(\frac{2m^* E}{\hbar^2} \right) dE \frac{\partial f_0}{\partial E}$$

$$= -\frac{2e^2 \varepsilon_z}{3m^*} \int \tau \rho(E) E \frac{\partial f_0}{\partial E} dE \tag{15.53}$$

where $\rho(E)$ is the density of states given by (2.36) with $\Omega = 1$.

One supposes now that the electric field is weak, so τ does not depend on the energy E of the electron. One can replace τ by an average value $<\tau>$. Using the expression (2.36) for $\rho(E)$, and integrating by parts, one has

$$j_z = -\frac{e^2 \varepsilon_z}{3\pi^2} \frac{<\tau>}{m^*} \left(\frac{2m^*}{\hbar^2} \right)^{3/2} \frac{3}{2} \left[f_0 E^{3/2} |_0^\infty - \int_0^\infty f_0 E^{1/2} dE \right]$$

$$= \frac{e^2 \varepsilon_z}{m^*} \frac{<\tau>}{2\pi^2} \frac{1}{2\pi^2} \left(\frac{2m^*}{\hbar^2} \right)^{3/2} \int_0^\infty f_0 E^{1/2} dE$$

$$= \frac{e^2 \varepsilon_z}{m^*} <\tau> \int_0^\infty f_0 \rho(E) dE \tag{15.54}$$

where $f_0 E^{3/2}|_0^\infty = 0$ (f_0 tends to 0 faster than $E^{3/2}$ when $E \to \infty$). The last integral is the total number of electrons n per volume unit [cf. (5.13)]. One has finally

$$j_z = \frac{e^2 \varepsilon_z <\tau>}{m^*} n = \sigma \varepsilon_z \qquad (15.55)$$

where

$$\sigma \equiv \frac{e^2 <\tau>}{m^*} \qquad (15.56)$$

σ is called "electric conductivity". Equation (15.55) is the "Ohm's law" which can be found by a simpler calculation. However, the method using the solution of the Boltzmann's equation allows us to treat the case of strong fields or when τ depends on E. In these cases, the Ohm's law is no more valid.

In the general case when the electric field $\vec{\varepsilon}$ is applied in an arbitrary direction the conductivity is a tensor. One writes the current density in the i direction as

$$j_i = -e \int \varphi v_i d^3 v = e \int \phi \frac{\partial f_0}{\partial E} v_i d^3 v$$

$$= -e^2 \int d^3 v \tau v_i \sum_j v_j \varepsilon_j \frac{\partial f_0}{\partial E}$$

$$\equiv \sum_{j=x,y,z} \sigma_{ij} \varepsilon_j \qquad (15.57)$$

where

$$\sigma_{ij} = -\frac{e^2}{4\pi^3} \int \frac{\partial f_0}{\partial E} \tau v_i v_j d\vec{k} \qquad (15.58)$$

15.7   Transport in strong electric fields — Hot electrons

Let us consider the Boltzmann's equation in the case of electrons moving in a strong electric field. We will see below that the Ohm's law is no more valid.

One applies an electric field $\vec{\varepsilon}$ parallel to the $\vec{Oz}$ axis. The distribution function f depends on the electron energy E and the angle θ between the wave vector $\vec{k}$ and $\vec{\varepsilon}$. One expresses f as a series of the Legendre polynomials P_n as follows

$$f(E, \cos \theta) = \sum_{n=0}^{\infty} g_n(E) P_n(\cos \theta).$$

One retains only the first two terms, namely

$$f(E, \cos\theta) = g_0(E) + g_1(E)\cos\theta.$$

• Field term:
Since $\cos\theta = k_z/k$, one has

$$f(E, \cos\theta) = g_0(E) + \frac{g_1(E)}{k}k_z$$

$$\frac{\partial f(E, \cos\theta)}{\partial k_z} = \frac{\partial g_0}{\partial k_z} + g + k_z\frac{\partial g}{\partial k_z} \quad (g \equiv g_1/k)$$

$$= \frac{\hbar^2 k_z}{m^*}\frac{dg_0}{dE} + g + \frac{\hbar^2 k_z^2}{m^*}\frac{dg}{dE}$$

$$= \frac{\hbar^2 k_z}{m^*}g_0' + g + \frac{\hbar^2 k_z^2}{m^*}g' \qquad (15.59)$$

Note that

$$\frac{\hbar^2 k_z^2}{m^*} = \frac{\hbar^2 k^2 \cos^2\theta}{m^*} = \frac{\hbar^2 k^2}{m^*}\frac{1}{3} = \frac{2}{3}E$$

where $\cos^2\theta$ was replaced by $1/3$ because we have set $P_2 = 0$, the field term in the Boltzmann's equation is thus

$$-\left(\frac{\partial f}{\partial t}\right)_F = \frac{q\varepsilon}{\hbar}\frac{\partial f(E, \cos\theta)}{\partial k_z}$$

$$= \frac{q\varepsilon}{\hbar}\left(g + \frac{2}{3}Eg' + \frac{\hbar^2 k_z}{m^*}g_0'\right) \qquad (15.60)$$

where $g = g_1/k$ ($k = k_z\cos\theta$), g' and g_0' are the first derivatives with respect to E.

• Collision term:
The collision term of the Boltzmann's equation is written as [see the right-hand side of Eq. (15.9)]

$$-\left(\frac{\partial f}{\partial t}\right)_C = \frac{\Omega}{(2\pi)^3}\int\left[w(\vec{k}, \vec{k}')f(\vec{k}') - w(\vec{k}', \vec{k})f(\vec{k})\right]d^3k' \qquad (15.61)$$

where one took $1 - f \simeq 1$ for high temperatures. Replacing f by (15.59), one has

$$\left(\frac{\partial f}{\partial t}\right)_C = \phi_0 + \phi_1$$

where

$$\phi_0 = \frac{\Omega}{(2\pi)^3} \int \left[w(\vec{k}, \vec{k}') g_0(E') - w(\vec{k}', \vec{k}) g_0(E) \right] d^3 k'$$

$$\phi_1 = \frac{\Omega}{(2\pi)^3} \int \left[w(\vec{k}, \vec{k}') k_z' g(E') - w(\vec{k}', \vec{k}) k_z g(E) \right] d^3 k'$$

$w(\vec{k}, \vec{k}')$ is the probability per time unit of the transition $\vec{k} \to \vec{k}'$. Taking into account the probability of an electron in the initial state $\vec{k}$ and that of the empty final state $\vec{k}'$, the transition probability is written as

$$\bar{w} = w(\vec{k}, \vec{k}') f(E_k)[1 - f(E_{k'})]$$

with

$$w(\vec{k}, \vec{k}') = \frac{\pi}{NM} \frac{|I|^2}{w_{\vec{q}_j}} \left[s(\omega_{\vec{q}_j}) + \frac{1}{2} \pm \frac{1}{2} \right] \delta(E_{k'} - E_k \pm \hbar\omega_{\vec{q}_j})$$

where I is the electron-phonon coupling, N the total number of electrons and M the mass of ions. The signs $\pm$ correspond respectively to emission and absorption of a phonon of energy $\hbar\omega_{\vec{q}_j}$ of the phonon branch j, of wave vector $\vec{q}$. $s(\omega_{\vec{q}_j})$ is the occupation number of mode $\omega_{\vec{q}_j}$. One supposes that $f[1 - f] \simeq f$ (high temperatures). For acoustic phonons, one shall use $\omega_{(\vec{q})} = v_s q$ (v_s: sound velocity) and $I = \pm iCq$ (C: deformation potential). After a lengthy calculation, one arrives at the following result (see Ref. [77] for details):

$$\phi_0 = \frac{2m^* v_s^2}{\tau} \left[E g_0'' + \left(\frac{E}{k_B T} + 2 \right) g_0' + \frac{2g_0}{k_B T} \right] \tag{15.62}$$

$$\phi_1 = -\frac{k_z g}{\tau} \tag{15.63}$$

where τ is the relaxation time calculated for weak fields:

$$\tau = \frac{2\pi M}{\Omega} \frac{\hbar v_s^2}{C^2} \left(\frac{\hbar^2}{2m^*} \right)^{3/2} \frac{1}{\sqrt{E}} \frac{1}{k_B T} \tag{15.64}$$

With Eqs. (15.60), (15.62), (15.63) and (15.64), one has

$$2m^* v_s^2 \left[E g_0'' + \left(\frac{E}{k_B T} + 2 \right) g_0' + \frac{2g_0}{k_B T} \right] - k_z g$$

$$= \tau \frac{q\varepsilon}{\hbar} \left(g + \frac{2}{3} E g' + \frac{\hbar^2 k_z}{m^*} g_0' \right)$$

$$= \frac{\ell}{v} \frac{q\varepsilon}{\hbar} \left(g + \frac{2}{3} E g' + \frac{\hbar^2 k_z}{m^*} g_0' \right)$$

$$= \ell \sqrt{\frac{m^*}{2E}} \frac{q\varepsilon}{\hbar} \left(g + \frac{2}{3} E g' + \frac{\hbar^2 k_z}{m^*} g_0' \right)$$

$$\tag{15.65}$$

where one replaced τ by ℓ/v because $\ell = v\tau \propto \sqrt{2E/m^*}/\sqrt{E}$ is independent of E. To solve (15.65), one equalizes the coefficients of k_z and the coefficients independent of k_z on the two sides of the above equation, one gets

$$g = -\frac{q\varepsilon\hbar\ell}{\sqrt{2m^*E}}g_0' \tag{15.66}$$

$$g' = -\frac{q\varepsilon\hbar\ell}{\sqrt{2m^*}}\left(\frac{g_0''}{\sqrt{E}} - \frac{1}{2}\frac{g_0'}{E^{3/2}}\right) \tag{15.67}$$

Replacing these expressions in (15.65), one obtains the following differential equation for g_0

$$(E + \lambda k_B T)g_0'' + \left(\frac{E}{k_B T} + 2 + \frac{\lambda k_B T}{E}\right)g_0' + \frac{2}{k_B T}g_0 = 0 \tag{15.68}$$

where

$$\lambda = \frac{q^2\ell^2}{6m^*v_s^2 k_B T}\varepsilon^2 \equiv \frac{\varepsilon^2}{\varepsilon_0^2} \tag{15.69}$$

- Strong field limit:
In the strong-field limit $\lambda = \varepsilon^2/\varepsilon_0^2 \gg E/k_B T$, Eq. (15.68) becomes

$$\lambda k_B T g_0'' + \left(\frac{E}{k_B T} + 2 + \frac{\lambda k_B T}{E}\right)g_0' + \frac{2}{k_B T}g_0 = 0 \tag{15.70}$$

where 2 was neglected when compared to $\frac{\lambda k_B T}{E}$. One can verify by substitution that the following expression is the solution of (15.70)

$$g_0 = Ae^{-E^2/2\lambda k_B^2 T^2}$$

where A is a constant. Taking the derivative of g_0, one has from (15.66)

$$g = \frac{q\varepsilon\hbar\ell}{\lambda k_B^2 T^2}\sqrt{\frac{E}{2m^*}}Ae^{-E^2/2\lambda k_B^2 T^2}$$

To calculate A, one can calculate the number of electrons by integrating

$f = g_0 + k_z g$ as follows

$$
\begin{aligned}
N &= \frac{\Omega}{(2\pi)^3} \int (g_0 + k_z g) d^3 \vec{k} \\
&= \frac{\Omega}{(2\pi)^3} \int g_0 d^3 \vec{k} \quad \text{(second term=0 because odd in } k_z) \\
&= A \frac{\Omega}{2\pi^2} \left(\frac{2m^*}{\hbar^2} \right)^{3/2} \int_0^\infty \sqrt{E} e^{-E^2/2\lambda k_B^2 T^2} dE \\
&= A \frac{\Omega}{2\pi^2} \left(\frac{2m^*}{\hbar^2} \right)^{3/2} \frac{1}{2} (2\lambda k_B^2 T^2)^{3/4} \Gamma 3/4
\end{aligned}
$$

$$
A = \frac{4\pi^2}{\Gamma(3/4)} \frac{N}{\Omega(2\lambda)^{3/4}} \left(\frac{\hbar^2}{2m^* k_B T} \right)^{3/2} \tag{15.71}
$$

where one has transformed the k integral into an E integral with the help of the density of states $\rho \propto \sqrt{E}$ [see (2.36)]. Only the term $k_z g$ of f contributes to the current j. Therefore, one has

$$
\begin{aligned}
j &= q \int v_z \, k_z g \, d^3 \vec{k} \\
&= \frac{q\hbar}{4\pi^3 m^*} \int g \, k_z^2 \, d^3 \vec{k} \\
&= \frac{q^2 \varepsilon \ell \hbar^2 A}{2\pi^3 \lambda k_B^2 T^2 (2m^*)^{3/2}} \\
&\quad \times \int \sqrt{E} e^{-E^2/2\lambda k_B^2 T^2} k^2 \cos^2 \theta \, k^2 dk \sin \theta d\theta d\phi \tag{15.72}
\end{aligned}
$$

The angular part gives $4\pi/3$. For the radial part, putting $E^2 = (\hbar^2 k^2/2m^*) = x$, one has

$$
dk = \frac{1}{4} \sqrt{\frac{2m^*}{\hbar^2}} \frac{dx}{x^{3/4}}
$$

so that

$$
\begin{aligned}
j &= \frac{q^2 \varepsilon \ell \hbar^2 A}{2\pi^3 \lambda k_B^2 T^2 (2m^*)^{3/2}} \frac{4\pi}{3} \left(\frac{2m^*}{\hbar^2} \right)^{5/2} \\
&\quad \times \frac{1}{4} \int_0^\infty \sqrt{x} e^{-x/2\lambda k_B^2 T^2} dx \\
&= \frac{q^2 \varepsilon \ell A m^*}{3\pi^2 \lambda k_B^2 T^2 \hbar^3} \frac{\sqrt{\pi}}{2} (2\lambda k_B^2 T^2)^{3/2} \tag{15.73}
\end{aligned}
$$

Replacing q by $-e$, A and λ by (15.71) and (15.69), one has

$$j = \gamma\sqrt{\varepsilon} \qquad (15.74)$$

where

$$\gamma = en\frac{\sqrt{e\ell v_s}}{(m^*k_BT)^{1/4}}\frac{\sqrt{2\pi}}{3^{3/4}\Gamma(3/4)} \qquad (15.75)$$

The expression of j is quite different from the Ohm's law: it is valid for electron-acoustic phonon collisions in strong fields. Electrons obeying Eq. (15.74) are called "hot electrons".

15.8 Transport in semiconductors

The origin of band structures has been presented in chapter 11. In semiconductors, the conduction band is separated from the valence band by an energy gap. The simplest picture of the mechanism of the conductivity in a pure, intrinsic semiconductor has been given in section 11.3.2: valence electrons jump across the gap to go to the conduction band under the effect of the temperature. They become conducting electrons which create a current under an applied electric field. To facilitate the jump, one can dope semiconductors with different kinds of impurities so as to create "impurity energy levels" inside the gap: electrons can then transit by these levels on their way.

In this section, one presents some principal properties of semiconductors and outline mechanisms of conductivity in doped semiconductors from statistical physics point of view. Technological details of semiconductor devices can be consulted in books such as Ref. [163].

15.8.1 *Motion of electrons in applied fields — Hall effect*

Let us first consider an electron of effective mass m^* under a magnetic field $\vec{B}$ applied in the z direction. The equation of motion in the $\vec{k}$ space is written as

$$\hbar\frac{d\vec{k}}{dt} = -e\vec{v}\wedge\vec{B} = -\frac{e\hbar}{m^*}\vec{k}\wedge\vec{B}$$

In Cartesian coordinates, one has

$$\frac{dk_x}{dt} = -\frac{eB}{m^*}k_y$$

$$\frac{dk_y}{dt} = -\frac{eB}{m^*}k_z$$

$$\frac{dk_z}{dt} = 0$$

from which one has $k_z =$ constant, and

$$k_x(t) = A\cos(\omega_c t + \varphi), \quad k_y(t) == A\sin(\omega_c t + \varphi) \quad (A : \text{constant}) \quad (15.76)$$

where

$$\omega_c = \frac{eB}{m^*} = \text{cyclotron frequency.}$$

One sees that

$$k_x(t)^2 + k_y(t)^2 = A^2 = \text{constant, independent of t} \quad (15.77)$$

This means that the electron moves on a circle of radius A in the xy plane perpendicular to the field $\vec{B}$, with the cyclotron frequency ω_c.

If one wants to see the trajectory in the real space, one writes

$$\vec{v} = \frac{1}{\hbar}\vec{\nabla}E(\vec{k}), \quad \text{with } E(\vec{k}) = \frac{\hbar^2 k^2}{2m^*}$$

Integrating this equation, one gets

$$x(t) = x(0) + \frac{\hbar}{m^*}\int_0^t k_x dt = x(0) + \frac{\hbar A}{m^*\omega_c}[\sin(\omega_c + \varphi) - \sin\varphi]$$

$$y(t) = y(0) + \frac{\hbar}{m^*}\int_0^t k_y dt = y(0) - \frac{\hbar A}{m^*\omega_c}[\cos(\omega_c + \varphi) - \cos\varphi]$$

$$z(t) = z(0) + \frac{\hbar}{m^*}\int_0^t k_z dt = z(0) + \frac{\hbar k_z}{m^*}t$$

where one has used (15.76), φ being a phase. The above equations describe in the real space a helicoidal trajectory with the revolution axis parallel to $\vec{B}$. The radius of the helix is

$$\rho_h = \frac{v_{xy}}{\omega_c} = \frac{1}{\omega_c}(v_x^2 + v_y^2)^{1/2}$$

$$= \frac{\hbar}{m^*\omega_c}(k_x^2 + k_y^2)^{1/2} = \frac{\hbar A}{m^*\omega_c}$$

where one has used the velocity in the xy plane $v_{xy} = \rho_h\omega_c$ and (15.77).

In a very strong magnetic field, one has to incorporate the potential vector into the Schrödinger equation. The field effect is described by the

Landau's levels which have been treated in details in Problem 5 of chapter 14.

In order to describe the Hall effect, let us consider a slab of semiconductor of dimension $L_x \times L_y \times L_z$ where $L_x >> Ly > L_z$. One applies a magnetic field $\vec{B}$ along the $\vec{Oz}$ axis perpendicular to the slab and an electric field $\vec{\varepsilon}$ along the $\vec{Ox}$ axis along the slab, as indicated in Fig. 15.1. The Lorentz force acting on an electron under the applied fields is written as

$$F = -e[\vec{\varepsilon} + \vec{v} \wedge \vec{B}] \tag{15.78}$$

At the beginning, electrons move under the actions of the force by $\vec{\varepsilon}$ and of the force by $\vec{v} \wedge \vec{B}$. The first force "pushes" electrons in the direction $-\vec{Ox}$ and the second force drives electrons to the left edge of the sample (see figure). Negative electron charges are thus accumulated on the left edge leaving a negative-charge deficit on the right edge. The slab is quickly polarized and gives rise to an induced electric field $\vec{\varepsilon}_y$ in the y direction. This field will put later-coming electrons on a straight trajectory in the $-\vec{Ox}$ as indicated by the thin arrow in Fig. 15.1. Thus, in the steady state, electrons go on a straight trajectory: the effect of the induced transverse field is to cancel the Lorentz force in the y direction.

Let us calculate the Hall coefficient at weak fields. The applied electric field $\vec{\varepsilon}$ is supposed to be parallel to $\vec{Ox}$ and $\vec{B}$ parallel to $\vec{Oz}$. Since $\vec{\varepsilon} \cdot \vec{B} = 0$, we have the geometry of the Hall experiment.

In the steady state, the Lorentz force is written as

$$\vec{F} = -e(\vec{E} + \vec{v} \wedge \vec{B}) \tag{15.79}$$

where $\vec{E}$ is the total electric field acting on the electron: $E_x = \varepsilon_x = \varepsilon$ (applied field), $E_y = \varepsilon_y$ (polarization-induced field), $E_z = \varepsilon_z = 0$. The equation of motion $\vec{F} = \frac{d\vec{k}}{dt}$ gives

$$m\dot{v}_x = -e(\varepsilon_x + Bv_y) \tag{15.80}$$

$$m\dot{v}_y = -e(\varepsilon_y - Bv_x) \tag{15.81}$$

$$m\dot{v}_z = -e\varepsilon_z \tag{15.82}$$

Note that ε_y is the electrostatic field induced by the polarization of the two edges of the sample (see Fig. 15.1): it cancels the initial Lorentz force due to $\vec{B}$ so that $v_y = 0$ in the steady state, namely there is no current in the y direction. Setting $v_y = 0$, the first equation then gives $v_x = -e\varepsilon_x\tau/m^*$ when integrating from 0 to τ. The second equation gives, with $\dot{v}_y = 0$ (no force in the y direction),

$$\varepsilon_y = Bv_x = B\frac{j_x}{ne} \tag{15.83}$$

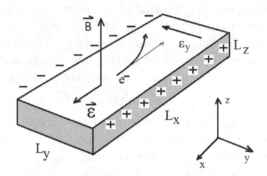

Fig. 15.1   Experiment setup for Hall effect: $\vec{B}$: magnetic field, $\vec{\varepsilon}$: applied electric field, ε_y: electric field in the y direction induced by the charge polarization, e^-: electron, thick arrow: electron trajectory if there is no charge polarization, thin arrow: electron trajectory in the steady state by effect of charge polarization field ε_y.

The Hall coefficient R_e is defined by (the index indicates the electron case):

$$R_e = \frac{\varepsilon_y}{j_x B} = -\frac{1}{ne} \qquad (15.84)$$

where one has used (15.83). For holes, one replaces the electron density n by the hole density p and the electron charge $-e$ by the hole charge e. One has

$$R_h = \frac{1}{pe}$$

The mobility μ_e for an electron is defined by

$$\vec{v} = -\mu_e \vec{\varepsilon} \qquad (15.85)$$

where the minus sign is used to make μ_e positive by convention (note that for electrons $\vec{v}$ is opposite to $\vec{\varepsilon}$). Using the Ohm's law $j = -nev = \sigma\varepsilon$, one has $v = -\sigma\varepsilon/ne$. Comparing this to (15.85), one has

$$\mu_e = \sigma_e/ne = -\sigma_e R_e, \quad \text{and similarly} \quad \mu_h = \sigma_h R_h.$$

From (15.83) one has $\varepsilon_y = Bv_x = -B\mu_e\varepsilon_x$. The Hall angle is then given by

$$\tan\theta_e = \varepsilon_y/\varepsilon_x = -\mu_e B = -\omega_c \tau.$$

When there are two types of carriers, ε_x is unchanged but the induced ε_y is changed. To calculate it one shall use $j_y = \sigma\varepsilon_y$ where $\sigma = \sigma_e + \sigma_h = ne\mu_e + pe\mu_h$. Note that j_y is the current due to the induced ε_y but it is canceled by the current due to the Lorentz term $[\vec{v} \wedge \vec{B}]_y$ because there is no total current in the y-direction as described earlier. One can decompose the current j_y into two currents j_{ye} and j_{yh}: using (15.83) one has

$$j_{ye} = \sigma_e\varepsilon_y = -\sigma_e\frac{m\omega_c}{e}\frac{j_x}{ne} = -\mu_e B j_{xe}.$$

and similarly,

$$j_{yh} = \mu_h B j_{xh}.$$

Using $j_x = \sigma\varepsilon_x$, one has

$$\begin{aligned}
j_y &= -\mu_e B j_{xe} + \mu_h B j_{xh} \\
&= -\mu_e B \sigma_e \varepsilon_x + \mu_h B \sigma_h \varepsilon_x \\
&= -ne\mu_e^2 B \frac{j_x}{\sigma} + pe\mu_h^2 B \frac{j_x}{\sigma}
\end{aligned}$$

One obtains

$$\varepsilon_y = \frac{j_y}{\sigma} = -\left(ne\mu_e^2 B\frac{j_x}{\sigma^2} - pe\mu_h^2 B\frac{j_x}{\sigma^2} \right)$$

$$\begin{aligned}
R_H = \frac{\varepsilon_y}{B j_x} &= -\left(\frac{ne\mu_e^2}{\sigma^2} - \frac{pe\mu_h^2}{\sigma^2} \right) \\
&= -\frac{e}{\sigma^2}(n\mu_e^2 - p\mu_h^2) = -\frac{e(n\mu_e^2 - p\mu_h^2)}{e^2(n\mu_e + p\mu_h)^2} \\
&= -\frac{1}{e}\frac{(n\mu_e^2 - p\mu_h^2)}{(n\mu_e + p\mu_h)^2}
\end{aligned} \tag{15.86}$$

where R_H is the Hall coefficient when there are two types of carriers. The Hall mobility is

$$\mu_H = \sigma R_H = \frac{(p\mu_h^2 - n\mu_e^2)}{(n\mu_e + p\mu_h)}.$$

One sees that R_H and μ_H are zero if $p\mu_h^2 = n\mu_e^2$.

The above results have been obtained for weak magnetic fields. Modifications of these results at stronger fields as well as with collisions are treated in Problem 11 below.

15.8.2   *Calculation of the diffusion coefficient by the Boltzmann's equation*

One considers a semiconductor (or a metal) in which one maintains a gradient of carrier concentration. One supposes that there is no other applied external force.

The Boltzmann's equation in the relaxation time approximation is

$$\vec{v} \cdot \vec{\nabla}_{\vec{r}} f(\vec{v}, \vec{r}) \simeq \vec{v} \cdot \vec{\nabla}_{\vec{r}} f_0(\vec{v}, \vec{r}) = -\frac{f(\vec{v}, \vec{r}) - f_0(E, \vec{r})}{\tau(E)} \tag{15.87}$$

In the stationary regime, one looks for the distribution function f in supposing a local equilibrium, namely in supposing that at each spatial point $\vec{r}$, f relaxes toward the Fermi-Dirac distribution $f_0(E, \vec{r})$ with a local chemical potential $\mu(\vec{r})$. Let $\tau(E)$ be the relaxation time supposed to be dependent on the energy E. The solution of the Boltzmann's equation (15.23) at the first order with respect to the gradient of carrier concentration is

$$f(\vec{v}, \vec{r}) = f_0(E, \vec{r}) - \tau(E)\vec{v}.\vec{\nabla}_{\vec{r}} f_0(E, \vec{r}) \tag{15.88}$$

As f_0 depends on $\vec{r}$ via μ, one rewrites (15.88) as

$$f(\vec{v}, \vec{r}) = f_0(E, \vec{r}) - \tau(E)\vec{v}.\vec{\nabla}_{\vec{r}} n \frac{\partial f_0}{\partial \mu} \frac{d\mu}{dn} \tag{15.89}$$

where $n(\vec{r})$ is the electron density at $\vec{r}$. One can use the formula (5.21) at $\vec{r}$ because one supposes a local equilibrium:

$$\mu(\vec{r}) = \frac{\hbar^2}{2m^*} (3\pi^2 n(\vec{r}))^{2/3} \tag{15.90}$$

The electron current, or diffusion current, is written as

$$\vec{j}_D(\vec{r}) = -2e \frac{1}{(2\pi)^3} \int d\vec{k}\vec{v}\varphi \quad \text{(factor 2 is due to spin degeneracy)}$$

$$= 2e \frac{1}{(2\pi)^3} \int d\vec{k}\vec{v}\tau(E)\vec{v}.\vec{\nabla}_{\vec{r}} n \frac{\partial f_0}{\partial \mu} \frac{d\mu}{dn}$$

$$= 2e \frac{1}{(2\pi)^3} \int d\vec{k}\vec{v}\tau(E)\vec{v}.\vec{\nabla}_{\vec{r}} n \left(-\frac{\partial f_0}{\partial E}\right) \frac{d\mu}{dn}$$

$$\equiv e\bar{\bar{D}}\vec{\nabla}_{\vec{r}} n \tag{15.91}$$

where $\bar{\bar{D}}$ is the diffusion tensor defined by

$$D_{\alpha\beta} = 2\frac{d\mu}{dn}\frac{1}{(2\pi)^3}\int d\vec{k}v_\alpha v_\beta\left(-\frac{\partial f_0}{\partial E}\right)\tau(E) \tag{15.92}$$

where $\frac{d\mu}{dn}$ is taken out of the integral because it does not depend on $\vec{k}$ as seen in (15.90).

In the general case with an arbitrary energy band structure, $D_{\alpha\beta}$ is not diagonal. However, in the effective-mass approximation, one has $\vec{v} = \frac{\hbar\vec{k}}{m^*}$, from which one sees that $D_{\alpha\beta} = 0$ if $\alpha \neq \beta$. One puts $D_{xx} = D_{yy} = D_{zz} \equiv D$, so that $D = \frac{1}{3}(D_{xx} + D_{yy} + D_{zz})$. One rewrites (15.92) as

$$
\begin{aligned}
D &= \frac{2}{3}\frac{d\mu}{dn}\frac{1}{(2\pi)^3}\int \vec{k}(v_x^2 + v_y^2 + v_z^2)(-\frac{\partial f_0}{\partial E})\tau(E) \\
&= \frac{4}{3}\frac{d\mu}{dn}\frac{1}{(2\pi)^3}\int \vec{k}\frac{E}{m^*}(-\frac{\partial f_0}{\partial E})\tau(E)
\end{aligned} \tag{15.93}
$$

Replacing $2\frac{1}{(2\pi)^3}\int d\vec{k}...$ by $\int dE\rho(E)...$, one has

$$
\begin{aligned}
D &= \frac{2}{3}\frac{d\mu}{dn}\int dE\rho(E)\frac{E}{m^*}\left(-\frac{\partial f_0}{\partial E}\right)\tau(E) \\
&= \frac{2}{3m^*}\frac{d\mu}{dn}n\frac{\int_0^\infty dE\rho(E)E\left(-\frac{\partial f_0}{\partial E}\right)\tau(E)}{\int_0^\infty dE\rho(E)f_0}
\end{aligned} \tag{15.94}
$$

where one has introduced $n = \int_0^\infty dE\rho(E)f_0$ in the numerator and denominator. Using $\rho(E) = AE^{1/2}$, one has

$$
\begin{aligned}
\int_0^\infty dE\rho(E)f_0 = A\int_0^\infty dEE^{1/2}f_0 &= A\frac{2}{3}\left[E^{3/2}f_0|_0^\infty - \int_0^\infty dEE^{3/2}(\frac{\partial f_0}{\partial E})\right] \\
&= A\frac{2}{3}\int_0^\infty dEE^{3/2}\left(-\frac{\partial f_0}{\partial E}\right)
\end{aligned} \tag{15.95}
$$

The term $E^{3/2}f_0|_0^\infty$ is zero because f_0 tends to 0 at the upper bound. One then has

$$
\begin{aligned}
D &= \frac{2}{3m^*}\frac{d\mu}{dn}n\frac{A\int_0^\infty dEE^{3/2}\left(-\frac{\partial f_0}{\partial E}\right)\tau(E)}{A\frac{2}{3}\int_0^\infty dEE^{3/2}\left(-\frac{\partial f_0}{\partial E}\right)} \\
&= \frac{n}{m^*}\frac{d\mu}{dn} <\tau(E)>
\end{aligned} \tag{15.96}
$$

where the mean value of $\tau(E)$ is defined using the distribution of the denominator.

One shows now that D is proportional to the electron mobility μ_e defined by

$$\vec{v} \equiv -\mu_e \vec{\mathcal{E}} \quad (\mu_e \text{ is positively defined}) \tag{15.97}$$

Using $\vec{j} = -ne\vec{v} = \sigma\vec{\mathcal{E}}$, one has

$$\vec{v} = -\frac{\sigma\vec{\mathcal{E}}}{ne} = -\mu_e\vec{\mathcal{E}} \tag{15.98}$$

from which one obtains

$$\mu_e = \frac{\sigma}{ne} = \frac{e<\tau>}{m^*} \tag{15.99}$$

using (15.56).

Equation (15.96) can be rewritten by

$$D = \frac{n}{e}\mu_e\frac{d\mu}{dn} \tag{15.100}$$

This relation is called "Einstein's relation".

In a metal, using (15.90) one has

$$\frac{d\mu}{dn} = \frac{2\hbar^2}{6m^*}\left(3\pi^2\right)^{2/3} n(\vec{r})^{-1/3} = \frac{2\mu}{3n(\vec{r})} \tag{15.101}$$

from which one obtains

$$\frac{D}{\mu_e} = \frac{2}{3}\frac{\mu}{e} \tag{15.102}$$

In a non degenerate semiconductor, one has (see paragraph 11.3.2)

$$n = 2\left(\frac{m^*k_BT}{2\pi\hbar^2}\right)e^{\beta\mu} \tag{15.103}$$

from which one gets $n\frac{d\mu}{dn} = k_BT$. One then has

$$\frac{D}{\mu_e} = \frac{k_BT}{e} \tag{15.104}$$

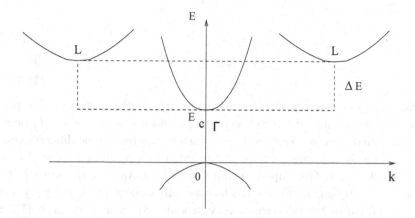

Fig. 15.2 Band structure with two valleys for GaAs and InP.

15.8.3 *Transport in semiconductors: Gunn's effect*

One considers a semiconductor with two valleys corresponding for example to GaAs and InP [163]. The band structures of GaAs and InP are schematically shown in Fig. 15.2.

The conduction band bottom E_c is situated at $k = 0$ (point Γ, center of the Brillouin zone), while the upper conduction band is composed of 8 parts centered at 8 points L at the border of the Brillouin zone in the 8 [111] directions, at the energy level $E_c + \Delta E$ where $\Delta E = 0.31$ eV for GaAs and 0.53 eV for InP. These valleys have very different effective masses: $m_1 \ll m_2$. Let μ_1 and μ_2 be their respective electron mobilities, and n_1 and n_2 the electron densities in the two valleys. The total concentration of electrons in the semiconductor is $n = n_1 + n_2$.

One applies an electric field ε on the sample of GaAs or InP. One calculates the current j as a function of n_1, n_2, μ_1 and μ_2 as follows

$$j = (\sigma_1 + \sigma_2)\varepsilon = -e(n_1\mu_1 + n_2\mu_2)\varepsilon \equiv -env \qquad (15.105)$$

The mean velocity v is given by

$$v = \frac{n_1\mu_1 + n_2\mu_2}{n_1 + n_2}\varepsilon \simeq \frac{\mu_1\varepsilon}{1 + \frac{n_2}{n_1}} \qquad (15.106)$$

where one has used $\mu_2 \ll \mu_1$ because $m_1 \ll m_2$.

One calculates now n_1 and n_2. At high temperatures, one has (see paragraph 11.3.2):

$$n_1 = g_1 e^{-\beta_e(E_1-\mu)} \tag{15.107}$$

$$n_2 = g_2 e^{-\beta_e(E_2-\mu)} \tag{15.108}$$

$$\tag{15.109}$$

where g_1 and g_2 are effective densities of states of the two valleys of respective energies E_1 and E_2, μ the chemical potential and $\beta_e = \frac{1}{k_B T_e}$, T_e being the so-called "electron temperature" which is supposed to be different from that of the lattice ions because of the extra energy given by the electric field to electrons. One supposes that the two valleys have the same T_e for simplicity. To be general, one has to take into account the degeneracy d in the definition of the effective density of states given in paragraph 11.3.2. One writes

$$g = d \left(\frac{m^* k_B T}{2\pi \hbar^2} \right)^{3/2} \tag{15.110}$$

In the case considered here, one has $d = 8$ for the upper band because there are 8 equivalent valleys in 8 [111] directions at the Brillouin zone border. In fact, these 8 points L belong half of each to the first Brillouin zone, in addition to a factor 2 due to the spin degeneracy. One has $d = 8$ valleys $\times \frac{1}{2} \times 2$ (spin) $= 8$. As for the lower band, it belongs entirely to the first Brillouin zone so that $d = 2$ (only spin degeneracy). One has

$$\frac{n_2}{n_1} = \frac{g_2}{g_1} e^{-\beta_e(E_2-E_1)} \tag{15.111}$$

$$= 4 \left(\frac{m_2^*}{m_1^*} \right)^{3/2} e^{-\beta_e \Delta E} \equiv R e^{-\beta_e \Delta E} \tag{15.112}$$

One sees that $n_2 \ll n_1$ because ΔE is not small. Equation (15.106) becomes

$$v = \mu_1 \varepsilon \left[1 + R e^{-\beta_e \Delta E} \right]^{-1} \tag{15.113}$$

This relation shows that the electron temperature T_e is related to the velocity v. If one supposes that the energy provided to electrons by the electric field is transferred to lattice ions during collisions so that temperatures T_e and T will become equal, the energy balance is written as

$$e\varepsilon v\tau = \frac{3k_B}{2}(T_e - T) \tag{15.114}$$

from which one has

$$T_e = T + \frac{2}{3}\frac{e\tau\mu_1}{k_B}\varepsilon^2\left[1 + Re^{-\frac{\Delta E}{k_B T_e}}\right]^{-1} \qquad (15.115)$$

This is an implicit equation for T_e. For weak fields one has

$$T_e \simeq T + \frac{2}{3}\frac{e\tau\mu_1}{k_B}\varepsilon^2 \qquad (15.116)$$

For strong fields, one solves (15.115) by a graphical method: one plots the curves

$$y_1 = T_e - T$$

and

$$y_2 = \frac{2}{3}\frac{e\tau\mu_1}{k_B}\varepsilon^2\left[1 + Re^{-\frac{\Delta E}{k_B T_e}}\right]^{-1}.$$

Their intersection gives T_e.

To get an idea about the order of magnitude of physical quantities in y_1 and y_2, one gives hereafter a table containing values used for the materials: for GaAs, one has $\tau = 10^{-12}$ sec, $m_1 = 0.067m_0$, $m_2 = 0.55m_0$, m_0 being the electron mass at rest, $\mu_1 = 8000$ cm^2/volt-sec. With these values, one has $R = 94$. For $\varepsilon = 5$kV/cm and $T = 300$ K, one obtains

$T_e(K)$	$k_B T_e$ (meV)	$\exp(-\frac{\Delta E}{k_B T_e})$	y_2
300	25.8	6×10^{-6}	1546
600	51.6	2.5×10^{-6}	1256
900	77.4	1.8×10^{-2}	570
1200	103.2	5×10^{-2}	273

Figure 15.3 shows the graphical solution for $\varepsilon = 5$ kV/cm: the solution is $T_e \simeq 890$ K.

The following table shows solutions for other values of ε:

ε(kV/cm)	1	2.5	5	10
$T_e(K)$	360	610	890	1240
v (cm/sec)	8×10^6	1.7×10^7	1.52×10^7	1.3×10^7

Figure 15.4 shows the variation of v as a function of ε.

One observes in this figure that $v = \mu_1\varepsilon$ when $\varepsilon < 2$ kV/cm. Above this value, one sees a non linear behavior of v. If one defines the differential resistance by

$$\frac{dj}{d\varepsilon} \propto \frac{dv}{d\varepsilon} \qquad (15.117)$$

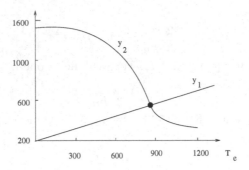

Fig. 15.3   Graphical solution for $\varepsilon = 5$ kV/cm in the case of GaAs.

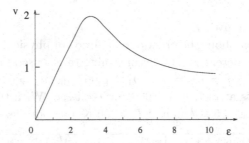

Fig. 15.4 v, in unit of 10^7 cm/sec, versus ε (kV/cm) in the case GaAs, at $T = 300$ K. One observes a negative differential resistance above the maximum of v which occurs at $\varepsilon \simeq 3.2$ kV/cm.

then the threshold field ε_s above which one observes a negative differential resistance at 300 K corresponds to the maximum electron velocity v_m, namely at the field $\varepsilon_s \simeq 3.2$ kV/cm. This negative differential resistance is called "Gunn's effect".

The variation of $v(\varepsilon)$ versus ε for GaAs at different temperatures shows that ε_s increases with increasing T but the maximum value v_m diminishes.

One shows in the following table the ratio n_2/n_1 at $T = 300$ K for GaAs. One observes that the upper valley starts to be filled for fields larger than $\varepsilon \simeq 2$ kV/cm.

T_e (K)	n_2/n_1	ε (kV/cm)
360	4×10^{-3}	1
610	0.255	2.5
890	1.63	5
1240	5.13	10

The Gunn's effect is interpreted as follows. A number of electrons of the first valley is excited to the second valley under a strong electric field. However, once they are in the second valley, these electrons become less mobile because they have a heavier effective mass. This explains the diminution of the velocity beyond the threshold field.

15.8.4 *Conductivity in extrinsic semiconductors — Doping effects*

Intrinsic (or pure) semiconductors have been described in section 11.3.2. In order to modify the number of charge carriers responsible for the conductivity in semiconductors, pure semiconductors can be doped with other kinds of atoms. In metals, the number of conducting electrons is $\simeq 10^{22}$ per cm^3, while in semiconductors it is between $\sim 10^{13}$ and $\sim 10^{17}$. There are two types of doping (see Fig. 15.5):

(i) n-type by replacing a number of atoms by "impurity" atoms having one more valence electron per atom. Extra electrons so provided by impurities increase the concentration of electrons in the conduction band. Such impurities are called "donors". For example, pure semiconductors Silicon and Germanium in the column IV of the periodic table have 4 valence electrons per atom. Dopant impurities of n-type have 5 valence electrons such as Phosphorus and Arsenic of column V.

(ii) p-type by replacing a number of atoms by "impurity" atoms having one valence electron less per atom. There are thus no complete bonding leaving a hole at each impurity site. Hole concentration in the valence band is therefore increased with such dopant impurities. These are called "acceptors". Acceptor impurities of Si and Ge are atoms from column III of the periodic table such as Boron and Aluminum.

Note that semiconducting III-V compounds such as Aluminum Phosphide, Aluminum Arsenide, Gallium Arsenide and Gallium Nitride have donors from atoms of columns IV and VI, depending on which atoms they replace: Selenium, Tellurium, Silicon and Germanium. The III-V semicon-

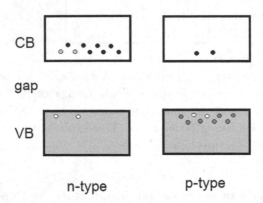

Fig. 15.5   Conduction band (CB) and valence band (VB) in doped semiconductors. In n-type semiconductor (left), extra electrons (black circles) from donors occupy the CB in addition to a much smaller number of native electrons represented by gray circles, white circles in the VB are holes. In p-type semiconductor (right), extra holes (gray circles) from acceptors occupy the VB in addition to a much smaller number of native holes represented by white circles, black circles in the CB are electrons.

ductors have acceptors from atoms of columns II and IV such as Beryllium, Zinc, Cadmium, Silicon and Germanium.

One sees that with just 0.001% doping of donors, one has $\sim 10^{17}$ extra conducting electrons per cm^3.

Doped semiconductors are called "extrinsic semiconductors". The conductivity in n-type semiconductors is mainly by electrons which are called "majority carriers". Holes are called in this case "minority carriers". In the case of p-type semiconductors, holes are majority carriers while electrons are minority ones. One has seen in section 11.3.2 that intrinsic semiconductors have the Fermi level at the middle of the band gap, assuming the same effective mass for the conduction and valence bands. In an n-type semiconductor, the Fermi level is greater than that of the intrinsic semiconductor and lies closer to the conduction band than the valence band. While, in a p-type semiconductor, the Fermi level lies closer to the valence band.

In low doping cases, the electron and hole concentrations are small. This allows us to use the Maxwell-Boltzmann statistics as we have done in intrinsic semiconductors shown in section 11.3.2.

Let us calculate n and p in doped semiconductors. The electric neutrality imposes that $n + N_a^- = p + N_d^+$ where N_a^- and N_d^+ are the numbers of ionized acceptors and donors, respectively, n and p being the numbers of free electrons and holes. If all impurities are ionized then

$$n + N_a = p + N_d$$

There are three types of doped semiconductors: apart from the n and p types mentioned above, there is the case where $N_a = N_d$ so that $n = p = n_i$ just like in the case of an intrinsic semiconductor. This case is called "compensated semiconductor". The Fermi level is at the middle of the gap. In the n type one has $N_d \gg N_a$, namely $n \gg p$. Since $np = n_i^2$, one has

$$n \simeq N_d - N_a, \quad \text{and} \quad p = \frac{n_i^2}{n} = \frac{n_i^2}{N_d - N_a}.$$

In the p type, one has $p \gg n$, thus

$$p \simeq N_a - N_d, \quad \text{and} \quad n = \frac{n_i^2}{p} = \frac{n_i^2}{N_a - N_d}.$$

The Fermi levels of the doped semiconductors are
-n type:

$$n = N_d - N_a = N_c \exp[-\beta(E_c - E_F^n)] \quad \Rightarrow \quad E_F^n = E_c - k_B T \ln \frac{N_c}{N_d - N_a}$$
$$(15.118)$$

-p type:

$$p = N_a - N_d = N_v \exp[\beta(E_v - E_F^p)] \quad \Rightarrow \quad E_F^p = E_v + k_B T \ln \frac{N_v}{N_a - N_d}$$
$$(15.119)$$

One sees that the Fermi level of the n type lies closer to the bottom of the conduction band, and that of the p type closer to the top of the valence band (see Fig. 15.6).

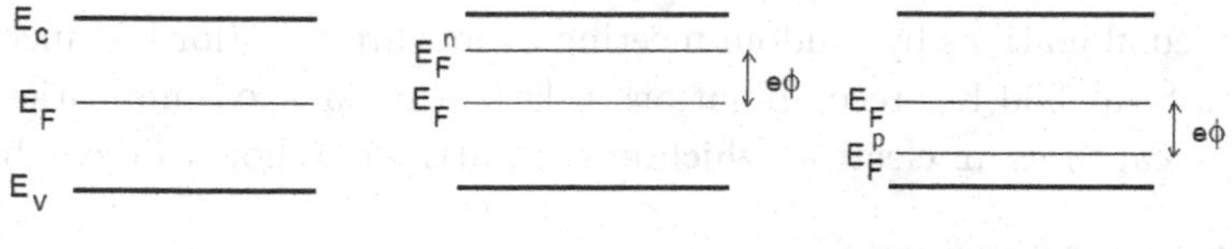

Fig. 15.6   Fermi level in an intrinsic semiconductor (E_F, left), an n-type semiconductor (E_F^n, middle) and a p-type semiconductor (E_F^p, right). E_c: bottom of the conduction band, E_v: top of the valence band.

In summary, one has

-intrinsic semiconductor:

$$n = p = n_i = N_c e^{-\beta(E_c - E_F)} = N_v e^{+\beta(E_v - E_F)}$$

-n-type semiconductor:

$$n = N_c e^{-\beta(E_c - E_F^n)}$$

-p-type semiconductor:

$$p = N_v e^{+\beta(E_v - E_F^p)}$$

Putting $e\phi = E_F^n - E_F$, one can rewrite the above equations as

$$n = n_i e^{+e\phi/k_B T} \tag{15.120}$$

$$p = n_i e^{-e\phi/k_B T} \tag{15.121}$$

If $e\phi < 0$, then $n > n_i$ and $p < n_i$: one has semiconductors of type n. If $e\phi > 0$, then $p > n_i$ and $n < n_i$: one has semiconductors of type p.

15.8.5 Doped semiconductors: generation, recombination, equation of continuity

In doped semiconductors, charge carriers (electrons and holes) are created and disappear with time evolution. The origin of carrier creation is numerous (thermal excitation, optical excitation, particle irradiation, intense applied electric field, carrier injection, ...), the carrier generation is characterized by a generation rate g (cm^{-3}s^{-1}). The carrier disappearance is mainly caused by electron-hole recombinations characterized by a recombination rate r. One has

$$\left[\frac{dn}{dt}\right]_{g,r} = g - r$$

For recombinations, there are two kinds: (i) direct recombination: one electron meets a hole, (ii) indirect recombination: in the case of low doping, direct recombinations by random meeting events are rare (for instance ~ 3 hours in Si at 300 K), recombinations take place via fixed impurities: an impurity captures an electron which in turn attracts a hole to recombine.

- Direct recombinations:
  One can show that (see Problem 7 below) the recombination rate in the case of weak doping is given by

$$r \simeq \frac{\Delta n}{\tau_{\Delta n}}, \quad \text{or} \quad r \simeq \frac{\Delta p}{\tau_{\Delta p}}, \tag{15.122}$$

where Δn (Δp) is the electron (hole) excess in an $n-$ ($p-$)doped semi-conductor and $\tau_{\Delta n}$ ($\tau_{\Delta p}$) the life-time of carriers.

• Indirect recombinations:

One can show that (see Problem 7 below) the indirect recombination rate is given by

$$r \simeq \frac{1}{\tau_{\Delta m}} \frac{pn - n_i^2}{2n_i + p + n} \tag{15.123}$$

where

$$\tau_{\Delta m} \equiv \tau_{n0} = \tau_{p0} = \frac{1}{CN_R} \tag{15.124}$$

C is the carrier-capture coefficient of an impurity serving as a recombination center and N_R the number of such centers. Of course, in an n-type semiconductor where $n_0 \gg n_i \gg p_0$ one has

$$r \simeq \frac{1}{\tau_{\Delta m}} \frac{pn - n_i^2}{n} \simeq \frac{1}{\tau_{\Delta m}} \left(p - \frac{n_i^2}{n_0} \right) = \frac{p - p_0}{\tau_{\Delta m}} \tag{15.125}$$

It is noted that r is proportional to the minority concentration. A similar expression is obtained for a p-type semiconductor by replacing $p - p_0$ by $n - n_0$.

• Return to equilibrium:

The material reacts to artificial creations or extractions of carriers by carrier recombinations or generations. The return to equilibrium is calculated as follows.

One considers an isolated semiconductor. It is excited by a photon beam. The variations of carrier densities are

$$\frac{dn}{dt} = g_n - r_n, \quad \text{and} \quad \frac{dp}{dt} = g_p - r_p \tag{15.126}$$

For a p-type semiconductor with a weak excitation, one has $p = p_0 + \Delta p \simeq p_0$, $n = n_0 + \Delta n \ll p_0$. Thus, from (15.125) one writes

$$\frac{dn}{dt} = g_n - \frac{n - n_0}{\tau_n} \tag{15.127}$$

so that in the stationary regime where $\frac{dn}{dt} = 0$ with continuing photon beam, one gets $n_1 - n_0 = g_n \tau_n$ (electron excess), and in the transitory regime (return to equilibrium) where $g_n = 0$ (no more photon beam) one obtains

$$\frac{dn}{dt} = -\frac{n - n_0}{\tau_n} \quad \Rightarrow \quad n - n_0 = (n_1 - n_0) \exp(-t/\tau_n) \tag{15.128}$$

where $n = n_1$ at $t = 0$.

- Equation of continuity:
  The variations of the carrier densities are given by the following balance:

$$\frac{dn}{dt} = \frac{1}{e}\vec{\nabla}\cdot\vec{j}_n + g_n - r_n \tag{15.129}$$

$$\frac{dp}{dt} = -\frac{1}{e}\vec{\nabla}\cdot\vec{j}_p + g_p - r_p \tag{15.130}$$

where $\vec{j}_n$ and $\vec{j}_p$ are the electron and hole currents each of which is composed of two contributions: the drift current caused by the electric field $\vec{\varepsilon}$ and the diffusion current due to the concentration gradient. One writes

$$\vec{j}_n = ne\mu_n\vec{\varepsilon} + eD_n\vec{\nabla}n$$

$$\vec{j}_p = pe\mu_p\vec{\varepsilon} - eD_p\vec{\nabla}p$$

where the electron mobility μ_n is defined in (15.97). A similar definition is used for holes. D_n and D_p are electron and hole diffusion coefficients (cf. paragraph 15.8.2). For a weak carrier injection, one explicitly rewrites (15.129) and (15.130) as

$$\frac{dn}{dt} = n\mu_n\frac{d\varepsilon}{dx} + \mu_n\varepsilon\frac{\partial n}{\partial x} + D_n\frac{\partial^2 n}{\partial x^2} + g_n - \frac{n - n_0}{\tau_n} \tag{15.131}$$

$$\frac{dp}{dt} = -p\mu_p\frac{d\varepsilon}{dx} - \mu_p\varepsilon\frac{\partial p}{\partial x} + D_p\frac{\partial^2 p}{\partial x^2} + g_p - \frac{p - p_0}{\tau_p} \tag{15.132}$$

where one has supposed that the electric field as well as the concentration gradient are along the x axis and one has used (15.125).

One considers a particular case where a p-type semiconductor receives at its surface ($x = 0$) a photon beam not deeply penetrating. One supposes there is no electric field. The electrons excited by the beam diffuse toward inside ($x > 0$): the resulting current is

$$j_{nx} = eD_n\frac{\partial n}{\partial x}$$

One has from (15.131)

$$\frac{dn}{dt} = D_n\frac{\partial^2 n}{\partial x^2} + g_n - \frac{n - n_0}{\tau_n} \tag{15.133}$$

Inside the material, $g_n = 0$ so that if the beam continues then one has, in the stationary regime, the following expression

$$\frac{d^2(n - n_0)}{dx^2} - \frac{n - n_0}{L_n^2} = 0 \tag{15.134}$$

where $L_n^2 \equiv D_n\tau_n$. L_n is called "diffusion length". If at $x = 0$ (surface), $n = n_1$ then the solution of the above equation is

$$n - n_0 = (n_1 - n_0)\exp(-x/L_n) \tag{15.135}$$

The charge excess decays exponentially from the surface toward inside.

15.8.6 *p − n junctions — Diodes*

In this paragraph, one shows the conductivity across a $p - n$ junction which is the essential element in electronic devices. One puts a slab of a p-type semiconductor in contact with a slab of an n-type semiconductor as shown in Fig. 15.7. After putting in contact, electrons of the n side diffuse into the p region leaving behind, near the contact interface, positive ion charges. Inversely, holes of the p side diffuse into the n region creating near the interface a zone of negative charges. There is thus a "space charge region" between $-x_a$ and x_d around the interface (see Fig. 15.7) where there is no mixing of charges. This charge polarization creates between these two points an induced electric field leading to a so-called "built-in voltage" V_D. The space charge region is also called the "depletion zone". Outside the depletion zone, namely $x < -x_a$ and $x > x_d$, one has the charge neutral zones.

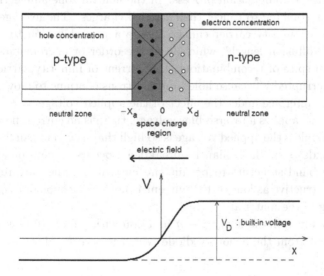

Fig. 15.7 Top: A $p - n$ junction with electron and hole concentration profiles across the junction. Positive and negative charges, accumulated on the two sides of the interface, are shown by white and black circles, respectively. The depletion zone is between $-x_a$ and x_d where there is an induced electric field oriented from right to left. Bottom: built-in voltage V_D due to the induced electric field.

Note that when they are alone, both p- and n-types are conductive. However, when they are put in contact, there is no current flow unless one applies a voltage as seen in what follows.

If a forward bias voltage is applied, namely the p side is connected with the positive electrode of a battery and the n site with the negative one, then holes of the p side rush to the junction, and electrons of the n side run also toward the junction. As a consequence, the width of the depletion zone is reduced, reducing therefore the potential barrier between the two sides of the interface. With increasing forward-bias voltage, the width of the depletion zone becomes so thin that the induced electric field becomes so weak to stop electrons and holes from crossing the interface. As a consequence, electrons diffuse into the neutral p zone and holes diffuse into the n zone. Electrons in the p zone and holes in the n zone are called "minority carriers". Using a forward-bias voltage larger than a threshold value, one can thus make minority carriers to cross the junction, creating thus a current. Note that minority carriers cannot travel very far from the interface because of the recombination process: in the neutral zone an electron and a hole can recombine to cancel their carried charges. The average length over which a minority carrier can go before a recombination to occur is called the "diffusion length" which is of the order of micrometers. Note also that, in spite of recombinations, the current of minority carriers continue uninterruptedly because holes and electrons continue to move toward the junction interface under the effect of the applied voltage.

Now, if one applies a reverse-bias voltage, there is no current flow across the contact unless the applied voltage is so high that a current can flow if the Zener breakdown or the avalanche breakdown occurs. These breakdowns change the band structure to permit the current but they are reversible and non destructive as long as the current is not strong enough to cause an overheating of the material.

A $p - n$ junction diode is a $p - n$ junction with a forward-bias voltage. It is different from the Schottky diode which uses a metal-semiconductor junction.

- Calculation of the built-in potential:
  Let us calculate the built-in potential and the width of the depletion zone in a non polarized $p - n$ junction (no applied voltage). Suppose that the electron concentration n on the n-type side is much larger than the concentration of the minority hole carriers p. In the p-type side, the reverse is true, namely $p >> n$. Suppose also that concentrations

of ionized donors and acceptors are N_d and N_a. Let n_{n0} and n_{p0} be the electron concentrations at equilibrium on the n and p sides, respectively. Similarly, let p_{p0} and p_{n0} be the hole concentrations at equilibrium on the p and n sides, respectively.

Far from the depletion zone, the Fermi level in the p zone lies near the top of the valence band, while the Fermi level of the n side lies closer to the bottom of the conduction band. However, at equilibrium the Fermi level should be the same in all the system. It means that the bottom of the conduction band is not the same in the two zones: the difference is due to the built-in voltage V_D. Since the energy of an electron in a potential $V(x)$ is $-eV(x)$, the difference in energy of an electron when it is in the p zone and when it is in the n zone is thus eV_D. One has (see section 11.3.2):

$$n_{n0} = N_c e^{-\beta(E_{cn}-E_F)}$$
$$n_{p0} = N_c e^{-\beta(E_{cp}-E_F)}$$

where E_{cn} and E_{cp} are the energies of the bottom of the conduction band in the n and p zones, respectively. One deduces

$$\frac{n_{n0}}{n_{p0}} = \exp\left[\frac{E_{cp} - E_{cn}}{k_B T}\right] = \exp\left[\frac{eV_D}{k_B T}\right] \tag{15.136}$$

One obtains

$$V_D = \frac{k_B T}{e} \ln\left(\frac{n_{n0}}{n_{p0}}\right) \tag{15.137}$$

One assumes that the two sides of the junction are obtained by doping the same pure material with p and n impurities. According to the equilibrium condition in each zone, one has

$$n_{n0}p_{n0} = n_{p0}p_{p0} = n_i^2 \tag{15.138}$$

where n_i is the intrinsic carrier concentration [see (11.73)]. Using this relation and noticing that $n_{n0} \sim N_d$ and $p_{p0} \sim N_a$, one rewrites (15.137) as

$$V_D = \frac{k_B T}{e} \ln\left(\frac{n_{n0}p_{p0}}{n_{p0}p_{p0}}\right) = \frac{k_B T}{e} \ln\left(\frac{N_d N_a}{n_i^2}\right) \tag{15.139}$$

Using typical values of parameters $N_d = N_a = 10^{16}$ cm^{-3}, $n_i = 10^{13}$ cm^{-3} at $T = 300$ K for Ge, one has $V_D \sim 0.34$ V. For Si with $n_i = 3.10^{10}$ cm^{-3} at $T = 300$ K, one obtains $V_D \sim 0.64$ V.

- Calculation of the width of the depletion zone:
One uses the Poisson equation

$$\vec{\nabla} \cdot \vec{E} = \frac{\rho}{\varepsilon}, \quad \text{and} \quad \vec{E} = -\vec{\nabla}V(x),$$

where $\vec{E}$ is the electric field, ρ denotes the charge density, and ε the dielectric constant. One has

$$0 \le x \le x_d: \quad \frac{dE}{dx} = \frac{eN_d}{\varepsilon} \Rightarrow E(x) = \frac{eN_d}{\varepsilon}(x - x_d)$$

$$-x_a \le x \le 0: \quad \frac{dE}{dx} = \frac{eN_d}{\varepsilon} \Rightarrow E(x) = -\frac{eN_a}{\varepsilon}(x + x_a)$$

Since the induced electric field should be continuous at $x = 0$, one has

$$N_a x_a = N_d x_d \tag{15.140}$$

This relation expresses the charge neutrality in the depletion zone: the left-hand side is the charges on the p side and the right-hand side is that on the n side. Using $E(x) = -dV(x)/dx$ one has

$$0 \le x \le x_d: \quad V(x) = -\frac{eN_d}{2\varepsilon}(x - x_d)^2 + A$$

$$-x_a \le x \le 0: \quad V(x) = \frac{eN_a}{2\varepsilon}(x + x_a)^2 + B$$

where A and B are determined by the continuity condition at $x = 0$ and a limit condition, say the value of $V(x = 0)$. Taking the latter to be zero, the continuity at $x = 0$ gives $A = \frac{eN_d}{2\varepsilon}x_d^2$ and $B = -\frac{eN_a}{2\varepsilon}x_a^2$. Replacing these into the above equations, one obtains

$$0 \le x \le x_d: \quad V(x) = -\frac{eN_d}{2\varepsilon}x(x - 2x_d)$$

$$-x_a \le x \le 0: \quad V(x) = \frac{eN_a}{2\varepsilon}x(x + 2x_a).$$

This potential is shown in Fig. 15.7. One sees that one can express V_D as

$$V_D = V(x_d) - V(x_a) = \frac{eN_d}{2\varepsilon}x_d^2 + \frac{eN_a}{2\varepsilon}x_a^2$$

$$= \frac{eN_a}{2\varepsilon}(x_a^2 + x_d^2)$$

Since $x_d = \sqrt{2V_D\varepsilon/eN_d}$ and $x_a = \sqrt{2V_D\varepsilon/eN_a}$, one has

$$X_D = x_a + x_d = \left[\frac{2V_D\varepsilon}{e}\right]^{1/2}\left[\frac{1}{N_a} + \frac{1}{N_d}\right]^{1/2} \quad (15.141)$$

The width X_D of the depletion zone is a fraction of a micrometer. For example, in the case of Silicon with $V_D \sim 0.65$ V, $\varepsilon = 11.5\varepsilon_0$ and $N_a = N_d = 10^{17}$ cm^{-3}, one has $X_D \sim 0.13$ μm which is in general much smaller than the diffusion length $L_D \sim 50$ μm.

The case of a polarized $p - n$ junction is studied in Problem 9 given below together with other problems on doping which provide details and orders of magnitude of physical quantities in some frequently used extrinsic semiconductors.

15.9 Spin transport in magnetic materials

The theory of spin resistivity R_m in magnetic materials has been studied since the 50's [95, 40, 62]. Most of these early works have focused on the motion of an itinerant spin within a magnetically ordered lattice. The main interaction is between spins of "free" conducting s electrons and spins of localized d or f electrons of lattice ions. It has been shown that the phase transition of the lattice spin system causes a peak in the electron-spin resistivity. At low temperatures, the scattering of itinerant-electron spins by lattice spin waves dominates, leading to a T^2 dependence as seen in Eq. (14.1). At higher temperatures and near the phase transition, there is no theory which adequately describes the behavior of the peak of the spin resistivity, although it is known that its shape is closely related to the nature of the phase transition. More recent works have provided detailed effects on R_m from various physical parameters such as localization length of impurities [187], spin-spin correlation length or competing interaction strength [96]. Note that Monte Carlo simulations provide an interesting technique to probe the spin resistivity [124].

The spin transport in magnetic materials and magnetic devices has become now a subject of intensive research activities due to numerous applications in spintronics (or spin electronics) [43, 154, 191]. The main principle is to take advantage of the spin degree of freedom to manipulate the electron current in artificially created systems such as multilayers, superlattices and wires. One of the remarkable effects is the giant magnetoresistance which results from the fact that the spin resistivity depends on the relative orientation of the magnetizations in layers coupled in a sandwich structure of alternating ferromagnetic and nonmagnetic thin films. Depending on the relative orientation of the magnetizations in the magnetic layers, the device resistance changes from small (parallel magnetizations) to large (antiparallel magnetizations) values. This kind of devices is already in use in industry as a read head and a memory-storage cell since its discovery [14, 74]. Since this is a subject of rapid development, the reader is referred to research papers for ongoing progress.

15.10   Conclusion

In this chapter, the Boltzmann's equation has been introduced to study systems out of equilibrium. For systems not far from equilibrium, a linearization of this equation together with the relaxation-time approximation can be used. One has treated some concrete examples of transport in metals and semiconductors. In particular, effects of magnetic and electric fields on the electron conductivity have been calculated at weak and strong limits. A special attention has been paid to semiconductors, due to their numerous applications in electronics. Properties of doped semiconductors, principles of functionality and characteristics of $p - n$ junctions have been treated to a certain extent. There are many applications using $p - n$ junctions: the most used device is no doubt the transistor which is a combination in series of two junctions as $p - n - p$ and $n - p - n$. Other applications include tunneling diodes and non homogenously doped junctions. Note that the above treatment of minority carrier current in semiconductors can be applied to the case of charge excess created by sending photons to a surface of a semiconductor. Applications of such phenomena are numerous in optoelectronics.

The purpose of this chapter was to outline principal mechanisms which dominate transport properties in some selected topics, leaving aside a large number of subjects related to the transport of electron charges and spins which can be found in textbooks on semiconductors [163].

15.11   Problems

Problem 1. Effect of magnetic field:
        Demonstrate Eq. (15.24).

Problem 2. Electrons in a strong electric field: an approximation

One has studied above the deviation of the Ohm's law due to a strong electric field. In this exercise, one studies again this problem using a simple approximation which leads to the same result. In weak fields, during collisions with ions, electrons transfer immediately to phonons the energy they received from the electric field. Thus, the electron energy is mainly the thermal energy. In strong fields, that picture is not the case. To get quickly the result (15.74), one supposes that the energy loss by an electron to a phonon during a collision is proportional to its kinetic energy, namely $\Delta E = -\alpha^2 m^* v^2$ where α^2 is a coefficient.

Show that

$$|j| = \sigma_0 \varepsilon (1 + a\varepsilon^2)$$

for intermediate fields and

$$|j| = en\sqrt{\frac{el}{\alpha m^*}} \sqrt{\varepsilon}$$

for strong fields. l is the mean free path and n the electron density.

Problem 3. Semiconductors: effect of temperature on conductivity

Show that in order to increase the conductivity σ in an intrinsic semiconductor, one can increase the temperature by an amount ΔT. Calculate ΔT to increase twice σ in Ge and in Si at $T = 300$ K. One shall use the following gap values $E_g = 0.6$ eV for Ge and $E_g = 1$ eV for Si.

Problem 4. Semiconductor: effect of magnetic field on the gap

Calculate the variation of the forbidden energy gap in a semiconductor under a strong applied magnetic field $\vec{B}$, supposing the effective mass m^* equal to rest mass.

Problem 5. Effect of doping in semiconductors:

Consider a semiconductor which contains donor impurities of concentration N_d. Using the condition of electric neutrality, write the equation which allows for the determination of E_F at the temperature T as functions of effective densities of states N_c and N_v of conduction and valence bands, of N_d, E_c (bottom of the conduction band), E_v (top of the valence band) and E_D the energy level of impurities. Solve this equation in the case of Si at 300 K with $N_d = 10^{14}$ cm^{-3}, $N_c = N_v = 10^{19}$ cm^{-3}, the forbidden band gap is 1.12 eV.

Problem 6. Swallow impurity states in semiconductors: an approximation

Consider the interaction between the extra electron brought about by an ionized donor impurity. The ionized donor creates around it an electrostatic potential $U(r) = e/(\varepsilon r)$ where ε is the static dielectric constant. The extra electron stays in a large orbit around the ion to form with it a bound state. Assuming that this is equivalent to a hydrogen atom, find the location of impurity energy levels with respect to the bottom of the conduction band energy.

Problem 7. Recombinations in semiconductors:

Calculate the recombination rate in a doped semiconductor.

Problem 8. Dielectric relaxation:

Show that any charge excess artificially created in a semiconductor tends exponentially to zero with a dielectric relaxation time τ_d to be determined.

Problem 9. Polarized $p - n$ junction: direct current

In section 15.8.6 a non polarized $p - n$ junction has been studied. Consider in this exercise a $p - n$ junction under an applied forward-bias voltage V_0. One supposes:

(i) the hole concentration in the n side (minority carriers) is very small, namely $\delta p(x_d) << n_0$ (very weak injection, n_0 being electron concentration at equilibrium)

(ii) the recombinations in the depletion zone are negligible due to a short time needed to cross this narrow zone ($X_D \sim 0.4$ μm) compared to a long diffusion length of minority carriers ($L_D \sim 50$ μm)

(iii) the quasi Fermi level of each kind of majority carriers is conserved until the entrance of the other majority zone, namely $E_{Fp}(x) = E_{Fp0}$ for $x \leq x_d$, and $E_{Fn}(x) = E_{Fn0}$ up to $x \geq -x_a$.

Calculate the direct current which flows across the junction.

Problem 10. Transport in a superlattice:

One considers a superlattice composed of alternate well-barrier semiconducting layers stacked in the x direction (sandwich structure) as shown in Fig. 15.8. The layers are yz planes. One can use the Kronig-Penney model (see Problem 4 of chapter 11) for the well-barrier in the x direction. Let a and b be the thickness of the well and barrier layers, respectively. The spatial periodicity c of the superlattice is thus $c = a + b$.

One admits that the electron energy in the x direction is written

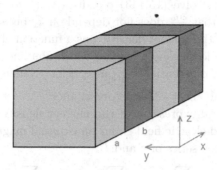

Fig. 15.8 Superlattice composed of alternate well and barrier layers stacked in the x direction.

as

$$E(k_x) = n^2 E_0 + t_n \cos(k_x b) \qquad (15.142)$$

where n is the band index, $E_0 = \frac{\hbar^2}{2m} \frac{\pi^2}{a^2}$ and $t_n \ll E_0$ (the energy in the yz plane is just the kinetic energy $E(k_y, k_z) = \frac{\hbar^2}{2m}(k_y^2 + k_z^2)$). The second term in (15.142) results from a weak coupling between neighboring wells if b is not small. Hereafter, one considers only a miniband $n = 1$.

a) Calculate the velocity in the x direction and the effective mass of the electron in the superlattice. Calculate the density of states near $k_x \simeq 0$.

b) One applies an electric field ε along the x direction. Using the linearized Boltzmann's equation in the relaxation-time approximation, calculate the conductivity σ. One shall use a constant C to represent $\int \sin^2 k_x b \frac{\partial f_0}{\partial E} dk_x$ (f_0: Fermi-Dirac distribution at equilibrium). Without integrating show that $C \neq 0$.

c) One takes into account non linear terms of the Boltzmann's equation. Show that $g \equiv f - f_0$ satisfies

$$\frac{\partial g}{\partial k_x} + Ag = B \sin(k_x b)\left(-\frac{\partial f_0}{\partial E}\right) \qquad (15.143)$$

where A and B are constants to be determined. Find a solution of the form

$$g = g_0 \exp\left(-\frac{\hbar k_x}{e\varepsilon\tau}\right).$$

Find the equation which is satisfied by $\frac{\partial g_0}{\partial k_x}$. Deduce g.

Guide: To solve (15.143) one uses $\sin(k_x b) = \frac{e^{ik_x b} - e^{-ik_x b}}{2i}$ and supposes that $\frac{\partial f_0}{\partial E}$ does not depend on k_x because $t \ll E_0$.

d) Calculate the current density j as a function of ε. Show that there exists a field threshold beyond which non linear effects set in with a negative differential resistance. Plot j versus ε.

Problem 11. Hall effect - Magnetoresistance:

The general expression of the current density in a system under an applied electric field $\vec{\varepsilon}$ and an external magnetic field $\vec{B}$ can be written as a series of $\vec{\varepsilon}$ and $\vec{B}$:

$$j_i = \sum_j \sigma_{ij}\varepsilon_j + \sum_{j,l} \sigma_{ijl}\varepsilon_j B_l + \sum_{j,l,m} \sigma_{ijlm}\varepsilon_j B_l B_m$$

where σ_{ij} is the "normal" or "ordinary" electric conductivity tensor, and σ_{ijl} denotes the conductivity tensor due to the interaction between $\vec{\varepsilon}$ and $\vec{B}$. When $\vec{\varepsilon} \cdot \vec{B} = 0$, we have the geometry of the Hall effect. σ_{ijlm} is the conductivity tensor due to the interaction between $\vec{\varepsilon}$ and $\vec{B}$ at the second order. This is at the origin of the magneto-resistance.

The case of weak magnetic fields has been treated in section 15.8.1. with a linear approximation. This approximation is used when $\omega_c \tau \ll 1$ where ω_c is the cyclotron frequency and τ the relaxation time. One supposes that $\vec{B}$ is parallel to $\vec{Oz}$ and $\vec{\varepsilon}$ parallel to $\vec{Ox}$. One supposes in addition that τ is independent of the electron energy and that the electron effective mass is isotropic.

a) Moderate fields:

i) Write down the equations of motion of an electron in the x and y directions. One introduces the following complex quantity $Z = v_x + iv_y$ where v_x and v_y are the components of the electron velocity. Show that the solution of the equations of motion is of the form

$$Z(t) = Z_0 + \frac{e}{i\omega_c m_e}(\varepsilon_x + i\varepsilon_y)(1 - e^{i\omega_c t})$$

where $Z_0 = Z(t = 0)$.

ii) Show that the average value of Z taken over all collisions is given by

$$\overline{Z} = \frac{e}{m_e} \frac{\tau}{i\omega_c \tau - 1}.$$

Deduce that

$$j_x = \frac{ne^2}{m_e}\left[\frac{\tau}{1+\omega_c^2\tau^2}\varepsilon_x - \frac{\omega_c\tau^2}{1+\omega_c^2\tau^2}\varepsilon_y\right]$$

$$j_y = \frac{ne^2}{m_e}\left[\frac{\tau}{1+\omega_c^2\tau^2}\varepsilon_y + \frac{\omega_c\tau^2}{1+\omega_c^2\tau^2}\varepsilon_x\right]$$

iii) Show that when both electrons and holes participate in the conduction, the Hall coefficient is given by

$$R = \frac{\sigma_e^2 R_e + \sigma_h^2 R_h + \sigma_e^2\sigma_h^2 R_e R_h(R_e + R_h)B^2}{(\sigma_e + \sigma_h)^2 + \sigma_e^2\sigma_h^2(R_e + R_h)^2 B^2}.$$

Give comments on the case where the conduction is due to only one type of impurity (p or n) and on the case of an intrinsic conductor ($p = n$).

iv) The above results show that the presence of a magnetic field has no effect on the resistance when the conduction is due to only one type of carriers (n or p) and when the relaxation time is constant and when the effective mass is isotropic (spherical iso-energy surfaces). If one of these three conditions is not fulfilled, there is a correction to the initial resistivity ρ_0. Show that when both electrons and holes participate to the conduction, keeping isotropic effective mass and constant τ, the correction is given by

$$\frac{\Delta\rho}{\rho_0} = \frac{np\mu_e\mu_h(\mu_e + \mu_h)^2 B^2}{(n\mu_e + p\mu_h)^2} \equiv \xi R_0^2\sigma_0^2 B^2$$

where the coefficient of the transverse magneto-resistance ξ is

$$\xi = \frac{np\mu_e\mu_h(\mu_e + \mu_h)^2}{(p\mu_h^2 - n\mu_e^2)^2}$$

and $\sigma_0 \equiv \sigma_e + \sigma_h$ when $B = 0$, R_0 being given by the above weak-field approximation. Numerical application: in order to have $\frac{\Delta\rho}{\rho_0} \simeq 0.1$ in an applied field of 10^3 Gauss, what is the total mobility $\mu_e + \mu_h$?

b) Effects of collisions:

One supposes again that the effective mass m is isotropic, but the relaxation time τ depends on the electron energy under the form $\tau = aE^{-s}$. This form represents several types of collision. The results for j_x and j_x in the weak-field approximation shown above are still valid provided that the quantities $\frac{\tau}{1+\omega_c^2\tau^2}$ and $\frac{\omega_c\tau^2}{1+\omega_c^2\tau^2}$ are replaced by their values averaged over all energies.

i) Show that in the present case the Hall coefficient is written as $R = -\frac{K}{ne}$ where K is given by

$$K = \frac{< \frac{\tau^2}{1+\omega_c^2\tau^2} >}{< \frac{\tau}{1+\omega_c^2\tau^2} >^2 + \omega_c^2 < \frac{\tau^2}{1+\omega_c^2\tau^2} >^2} \qquad (15.144)$$

Show that $K = \frac{<\tau^2>}{<\tau>^2}$ for weak fields and strong enough collisions.

ii) Calculate $< \tau^2 >$ and $< \tau >$. Show that

$$< \tau >= a(k_BT)^{-s}\frac{\Gamma(5/2 - s)}{\Gamma(5/2)}$$

and

$$< \tau^2 >= a^2(k_BT)^{-2s}\frac{\Gamma(5/2 - 2s)}{\Gamma(5/2)^2}.$$

Estimate K for $s = 1/2$ (collisions with phonons) and $s = -3/2$ (collisions with impurities).

iii) One supposes there is only one kind of carriers, calculate $\frac{\Delta\rho}{\rho_0}$.

PART 3
Solutions of Problems

Solutions of Problems of Part 1

16.1 Solutions of problems of chapter 1

Problem 1. Central limit theorem:

One shows first that the mean value $\bar{n}$ of the binomial law corresponds to the most probable value n_0 when $N, n \gg 1$. The most probable value n_0 is the value at the maximum of the probability, namely when $dP(N,n)/dn = 0$ or equivalently, $d\ln P(N,n)/dn = 0$. One has

$$\frac{d\ln P(N,n)}{dn} = \frac{d}{dn} \ln \frac{N!}{n!(N-n)!} P_A^n P_B^{N-n}$$

$$= \frac{d}{dn}[\ln N! - \ln n! - \ln(N-n)! + \ln P_A^n + \ln P_B^{N-n}]$$

$$= \frac{d}{dn}[-\ln n! - \ln(N-n)!] + \ln \frac{P_A}{P_B}$$

Using the Stirling formula $\ln n! \simeq n\ln n - n$ (see Appendix A.1), one obtains

$$\frac{d}{dn}\ln n! \simeq \frac{d}{dn}(n\ln n - n) = \ln n$$

$$\frac{d}{dn}\ln(N-n)! \simeq \frac{d}{dn}[(N-n)\ln(N-n) - (N-n)] = -\ln(N-n)$$

from which one gets

$$\frac{d\ln P(N,n)}{dn} = \ln \frac{N-n}{n} + \ln \frac{P_A}{P_B} \qquad (16.1)$$

At $n = n_0$, one has $\frac{d\ln P(N,n)}{dn} = 0$. One finds

$$\ln \frac{n_0}{N - n_0} = \ln \frac{P_A}{P_B}$$

$$n_0 \simeq N P_A \tag{16.2}$$

where $N \gg n \gg 1$. One sees under this condition $n_0 = \bar{n}$ [see (1.23)].

One expands now $\ln P(N, n)$ around n_0:

$$\ln P(N, n) = \ln P(N, n_0) + \left[\frac{d \ln P(N, n)}{dn} \right]_{n=n_0} (n - n_0)$$

$$+ \frac{1}{2} \left[\frac{d^2 \ln P(N, n)}{dn^2} \right]_{n=n_0} (n - n_0)^2 + \dots$$

$$\simeq \ln P(N, n_0) + \frac{1}{2} \left[\frac{d^2 \ln P(N, n)}{dn^2} \right]_{n=n_0} (n - n_0)^2 + \dots \tag{16.3}$$

where the second term of the first line corresponding to the maximum of $P(N, n)$ is thus zero. Taking the derivative of (16.1) and replacing n by $n_0 = N P_A$, one has

$$\left[\frac{d^2 \ln P(N, n)}{dn^2} \right]_{n=n_0} = -\frac{1}{N P_A P_B} \tag{16.4}$$

Using (1.25) one has

$$\left[\frac{d^2 \ln P(N, n)}{dn^2} \right]_{n=n_0} = -\frac{1}{(\Delta n)^2} \tag{16.5}$$

Neglecting terms of order higher than 2 in (16.3), one obtains

$$\ln P(N, n) = \ln P(N, n_0) - \frac{1}{2(\Delta n)^2} (n - n_0)^2 \tag{16.6}$$

from which

$$\frac{P(N, n)}{P(N, n_0)} = \exp \left[-\frac{1}{2(\Delta n)^2} (n - n_0)^2 \right] \tag{16.7}$$

This relation is the Gaussian density of probability where $x_0 = n_0$ and $\Delta n = \sigma$.

One concludes that the binomial law becomes the Gaussian law when $N \gg n \gg 1$.

Remark: One can show that terms of higher orders of (16.3) are negligible when $N \gg n \gg 1$ by taking successive derivatives of (16.1) at $n = n_0$.

Problem 2. Poisson law:

a) One has

$$\bar{n} = \sum_{n=0}^{\infty} nP(n) = \sum_{n=0}^{\infty} n \frac{\mu^n \exp(-\mu)}{n!} = \exp(-\mu) \sum_{n=1}^{\infty} \frac{n\mu^n}{n!}$$

$$= \exp(-\mu) \sum_{n=1}^{\infty} \frac{\mu^n}{(n-1)!} = \mu \exp(-\mu) \sum_{n=1}^{\infty} \frac{\mu^{n-1}}{(n-1)!}$$

$$= \mu \exp(-\mu) \sum_{n'=0}^{\infty} \frac{\mu^{n'}}{(n')!} = \mu \exp(-\mu) \exp(\mu) = \mu \qquad (16.8)$$

where n in the sum of the last equality in the first line starts with $n = 1$ because the $n = 0$ term has a zero contribution to the sum. Note that in the last line one used the series expansion of the exponential (see Appendix A).

Remark: It can be shown that the variance of the Poisson law is also equal to μ.

b) One shows now that the binomial law becomes the Poisson law when $P_A \ll P_B$ and $N \gg n \gg 1$:

$$\ln P(N, n) = \ln \left[\frac{N!}{n!(N-n)!} P_A^n P_B^{N-n} \right]$$

$$= \ln \left[\frac{N!}{n!(N-n)!} \right] + \ln \left(P_A^n P_B^{N-n} \right) \qquad (16.9)$$

Using the Stirling formula, one has

$$\ln \left[\frac{N!}{(N-n)!} \right] \simeq N \ln N - N - (N-n) \ln(N-n) + N - n$$

$$\simeq n \ln N = \ln N^n \qquad (16.10)$$

where one has neglected n in front of N. One gets

$$\frac{N!}{(N-n)!} \simeq N^n \qquad (16.11)$$

Replacing it in $P(N, n)$ one obtains

$$P(N, n) = \frac{1}{n!} N^n P_A^n P_B^{N-n} = \frac{1}{n!} (N P_A)^n (1 - P_A)^{N-n}$$

$$= \frac{(\bar{n})^n}{n!} \left(1 - \frac{\bar{n}}{N}\right)^{N-n}$$

$$= \frac{(\bar{n})^n}{n!} \left[1 - (N-n)\frac{\bar{n}}{N} + \frac{(N-n)(N-n-1)}{2!}(\frac{\bar{n}}{N})^2 + ..\right]$$

$$= \frac{(\bar{n})^n}{n!} \left[1 - \bar{n} + \frac{\bar{n}^2}{2!} + ...\right]$$

$$= \frac{(\bar{n})^n}{n!} \exp(-\bar{n}) \tag{16.12}$$

where one has taken $n/N, 1/N, 1/N^2, ... \simeq 0$ before the last line. The relation (16.12) is the Poisson law with $\mu = \bar{n}$.

Problem 3. Demonstration of (1.34) and (1.35):

a) Formula (1.34): One considers N indiscernible particles.

- There are N ways to choose the first particle for the level 0
- There are $N - 1$ ways to choose the second particle for the level 0
- There are $N - n_0 + 1$ ways to choose the n_0-th particle for the level 0

Therefore, there are a total of $\omega_0 = N(N-1)...(N - n_0 + 1)$ ways to choose n_0 particles for the level 0. One can write ω_0 as

$$\omega_0 = \frac{N!}{(N - n_0)!} \tag{16.13}$$

Note that permutations between particles belonging to the same level 0 do not add new states so that

$$\omega_0 = \frac{N!}{n_0!(N - n_0)!} = C_N^{n_0} \tag{16.14}$$

One proceeds in the same manner to choose n_1 particles for the level 1 among the remaining $(N - n_0)$ particles. One has then

$$\omega_1 = \frac{(N - n_0)!}{n_1!(N - n_0 - n_1)!} = C_{N-n_0}^{n_1} \tag{16.15}$$

Finally, the total number of ways to distribute $n_0, n_1, n_2, ...$ particles on levels 0, 1, 2, ... is given by

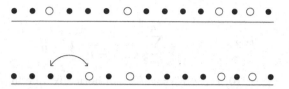

Fig. 16.1 Two examples of repartition of 11 energy units on 5 particles. The particles are separated by walls (white balls). The number of energy units of a particle is the number of black balls between two walls. The example shown on the top schema is a repartition from left to right 2, 3, 4, 1 and 1. A permutation of a white ball and a black ball gives the new repartition shown on the bottom: 4, 1, 4, 1 and 1.

$$\omega = \omega_0\omega_1\omega_2... = \frac{N!}{n_0!(N-n_0)!}\frac{(N-n_0)!}{n_1!(N-n_0-n_1)!}$$
$$\times \frac{(N-n_0-n_1)!}{n_2!(N-n_0-n_1-n_2)!}...$$
$$= \frac{N!}{n_0!n_1!n_2!...} \tag{16.16}$$

b) Formula (1.35):

The number of ways to distribute E energy units on N particles is given by (1.35). An easy way to demonstrate this formula is to consider the following schema: each energy unit is represented by a black ball. N particles are represented by a row of N "cases" each of which contains the number of black balls of the particle. Two particle cases are separated by a wall represented by a white ball: there are $N-1$ walls (see Fig. 16.1). The total number of energy repartitions corresponds to the number of permutations of $(E+N-1)$ black balls and white balls. However, one has to discard permutations between white balls and permutations between black balls because these permutations do not create new energy repartitions.

Finally, one has

$$\Omega(E) = \frac{(N+E-1)!}{(N-1)!E!} \tag{16.17}$$

Problem 4. Application of the binomial law:

One uses the binomial law. The probability to find the number of heads between 3 and 6 (including boundaries) when one flips 10 times a coin is given by

$$P(3 \le n \le 6) = \sum_{n=3}^{6} \frac{N!}{n!(N-n)!} P_A^n P_B^{N-n} \qquad (16.18)$$

where $N = 10$, $P_A = P_B = 1/2$. Numerically, one finds $P(3 \le n \le 6) = 99/128$.

Problem 5. Random walk in one dimension:

a) The probability for the particle to make n steps to the right and n' steps to the left after N steps is given by the binomial law:

$$P(N, n) = \frac{N!}{n!(N-n)!} P_A^n P_B^{N-n} \qquad (16.19)$$

where $P_A = P_B = 1/2$ and $n' = N - n$.

b) The binomial law gives $\bar{n} = NP_A$ and $\bar{n'} = N - \bar{n} = N(1 - P_A) = NP_B$. One sees that $\bar{n} = \bar{n'}$. The average position after N steps is thus at the origin. The variance is $(\Delta n)^2 = NP_A P_B$ and the relative uncertainty on $\bar{n}$ is proportional to $1/\sqrt{N}$ [see (1.26)].

c) The particle is at the position $x = ml$ from the origin after N steps: one has $x = ml = nl - (N - n)l = (2n - N)l$. Therefore,

$$n = \frac{N+m}{2} \qquad (16.20)$$

from which $n' = N - n = \frac{N-m}{2}$. The corresponding probability is

$$P(N, n) = \frac{N!}{[\frac{N+m}{2}]![\frac{N-m}{2}]!} P_A^{\frac{N+m}{2}} P_B^{\frac{N-m}{2}} \qquad (16.21)$$

One has $\bar{x} = \overline{ml} = 2\bar{n} - N = 2NP_A - N = 0$ (because $P_A = 1/2$). One obtains after a straightforward calculation

$$(\Delta x)^2 = \overline{x^2} - (\bar{x})^2 = l^2 \overline{(2n - N)^2} - 0$$
$$= 4NP_A P_B \qquad (16.22)$$

d) One supposes now that the step length x is variable and N is very large. The probability for x to be in the interval x and $x + dx$ is $W(x)dx$ where $W(x)$ is a Gaussian density of probability given by

$$W(x) = A \exp\left[-\frac{(x - \bar{x})^2}{2\sigma^2}\right] \qquad (16.23)$$

$\bar{x}$ being the average step length, $\sigma^2 = (\Delta x)^2$ and $A = 1/\sqrt{2\pi\sigma^2}$.

After N steps, the position of the particle is $X = \sum_{i=1}^{N} x_i$, the average position is $\overline{X} = \sum_{i=1}^{N} \overline{x}_i = N\overline{x}$ and the variance is

$$(\Delta X)^2 = \sum_{i=1}^{N} (\Delta x_i)^2 = N(\Delta x)^2 = N\sigma^2$$

The density of probability to find the particle at X after N steps is thus

$$W_N(X) = B \exp\left[-\frac{(X - N\overline{x})^2}{2N\sigma^2}\right] \qquad (16.24)$$

where $B = 1/\sqrt{2\pi N\sigma^2}$.

Problem 6. Random walk in three dimensions:

a) The Cartesian components (X, Y, Z) of the particle position after N steps are $X = \sum_{i=1}^{N} x_i$, $Y = \sum_{i=1}^{N} y_i$, $Z = \sum_{i=1}^{N} z_i$. For large N, the average $(\overline{X}, \overline{Y}, \overline{Z})$ are zero because the probabilities are homogeneous in all directions.

The variance $(\Delta X)^2$ is $(\Delta X)^2 = \overline{X^2} - (\overline{X})^2 = \overline{(\sum_{i=1}^{N} x_i)^2} - 0 = \overline{\sum_{i=1}^{N} x_i^2 + 2\sum_{i\neq j} x_i x_j} = \overline{\sum_{i=1}^{N} x_i^2} = \sum_{i=1}^{N} \overline{x_i^2} = N\overline{x_i^2}$ (the average value of $x_i x_j$ is zero because x_i and x_j are independent). One has the similar expressions for $(\Delta Y)^2$ and $(\Delta Z)^2$.

Note that the average value of the square length step is $\overline{l^2} = \overline{x_i^2 + y_i^2 + z_i^2} = 3\overline{x_i^2}$ because the space is homogeneous. Therefore $(\Delta X)^2 = N\overline{l^2}/3$.

b) The density of probability to find the particle at X after N steps is given by the Gaussian law:

$$W_N(X) = B \exp\left[-\frac{(X - \overline{X})^2}{2(\Delta X)^2}\right] = B \exp\left[-\frac{3X^2}{2N\overline{l^2}}\right] \qquad (16.25)$$

where $B = 1/\sqrt{2\pi N\overline{l^2}/3}$.

c) One puts $(\Delta X)^2 = 2Dt$ where D is the diffusion coefficient and t the lapse of time of the particle displacement. The number of steps is given by the number of collisions during that time: $N = t/\tau$. One has $D = (\Delta X)^2/(2t) = N\overline{l^2}/(6t) = \frac{1}{6}\frac{\overline{l^2}}{\tau}$. Similar expressions are obtained for two other directions.

Problem 7. Exchange of energy:

a) The number of microstates of system 1: one uses the formula (1.35):

$$\Omega_1(E_1) = \frac{(N_1 + E_1 - 1)!}{(N_1 - 1)!E_1!} = \frac{4!}{1!3!} = 4 \qquad (16.26)$$

For system 2,

$$\Omega_2(E_2) = \frac{(N_2 + E_2 - 1)!}{(N_2 - 1)!E_2!} = \frac{8!}{3!5!} = 56 \qquad (16.27)$$

The total number of microstates of systems 1 and 2 is thus $\Omega = \Omega_1\Omega_2 = 224$.

b) One removes the wall. The total system is always isolated. However, there is an energy exchange between systems 1 and 2. The total number of microstates of the total system in this new situation is equal to the number of ways to distribute the total energy $E = E_1 + E_2 = 8$ on $N = N_1 + N_2 = 6$ particles. One has

$$\Omega(E) = \frac{(N + E - 1)!}{(N - 1)!E!} = \frac{13!}{5!8!} = 1287 \qquad (16.28)$$

One sees that $\Omega(E)$ is much larger than in the previous case where there is no energy exchange between the two subsystems.

c) The number of ways to choose 2 particles among 6 to receive each 2 energy units is given by the formula (1.34):

$$W = \frac{6!}{2!4!} = 15 \qquad (16.29)$$

These 2 particles have 4 energy units. The best way to obtain the number of possibilities to give 4 energy units to these 2 particles is to calculate the number of possibilities to distribute the remaining 4 energy units on the 4 remaining particles. This number is given by (1.35):

$$\Omega(E = 4, N = 4) = \frac{(4 + 4 - 1)!}{(4 - 1)!4!} = \frac{7!}{3!4!} = 35 \qquad (16.30)$$

Thus, the number of microstates in which at least 2 particles have each 2 energy units is $W \times \Omega(E = 4, N = 4) = 15 \times 35 = 525$. The probability of this situation is $525/1287$.

Problem 8. Statistical entropy:

One shows that the statistical entropy $S = -k_B \sum_l P_l \ln P_l$ is maximum when all probabilities P_l are equal. One has to maximize the many-variable function S under the constraint of probability normalization $\sum_l P_l = 1$. Using the Lagrange variational method, one has to maximize $S - \lambda \sum_l P_l$ where λ is Lagrange multiplier. One has

$$\frac{\partial}{\partial P_m}\left[S - \lambda \sum_l P_l\right] = \frac{\partial}{\partial P_m}\left[-k_B \sum_l P_l \ln P_l - \lambda \sum_l P_l\right] = 0$$
(16.31)

where P_m indicates the variable to make vary. One has Ω of such equations because $m = 1, ..., \Omega$. One concentrates oneself in the case of a particular variable P_k. One has

$$\frac{\partial}{\partial P_k}[-k_B P_k \ln P_k - \lambda P_k] = 0$$

$$-k_B(\ln P_k + 1) - \lambda = 0$$

$$\ln P_k = (-\lambda - k_B)/k_B$$

The last line shows that P_k is independent of k. In other words, all events have the same probability. Putting $P_k = C$ where C is a constant independent of k. The normalization of probabilities $\sum_{l=1}^{\Omega} P_l = 1$ gives $\sum_l C = 1$ from which $C^{-1} = \sum_{l=1}^{\Omega} 1 = \Omega$, namely $P_k = 1/\Omega$.

16.2 Solutions of problems of chapter 2

Problem 1. Joule expansion:

a) The molecules occupy the total volume, namely $2V_0$.

b) The probability to find a molecule in the compartment A in the final state is $P(1) = V_A/V_{A+B} = 1/2$. The probability to find n molecules in the compartment A in the final state is thus $P(n) = C_N^n (1/2)^n (1/2)^{N-n}$. This is the binomial law (a molecule in A or not in A). The most probable value of n is $NP(1) = N/2$. This corresponds to the maximum of $P(n)$. The number of microstates is therefore $\Omega(n = N/2) = C_N^{N/2} = \frac{N!}{(N/2)!(N/2)!}$. When all molecules find themselves in A, the probability of such situation corresponds to $n = N$, the number of microstates is thus $\Omega(n = N) = C_N^N = 1$. The variation of the statistical entropy between these situations is

$$\Delta S = S(n = N/2) - S(N) = k_B[\ln \Omega(n = N/2) - \ln \Omega(n = N)]$$

$$= k_B \left(\ln[\frac{N!}{(N/2)!(N/2)!}] - \ln 1 \right)$$

$$\simeq k_B[N \ln N - N - (N/2) \ln(N/2) + N/2$$

$$-(N/2) \ln(N/2) + N/2]$$

$$= k_B[N \ln N - N \ln(N/2)]$$

$$= k_B N \ln 2 > 0 \tag{16.32}$$

This shows that in a Joule expansion which is a spontaneous evolution of the system when an external constraint (wall) is removed,
the micro-canonical entropy increases. The new equilibrium state
corresponds to the maximum of the entropy.

Problem 2. Exchange of heat:

a) One enumerates the microstates of the energy levels accessible for
each particle:

- fundamental level: $(n_x = n_y = 1)$, $\epsilon = 2\epsilon_0$, degeneracy $g_1 = 1$
- 2nd level: states $(n_x = 2, n_y = 1)$ and $(n_x = 1, n_y = 2)$
  (permutation of n_x and n_y), $\epsilon = 5\epsilon_0$, degeneracy $g_2 = 2$
- 3rd level: $(n_x = n_y = 2)$, $\epsilon = 8\epsilon_0$, degeneracy $g_3 = 1$
- 4th level: $(n_x = 3, n_y = 1)$ + one permutation, $\epsilon = 10\epsilon_0$,
  degeneracy $g_4 = 2$
- 5th level: $(n_x = 3, n_y = 2)$+ one permutation, $\epsilon = 13\epsilon_0$,
  degeneracy $g_5 = 2$
- 6th level: $(n_x = 4, n_y = 1)$ + one permutation, $\epsilon = 17\epsilon_0$,
  degeneracy $g_6 = 2$
- 7th level: $(n_x = 3, n_y = 3)$, $\epsilon = 18\epsilon_0$, degeneracy $g_7 = 1$
- 8th level: $(n_x = 4, n_y = 2)$, $\epsilon = 20\epsilon_0$, degeneracy $g_8 = 2$
- 9th level: $(n_x = 4, n_y = 3)$, $\epsilon = 27\epsilon_0$, degeneracy $g_9 = 2$

b) At $t = 0$, $E_1 = 15\epsilon_0$. The repartitions of this energy on two
discernible particles are (the first value is the energy of the first
particle, the second value is that of the second particle):

- $(2\epsilon_0, 13\epsilon_0)$: degeneracy $g_1 \times g_5 = 1 \times 2 = 2$
- $(13\epsilon_0, 2\epsilon_0)$: degeneracy $g_5 \times g_1 = 2$
- $(5\epsilon_0, 10\epsilon_0)$: degeneracy $g_2 \times g_4 = 2 \times 2 = 4$
- $(10\epsilon_0, 5\epsilon_0)$: degeneracy $g_4 \times g_2 = 4$

The total degeneracy, namely the total number of microstates, of E_1 is thus $\Omega_1 = 2 + 2 + 4 + 4 = 12$.

For subsystem 2, $E_2 = 10\epsilon_0$ is to be distributed on two discernible particles. The repartitions are

- $(5\epsilon_0, 5\epsilon_0)$: degeneracy $g_2 \times g_2 = 4$.
- $(2\epsilon_0, 8\epsilon_0)$: degeneracy $g_1 \times g_3 = 1$.
- $(8\epsilon_0, 2\epsilon_0)$: degeneracy $g_3 \times g_1 = 1$.

The total number of microstates of E_2 is thus $\Omega_2 = 4 + 1 + 1 = 6$. The total number of microstates of the total system when two subsystems are independent is $\Omega = \Omega_1 \Omega_2 = 12 \times 6 = 72$.

c) The separating wall becomes now diathermic. It allows an exchange of heat between subsystems. The total system evolves toward a new equilibrium.

- The total energy is conserved with the evolution: $E_t = E_1 + E_2 = 25\epsilon_0$.
- The total number of microstates of the total system is the number of repartitions of E_t on the 4 discernible particles. In the first question, one has enumerated the accessible levels for each particle. Using this result, one has the following energy repartitions (the first number is for the first particle etc.):

    - $(2\epsilon_0, 2\epsilon_0, 8\epsilon_0, 13\epsilon_0)$: degeneracy $g_1 \times g_1 \times g_3 \times g_5 = 2$. There are $A_4^2 = 12$ arrangements of these energies on 4 particles, total degeneracy $= 12 \times 2 = 24$
    - $(2\epsilon_0, 5\epsilon_0, 5\epsilon_0, 13\epsilon_0)$: degeneracy $g_1 \times g_2 \times g_2 \times g_5 = 8$. There are $A_4^2 = 12$ arrangements, total degeneracy $= 12 \times 8 = 96$
    - $(2\epsilon_0, 5\epsilon_0, 8\epsilon_0, 10\epsilon_0)$: degeneracy $g_1 \times g_2 \times g_3 \times g_4 = 4$. There are $4! = 24$ arrangements (permutations of these energies on 4 particles), total degeneracy $= 24 \times 4 = 96$
    - $(5\epsilon_0, 5\epsilon_0, 5\epsilon_0, 10\epsilon_0)$: degeneracy $g_2 \times g_2 \times g_2 \times g_4 = 16$. There are $A_4^1 = 4$ arrangements, total degeneracy $= 4 \times 16 = 64$

Thus, there are in all $24 + 96 + 96 + 64 = 280$ microstates. One sees that the total number of microstates increases considerably from 72 to 280 with the evolution.

- Using the accessible levels for each particle $(2, 5, 8, 10, 13, 17, 18, 20, ...\epsilon_0)$, and taking into account the energy conservation, one has the following repartitions between two subsystems (using the same analysis as above):

$E_1 = 4 \ (=2+2), \ \Omega_1 = 1, \ E_2 = 21 \ (=8+13), \ \Omega_2 = 4 \to \Omega = \Omega_1\Omega_2 = 1 \times 4 = 4$

$E_1 = 7 \ (=2+5), \ \Omega_1 = 4, \ E_2 = 18 \ (=8+10 \text{ and } 5+13), \ \Omega_2 = 12 \to \Omega = 4 \times 12 = 48$

$E_1 = 10 \ (=2+8), \ \Omega_1 = 2, \ E_2 = 15 \ (=5+10 \text{ and } =2+13), \ \Omega_2 = 12 \to \Omega = 2 \times 12 = 24$

$E_1 = 10 \ (=5+5), \ \Omega_1 = 4, \ E_2 = 15 \ (=5+10 \text{ and } 2+13), \ \Omega_2 = 12 \to \Omega = 4 \times 12 = 48$

$E_1 = 12 \ (=2+10), \ \Omega_1 = 4, \ E_2 = 13 \ (=5+8), \ \Omega_2 = 4 \to \Omega = 4 \times 4 = 16$

$E_1 = 13 \ (=5+8), \ \Omega_1 = 4, \ E_2 = 12 \ (=2+10), \ \Omega_2 = 4 \to \Omega = 4 \times 4 = 16$

$E_1 = 15 \ (=5+10), \ \Omega_1 = 8, \ E_2 = 10 \ (=2+8 \text{ and } 5+5), \ \Omega_2 = 6 \to \Omega = 8 \times 6 = 48$

$E_1 = 15 \ (=2+13), \ \Omega_1 = 4, \ E_2 = 10 \ (=2+8 \text{ and } 5+5), \ \Omega_2 = 6 \to \Omega = 4 \times 6 = 24$

$E_1 = 18 \ (=8+10), \ \Omega_1 = 4, \ E_2 = 7 \ (=2+5), \ \Omega_2 = 4 \to \Omega = 4 \times 4 = 16$

$E_1 = 18 \ (=5+13), \ \Omega_1 = 8, \ E_2 = 7 \ (=2+5), \ \Omega_2 = 4 \to \Omega = 8 \times 4 = 32$

$E_1 = 21 \ (=8+13), \ \Omega_1 = 4, \ E_2 = 4 \ (=2+2), \ \Omega_2 = 1 \to \Omega = 4 \times 1 = 4$

One sees that the total number of microstates is 280 as found in the previous question.

Problem 3. Distribution of energy on particles:

a) One has $E = \sum_i \epsilon_i = u[n_1 + n_2 + ... + n_N] \equiv Mu$. The number of microstates is equal to the number of ways to distribute M energy units on N particles. One uses therefore the formula (1.35)
$$\Omega(E) = \frac{(M+N-1)!}{(N-1)!M!}$$

b) The micro-canonical entropy is $S = k_B \ln \Omega = k_B \ln[\frac{(M+N-1)!}{(N-1)!M!}]$ with $M = E/u$. Assuming that M and N are very large, one takes $(M + N - 1)! \simeq (M + N)!$ and $(N - 1)! \simeq N!$. One uses next the Stirling formula (2.47), one obtains
$$S \simeq k_B[(E/u + N) \ln(E/u + N) - (E/u) \ln(E/u) - N \ln N].$$

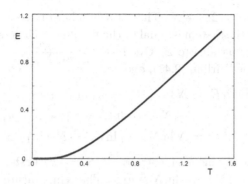

Fig. 16.2   Energy E versus $k_B T$ with $u = 1$ obtained in Problem 3.

c) The micro-canonical temperature is given by

$$T^{-1} = \left(\frac{\partial S}{\partial E}\right)_{V,N}$$

$$= k_B \left[\frac{1}{u}\ln(E/u + N) + (E/u + N)\frac{1/u}{E/u + N}\right.$$

$$\left. -\frac{1}{u}\ln(E/u) - (E/u)\frac{1/u}{E/u}\right]$$

$$= k_B \left[\frac{1}{u}\ln\frac{E/u + N}{E/u}\right]$$

from which one has

$$\frac{u}{k_B T} = \ln\left(\frac{E/u + N}{E/u}\right)$$

$$\exp\left[\frac{u}{k_B T}\right] = \frac{E/u + N}{E/u}$$

$$E = \frac{Nu}{\exp\left[\frac{u}{k_B T}\right] - 1} \tag{16.33}$$

At low temperatures, $E \simeq Nu\exp[-\frac{u}{k_B T}]$ and at high temperatures, $E \simeq Nk_B T$. The low-temperature result describes the energy of N quantum harmonic oscillators. The high-temperature result is the energy of N classical harmonic oscillators. The curve E/N versus $k_B T/u$ is shown in Fig. 16.2.

Problem 4. System of magnetic moments in a field:

a) E is given, n_1 and n_2 are determined because $E = -(n_1 - n_2)\mu B$ and $n_1 + n_2 = N$. One can rewrite E as $E = -[n_1 - (N - n_1)]\mu B$ or

$E/(\mu B) = -2n_1 + N$. The total number of accessible microstates $\Omega(E)$ of the system is equal to the number of choices of n_1 magnetic moments parallel to B. One has $\Omega = \frac{N!}{n_1!(N-n_1)!}$.

b) Using the Stirling (2.47), one has

$$\ln \Omega(E) \simeq N \ln N - N - n_1 \ln n_1 + n_1$$
$$-(N-n_1)\ln(N-n_1) + N - n_1$$
$$= N \ln N - n_1 \ln n_1 - (N - n_1)\ln(N-n_1)$$

c) The micro-canonical entropy is $S = k_B \ln \Omega = k_B[N \ln N - n_1 \ln n_1 - (N-n_1)\ln(N-n_1)]$. The temperature T is par

$$T^{-1} = \frac{\partial S}{\partial E} = \frac{\partial S}{\partial n_1}\frac{\partial n_1}{\partial E}$$

$$= -k_B \frac{1}{2\mu B}\left[-\ln n_1 - \frac{n_1}{n_1} + \ln(N-n_1) + \frac{N-n_1}{N-n_1}\right]$$

$$= -\frac{k_B}{2\mu B}\left[\ln \frac{N-n_1}{n_1}\right]$$

$$-\frac{2\mu B}{k_B T} = \ln \frac{N-n_1}{n_1}$$

$$\exp\left(-\frac{2\mu B}{k_B T}\right) = \frac{N-n_1}{n_1}$$

where one has used $\partial n_1/\partial E = -1/(2\mu B)$ (see the first question). One finds

$$n_1 = \frac{N}{\exp(-\frac{2\mu B}{k_B T}) + 1} \tag{16.34}$$

If $\frac{\mu B}{k_B T} \ll 1$ (weak field and/or high temperature), then $\exp(-\frac{2\mu B}{k_B T}) \simeq 1 - \frac{2\mu B}{k_B T}$, from which one gets $n_1 \simeq N(1 + \frac{\mu B}{k_B T})/2$: n_1 is slightly larger than $N/2$.

If $\frac{\mu B}{k_B T} \gg 1$ (strong field and/or low temperature), then $n_1 \simeq N$: all magnetic moments align themselves in the field direction.

Remark: The system studied here is a version of a two-level system.

Problem 5. Density of states:

One uses the same method as in the case of three dimensions [see (2.36)]: in one dimension, for a given E, the number of states of energy less than or equal to E, called $\phi(E)$, is equal to the segment $2k$ (segment from $-k$ to k) divided by the size of a microstate in one dimension, namely $2\pi/L$. One has

$$\phi(E) = \frac{2k}{\frac{2\pi}{L}}$$

$$= \frac{L}{\pi}k = \frac{L}{\pi}\left[\frac{2m}{\hbar^2}\right]^{1/2} E^{1/2} \tag{16.35}$$

The density of states is obtained, using (2.27),

$$\rho(E) = \frac{L}{2\pi}\left[\frac{2m}{\hbar^2}\right]^{1/2} E^{-1/2} \tag{16.36}$$

In two dimensions, $\phi(E)$ is equal to the surface of the circle of radius k divided by the size of a microstate, namely $(2\pi/L)^2$. One has then

$$\phi(E) = \frac{\pi k^2}{(\frac{2\pi}{L})^2}$$

$$= \frac{L^2}{4\pi}k^2 = \frac{L^2}{4\pi^2}\frac{2m}{\hbar^2}E \tag{16.37}$$

The density of states in two dimension is thus

$$\rho(E) = \frac{L^2}{4\pi^2}\frac{2m}{\hbar^2} \tag{16.38}$$

which is independent of E. One notes that the power of E in the density of states decreases by $1/2$ when the dimension decreases by 1.

Note that when the particle has a spin of amplitude S, one should add a factor $(2S+1)$ to take into account the spin degeneracy [see (2.39)].

Problem 6. Classical ideal gas in one and two dimensions:

The densities of states calculated in the previous problem are for one free particle. For a system of N free particles, the k-space has N degrees of freedom in one dimension and $2N$ degrees of freedom in two dimensions. The "volume" of of a "sphere" of radius r in the k-space is equal to Ar^N and Br^{2N}, for one and two dimensions, respectively. On has $r = k = \sqrt{2mE/\hbar^2}$. The coefficients A and B can be calculated as in Appendix A.1. One has in one dimension

$$\phi(E) = \left[\frac{2m}{\hbar^2}\right]^{N/2} \frac{AE^{N/2}}{(\frac{2\pi}{L})^N} \propto L^N E^{N/2} \tag{16.39}$$

and in two dimensions

$$\phi(E) = \left[\frac{2m}{\hbar^2}\right]^{N} \frac{BE^N}{(\frac{2\pi}{L})^{2N}} \propto [L^2]^N E^N \tag{16.40}$$

The density of states of a system of N free particles in one dimension is thus

$$\rho(E) \propto L^N E^{N/2-1} \simeq L^N E^{N/2} \tag{16.41}$$

and that in two dimensions is

$$\rho(E) \propto [L^2]^N E^{N-1} \simeq L^{2N} E^N \tag{16.42}$$

since $N \gg 1$.

Using $S = k_B \ln \rho(E)$ one obtains in one dimension $T^{-1} = (\partial S/\partial E)_{V,N} = k_B N/(2E)$ or

$$E = \frac{Nk_B T}{2} \tag{16.43}$$

Similarly, in two dimensions, one has

$$E = Nk_B T \tag{16.44}$$

Note that in three dimensions one has $E = \frac{3Nk_B T}{2}$

Using $p/T = (\partial S/\partial V)_{E,N}$ where $V = L$ and $V = L^2$ in one and two dimensions, one obtains the equations of state $pL = Nk_B T$ and $pL^2 = Nk_B T$, in one and two dimensions, respectively.

Remark: One can obtain the above results of $\rho(E)$ by using the method for (2.44).

Problem 7. If one uses (2.46) for a molecule in three dimensions, with the density of states (2.36), one has $S = k_B \ln \rho(E) = k_B \ln AE^{1/2} = k_B \ln E^{1/2} + k_B \ln A$ [A: constant given by (2.36)]. One then has

$$T^{-1} = \frac{\partial S}{\partial E} = \frac{k_B}{2E} \tag{16.45}$$

from which $E = k_B T/2$ in contradiction with the result $E = 3k_B T/2$ for a particle in three dimensions. The reason of this error is that the formula (2.46) is valid only for a grand system (see also the case of a harmonic oscillator in Problem 8).

Problem 8. Classical harmonic oscillator:

The energy of a classical harmonic oscillator is written as $E = p^2/2m + Kr^2/2$ where K is the restoring-force constant, p the modulus of the momentum and m the mass. Following the calculation of (2.41), one calculates $\phi(E)$ which is the number of states of energy less than or equal to E for just one particle. There are 6 degrees of freedom: one has

$$\phi(E) = \frac{1}{h^3} \int d\vec{p} \int d\vec{r} \tag{16.46}$$

where the integration domain is given by the condition

$$\frac{p^2}{2m} + Kr^2/2 \leq E \tag{16.47}$$

Putting $X = p/\sqrt{2m}$ and $Y = \sqrt{K/2}r$, the integration domain is

$$X^2 + Y^2 \leq E \tag{16.48}$$

The integration domain is thus a sphere of radius $\sqrt{E}$. Expression (16.46) becomes

$$\phi(E) = \frac{1}{h^3} \int 2\pi p^2 dp \int 2\pi r^2 dr$$
$$= \frac{1}{h^3}(2m)^{3/2}(K/2)^{3/2} \int\int 2\pi X^2 dX 2\pi Y^2 dY \tag{16.49}$$

The above integrals give the volume of a sphere in a space of 6 degrees of freedom. One obtains

$$\phi(E) = \frac{1}{h^3}(2m)^{3/2}(K/2)^{3/2}A(\sqrt{E})^6 \tag{16.50}$$

where A is a constant (see Appendix A.1). One gets the following density of states

$$\rho(E) = \frac{d\phi(E)}{dE} = \frac{3}{h^3}(2m)^{3/2}(K/2)^{3/2}AE^2 \tag{16.51}$$

If one uses (2.46), the micro-canonical entropy is $S = k_B \ln \rho(E) = k_B \ln E^2 +$ constant, and the micro-canonical temperature is $T^{-1} = \partial S/\partial E = 2k_B/E$. One obtains $E = 2k_BT$ for a classical harmonic oscillator in three dimensions (6 degrees of freedom). This expression is wrong since it is known that the energy corresponding to 6 degrees of freedom in classical mechanics is (theorem of energy equipartition) $E = 6 \times \frac{1}{2}k_BT = 3k_BT$. As in the previous exercise, this error comes from the fact that the formula (2.46) has been demonstrated for a large system (see its demonstration).

Problem 9. System of classical harmonic oscillators:

The energy of the system is

$$E = \frac{1}{2m}(p_1^2 + p_2^2 + ... + p_N^2) + \frac{K}{2}(r_1^2 + r_2^2 + ... + r_N^2)$$

where K is the restoring constant, p_i the modulus of the momentum of the i-th oscillator and m the oscillator mass. One calculates $\phi(E)$ the number of microstates of energy from 0 to E. There are $6N$ degrees of freedom in the phase space. One has

$$\phi(E) = \frac{1}{h^{3N}} \int d\vec{p}_1 \int d\vec{p}_2 ... \int d\vec{p}_N \int d\vec{r}_1 \int d\vec{r}_2 ... \int d\vec{r}_N \quad (16.52)$$

where the integration domain is given by

$$\frac{1}{2m}(p_1^2 + p_2^2 + ... + p_N^2) + \frac{K}{2}(r_1^2 + r_2^2 + ... + r_N^2) \leq E \quad (16.53)$$

One puts $X_i = p_i/\sqrt{2m}$ and $Y_i = \sqrt{K/2}r_i$. The integration domain in the space $(X_i, Y_i; i = 1, ..., N)$ is a hypersphere of radius $\sqrt{E}$ in the space of $6N$ degrees of freedom. Its volume V is given by $V = A(\sqrt{E})^{6N}$ where $A = \pi^{6N/2}/\Gamma(6N/2 + 1)$. One obtains $\phi(E) = CE^{3N}$ where C includes all constants of E. The density of states is thus $\rho(E) = 3NCE^{3N-1} \simeq 3NCE^{3N}$ with $N >> 1$. The micro-canonical entropy is obtained using (2.46): $S = k_B \ln \rho(E) = k_B \ln E^{3N}$ + constant. Temperature T is $T^{-1} = \partial S/\partial E = 3Nk_B/E$. One obtains $E = 3Nk_BT$ for N classical harmonic oscillators in three dimensions as expected from the theorem of energy equipartition in classical mechanics.

Problem 10. System of quantum harmonic oscillators:

a) The energy of the system is

$$E = \sum_{i=1}^{N} \epsilon_i = 3Nh\nu/2 + (n_1^x + n_1^y + n_1^z + n_2^x + n_2^y + n_2^z... + n_N^z)h\nu$$

$$= 3Nh\nu/2 + Mh\nu$$

where $M = n_1^x + n_1^y + n_1^z + n_2^x + n_2^y + n_2^z... + n_N^z$. As in Problem 3, the number of microstates is equal to the number of repartitions of M energy units on $3N$ objects [$3N$ is the number of $n_i^\alpha (\alpha = x, y, z)$]. Using (1.35), one has

$$\Omega(E) = \frac{(M + 3N - 1)!}{(3N - 1)!M!}$$

b) The micro-canonical entropy is $S = k_B \ln \Omega = k_B \ln[\frac{(M+3N-1)!}{(3N-1)!M!}]$ where $M = U/(h\nu)$ and $U \equiv E - 3Nh\nu/2$. Assuming $M, N \gg 1$ so that $(M + 3N - 1)! \simeq (M + 3N)!$ and $(3N - 1)! \simeq (3N)!$. Using now the Stirling formula (2.47), one obtains

$$S \simeq k_B[(M + 3N)\ln(M + 3N) - M \ln M - 3N \ln 3N].$$

The temperature is

$$T^{-1} = \frac{\partial S}{\partial E} = \frac{\partial S}{\partial M}\frac{\partial M}{\partial U}$$

$$= \frac{k_B}{h\nu}[\ln(M + 3N) + 1 - \ln M - 1]$$

$$= \frac{k_B}{h\nu}\ln\frac{M + 3N}{M}$$

from which

$$\exp(\frac{h\nu}{k_B T}) = \frac{M + 3N}{M}$$

$$M = \frac{3N}{\exp(\frac{h\nu}{k_B T}) - 1} \tag{16.54}$$

c) One has in the first question $M = E/(h\nu) - 3N/2$. Thus

$$E = 3N\left[\frac{h\nu}{\exp(\frac{h\nu}{k_B T}) - 1} + \frac{h\nu}{2}\right] \tag{16.55}$$

This is the expression of E as a function of T for N harmonic oscillators in three dimensions.

The calorific capacity C_v is

$$C_v = \frac{dE}{dT} = \frac{3N}{k_B T^2}\frac{(h\nu)^2 \exp(\frac{h\nu}{k_B T})}{\left[\exp(\frac{h\nu}{k_B T}) - 1\right]^2} \tag{16.56}$$

The curves representing $\frac{E(T)}{3N}$ and $\frac{C_v(T)}{3N}$ are shown in Fig. 16.3.

d) At high temperatures,

$$e^{\frac{h\nu}{k_B T}} \simeq 1 + \frac{h\nu}{k_B T}$$

One has $E \to 3Nk_B T$. One recovers the result of a system of classical harmonic oscillators studied in Problem 9.

Problem 11. System composed of subsystems of quantum harmonic oscillators:

One deals with the case of one dimension.

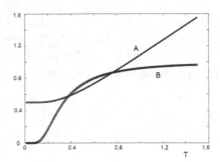

Fig. 16.3   Energy $E/(3N)$ (curve A) and calorific capacity $C_v/(3N)$ (curve B) studied in Problem 10 versus T with $h\nu = 1$ and $k_B = 1$.

a) The two subsystems are separated by a fixed insulating wall. The number of microstates of each subsystem is calculated as in Problem 10. In one dimension, one has

$$\Omega_1(E_1) = \frac{(M_1 + N_1 - 1)!}{(N_1 - 1)!M_1!}$$

$$\Omega_2(E_2) = \frac{(M_2 + N_2 - 1)!}{(N_2 - 1)!M_2!}$$

The total number of microstates is $\Omega(E_1, E_2) = \Omega_1(E_1)\Omega_2(E_2)$. The micro-canonical entropy is

$$\begin{aligned}
S_0 &= k_B \ln \Omega(E_1, E_2) = k_B \ln \Omega_1(E_1) + k_B \ln \Omega_2(E_2) \\
&= S_1 + S_2 \simeq k_B[(M_1 + N_1)\ln(M_1 + N_1) - M_1 \ln M_1 \\
&\quad - N_1 \ln N_1] + k_B[(M_2 + N_2)\ln(M_2 + N_2) \\
&\quad - M_2 \ln M_2 - N_2 \ln N_2]
\end{aligned}$$

b) The wall is now diathermic but impermeable to oscillators. The energy of the system is $E = E_1 + E_2 = Nh\nu/2 + Mh\nu$ where $M = M_1 + M_2$ and $N = N_1 + N_2$.

- The total number of microstates of the system is

$$\Omega(E) = \frac{(M + N - 1)!}{(N - 1)!M!}$$

Replacing $M = E/h\nu - N/2$, one gets $\Omega(E)$ as a function of E, N_1 and N_2.

- Micro-canonical entropy S of the system, as in Problem 10, is given by

$$S \simeq k_B[(M+N)\ln(M+N) - M\ln M - N\ln N].$$

One has $S > S_0$ because $\Omega(E) > \Omega(E_1, E_2)$: one sees that the spontaneous evolution accompanied by an exchange of energy between the subsystems increases the total micro-canonical entropy.

- The relation between the total energy and T is

$$E = N\left[\frac{h\nu}{\exp(\frac{h\nu}{k_B T}) - 1} + \frac{1}{2}\right]$$

$$= (N_1 + N_2)\left[\frac{h\nu}{\exp(\frac{h\nu}{k_B T}) - 1} + \frac{1}{2}\right]$$

$$= E_1 + E_2 \tag{16.57}$$

where

$$E_1 = N_1\left[\frac{h\nu}{\exp(\frac{h\nu}{k_B T}) - 1} + \frac{1}{2}\right]$$

$$E_2 = N_2\left[\frac{h\nu}{\exp(\frac{h\nu}{k_B T}) - 1} + \frac{1}{2}\right]$$

$$\tag{16.58}$$

One has $E_1/N_1 = E_2/N_2$: energy per oscillator is thus the same in the two subsystems.

Problem 12. Frenkel's defects:

a) One has $E - E_0 = n\epsilon$.

b) The number of microstates of the system Ω is equal to the number of ways to choose n sites among N regular sites and to distribute them in a random manner on n sites among N' interstices. One has thus

$$\Omega = C_N^n C_{N'}^n = \frac{N!}{n!(N-n)!} \frac{N'!}{n!(N'-n)!} \tag{16.59}$$

c) The micro-canonical entropy is $S = k_B \ln \Omega$. Assuming $N, N' >> n >> 1$, and using the Stirling formula (2.47) one gets

$$S \simeq k_B[N \ln N - N - n \ln n + n - (N - n) \ln(N - n) + (N - n)$$
$$+N' \ln N' - N' - n \ln n + n - (N' - n) \ln(N' - n)$$
$$+(N' - n)]$$
$$= k_B[N \ln N + N' \ln N' - 2n \ln n - (N - n) \ln(N - n)$$
$$-(N' - n) \ln(N' - n)]$$

The temperature T is given by

$$T^{-1} = = \left(\frac{\partial S}{\partial E}\right)_{V,N} = \frac{\partial S}{\partial n} \frac{\partial n}{\partial E}$$
$$= \frac{k_B}{\epsilon}[-2 \ln n - 2 + \ln(N - n) + 1 + \ln(N' - n) + 1]$$
$$\frac{\epsilon}{k_B T} = \ln \left[\frac{(N - n)(N' - n)}{n^2}\right]$$
$$\exp \left(\frac{\epsilon}{k_B T}\right) = \frac{(N - n)(N' - n)}{n^2}$$

where one used $\partial n/\partial E = 1/\epsilon$ (see the first question).

d) If $n << N, N'$, one obtains

$$n^2 \simeq NN' \exp \left(-\frac{\epsilon}{k_B T}\right)$$
$$n \simeq \sqrt{NN'} \exp \left(-\frac{\epsilon}{2k_B T}\right) \tag{16.60}$$

At low T, one sees that n diminishes exponentially. However, one should not go down to $T = 0$ to keep the validity of (16.60), namely $n >> 1$ (otherwise $n \to 0 \to$ no more defects). At high T, n increases but one should not go up to $T = \infty$, otherwise $n \xrightarrow{T \to \infty} \sqrt{NN'} \simeq N$ (if $N \simeq N'$) in contradiction with the hypothesis $N, N' >> n$.

Problem 13. Schottky's defects:

The energy of the system having n empty sites is $E = n\epsilon_0$. The empty sites are created by the displacement of interior atoms to the crystal surface leaving behind vacant sites.

a) The number of microstates Ω of the system is equal to the number of random choices of n empty sites among $N + n$ sites. One has $\Omega = C_{N+n}^n$. The micro-canonical entropy is thus

$$S = k_B \ln \Omega = k_B \ln \frac{(N+n)!}{n!N!} = k_B[\ln(N+n)! - \ln n! - \ln N!]$$
$$\simeq k_B[(N+n)\ln(N+n) - N - n - n\ln n + n - N\ln N + N]$$
$$= k_B[(N+n)\ln(N+n) - n\ln n - N\ln N] \qquad (16.61)$$

where one has used the Stirling formula for N and n.

b) The micro-canonical temperature is given by

$$T^{-1} = \frac{\partial S}{\partial E} = \frac{\partial S}{\partial n}\frac{\partial n}{\partial E}$$

$$T^{-1} = k_B[\ln(N+n) - \ln n]\frac{1}{\epsilon}$$

$$\frac{\epsilon}{k_B T} = \ln \frac{N+n}{n}$$

$$\frac{N+n}{n} = \exp\left(\frac{\epsilon}{k_B T}\right)$$

$$n \simeq N \exp\left(-\frac{\epsilon}{k_B T}\right) \qquad (16.62)$$

where one has used $N >> n$.

When $T \to 0$ (but remains non zero), n/N diminishes exponentially: there are fewer vacant sites at low T. When T increases (but remains finite), n increases: there are more and more empty sites created.

Remark: In the enumeration of vacant sites one has included those on the surface. This is not correct by definition (a vacant site should be surrounded by atoms). However, this approximation is not bad if $n << N$ so that the number of surface vacancies is still small with respect to n and N.

Problem 14. Exchange of heat in three dimensions:

a) The energy levels for each particle are

- fundamental level ($n_x = n_y = n_z = 1$): $\epsilon = 3\epsilon_0$, degeneracy $g_1 = 1$
- 2nd level ($n_x = 2, n_y = n_z = 1 \to 3$ permutations): $\epsilon = 6\epsilon_0$, degeneracy $g_2 = 3$

- 3rd level ($n_x = n_y = 2, n_z = 1 \to$ 3 permutations): $\epsilon = 9\epsilon_0$, degeneracy $g_3 = 3$
- 4th level ($n_x = 3, n_y = n_z = 1 \to$ 3 permutations): $\epsilon = 11\epsilon_0$, degeneracy $g_4 = 3$
- 5th level ($n_x = n_y = n_z = 2$): $\epsilon = 12\epsilon_0$, degeneracy $g_5 = 1$
- 6th level ($n_x = 3, n_y = 2, n_z = 1 \to 3! = 6$ permutations): $\epsilon = 14\epsilon_0$, degeneracy $g_6 = 6$
- 7th level ($n_x = 3, n_y = n_z = 2 \to$ 3 permutations): $\epsilon = 17\epsilon_0$, degeneracy $g_7 = 3$

At $t = 0$, $E_1 = 9\epsilon_0$. The repartitions of E_1 on two discernible particles (the first value in the parentheses is for the first particle) are:

- $(3\epsilon_0, 6\epsilon_0)$: degeneracy $g_1 \times g_2 = 3$
- $(6\epsilon_0, 3\epsilon_0)$: degeneracy $g_2 \times g_1 = 3$

The total degeneracy (total number of microstates) of E_1 is thus $\Omega_1 = 3 + 3 = 6$.

For the second box, $E_2 = 12\epsilon_0$ is to be distributed on the two discernible particles. The repartitions are

- $(3\epsilon_0, 9\epsilon_0)$: degeneracy $g_1 \times g_3 = 3$
- $(9\epsilon_0, 3\epsilon_0)$: degeneracy $g_3 \times g_1 = 3$
- $(6\epsilon_0, 6\epsilon_0)$: degeneracy $g_2 \times g_2 = 9$

The total number of microstates of E_2 is $\Omega_2 = 3 + 3 + 9 = 15$. The total number of microstates of the system is thus $\Omega = \Omega_1 \Omega_2 = 6 \times 15 = 90$.

b) The wall is now diathermic to allow for a heat exchange. The total system evolves toward a new equilibrium:

- The total energy is conserved during the evolution $E_t = E_1 + E_2 = 21\epsilon_0$.
- The total number of microstates of the total system is equal th the number of repartitions of E_t on 4 discernible particles:

    - $(3\epsilon_0, 3\epsilon_0, 3\epsilon_0, 12\epsilon_0)$: there are 4 arrangements, the degeneracy is thus $4g_1 \times g_1 \times g_1 \times g_5 = 4$
    - $(3\epsilon_0, 3\epsilon_0, 6\epsilon_0, 9\epsilon_0)$: there are $A_4^2 = 12$ arrangements, degeneracy $= 12g_1 \times g_1 \times g_2 \times g_3 = 12 \times 1 \times 1 \times 3 \times 3 = 108$
    - $(3\epsilon_0, 6\epsilon_0, 6\epsilon_0, 6\epsilon_0)$: $A_4^1 = 4$ arrangements, degeneracy $= 4g_1 \times g_2 \times g_2 \times g_2 = 4 \times 1 \times 3 \times 3 \times 3 = 108$

There are in all $4 + 108 + 108 = 220$ microstates.

- One sees that the total number of microstates increases from 90 to 220 during the evolution.

c) One considers the total system at its new equilibrium:

- the system is isolated, the probability of a microstate is $P = 1/220$.
- one calculates the number of microstates when the subsystem 1 has a given energy:

- If $E_1 = 6\epsilon_0$ (non degenerate): the degeneracy comes from subsystem 2 with $E_2 = 15\epsilon_0$:

$(3\epsilon_0, 12\epsilon_0)$, degeneracy $= 1$,
$(12\epsilon_0, 3\epsilon_0)$, degeneracy $= 1$,
$(6\epsilon_0, 9\epsilon_0)$, degeneracy $= 3 \times 3$,
$(9\epsilon_0, 6\epsilon_0)$, degeneracy $= 3 \times 3$.

Subsystem 2 has $1+1+9+9 = 20$ microstates: Thus, $\Omega_1 = 20$ microstates when $E_1 = 6\epsilon_0$, $E_2 = 15\epsilon_0$. The corresponding probability of the macrostate when $E_1 = 6\epsilon_0$ is $20/220$.

- If $E_1 = 9\epsilon_0$, this case corresponds to the following repartitions within subsystem 1: $(3\epsilon_0, 6\epsilon_0)$ and $(6\epsilon_0, 3\epsilon_0)$. The degeneracy of subsystem 1 is thus $g(E_1) = 3 + 3 = 6$. Subsystem 2 has $E_2 = 12\epsilon_0$ which corresponds to the following repartitions: $(6\epsilon_0, 6\epsilon_0)$ $(3 \times 3$ microstates$)$, $(3\epsilon_0, 9\epsilon_0)$ $(1 \times 3$ microstates$)$ and $(9\epsilon_0, 3\epsilon_0)$ $(3 \times 1$ microstates$)$. The number of microstates of subsystem 2 is therefore $g(E_2) = 9 + 3 + 3 = 15$. In summary, there are $\Omega_1 = g(E_1)g(E_2) = 6 \times 15 = 90$ microstates in which $E_1 = 9\epsilon_0$. The corresponding probability is $90/220$.

- If $E_1 = 12\epsilon_0$, the repartitions are $(3\epsilon_0, 9\epsilon_0)$ $(1x3$ microstates$)$, $(9\epsilon_0, 3\epsilon_0)$ $(3 \times 1$ microstates$)$, $(6\epsilon_0, 6\epsilon_0)$ $(3 \times 3$ microstates$)$. The degeneracy of subsystem 1 is thus $g(E_1) = 3 + 3 + 9 = 15$. System 2 has $E_2 = 9\epsilon_0$ which corresponds to the repartitions $(3\epsilon_0, 6\epsilon_0)$ and $(6\epsilon_0, 3\epsilon_0)$. Its degeneracy is $g(E_2) = 3 + 3 = 6$. In summary, there are $\Omega_1 = g(E_1)g(E_2) = 15 \times 6 = 90$ microstates in which $E_1 = 12\epsilon_0$. The corresponding probability is $90/220$.

- one calculates the average energy of subsystem 1 at equilibrium. This is given by

$$\overline{E}_1 = \frac{6 \times 20 + 9 \times 90 + 12 \times 90}{220}\epsilon_0 = 10.5\epsilon_0 \qquad (16.63)$$

One deduces $\overline{E}_2 = E_t - \overline{E}_1 = 10.5\epsilon_0$. Thus, the total energy is identically shared by two subsystems.

. The most probable energy of subsystem 1 corresponds, by definition, to the maximum of the probability of E_1. From the above calculation, one has
- $E_1 = 6\epsilon_0$ with probability 20/220,
- $E_1 = 9\epsilon_0$ with probability 90/220,
- $E_1 = 12\epsilon_0$ with probability 90/220,
- $E_1 = 15\epsilon_0$ with probability 20/220,

The probability is maximum at $E_1 = 9\epsilon_0$ and $E_1 = 12\epsilon_0$. This yields the average value at $(12 + 9)\epsilon_0/2 = 10.5\epsilon_0$ as found above.

One shows that the micro-canonical temperatures of two subsystems are equal when $\Omega = \Omega_1\Omega_2$, or $\ln\Omega$ (namely the micro-canonical entropy) is maximum:

$$0 = \frac{\partial ln\Omega}{\partial E_1} = \frac{\partial ln\Omega_1}{\partial E_1} + \frac{\partial ln\Omega_2}{\partial E_1}$$

$$= \frac{\partial ln\Omega_1}{\partial E_1} - \frac{\partial ln\Omega_2}{\partial E_2}$$

$$= \left(\frac{1}{T_1} - \frac{1}{T_2}\right) \tag{16.64}$$

from which $T_1 = T_2$.

• The average energy gain by subsystem 1 between $t = 0$ and the new equilibrium is $1.5\epsilon_0$.

Problem 15. Binary alloy:

a) Since $\epsilon > \phi$, the energy of an atom is minimum when it is surrounded by atoms of the other kind. This defines the ground-state configuration of a binary alloy.

Remark: This configuration is possible only when the lattice structure allows for such an arrangement, as the case shown in Fig.2.5, for example. There are other crystalline structures in which such alternate arrangements are impossible as in the triangular lattice.

b)   • One has $0 \leq N_{A,I} = N(1+x)/4 \leq N/2$ from which $-1 \leq x \leq 1$. When $x = 0$, $N_{A,I} = N/4$: the system is in the disordered phase.

The number of A atoms occupying the sites II is $N_{A,II} = N/2 - N_{A,I} = N(1-x)/4$.

For B atoms, one has $N_{B,II} + N_{A,II} = N/2$ from which $N_{B,II} = N(1+x)/4$. One finds $N_{B,I} = N/2 - N_{B,II} = N(1-x)/4$.

One limits oneself in the case $x > 0$ in the following.

• The probability to find an A atom on a site of type I is

$$P(A,I) = N_{A,I}/(N/2) = (1+x)/2,$$

and that on a site of type II is

$$P(A,II) = N_{A,II}/(N/2) = (1-x)/2.$$

Similarly, for B atoms, one has

$$P(B,I) = N_{B,I}/(N/2) = (1-x)/2$$

and

$$P(B,II) = N_{B,II}/(N/2) = (1+x)/2.$$

• Let $N_{A,A}$, $N_{B,B}$, and $N_{A,B}$ be the numbers of nearest pairs AA, BB and AB, respectively. The probability to find an AA pair is

$$P(A,A) = P(A,I)P(A,II) + P(A,II)P(A,I) = (1-x^2)/2$$

from which

$$N_{A,A} = NP(A, A) = N(1 - x^2)/2.$$

Similarly, the probability to find an BB pair is $N_{B,B} = N(1 - x^2)/2$.

The probability to find an AB pair is

$$P(A, B) = P(A, I)P(B, II) + P(A, II)P(B, I)$$
$$+P(B, I)P(A, II) + P(B, II)P(A, I)$$
$$= (1 + x)^2/4 + (1 - x)^2/4 + (1 - x)^2/4 + (1 + x)^2/4$$
$$= 1 + x^2$$

Therefore, $N_{A,B} = NP(A, B) = N(1 + x^2)$.

- One has $E = [N_{A,A} + N_{B,B}]\epsilon + N_{A,B}\phi =$ from which,

$$E = N(\epsilon + \phi) - N(\epsilon - \phi)x^2 \qquad (16.65)$$

- When E is fixed, x^2 is fixed. One takes $x > 0$. Let $\Omega(E)$ be the number of microstates having energy E. $\Omega(E)$ is equal to the number of choices of $N(1+x)/4$ A atoms among $N/2$ sites of type I and to put them on $N(1 + x)/4$ sites among $N/2$ sites of type II. One has

$$\Omega(E) = \left[C_{N/2}^{N(1+x)/4} \right]^2 = \left[\frac{(N/2)!}{[N(1+x)/4]![N(1-x)/4]!} \right]^2$$

Using the Stirling formula, one gets

$$\ln \Omega(E) = 2 \left[\ln \frac{N}{2}! - \ln \frac{N(1 + x)}{4}! - \ln \frac{N(1 - x)}{4}! \right]$$
$$\simeq 2\{(N/2)\ln(N/2) - N/2 - [N(1 + x)/4]$$
$$\times \ln [N(1 + x)/4]$$
$$+N(1 + x)/4 - [N(1 - x)/4]$$
$$\times \ln [N(1 - x)/4] + N(1 - x)/4\}$$
$$= 2\{(N/2)\ln(N/2) - [N(1 + x)/4] \ln [N(1 + x)/4]$$
$$-[N(1 - x)/4] \ln [N(1 - x)/4]\} \qquad (16.66)$$

The micro-canonical entropy is given by $S = k_B \ln \Omega$.

- The temperature T is

$$T^{-1} = \frac{\partial S}{\partial E} = k_B \frac{\partial \ln \Omega(E)}{\partial x} \frac{\partial x}{\partial E}$$

$$= \frac{k_B}{4(\epsilon - \phi)x} \{\ln[N(1+x)/4] - \ln[N(1-x)/4]\}$$

$$= \frac{k_B}{4(\epsilon - \phi)x} \ln \frac{1+x}{1-x}$$

One obtains then

$$\ln \frac{1+x}{1-x} = \frac{4(\epsilon - \phi)x}{k_B T}$$

$$\frac{1+x}{1-x} = \exp \left[\frac{4(\epsilon - \phi)x}{k_B T} \right]$$

$$x = \tanh \left[\frac{2(\epsilon - \phi)x}{k_B T} \right] \tag{16.67}$$

One sees that $x = 1$ at $T = 0$ and $x = 0$ at $T = \infty$. Between these two limits, one shows that there exists a finite temperature far below $T = \infty$ where x becomes zero. To demonstrate this, one expands tanh in the expression of x:

$$x \simeq 2(\epsilon - \phi)x/(k_B T) - [2(\epsilon - \phi)x/(k_B T)]^3/3$$

$$x[1 - 2(\epsilon - \phi)/(k_B T)] = -[2(\epsilon - \phi)/(k_B T)]^3 x^3 \tag{16.68}$$

If $x \neq 0$, one can simplify it on both sides and one gets

$$2(\epsilon - \phi)/(k_B T) - 1 = [2(\epsilon - \phi)/(k_B T)]^3 x^2 \tag{16.69}$$

Since $2(\epsilon - \phi)/(k_B T) > 0$, the right-hand side is positive. The above relation is possible only if the left-hand side is also positive, namely

$$2(\epsilon - \phi)/(k_B T) - 1 > 0$$

from which

$$T < 2(\epsilon - \phi)/k_B \equiv T_c.$$

T_c is called "transition temperature" at which the system undergoes a transition from the ordered phase ($x \neq 0$) to the disordered phase ($x = 0$).

16.3   Solutions of problems of chapter 3

Problem 1. The calorific capacity C_V is given by (3.18):

$$C_V = \frac{\partial \overline{E}}{\partial T} = \frac{1}{k_B T^2} \frac{\partial^2 \ln Z}{\partial \beta^2} \qquad (16.70)$$

where

$$
\begin{aligned}
\frac{\partial^2 \ln Z}{\partial \beta^2} &= \frac{\partial}{\partial \beta}\left(\frac{\partial \ln Z}{\partial \beta}\right) = \frac{\partial}{\partial \beta}\left[-\frac{1}{Z}\sum_l E_l \exp(-\beta E_l)\right] \\
&= \frac{1}{Z^2}\frac{\partial Z}{\partial \beta}\sum_l E_l \exp(-\beta E_l) + \frac{1}{Z}\sum_l E_l^2 \exp(-\beta E_l) \\
&= \left[\frac{1}{Z}\cdot\frac{\partial Z}{\partial \beta}\right]\left[\frac{1}{Z}\sum_l E_l \exp(-\beta E_l)\right] + \overline{E^2} \\
&= -[\overline{E}]^2 + \overline{E^2} \qquad (16.71)
\end{aligned}
$$

from which one gets

$$C_V = \frac{1}{k_B T^2}[\overline{E^2} - (\overline{E})^2] = \frac{\overline{(E-\overline{E})^2}}{k_B T^2} \qquad (16.72)$$

Problem 2. Maxwell-Boltzmann approximation:

One shows below that the condition $\bar{N}_{\lambda_i} \ll 1$ for the Maxwell-Boltzmann approximation (3.49) corresponds to the high-temperature region.

One uses the formula (3.52) with Z given by (3.49) and $z = \sum_{\lambda_j} \exp(-\beta\epsilon_{\lambda_j})$:

$$
\begin{aligned}
\overline{N}_{\lambda_j} &= -k_B T \frac{\partial \ln Z}{\partial \epsilon_{\lambda_j}} = -k_B T N \frac{\partial \ln z}{\partial \epsilon_{\lambda_j}} \\
&= \frac{N}{z}\exp(-\beta\epsilon_{\lambda_j}) \qquad (16.73)
\end{aligned}
$$

If $\overline{N}_{\lambda_j} \ll 1$ one has $\frac{N}{z}\exp(-\beta\epsilon_{\lambda_j}) \ll 1$, namely $z\exp(\beta\epsilon_{\lambda_j}) \gg N$. If the lowest energy ϵ_0 satisfies this relation, then any other energy will do. So, one has to prove $z\exp(\beta\epsilon_0) \gg N$ or $z\exp(\beta\epsilon_0) = \sum_{\lambda_j}\exp[-\beta(\epsilon_{\lambda_j}-\epsilon_0)] \gg N$. One sees that the argument of the exponential is negative, therefore each term of the sum is smaller than 1. In order that the sum is much larger than N, there should be many states to contribute and/or the temperature should be high so that each term is close to 1. In summary, the condition $z\exp(\beta\epsilon_{\lambda_j}) \gg N$ is satisfied

- if T and V are fixed, then N should not be very large. This corresponds to a dilute gas: N/V is not very high
- if N and V are fixed, namely the density N/V is given, then T should be high.

Under these conditions, the Maxwell-Boltzmann approximation is valid.

Problem 3. Classical ideal gas in the gravitational field:

One considers a classical ideal gas of N particles of mass m, contained in a cylinder of section A, of infinite height. One puts the system in the gravitational field. The gas is at equilibrium.

The energy of a particle in the cylinder is $E = p^2/(2m) + mgl$ where l is the altitude with respect to the ground and g the gravity. The partition function of a particle is

$$
\begin{aligned}
z &= \frac{1}{h^3} \int_A dS \int_0^\infty dl \exp(-\beta mgl) \int_{-\infty}^\infty \exp[-\beta p_x^2/(2m)] dp_x \\
&\quad \times \int_{-\infty}^\infty \exp[-\beta p_y^2/(2m)] dp_y \int_{-\infty}^\infty \exp[-\beta p_z^2/(2m)] dp_z \\
&= \frac{A}{h^3} \frac{1}{\beta mg} \left[\sqrt{\frac{2\pi m}{\beta}} \right]^3 \\
&= \frac{A}{h^3 mg} (2\pi m)^{3/2} \beta^{-5/2}
\end{aligned}
\tag{16.74}
$$

The partition function of the gas is $Z = z^N/N!$. One has

$$
\begin{aligned}
\ln Z &= N \left[-\frac{5}{2} \ln \beta + \ln \frac{A}{h^3 mg} (2\pi m)^{3/2} \right] - \ln N! \\
&\simeq N \left[-\frac{5}{2} \ln \beta + \ln \frac{A}{h^3 mg} (2\pi m)^{3/2} \right] - N \ln N + N \\
\overline{E} &= -\frac{\partial \ln Z}{\partial \beta} = \frac{5N}{2\beta} = \frac{5}{2} N k_B T \\
C_V &= \frac{5}{2} N k_B
\end{aligned}
$$

The canonical entropy is calculated by $S = k_B(\ln Z + \beta \overline{E})$ [see (3.21)]. The free energy is $F = -k_B T \ln Z$.

Problem 4. Bi-dimensional classical ideal gas:

The bi-dimensional classical ideal gas of N particles on a surface S is maintained at a constant temperature T.

a) The density of states is $\rho(E) = A$ where the constant A was calculated in Problem 5 of chapter 2: $A = \frac{S}{4\pi^2} \frac{2m}{\hbar^2}$ [see (16.38)].

b) The partition function z of a particle can be calculated by (3.14) as follows:

$$z = \int_0^\infty \rho(E)\exp(-\beta E)dE = A\int_0^\infty \exp(-\beta E)dE$$
$$= A/\beta \tag{16.75}$$

For N particles, one has $Z = z^N/N! = (1/N!)(A/\beta)^N$.

c) The average energy of the gas is

$$\overline{E} = -\frac{\partial \ln Z}{\partial \beta} = -\frac{\partial}{\partial \beta}[N\ln z + \ln N!]$$
$$= N\frac{1}{\beta} = Nk_B T \tag{16.76}$$

d) One has

$$p = -\frac{\partial F}{\partial S} = -k_B T\frac{\partial \ln Z}{\partial S} = -k_B T\frac{\partial}{\partial S}[N\ln z + \ln N!]$$
$$= Nk_B T/S \tag{16.77}$$

where S is the "volume" (surface) of the system. One gets then the equation of state $pS = Nk_B T$.

Problem 5. Classical harmonic oscillators:

One consider a system of N classical harmonic oscillators of mass m which are supposed to be independent, indiscernible, maintained at a constant temperature T. The partition function is $Z = z^N$ where z is the partition function of an oscillator of energy $E_i = p^2/(2m) + Kr^2/2$ in three dimensions (K: constant of restoring force). One has

$$z = \sum_i \exp(-\beta E_i) = \frac{1}{h^3}\int_V d\vec{r}\int d\vec{p}\exp[-\beta(p^2/(2m) + Kr^2/2)]$$
$$= \frac{1}{h^3}\int_{-\infty}^\infty \exp[-\beta Kx^2/2]dx \int_{-\infty}^\infty \exp[-\beta Ky^2/2]dy$$
$$\times \int_{-\infty}^\infty \exp[-\beta Kz^2/2]dz \int_{-\infty}^\infty \exp[-\beta p_x^2/(2m)]dp_x$$
$$\times \int_{-\infty}^\infty \exp[-\beta p_y^2/(2m)]dp_y \int_{-\infty}^\infty \exp[-\beta p_z^2/(2m)]dp_z$$
$$= \frac{1}{h^3}\left(\sqrt{\frac{2\pi}{K\beta}}\right)^3\left(\sqrt{\frac{2\pi m}{\beta}}\right)^3$$
$$= \frac{1}{h^3}\left(\frac{2\pi}{\beta}\right)^3\left(\frac{m}{K}\right)^{3/2} \tag{16.78}$$

where one has used the Gauss integral I_0 of Appendix A. The average energy is

$$\overline{E} = -\frac{\partial \ln Z}{\partial \beta} = -N\frac{\partial \ln z}{\partial \beta}$$

$$\overline{E} = -N\frac{\partial}{\partial \beta}\left[-\ln(\beta^3) + \ln\frac{1}{h^3}(2\pi)^3\left(\frac{m}{K}\right)^{3/2}\right]$$

$$\overline{E} = 3N/\beta = 3Nk_BT \tag{16.79}$$

One can of course find this result using the theorem of energy equipartition shown in 3.6.4. The calorific capacity is $C_V = \frac{d\overline{E}}{dT} = 3Nk_B$.

Problem 6. Quantum harmonic oscillators:

The energy of a quantum harmonic oscillator is written as $E(n_x, n_y, n_z) = (n_x + n_y + n_z + 3/2)h\nu$ where n_x, n_y and n_z are positive integers or zero. The partition function Z of the system of N independent and indiscernible oscillators is $Z = z^N$ where z is the partition function of one oscillator. One has

$$z = \sum_{n_x, n_y, n_z} \exp[-\beta(n_x + n_y + n_z + 3/2)h\nu]$$

$$z = \exp(-3\beta h\nu/2) \sum_{n_x=0}^{\infty} \exp(-\beta n_x h\nu) \sum_{n_y=0}^{\infty} \exp(-\beta n_y h\nu) \times$$

$$\sum_{n_z=0}^{\infty} \exp(-\beta n_z h\nu)$$

$$z = \exp(-3\beta h\nu/2)\left[\frac{1}{1-\exp(-\beta h\nu)}\right]\left[\frac{1}{1-\exp(-\beta h\nu)}\right]$$

$$\times \left[\frac{1}{1-\exp(-\beta h\nu)}\right]$$

$$z = \left[\frac{\exp(-\beta h\nu/2)}{1-\exp(-\beta h\nu)}\right]^3$$

$$z = \left[\frac{1}{2\sinh(\beta h\nu/2)}\right]^3 \tag{16.80}$$

from which one gets

$$Z = z^N = \left[\frac{1}{2\sinh(\beta h\nu/2)}\right]^{3N} \tag{16.81}$$

The average energy is

$$\overline{E} = -\frac{\partial \ln Z}{\partial \beta} = -N \frac{\partial \ln z}{\partial \beta}$$

$$\overline{E} = \frac{3Nh\nu}{2} \frac{\cosh(\beta h\nu/2)}{\sinh(\beta h\nu/2)} = \frac{3Nh\nu}{2} \coth(\beta h\nu/2) \quad (16.82)$$

or under a more frequently used expression

$$\overline{E} = 3N \left[\frac{h\nu}{2} + \frac{h\nu}{\exp(\beta h\nu) - 1} \right] \quad (16.83)$$

where one has used the following identities

$$\coth(x) = \frac{\exp(x) + \exp(-x)}{\exp(x) - \exp(-x)} = \frac{1 + \exp(-2x)}{1 - \exp(-2x)}$$

$$= \frac{1 + 2\exp(-2x) - \exp(-2x)}{1 - \exp(-2x)} = 1 + \frac{2\exp(-2x)}{1 - \exp(-2x)}$$

$$= 1 + \frac{2}{\exp(2x) - 1} \quad (16.84)$$

with $x = \beta h\nu/2$. The calorific capacity is

$$C_V = \frac{\partial \overline{E}}{\partial T} = 3N k_B (\beta h\nu)^2 \frac{\exp(\beta h\nu)}{[\exp(\beta h\nu) - 1]^2} \quad (16.85)$$

Problem 7. Three-level system:

Each particle has three non degenerate states of energy $E_i = -\epsilon, 0, \epsilon$ $(\epsilon > 0)$.

a) The particles are supposed to be identical and independent. If they are discernible the partition function of the system is $Z = z^N$ [see (3.45)] where z, partition function of one particle, is calculated as follows:

$$z = \sum_i \exp(-\beta\epsilon_i) = \exp(\beta\epsilon) + 1 + \exp(-\beta\epsilon) = 2\cosh(\beta\epsilon) + 1 \quad (16.86)$$

The average energy of the system is thus

$$\overline{E} = -\frac{\partial \ln Z}{\partial \beta} = -N \frac{\partial \ln z}{\partial \beta}$$

$$= -N\epsilon \frac{2\sinh(\beta\epsilon)}{2\cosh(\beta\epsilon) + 1} \quad (16.87)$$

b) The probability of level E_i is $P_i = \exp(-\beta E_i)/z$. One has then
$$P_1 = \exp(\beta\epsilon)/z, \quad P_2 = 1/z, \quad P_3 = \exp(-\beta\epsilon)/z.$$
The numbers of particles on three levels are
$$N_1 = N\exp(\beta\epsilon)/z, \quad N_2 = NP_2 = N/z, \quad N_3 = NP_3 = \exp(-\beta\epsilon)/z.$$

c) At high temperatures, $z \to 3$, $N_1 \simeq N/3$, $N_2 \simeq N/3$ and $N_3 \simeq N/3$: the particles are equally distributed on three levels.

At low temperatures, $N_1 \simeq N$, $N_2 \simeq 0$ and $N_3 \simeq 0$: the particles occupy the level of lowest energy as expected for the ground state.

Problem 8. Frenkel's defects:

One studies Frenkel's defects in the following using the canonical method.

a) The energy of the system in a microstate with n interstices is written as $E_n = n\epsilon$. Let $g(E_n)$ be the degeneracy of the energy E_n. Note that $g(E_n)$ is equal to Ω, the number of microstates of energy E given in Problem 12 of chapter 2. The probability of energy E_n is thus

$$P(E_n) = g(E_n)\exp(-\beta E_n)/Z \qquad (16.88)$$

where

$$g(E_n) = C_N^n C_{N'}^n = \frac{N!}{n!(N-n)!}\frac{N'!}{n!(N'-n)!} \qquad (16.89)$$

b) The most probable energy E_M corresponds to the maximum of $P(E_n)$: one has

$$\frac{\partial P(E_n)}{\partial E_n}\Big|_{E_M} = 0, \quad \text{or} \quad \frac{\partial P(E_n)}{\partial n}\Big|_{E_M} = 0 \qquad (16.90)$$

Using the Stirling formula (see Appendix A.1) with $N, N' \gg n \gg 1$ one has

$$0 = \frac{\partial \ln P(E_n)}{\partial n}$$

$$0 \simeq \frac{\partial}{\partial n}[-\beta E_n + N\ln N - N - n\ln n + n - (N-n)\ln(N-n)$$
$$+(N-n) + N'\ln N' - N' - n\ln n + n$$
$$-(N'-n)\ln(N'-n) + (N'-n)]$$

$$0 = \frac{\partial}{\partial n}[-\beta n\epsilon + N\ln N + N'\ln N' - 2n\ln n - (N-n)\ln(N-n)$$
$$-(N'-n)\ln(N'-n)]$$

$$0 = \left[-\beta\epsilon + \ln\frac{(N-n)(N'-n)}{n^2}\right] \qquad (16.91)$$

from which one obtains

$$\exp(\beta\epsilon) = \frac{(N-n)(N'-n)}{n^2}$$

$$n \simeq \sqrt{NN'}\exp(-\beta\epsilon/2) \tag{16.92}$$

This value of n corresponds to the maximum of $P(E_n)$. The most probable energy is thus $E_M = \epsilon\sqrt{NN'}\exp(-\beta\epsilon/2)$.

Note that the result (16.92) is identical to (16.60) obtained by the micro-canonical method.

c) If one uses (3.13) for the partition function:

$$Z = \sum_{E_n} g(E_n)\exp(-\beta E_n) \tag{16.93}$$

Then the sum on E_n is impossible to perform for very large values of N and N' due to the formula (16.89) for $g(E_n)$. If $N, N' \to \infty$, Z is expressed as follows:

$$Z = \sum_{n=0}^{\infty}\exp(-\beta E_n) = \sum_{n=0}^{\infty}\exp(-\beta n\epsilon) = \frac{1}{1-\exp(-\beta\epsilon)} \tag{16.94}$$

Then the average energy $\overline{E}$ of the system is

$$\overline{E} = -\frac{\partial\ln Z}{\partial\beta}$$

$$\overline{E} = \frac{\epsilon}{\exp(\beta\epsilon)-1} \tag{16.95}$$

Since $\overline{E} = n\epsilon$, one finds $n = \frac{1}{\exp(\beta\epsilon)-1}$. This result is different from (16.92). It is not valid because the degeneracy $g(E_n)$ is not correctly taken into account in (16.94): one has taken $g(E_n) = 1$ when $N, N' \to \infty$ instead of $g(E_n) = \frac{1}{(n!)^2}$. If this is added in (16.94) then the sum cannot be done.

Problem 9. Schottky's defects:

One considers again Problem 13 of chapter 2 using the canonical method.

a) The probability to have n vacant sites is

$$P(E_n) = g(E_n)\exp(-\beta E_n)/Z \tag{16.96}$$

where $g(E_n)$ is the degeneracy of E_n, namely the number of microstates having energy E_n. $g(E_n)$ is equal to $\Omega = C_{N+n}^n$ [see Problem 13 of chapter 2]. One has

$$P(E_n) = C_{N+n}^n \exp(-\beta E_n)/Z = \frac{(N+n)!}{n!N!} \exp(-\beta E_n)/Z$$

(16.97)

b) The most probable energy of the system corresponds to the maximum of $P(E_n)$ or to the maximum of $\ln P(E_n)$. Using the Stirling formula for $N \gg n \gg 1$ one obtains

$$0 = \frac{\partial}{\partial n} \ln P(E_n) = \frac{\partial}{\partial n} \left[-\beta E_n + \ln \frac{(N+n)!}{n!N!} - \ln Z \right]$$

$$0 = \frac{\partial}{\partial n} \left[-\beta n\epsilon + \ln(N+n)! - \ln n! - \ln N! \right]$$

$$0 \simeq \frac{\partial}{\partial n} [-\beta n\epsilon + (N+n)\ln(N+n) - N - n - n\ln n + n$$
$$-N\ln N + N]$$

$$0 = [-\beta\epsilon + \ln(N+n) - \ln n] = \left[-\beta\epsilon + \ln \frac{N+n}{n} \right]$$

from which one gets

$$n \simeq N \exp(-\beta\epsilon)$$

(16.98)

This value of n corresponds to the maximum of $P(E_n)$. The most probable energy of the system is thus $E_M = n\epsilon = \epsilon N \exp(-\beta\epsilon)$. The result of n found above is the same as the one obtained by the micro-canonical method.

Problem 10. Velocity distribution in a classical ideal gas:

Let n be the density of molecules per volume unit of a classical ideal gas maintained at the temperature T. The canonical probability of a microstate of energy $E(v_x, v_y, v_z) = \frac{m}{2}(v_x^2 + v_y^2 + v_z^2)$ is

$$P(v_x, v_y, v_z) = A \exp[-\beta \frac{m}{2}(v_x^2 + v_y^2 + v_z^2)]$$

where A is a constant to be determined by the normalization of probabilities. One recalls that the sum on microstates is performed on the momentum $\vec{p}$, not on the velocity $\vec{v}$. One will work with $\vec{p}$ and one will transform the result in terms of $\vec{v}$ at the end. One has

$$A^{-1} = Z = \frac{V}{h^3} \int_{-\infty}^{\infty} dp_x \exp\left(-\frac{\beta}{2m} p_x^2 \right)$$

$$\times \int_{-\infty}^{\infty} dp_y \exp\left(-\frac{\beta}{2m} p_y^2 \right) \int_{-\infty}^{\infty} dp_z \exp\left(-\frac{\beta}{2m} p_z^2 \right)$$

$$= V \left(\frac{2\pi m k_B T}{h^2} \right)^{3/2}$$

(16.99)

a) The number of molecules with momentum components lying between p_x and $p_x + dp_x$, p_y and $p_y + dp_y$, p_z and $p_z + dp_z$ is

$$dN(p_x, p_y, p_z) = \frac{1}{h^3} N P(p_x, p_y, p_z) dp_x dp_y dp_z \qquad (16.100)$$

where the volume of a microstates is $\sqrt{h}$ per degree of freedom (see chapter 2). In spherical coordinates, one has

$$dN(p) = \frac{1}{h^3} N P(p) 4\pi p^2 dp \qquad (16.101)$$

from which

$$dN(v) = \frac{1}{h^3} N P(v) 4\pi m^3 v^2 dp$$

$$= \frac{N}{V} \left(\frac{m}{2\pi k_B T} \right)^{3/2} \exp\left(-\beta \frac{mv^2}{2} \right) 4\pi v^2 dv \quad (16.102)$$

Putting $\frac{dN(v)}{n} \equiv f(v) dv$ where $n = N/V$ one has

$$f(v) = \left(\frac{m}{2\pi k_B T} \right)^{3/2} \exp\left(-\beta \frac{mv^2}{2} \right) 4\pi v^2 \qquad (16.103)$$

$f(v)$ is called "velocity distribution" in a classical ideal gas.

b) • The most probable velocity v_p corresponds to the maximum of $f(v)$, namely $\frac{\partial f(v)}{\partial v} = 0$. One finds the following non trivial solution:

$$v_p^2 = \frac{2k_B T}{m} \qquad (16.104)$$

Numerical application: One has $k_B = R/N_A$, $m = M/N_A$ (mass of an atom or molecule), from which $k_B/m = R/M$. One gets thus $v_p = \sqrt{2RT/M} = 496.8$ m/s.

 • The average velocity $\bar{v}$ is

$$\bar{v} = \int_0^\infty v f(v) dv$$

$$= 4\pi \left(\frac{m}{2\pi k_B T} \right)^{3/2} \int_0^\infty \exp\left(-\beta \frac{mv^2}{2} \right) v^3 dv$$

$$= 4\pi \left(\frac{m}{2\pi k_B T} \right)^{3/2} \frac{1}{2} \left(\frac{2k_B T}{m} \right)^2$$

$$= \sqrt{\frac{8RT}{\pi M}} = 560.6 \text{ m/s} \qquad (16.105)$$

where one has used integral I_3 of Appendix A.

- The average of the square velocity is

$$\overline{v^2} = \int_0^\infty v^2 f(v) dv$$

$$= 4\pi \left(\frac{m}{2\pi k_B T} \right)^{3/2} \int_0^\infty \exp\left(-\beta \frac{mv^2}{2} \right) v^4 dv$$

$$= 4\pi \left(\frac{m}{2\pi k_B T} \right)^{3/2} I_4 = 3RT/M$$

$$\sqrt{\overline{v^2}} = 608.5 \ \text{m/s} \tag{16.106}$$

c) The proportion of molecules of a gas of Neon having the velocity lying inside the interval $v_p \pm 10\%$ m/s is given by $\frac{dN(v)}{n} = f(v_p)2\Delta v$ because $2\Delta v$ is the distance between $v_p - \Delta v$ and $v_p + \Delta v$. One has

$$\frac{dN(v)}{n} = f(v_p)2\Delta v = 4\pi \left(\frac{m}{2\pi k_B T} \right)^{3/2} \frac{2k_B T}{m} e^{-1} 2\Delta v$$

$$= 3.34\% \tag{16.107}$$

where numerical values have been used.

Problem 11. System of spins in an applied magnetic field:

The magnetic moment associated with a spin $\vec{S}$ is written as

$$\vec{M} = -g\mu_B \vec{S} \tag{16.108}$$

where g is the Landé factor and μ_B the Bohr magneton.

One considers a solid of N independent atoms each of which has a spin $\vec{S}$. The magnetic moment of the i-th atom is

$$\vec{M}_i = -g\mu_B \vec{S}_i \tag{16.109}$$

The effect of an applied magnetic field $\vec{B}$ is called Zeeman effect. If the field is applied along the z direction, the Zeeman energy is

$$\mathcal{H} = -\sum_{i=1}^N \vec{M}_i \cdot \vec{B} = \sum_{S_i^z} g\mu_B S_i^z B = \sum_{i=1}^N E_i \tag{16.110}$$

where $S_i^z = 3/2, 1/2, -1/2, -3/2$ and E_i is the Zeeman energy of the i-th atom. The solid is at equilibrium at the temperature T.

a) Since the particles are discernible, the partition function is

$$Z = z^N \tag{16.111}$$

where z is the partition function of an atom. One has

$$z = \sum_{M_i^z=-3/2}^{3/2} e^{-\beta E_i} = \sum_{M_i^z=-3/2}^{3/2} e^{-\beta g\mu_B B M_i^z}$$

$$= \frac{\sinh(2\alpha)}{\sinh(\alpha/2)} . \tag{16.112}$$

where one has used the formula for a geometric series of 4 terms, of ratio $e^{-\alpha}$ (see Appendix A) and one has put $\alpha = \beta g\mu_B B$.

b) The average energy of the system is

$$\overline{E} = -\frac{\partial \ln Z}{\partial \beta} = -\frac{\partial \ln Z}{\partial \alpha}\frac{\partial \alpha}{\partial \beta}$$

$$= -N g\mu_B B Q(\alpha) \tag{16.113}$$

where

$$Q(\alpha) = 2\coth(2\alpha) - \frac{1}{2}\coth(\frac{\alpha}{2}) \tag{16.114}$$

The calorific capacity is

$$C_V = Nk_B \left(\frac{g\mu_B B}{k_B T}\right)^2 \left[\frac{4}{\sinh^2(2\alpha)} - \frac{1}{4}\frac{1}{\sinh^2(\alpha/2)}\right] \tag{16.115}$$

c) The free energy F is calculated by

$$F = -k_B T \ln Z = -Nk_B T \ln z = -Nk_B T \ln\frac{\sinh(2\alpha)}{\sinh(\alpha/2)} \tag{16.116}$$

The canonical entropy is

$$S = \frac{\overline{E} - F}{T} \tag{16.117}$$

In the limit $\alpha \ll 1$, namely at high temperatures and/or weak field B, one expands $\coth(x) \to 1/x + x/3 - x^3/45 + ...$ when $x \to 0$. One has then

$$\overline{E} \to -N(g\mu_B B)^2\frac{1}{2k_B T} \ , \quad F \to -Nk_B T \ln 4$$

One deduces

$$S \to -N(g\mu_B B)^2\frac{1}{2k_B T^2} + Nk_B \ln 4 \to Nk_B \ln 4 \tag{16.118}$$

This value corresponds to the maximum of disorder as seen by the following argument: there are 4 spin states for each atom so that there are at maximum $\Omega = 4^N$ spin configurations for a system of N atoms. The canonical entropy is thus $S = Nk_B \ln 4$ which corresponds to its maximum where all configurations have the same probability.

At low temperatures and/or strong field, one has $\alpha \gg 1$, $\coth(x) \simeq 1$ so that $\overline{E} \to -\frac{3}{2}Ng\mu_B B$, $F \to -\frac{3}{2}Ng\mu_B B$, $S \to -\frac{3}{2T}Ng\mu_B B + \frac{3}{2T}Ng\mu_B B = 0$: the entropy is zero in agreement with the Nernst's principle in thermodynamics.

d) One considers the system in the microstate l of energy E_l given by

$$E_l = \sum_i E_i = -B \sum_i g\mu_B B S_i^z \equiv -BM_l \qquad (16.119)$$

where M_l is the magnetic moment of the system in the microstate l ($M_l = \sum_{i=1}^N S_i^z(l)$, $S_i^z(l)$: value of S_i^z in the state l). The average magnetic moment of the system is

$$
\begin{aligned}
\overline{M} &= \frac{1}{Z}\sum_l M_l e^{-\beta E_l} = \frac{1}{Z}\sum_l M_l e^{\beta B M_l} \\
&= \frac{1}{\beta}\frac{1}{Z}\frac{\partial}{\partial B}\sum_l e^{\beta B M_l} = \frac{1}{\beta}\frac{1}{Z}\frac{\partial Z}{\partial B} = \frac{1}{\beta}\frac{\partial \ln Z}{\partial B} \\
&= -\left[\frac{\partial F}{\partial B}\right]_{T,V} \qquad (16.120)
\end{aligned}
$$

Replacing (16.116) in the last equality, one obtains the magnetization $\overline{m} \equiv \overline{\mathcal{M}}/V$:

$$\overline{m} = \frac{Ng\mu_B}{V}Q(\alpha) \qquad (16.121)$$

The value of $\overline{m}$ at high temperatures and:or weak fields ($\alpha \ll 1$) is

$$\overline{m} = \frac{N}{V}(g\mu_B)^2\frac{B}{2k_B T} \qquad (16.122)$$

One finds the susceptibility

$$\chi = \frac{d\overline{m}}{dB} = \frac{N}{V}(g\mu_B)^2\frac{1}{2k_B T} \propto \frac{1}{T} \qquad (16.123)$$

This is known as the Curie law.

At $T = 0$ one has $\overline{m} = \frac{N}{V}3g\mu_B/2$. This is the saturated value of the magnetization.

Problem 12. Equilibrium of a vapor-solid system:

a) Atoms of the solid are considered as independent quantum harmonic oscillators with the same frequency ν (Einstein model of a solid). The partition function Z_s of the solid is $Z_s = z_s^{N_s}$ where z_s is the partition function of one atom. One has

$$z_s = \sum_{n_x,n_y,n_z} \exp[-\beta E(n_x, n_y, n_z)] \tag{16.124}$$

where $E(n_x, n_y, n_z) = -\phi + (n_x + n_y + n_z + 3/2)h\nu$ is the energy of an atom composed of the cohesion energy $-\phi$ which keeps the atom in the solid, and of the vibration energy dependent on 3 integers which are positive or zero. One has

$$z_s = \sum_{n_x,n_y,n_z} \exp[-\beta(-\phi + (n_x + n_y + n_z + 3/2)h\nu)]$$

$$z_s = \exp(\beta\phi)\exp(-3\beta h\nu/2)\sum_{n_x=0}^{\infty}\exp(-\beta n_x h\nu)\sum_{n_y=0}^{\infty}\exp(-\beta n_y h\nu)$$

$$\times \sum_{n_z=0}^{\infty}\exp(-\beta n_z h\nu)$$

$$z_s = \exp(\beta\phi)\exp(-3\beta h\nu/2)\left[\frac{1}{1-\exp(-\beta h\nu)}\right]\left[\frac{1}{1-\exp(-\beta h\nu)}\right]$$

$$\times \left[\frac{1}{1-\exp(-\beta h\nu)}\right]$$

$$z_s = \exp(\beta\phi)\left[\frac{\exp(-\beta h\nu/2)}{1-\exp(-\beta h\nu)}\right]^3$$

$$z_s = \exp(\beta\phi)\left[\frac{1}{2\sinh(\beta h\nu/2)}\right]^3 \tag{16.125}$$

from which one gets

$$Z_s = z_s^{N_s} = \exp(N_s\beta\phi)\left[\frac{1}{2\sinh(\beta h\nu/2)}\right]^{3N_s} \tag{16.126}$$

b) The vapor is considered as a classical ideal gas. The partition function Z_g is $Z_g = z_g^{N_g}/N_g!$ where N_g is the number of atoms in the gas and z_g the partition function of an atom given by (3.53):

$$z_g = \frac{V}{h^3}\left[\frac{2\pi m}{\beta}\right]^{3/2}$$

One has thus

$$Z_g = \frac{z_g^{N_g}}{N_g!} = \frac{V_g^N}{h^{3N_g}} \left[\frac{2\pi m}{\beta} \right]^{3N_g/2} \tag{16.127}$$

c) The global system conserves the total number of atoms $N_s + N_g = N = $ constant. There is thus only one parameter in the problem, either N_s or N_g. When the whole system is at equilibrium at T, this parameter is determined at the minimum of the system free energy F [see (3.34)]:

$$\frac{\partial F}{\partial N_g} = \frac{\partial (F_s + F_g)}{\partial N_g} = 0 \tag{16.128}$$

where

$$F_s = -k_B T \ln Z_s = -k_B T (N - N_g) \ln z_s$$
$$F_g = -k_B T \ln Z_g = -k_B T (N_g \ln z_g - \ln N_g!)$$
$$\simeq -k_B T (N_g \ln z_g - N_g \ln N_g + N_g)$$

Replacing these functions in (16.128) one obtains

$$\frac{\partial (F_s + F_g)}{\partial N_g} = 0 \quad \rightarrow \quad -k_B T \left[\ln \frac{z_g}{N_g} - \ln z_s \right] = 0 \tag{16.129}$$

from which one gets

$$N_g = \frac{z_g}{z_s} = \frac{\frac{V}{h^3} [\frac{2\pi m}{\beta}]^{3/2} [2 \sinh(\beta h\nu/2)]^3}{\exp(\beta\phi)} \tag{16.130}$$

using z_s and z_g given above.

When $T \rightarrow \infty$, one has $N_g \rightarrow \infty$ (in the limit of N obviously): all atoms are in the gaseous phase. When $T \rightarrow 0$, one has $N_g \rightarrow 0$: all atoms stay in the solid.

d) The gas pressure is given by the equation of state of a classical ideal gas $pV = N_g k_B T$ where N_g is the number of atoms in the vapor at equilibrium given by (16.130).

Problem 13. Harmonic oscillators:

a) • The partition function of a classical oscillator of mass m, of pulsation ω_j is calculated in Problem 5. The result is

$$z = \frac{1}{h^3} \left[\frac{2\pi}{\beta} \right]^3 \frac{1}{\omega_j^3} = \frac{1}{(\beta \hbar \omega_j)^3}$$

where one has replaced $h\nu$ by $\hbar\omega_j$ and K/m by ω_j^2 because one can write the potential energy as $m\omega_j^2 r^2/2$.

- The partition function of a quantum harmonic oscillator of mass m, of pulsation ω_j is calculated in Problem 6. The result is

$$z = \left[\frac{1}{2\sinh(\beta\hbar\omega_j/2)}\right]^3 \qquad (16.131)$$

The result in the classical case is valid at high temperatures. The result in the quantum case becomes at high temperatures $z \to \frac{1}{(\beta\hbar\omega_j)^3}$. The quantum case becomes the classical case at high temperatures.

b) If all pulsations are identical and equal to ω_E, one obtains the results of Problem 6 for the partition function of the system, the average energy and the calorific capacity. The free energy is calculated by $F = -k_B T \ln Z = -N k_B T \ln z$, and the entropy by $S = (\overline{E} - F)/T$.

c) In this question, one assumes that all pulsations ω_j are different.

- The energy of the system in microstate i is $E_i = \sum_j (n_j^x + n_j^y + n_j^z + 3/2)\hbar\omega_j$ where n_j^α ($\alpha = x, y, z$) are positive integers or zero. The sum is made over all pulsations. One has

$$Z = \sum_i \exp(-\beta E_i)$$

$$Z = \sum_{n_1^x, \dots, n_N^z} \exp\left[-\beta \sum_{j=1}^N (n_j^x + n_j^y + n_j^z + 3/2)\hbar\omega_j\right]$$

$$Z = \prod_{j=1}^N \exp(-3\beta\hbar\omega_j/2)$$

$$\times \sum_{n_j^x} \exp[-\beta n_j^x \hbar\omega_j] \sum_{n_j^y} \exp[-\beta n_j^y \hbar\omega_j] \sum_{n_j^z} \exp[-\beta n_j^z \hbar\omega_j]$$

$$Z = \prod_{j=1}^N \exp(-3\beta\hbar\omega_j/2) \left[\frac{1}{1 - \exp(-\beta\hbar\omega_j)}\right]^3$$

$$Z = \prod_{j=1}^N \left[\frac{\exp(-\beta\hbar\omega_j/2)}{1 - \exp(-\beta\hbar\omega_j)}\right]^3$$

$$Z = \prod_{j=1}^N \left[\frac{1}{2\sinh(\beta\hbar\omega_j/2)}\right]^3 \qquad (16.132)$$

where one has factorized the sums and used the series (A.21) of Appendix A. The free energy $F = -k_B T \ln Z$ is given by

$$F = 3k_B T \sum_j \ln[2\sinh(\beta\hbar\omega_j/2)] \qquad (16.133)$$

- For $N \gg 1$, one can rewrite F as

$$F = -k_B T \ln Z = 3k_B T \int_0^\infty \ln[2\sinh(\beta\hbar\omega/2)]g(\omega)d\omega \qquad (16.134)$$

where $g(\omega)$ is the number of pulsations between ω and $\omega + d\omega$ and V the system volume. The average energy is

$$\overline{E} = -\frac{\partial \ln Z}{\partial \beta} = -\frac{\partial}{\partial\beta}3\int_0^\infty \ln[2\sinh(\beta\hbar\omega/2)]g(\omega)d\omega$$

$$= 3\int_0^\infty \frac{\hbar\omega}{2}\coth(\beta\hbar\omega/2)g(\omega)d\omega$$

$$= \int_0^\infty \epsilon(\omega,T)g(\omega)d\omega \qquad (16.135)$$

where

$$\epsilon(\omega,T) \equiv \frac{3\hbar\omega}{2}\coth(\beta\hbar\omega/2) = 3\left[\frac{1}{2} + \frac{1}{\exp(\beta\hbar\omega)-1}\right]\hbar\omega \qquad (16.136)$$

where one has used (16.84) in the last equality.

Remark: Factor 3 corresponds to the case of three dimensions. For one dimension, one replaces it by 1.

$\epsilon(\omega,T)$ represents the energy of a quantum harmonic oscillator of pulsation ω at the temperature T. One will see in chapter 6 that the second term in the parentheses of (16.136) is the number of occupation of the state of energy $\hbar\omega$ at T (Bose-Einstein distribution). When $T = 0$, one has $\epsilon = 3\hbar\omega/2$ which is the "zero-point" energy of the quantum oscillator. The calorific capacity C of the system is

$$C = \frac{\partial \overline{E}}{\partial T} = \frac{\partial}{\partial T}\int_0^\infty 3\left[\frac{1}{2} + \frac{1}{\exp(\beta\hbar\omega)-1}\right]\hbar\omega g(\omega)d\omega$$

$$C = 3k_B \int_0^\infty \left(\frac{\hbar\omega}{k_B T}\right)^2 \frac{\exp(\beta\hbar\omega)}{[\exp(\beta\hbar\omega)-1]^2}g(\omega)d\omega \qquad (16.137)$$

d) The function $g(\omega)$ in the above expressions is called "density of modes" in analogy with $\rho(E)$ the density of states. One supposes $g(\omega) = 3\omega^2/\omega_D$ if $\omega < \omega_D$ and $g(\omega) = 0$ if $\omega > \omega_D$ where ω_D is a constant called "Debye pulsation".

In the case $\hbar\omega_j/k_B T \ll 1$ (high temperatures), one has

$$\frac{\exp(\beta\hbar\omega)}{[\exp(\beta\hbar\omega) - 1]^2} \simeq \frac{1}{[1 + \beta\hbar\omega - 1]^2} = \frac{1}{[\beta\hbar\omega]^2} \qquad (16.138)$$

from which

$$C \simeq 3k_B \int_0^{\omega_D} \frac{3\omega^2}{\omega_D} d\omega = 3k_B \qquad (16.139)$$

This is the result of a classical harmonic oscillator (see Problem 5) as expected at high temperatures.

In the case $\hbar\omega_j/k_B T \gg 1$ (low temperatures), one has $\exp(\beta\hbar\omega) \gg 1$ so that

$$\frac{\exp(\beta\hbar\omega)}{[\exp(\beta\hbar\omega) - 1]^2} \simeq \exp(-\beta\hbar\omega) \qquad (16.140)$$

from which

$$C = 3k_B \int_0^{\omega_D} [\beta\hbar\omega]^2 \exp(-\beta\hbar\omega) \frac{3\omega^2}{\omega_D} d\omega$$

$$C = 9k_B \frac{1}{\omega_D \hbar^3 \beta^3} \int_0^{x_D} x^4 \exp(-x) dx \qquad (16.141)$$

where $x \equiv \beta\hbar\omega$ and $x_D = \beta\hbar\omega_D$. At low temperatures, the upper bound $x_D \to \infty$. Using the $\Gamma(5)$ function of Appendix A one obtains

$$C = 9k_B \frac{1}{\omega_D \hbar^3 \beta^3} \Gamma(5) \propto T^3 \qquad (16.142)$$

The T^3 dependence at low temperatures of C is experimentally verified in solid crystals. The model studied in this question is a simplified version of the Debye model which will be developed in chapter 9.

16.4   Solutions of problems of chapter 4

Problem 1. Fluctuations de N:

Fluctuations of a physical quantities are represented by its variance. One calculates the variance of N, the number of particles of a system in the grand-canonical situation.

The grand partition function is written as

$$\mathcal{Z} = \sum_N \exp(\beta\mu N)Z(N) \qquad (16.143)$$

where $Z(N)$ is the partition function for a fixed N. One has

$$\overline{N} = \sum_N N\frac{\exp(\beta\mu N)Z(N)}{\mathcal{Z}} \qquad (16.144)$$

$$\frac{\partial\overline{N}}{\partial\mu} = \beta\sum_N N^2\frac{\exp(\beta\mu N)Z(N)}{\mathcal{Z}} - \frac{1}{\mathcal{Z}^2}\sum_N N\exp(\beta\mu N)Z(N)\frac{\partial\mathcal{Z}}{\partial\mu}$$

$$= \beta\overline{N^2} - \frac{1}{\mathcal{Z}^2}\left[\sum_N N\exp(\beta\mu N)Z(N)\right]$$

$$\times\left[\beta\sum_N N\exp(\beta\mu N)Z(N)\right]$$

$$= \beta\left[\overline{N^2} - (\overline{N})^2\right] = \beta\overline{(N - (\overline{N})^2} \qquad (16.145)$$

from which

$$\Delta N^2 = \overline{(N - \overline{N})^2} = k_B T\frac{\partial\overline{N}}{\partial\mu} \qquad (16.146)$$

Problem 2. Classical ideal gas in the gravitational field:

One calculates the variation of the atmosphere pressure as a function of altitude y at the temperature T using the grand-canonical method. The atmosphere is considered as a classical ideal gas. The energy of a particle at altitude y is $E = p^2/(2m) + mgy$ where g is the gravity. One uses (4.52):

$$\mathcal{Z} = \exp(\lambda z) \qquad (16.147)$$

where $\lambda \equiv e^{\beta\mu}$ and the partition function for a particle in a cylinder of section S, of height dy at altitude y is given by [see (3.53)]

$$z = \frac{1}{h^3}\exp(-\beta mgy)dy\int_S ds\int_{-\infty}^{\infty}\exp[-\beta p_x^2/(2m)]dp_x$$

$$\times\int_{-\infty}^{\infty}\exp[-\beta p_y^2/(2m)]dp_y\int_{-\infty}^{\infty}\exp[-\beta p_z^2/(2m)]dp_z$$

$$= \frac{Sdy}{h^3}\exp(-\beta mgy)\left[\sqrt{\frac{2\pi m}{\beta}}\right]^3$$

$$= \frac{V}{h^3}\exp(-\beta mgy)\left[\frac{2\pi m}{\beta}\right]^{3/2} \qquad (16.148)$$

where $V = Sdy$ and the Gauss integral for integrals on p_x, p_y and p_z has been used (see Appendix A).

The grand-canonical pressure P is

$$P = -\frac{\partial J}{\partial V} = k_B T \frac{\partial \ln \mathcal{Z}}{\partial V}$$

$$= k_B T \frac{\partial (\lambda z)}{\partial V}$$

$$= \frac{k_B T \lambda z}{V} \qquad (16.149)$$

from which one gets

$$P = k_B T \left(\frac{2\pi m}{h^2 \beta}\right)^{3/2} \lambda \exp(-\beta m g y)$$

$$= k_B T \left(\frac{2\pi m k_B T}{h^2}\right)^{3/2} \exp[\beta(\mu - m g y)] \qquad (16.150)$$

The pressure diminishes with increasing y.

Problem 3. Two-level system:

One considers the two-level system in section 4.7.2 using formulas (4.41) and (4.42) which are more difficult to handle.

a) The partition function z of a particle is calculated by making it visit all individual states of different energies. While, the function z_k in (4.42) is calculated by taking a state of energy ϵ_k and making all particles to go through it.

b) One has $\mathcal{Z} = \prod_k z_k$ [see (4.41)] where z_k is given by (4.42). One supposes that permutations between particles on the same level ϵ_k do not yield new microstates. One has thus

$$z_k = \sum_{n_k=0}^{\infty} \exp[-\beta n_k(\epsilon_k - \mu)]/n_k! = \exp[\exp[-\beta(\epsilon_k - \mu)]] \quad (16.151)$$

For $\epsilon_1 = -\epsilon$ and $\epsilon_2 = \epsilon$, one has

$$z_1 = \exp\{\exp[-\beta(-\epsilon - \mu)]\}$$

$$z_2 = \exp\{\exp[-\beta(\epsilon - \mu)]\} \qquad (16.152)$$

from which one gets

$$\mathcal{Z} = \prod_i z_i = z_1 z_2 = \exp\{\exp[-\beta(-\epsilon - \mu)] + \exp[-\beta(\epsilon - \mu)]\}$$

$$= \exp\{\exp(\beta\mu)[\exp(-\beta\epsilon) + \exp(\beta\epsilon)]\}$$

$$= \exp[\exp(\beta\mu)2\cosh(\beta\epsilon)] \qquad (16.153)$$

One recovers here $\mathcal{Z}$ of section 4.7.2.

c) The calculations of the average energy and the average numbers of particles on the two levels are identical to those in section 4.7.2.

Remark: It is more delicate to use the formulas (4.41) and (4.42) because if one does not divide z_k by $n!$ as in (16.151), one obtains

$$z_k = \sum_{n_k=0}^{\infty} \exp[-\beta n_k(\epsilon_k - \mu)] = \frac{1}{1 - \exp[-\beta(\epsilon_k - \mu)]} \qquad (16.154)$$

One has thus

$$z_1 = \frac{1}{1 - \exp[-\beta(-\epsilon - \mu)]}$$

$$\bar{n}_1 = k_B T \frac{\partial \ln z_1}{\partial \mu} = \frac{1}{\exp[\beta(-\epsilon - \mu)] - 1}$$

$$z_2 = \frac{1}{1 - \exp[-\beta(\epsilon - \mu)]}$$

$$\bar{n}_2 = k_B T \frac{\partial \ln z_2}{\partial \mu} = \frac{1}{\exp[\beta(\epsilon - \mu)] - 1} \qquad (16.155)$$

The above results of z_k and $\bar{n}_k$ are identical to the case of bosons shown in (4.44) and (4.49). A question arises on the value of μ: in order to have $\bar{n}_k \geq 0$, one should have $\beta(\epsilon_k - \mu) \geq 0$, namely the lowest value of ϵ_k should be larger than μ. One should impose $-\epsilon - \mu \geq 0$. This question will be discussed in chapter 6 when one studies the Bose condensation which takes place at very low temperatures. For the two-level system considered here, one supposes that the temperature is not too low so that $\exp[\beta(\pm\epsilon - \mu)] >> 1$. With this hypothesis, one has

$$\ln z_1 = -\ln(1 - \exp[-\beta(-\epsilon - \mu)]) \simeq \exp[-\beta(-\epsilon - \mu)]$$

$$\ln z_2 = -\ln(1 - \exp[-\beta(\epsilon - \mu)]) \simeq \exp[-\beta(\epsilon - \mu)] \qquad (16.156)$$

from which one obtains

$$\ln \mathcal{Z} = \ln z_1 + \ln z_2 = \exp[-\beta(-\epsilon - \mu)] + \exp[-\beta(\epsilon - \mu)]$$

$$= \exp(\beta\mu) 2 \cosh(\beta\epsilon) \qquad (16.157)$$

This result is the same as that obtained in section 4.7.2.

Problem 4. Degeneracy in the case of fermions:

In the case of fermions, if an energy level ϵ_k in the expression (4.45) is g_k-fold degenerate, then (4.45) is rewritten as

$$z_k^{FD} = \sum_{n_k=0}^{1} g_k^{n_k} e^{-\beta n_k(\epsilon_k - \mu)}$$

$$= 1 + g_k e^{-\beta(\epsilon_k - \mu)}$$

The average number of particles is

$$\overline{n}_k^{FD} = k_B T \frac{\partial \ln z_k^{FD}}{\partial \mu} = k_B T \frac{\beta g_k e^{-\beta n_k(\epsilon_k - \mu)}}{1 + g_k e^{-\beta n_k(\epsilon_k - \mu)}}$$

$$= \frac{1}{\frac{1}{g_k} e^{\beta n_k(\epsilon_k - \mu)} + 1} \tag{16.158}$$

Problem 5. Particle trap:

a) Each site is considered as a subsystem which can have a variable number of particles (0 or 1). Let ϕ be the grand partition function of one site. The sites are supposed to be independent. The grand partition function of the system is factorizable: one has $\mathcal{Z} = \phi^N$ [see (4.42)]. To calculate ϕ, one observes that there are 3 energy levels 0, $-\epsilon_1$ and $-\epsilon_2$ corresponding to the numbers of particles $n_k = 0, 1, 1$, respectively. One has

$$\phi = \sum_{n_k} \exp[-n_k \beta(\epsilon_k - \mu)] = 1 + \exp[-\beta(-\epsilon_1 - \mu)] + \exp[-\beta(-\epsilon_2 - \mu)]$$
$$\tag{16.159}$$

b) The average number of captured particles $\overline{n}$ is given by [see (4.16)]

$$\overline{n} = \frac{1}{\beta} \frac{\partial \ln \mathcal{Z}}{\partial \mu}$$

$$= N \frac{\exp[-\beta(-\epsilon_1 - \mu)] + \exp[-\beta(-\epsilon_2 - \mu)]}{1 + \exp[-\beta(-\epsilon_1 - \mu)] + \exp[-\beta(-\epsilon_2 - \mu)]} \tag{16.160}$$

The average energy of the solid is calculated by

$$\overline{E} = -\frac{\partial \ln \mathcal{Z}}{\partial \beta} + \mu \overline{n}$$

$$= -N \frac{(\epsilon_1 + \mu) \exp[\beta(\epsilon_1 + \mu)] + (\epsilon_2 + \mu) \exp[\beta(\epsilon_2 + \mu)]}{1 + \exp[\beta(\epsilon_1 + \mu)] + \exp[\beta(\epsilon_2 + \mu)]}$$

$$+ \mu \overline{n} \tag{16.161}$$

c) $\bar{n}_1$, $\bar{n}_2$ and $\bar{n}_0$ are given by N multiplied by the respective probabilities. One has

$$\bar{n}_1 = N \frac{\exp[\beta(\epsilon_1 + \mu)]}{1 + \exp[\beta(\epsilon_1 + \mu)] + \exp[\beta(\epsilon_2 + \mu)]} \quad (16.162)$$

$$\bar{n}_2 = N \frac{\exp[\beta(\epsilon_2 + \mu)]}{1 + \exp[\beta(\epsilon_1 + \mu)] + \exp[\beta(\epsilon_2 + \mu)]} \quad (16.163)$$

$$\bar{n}_0 = N \frac{1}{1 + \exp[\beta(\epsilon_1 + \mu)] + \exp[\beta(\epsilon_2 + \mu)]} \quad (16.164)$$

d) Replacing $-\epsilon_1 = -\epsilon_0 + mB$ and $-\epsilon_2 = -\epsilon_0 - mB$ in (16.162) and (16.163), one obtains

$$\begin{aligned}
\bar{n}_1 &= N \frac{\exp[\beta(\epsilon_0 - mB + \mu)]}{1 + \exp[\beta(\epsilon_0 - mB + \mu)] + \exp[\beta(\epsilon_0 + mB + \mu)]} \\
&= N \frac{\exp[\beta(\epsilon_0 - mB + \mu)]}{1 + 2\exp[\beta(\epsilon_0 + \mu)]\cosh(\beta mB)} \\
\bar{n}_2 &= N \frac{\exp[\beta(\epsilon_0 + mB + \mu)]}{1 + \exp[\beta(\epsilon_0 - mB + \mu)] + \exp[\beta(\epsilon_0 + mB + \mu)]} \\
&= N \frac{\exp[\beta(\epsilon_0 + mB + \mu)]}{1 + 2\exp[\beta(\epsilon_0 + \mu)]\cosh(\beta mB)} \quad (16.165)
\end{aligned}$$

The average magnetic moment of the traps is

$$\begin{aligned}
\overline{M} &= m(\bar{n}_2 - \bar{n}_1) \\
&= mN \frac{2\exp[\beta(\epsilon_0 + \mu)]\sinh(\beta mB)}{1 + 2\exp[\beta(\epsilon_0 + \mu)]\cosh(\beta mB)} \quad (16.166)
\end{aligned}$$

At high temperatures, $\sinh[\beta mB] \to 0$, $\overline{M} \to 0$. This shows that half of the captured electrons have up spins, the other half have down spins.

At low temperatures, one can neglect 1 in the denominator. One has then

$$\begin{aligned}
\overline{M} &\simeq mN \frac{2\exp[\beta(\epsilon_0 + \mu)]\sinh(\beta mB)}{2\exp[\beta(\epsilon_0 + \mu)]\cosh(\beta mB)} \\
&= mN \tanh(\beta mB) \to mN \quad (16.167)
\end{aligned}$$

This means that all spins are parallel to the field $\vec{B}$ at low temperatures. The total magnetic moment is maximum.

Problem 6. Poisson law by the grand-canonical description:

One recalls that for a classical ideal gas, the grand partition function is given by (4.52): $\mathcal{Z} = \exp(\lambda z)$, and the average number of particles by (4.53): $\bar{N} = \lambda z$ where $\lambda = \exp(\beta\mu)$. The partition function z of a particle is given by (3.53): $z = V(2\pi m k_B T)^{3/2}/h^3$. One applies these results to a very small volume v: let n be the number of particles which find themselves in v. The grand partition function of v is written as $\mathcal{Z} = \exp(\lambda z) = \sum_n \lambda^n Z(n)$ where $Z(n) = z^n/n!$. The probability to find n particles in v is thus

$$
\begin{aligned}
P_n &= \frac{\lambda^n Z(n)}{\mathcal{Z}} = \frac{\lambda^n Z(n)}{\exp(\lambda z)} \\
&= \frac{\lambda^n z^n/n!}{\exp(\lambda z)} = \frac{\lambda^n z^n}{n!}\exp(-\lambda z)
\end{aligned}
\tag{16.168}
$$

Using $\bar{n} = \lambda z$ one gets

$$
P_n = \frac{(\bar{n})^n}{n!}\exp(-\bar{n})
\tag{16.169}
$$

This is the Poisson law.

Problem 7. System of interacting electrons:

A metal can be considered in a first approximation as a gas of free conducting electrons (chapter 5). If there are N impurities in the system, then electrons may be captured by these impurities. This is the subject of the present exercise. The steps to follow are similar to those in Problem 5.

Each impurity is supposed to be an independent subsystem able to capture 0, 1 or 2 electrons. Let ϕ be the grand partition function of an impurity. The grand partition function of the system is $\mathcal{Z} = \phi^N$ [see (4.42)]. There are 3 energy levels $\epsilon_k = 0$, ϵ_0 and $2\epsilon_0 + g$ corresponding to the capture of $n_k = 0, 1, 2$ electrons. The impurity which captures one electron is two-fold degenerate, due to two possibilities ↑ and ↓ spins. One has

$$
\begin{aligned}
\phi &= \sum_{n_k} \exp[-n_k\beta(\epsilon_k - \mu)] = 1 + 2\exp[-\beta(\epsilon_0 - \mu)] \\
&\quad + \exp[-\beta(2\epsilon_0 + g - 2\mu)]
\end{aligned}
\tag{16.170}
$$

a) The average number of captured electrons $\bar{n}$ is given by [see (4.16)]

$$\bar{n} = \frac{1}{\beta}\frac{\partial \ln \mathcal{Z}}{\partial \mu}$$

$$= 2N\frac{\exp[-\beta(\epsilon_0 - \mu)] + \exp[-\beta(2\epsilon_0 + g - 2\mu)]}{1 + 2\exp[-\beta(\epsilon_0 - \mu)] + \exp[-\beta(2\epsilon_0 + g - 2\mu)]}$$

(16.171)

The average energy of captured electrons is

$$\overline{E} = -\frac{\partial \ln \mathcal{Z}}{\partial \beta} + \mu\bar{n}$$

$$= -N\frac{2(\epsilon_0 - \mu)e^{-\beta(\epsilon_0 - \mu)} + (2\epsilon_0 + g - \mu)e^{-\beta(2\epsilon_0 + g - \mu)}}{1 + 2e^{-\beta(\epsilon_0 - \mu)} + e^{-\beta(2\epsilon_0 + g - \mu)}}$$

$$+ \mu\bar{n}$$

(16.172)

b) One supposes that $\bar{n}$ is fixed and it is taken to be equal to N. One has

$$\bar{n} = N = 2N\frac{\exp[-\beta(\epsilon_0 - \mu)] + \exp[-\beta(2\epsilon_0 + g - 2\mu)]}{1 + 2\exp[-\beta(\epsilon_0 - \mu)] + \exp[-\beta(2\epsilon_0 + g - 2\mu)]}$$

$$1 = 2\frac{\exp[-\beta(\epsilon_0 - \mu)] + \exp[-\beta(2\epsilon_0 + g - 2\mu)]}{1 + 2\exp[-\beta(\epsilon_0 - \mu)] + \exp[-\beta(2\epsilon_0 + g - 2\mu)]}$$

$$1 = \exp[-\beta(2\epsilon_0 + g - 2\mu)]$$

$$0 = 2\epsilon_0 + g - 2\mu$$

$$\mu = \epsilon_0 + \frac{g}{2}$$

(16.173)

c) The average numbers of impurities which capture zero, one and two electrons are equal to N multiplied by the respective probabilities. One has

$$\bar{n}_0 = N\frac{1}{1 + 2\exp[-\beta(\epsilon_0 - \mu)] + \exp[-\beta(2\epsilon_0 + g - 2\mu)]}$$

$$\bar{n}_1 = N\frac{2\exp[-\beta(\epsilon_0 - \mu)]}{1 + 2\exp[-\beta(\epsilon_0 - \mu)] + \exp[-\beta(2\epsilon_0 + g - 2\mu)]}$$

$$\bar{n}_2 = N\frac{\exp[-\beta(2\epsilon_0 + g - \mu)]}{1 + 2\exp[-\beta(\epsilon_0 - \mu)] + \exp[-\beta(2\epsilon_0 + g - 2\mu)]}$$

Since $\bar{n} = N$, one replaces μ by (16.173):

$$\bar{n}_0 = \frac{N}{2(1 + \exp[\beta g/2])}$$

$$\bar{n}_1 = N\frac{\exp[\beta g/2]}{1 + \exp[\beta g/2]}$$

$$\bar{n}_2 = \frac{N}{2(1 + \exp[\beta g/2])}$$

At high temperatures, $\overline{n}_0 \to N/4$, $\overline{n}_1 \to N/2$ and $\overline{n}_2 \to N/4$.

At low temperatures, $\overline{n}_0 \to 0$, $\overline{n}_1 \to N$ and $\overline{n}_2 \to 0$: only the level of lowest energy is occupied (one recalls that $\epsilon_0 < 2\epsilon_0 + g$).

d) The system is put in a magnetic field $\vec{B}$. The energy of impurities having two antiparallel spins does not change with the field since the Zeeman energies mutually cancel out. The degeneracy of impurities having one electron is removed (spins parallel to $\vec{B}$ are favored by negative Zeeman energy). One has

$$\phi = 1 + e^{-\beta(\epsilon_0 - mB - \mu)} + e^{-\beta(\epsilon_0 + mB - \mu)}$$
$$+ e^{-\beta(2\epsilon_0 + g - 2\mu)}$$
$$= 1 + 2\cosh(\beta mB)e^{-\beta(\epsilon_0 - \mu)} + e^{-\beta(2\epsilon_0 + g - 2\mu)} \quad (16.174)$$

The average magnetic moment of impurities is

$$\overline{M} = m(\overline{n}_2 - \overline{n}_1)$$
$$= mN \frac{e^{\beta(\epsilon_0 - mB - \mu)} - e^{-\beta(\epsilon_0 + mB - \mu)}}{1 + 2\cosh(\beta mB)e^{-\beta(\epsilon_0 - \mu)} + e^{-\beta(2\epsilon_0 + g - 2\mu)}}$$
$$= mN \frac{2\sinh(\beta mB)e^{-\beta(\epsilon_0 - \mu)}}{1 + 2\cosh(\beta mB)e^{-\beta(\epsilon_0 - \mu)} + e^{-\beta(2\epsilon_0 + g - 2\mu)}}$$

At high temperatures, $\sinh(\beta mB) \to 0$, $\overline{M} \to 0$: half of captured electrons have up spins, the other half have down spins.

A low temperatures, one neglects 1 in the denominator. One observes that $\exp[-\beta(\epsilon_0 - \mu)] \gg \exp[-\beta(2\epsilon_0 + g - 2\mu)]$ (one has used the assumption $g > 0$ and $g < |\epsilon_0|$). One has then

$$\overline{M} \simeq mN \frac{2\exp[\beta(\epsilon_0 + \mu)]\sinh(\beta mB)}{2\cosh(\beta mB)\exp[-\beta(\epsilon_0 - \mu)]}$$
$$= mN\tanh(\beta mB) \to mN \quad (16.175)$$

All spins are thus parallel to $\vec{B}$ at low temperatures, giving rise to the maximum of the magnetic moment.

Problem 8. Lattice model for an ideal gas:

a) Definition of the grand partition function $\mathcal{Z}$: see lecture. For a simple fluid, the grand potential and other physical quantities in the grand-canonical situation depend on T, μ and V. One has $J = J(T, \mu, V)$. If one increases the system volume from V to αV, without changing T and μ, one has $J' = J'(T, \mu, \alpha V) = \alpha J(T, \mu, V)$ because J is additive (see lecture). The pressure when the volume is V is $p = -\partial J/\partial V$ and when the volume is αV is $p' = -\partial J'/\partial V' = -\alpha(\partial J/\partial V)(\partial V/\partial V') = -\partial J/\partial V = p$. The pressure does not change when the volume changes. This implies that the pressure does not depend on V. Integrating $(\partial J/\partial V)dV = -pdV$, one obtains $J = -pV$.

b) Each small cube is an independent subsystem, its grand partition function is

$$\mathcal{Z}_i = \sum_{n_k=0,1} \exp[-n_k\beta(E_i - \mu)] = 1 + \exp[-\beta(E_i - \mu)] \quad (16.176)$$

where E_i is the energy of the molecule i. The grand partition function of the system is $\mathcal{Z} = \prod_{i=1}^{N} \mathcal{Z}_i$.

c) The average number of particles in the cube i is

$$\overline{N}_i = \frac{1}{\beta}\frac{\partial \ln \mathcal{Z}_i}{\partial \mu}$$

$$= \frac{\exp[-\beta(E_i - \mu)]}{1 + \exp[-\beta(E_i - \mu)]}$$

$$= \frac{1}{\exp[\beta(E_i - \mu)] + 1} \quad (16.177)$$

One recognizes that the above expression is the Fermi-Dirac distribution. The reason of this finding is that one has assumed only one particle at most can occupy a "site": the Pauli principle was implicitly imposed.

From this point, the same results are obtained using the model of a free Fermi gas shown in chapter 5: the total number of particles is

$$\overline{N} = \sum_i \overline{N}_i \quad (16.178)$$

For a large system, he sum can be replaced by an integral:

$$\overline{N} = \int_0^{\infty} dE\rho(E)\frac{1}{\exp[\beta(E - \mu)] + 1}$$

$$= A \int_0^{\infty} dE E^{1/2}\frac{1}{\exp[\beta(E - \mu)] + 1} \quad (16.179)$$

where one has used $\rho(E) = AE^{1/2}$ [see (2.36)].

The grand potential is

$$J = -k_B T \ln \mathcal{Z} = -k_B T \sum_i \ln(1 + \exp[-\beta(E_i - \mu)])$$

$$= -k_B T A \int_0^{\infty} dE E^{1/2} \ln(1 + \exp[-\beta(E - \mu)]) \quad (16.180)$$

Integrating by parts gives

$$J = -k_B T A \{ \left[\frac{2}{3} E^{3/2} \ln\left(1 + \exp[-\beta(E-\mu)]\right) \right]_0^\infty$$

$$+ \beta \frac{2}{3} \int_0^\infty dE E^{3/2} \frac{\exp[-\beta(E-\mu)]}{1 + \exp[-\beta(E-\mu)]} \}$$

$$= -k_B T A \left\{ 0 + \beta \frac{2}{3} \int_0^\infty dE E^{3/2} \frac{1}{1 + \exp[\beta(E-\mu)]} \right\}$$

$$= -\frac{2}{3} \overline{E} \tag{16.181}$$

where one has used for $\overline{E}$ the following result:

$$\overline{E} = \sum_i \overline{N}_i E_i = A \int_0^\infty dE E^{3/2} \frac{1}{\exp[\beta(E-\mu)] + 1} \tag{16.182}$$

The pressure is

$$p = -\frac{\partial J}{\partial V} = -\frac{J}{V} = \frac{2}{3V} \overline{E}$$

Thus,

$$pV = \frac{2}{3} \overline{E}$$

This is the equation of state of the lattice model for the gas defined in this exercise.

In the limit of strong dilution, i. e. $\overline{N}v \ll V$, the particle density is small so that $\exp[\beta(E-\mu)] \gg 1$. One has then

$$\overline{N}_i \simeq N \exp[-\beta(E_i - \mu)] \tag{16.183}$$

One recovers here the Maxwell-Boltzmann distribution. One has

$$\overline{N} = \sum_i \overline{N}_i$$

$$\simeq \int_0^\infty dE \rho(E) \exp[-\beta(E-\mu)]$$

$$= A \int_0^\infty dE E^{1/2} \exp[\beta(E-\mu)] \tag{16.184}$$

$$\overline{E} = \sum_i \overline{N}_i E_i$$

$$\simeq A \int_0^\infty dE \, E^{3/2} \exp[-\beta(E - \mu)]$$

$$= A \left\{ \left[-\frac{1}{\beta} \exp[-\beta(E - \mu)] E^{3/2} \right]_0^\infty \right.$$

$$\left. + \frac{3}{2\beta} \int_0^\infty dE \, E^{1/2} \exp[-\beta(E - \mu)] \right\}$$

$$= A \left\{ 0 + \frac{3}{2\beta} \int_0^\infty dE \, E^{1/2} \exp[-\beta(E - \mu)] \right\}$$

$$= \frac{3}{2\beta} \overline{N} = \frac{3}{2} \overline{N} k_B T \qquad (16.185)$$

where (16.184) has been used. Therefore, in the case of strong dilution one has

$$pV = \frac{2\overline{E}}{3} = \overline{N} k_B T$$

This is the equation of state of a classical ideal gas.

Problem 9. Adsorption:

The surface of the solid having N_0 sites is in contact with a classical ideal gas playing the role of a reservoir of heat and particles providing fixed temperature T and chemical potential μ. Each molecule adsorbed on the surface has an energy $-\epsilon$ ($\epsilon > 0$).

a) One considers each site of the surface as an independent subsystem, as in Problems 5 and 7. Its grand partition function is

$$\mathcal{Z}_i = \sum_{n_k = 0,1} \exp[-n_k \beta(E_i - \mu)] = 1 + \exp[-\beta(-\epsilon - \mu)] \quad (16.186)$$

where E_i is the energy of the molecule i. The grand partition function of the system is $\mathcal{Z} = \prod_{i=1}^N \mathcal{Z}_i$.

b) The average number of particles at the site i is

$$\overline{N}_i = \frac{1}{\beta} \frac{\partial \ln \mathcal{Z}_i}{\partial \mu}$$

$$= \frac{\exp[\beta(\epsilon + \mu)]}{1 + \exp[\beta(\epsilon + \mu)]}$$

$$= \frac{1}{\exp[\beta(-\epsilon - \mu)] + 1} \qquad (16.187)$$

As in Problem 8, one sees that the above expression is the Fermi-Dirac distribution which is due to the assumption of a single particle at most per site.

The average total number of particles adsorbed on the surface is

$$\overline{N}_c = \sum_i \overline{N}_i$$

$$= N_0 \frac{1}{\exp[-\beta(\epsilon + \mu)] + 1} \tag{16.188}$$

The corresponding energy is

$$\overline{E} = \sum_i \overline{N}_i E_i = \frac{-N_0 \epsilon}{\exp[-\beta(\epsilon + \mu)] + 1} \tag{16.189}$$

c) The grand potential is written as

$$J = -k_B T \ln \mathcal{Z} = -k_B T \sum_i \ln(1 + \exp[\beta(\epsilon + \mu)])$$

$$= -k_B T N_0 \ln(1 + \exp[\beta(\epsilon + \mu)]) \tag{16.190}$$

The entropy S of adsorbed molecules is calculated by

$$J = \overline{E} - \mu \overline{N}_c - TS$$

$$S = k_B N_0 \ln(1 + \exp[\beta(\epsilon + \mu)]) - N_0 \frac{\epsilon + \mu}{T} \tag{16.191}$$

where one has used the results of J, $\overline{E}$ and $\overline{N}_c$. From (16.188), one has

$$\beta(\epsilon + \mu) = \ln \frac{\overline{N}_c}{N_0 - \overline{N}_c} \tag{16.192}$$

One can express S as a function of $\overline{N}_c$ as follows:

$$S = k_B[N_0 \ln N_0 - \overline{N}_c \ln \overline{N}_c - (N_0 - \overline{N}_c) \ln(N_0 - \overline{N}_c)] \tag{16.193}$$

In the case of N_c molecules captured by N_0 surface sites studied by the micro-canonical description, the micro-canonical entropy is

$$S^* = k_B \ln \Omega = k_B \ln \frac{N_0!}{N_c!(N_0 - N_c)!} \tag{16.194}$$

If one uses the Stirling formula for the above expression, one recovers the result given by (16.193).

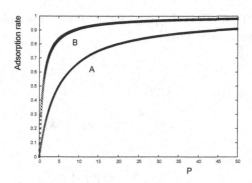

Fig. 16.4 Langmuir isotherms: adsorption rate θ versus pressure p, for two temperatures. Curve A corresponds to the higher temperature.

d) The chemical potential μ is related to the pressure p via (4.55). One has

$$\lambda \equiv \exp(\beta\mu) = \frac{pV}{k_B T z} = \frac{p}{k_B T} \left(\frac{2\pi\hbar^2}{mk_B T} \right)^{3/2} \tag{16.195}$$

The adsorption rate $\theta = \overline{N}_c/N_0$ is obtained from (16.188):

$$\theta = \frac{1}{\exp[-\beta(\epsilon + \mu)] + 1} \tag{16.196}$$

Replacing (16.195) in this equation, one obtains

$$\theta = \frac{1}{\frac{k_B T}{p} \left(\frac{mk_B T}{2\pi\hbar^2} \right)^{3/2} \exp(-\beta\epsilon) + 1}$$

$$= \frac{p}{p + p_0} \tag{16.197}$$

where one has put

$$p_o \equiv k_B T \left(\frac{mk_B T}{2\pi\hbar^2} \right)^{3/2} \exp(-\beta\epsilon) \tag{16.198}$$

Curves θ versus p at several temperatures, called "Langmuir isotherms", are schematically shown in Fig. 16.4. One sees that the adsorption rate decreases with increasing temperature.

Problem 10. Adsorption of an ideal gas on the surface of a solid:

a) One considers adsorbed molecules on the surface as a bidimensional classical ideal gas in the grand-canonical situation. The grand partition function is given by (4.10) in two dimensions:

$$\mathcal{Z} = \sum_n e^{\beta\mu n} Z(n, T, S) \tag{16.199}$$

where $Z(n, T, S)$ is the partition function of n molecules adsorbed on the surface S. One has

$$Z = \frac{z^n}{n!} \tag{16.200}$$

where z is given by (3.53)

$$z = \frac{S}{h^2} \exp(\beta \epsilon_0) 2\pi m k_B T \tag{16.201}$$

Thus,

$$\mathcal{Z} = \sum_n (\lambda z)^n / n! = e^{\lambda z} \tag{16.202}$$

where $\lambda = e^{\beta \mu}$. The average number of molecules adsorbed on the surface is

$$\overline{N}_a = \frac{1}{\beta} \frac{\partial \ln \mathcal{Z}}{\partial \mu}$$

$$= \frac{1}{\beta} \frac{\partial (\lambda z)}{\partial \mu} = \lambda z \tag{16.203}$$

$$= \frac{S}{h^2} \exp[\beta(\epsilon_0 + \mu)] 2\pi m k_B T \tag{16.204}$$

One expresses now $\overline{N}_a$ as a function of pressure of the ideal gas serving as the reservoir of molecules by replacing (16.195) in the above expression:

$$\overline{N}_a = \frac{pS}{k_B T} \exp(\beta \epsilon_0) \left(\frac{h^2}{2\pi m k_B T} \right)^{1/2} \tag{16.205}$$

b) The average energy of adsorbed molecules is given by

$$\overline{E} - \mu \overline{N}_a = -\frac{\partial \ln \mathcal{Z}}{\partial \beta} = -\frac{\partial (\lambda z)}{\partial \beta}$$

$$= -(\epsilon_0 + \mu)\lambda z + \lambda z / \beta$$

$$= -(\epsilon_0 + \mu)\overline{N}_a + k_B T \overline{N}_a \tag{16.206}$$

from which one gets

$$\overline{E} = \overline{N}_a (k_B T - \epsilon_0) \tag{16.207}$$

Let $\overline{N}_g$ be the average number of molecules of the reservoir. One has $N = \overline{N}_a + \overline{N}_g$=constant. The total energy of the whole system "adsorbed molecules + gas of reservoir" is

$$\overline{E}_T = \overline{E} + \frac{3}{2}\overline{N}_g k_B T$$

$$= \overline{N}_a(k_B T - \epsilon_0) + \frac{3}{2}(N - \overline{N}_a)k_B T$$

$$= \frac{3}{2}N k_B T - \overline{N}_a \epsilon_0 - \frac{1}{2}\overline{N}_a k_B T \qquad (16.208)$$

The calorific capacity of the total system is thus

$$C_V = \frac{3}{2}N k_B - \epsilon_0 \frac{\partial \overline{N}_a}{\partial T} - \frac{1}{2}\overline{N}_a k_B$$

$$- \frac{1}{2}k_B T \frac{\partial \overline{N}_a}{\partial T} \qquad (16.209)$$

To calculate $\frac{\partial \overline{N}_a}{\partial T}$, one can use the relation $\overline{N}_a = \lambda z$. The final expression of C_V is cumbersome to write down here.

c) By the canonical method, one considers the total system composed of a gas of N_g molecules in three dimensions and a gas of N_a molecules in two dimensions. The latter has an extra term due to interaction with surface atoms, namely $-N_a\epsilon_0$. One has thus

$$\overline{E}_T = \frac{3}{2}N_g k_B T + N_a k_B T - N_a \epsilon_0 \qquad (16.210)$$

$N_g + N_a = N$ being a constant. The remaining part of the present question can be done following the steps in Problem 12 of chapter 3: one writes the free energies F_g and F_a, then one minimizes $F = F_g + F_a$ with respect to N_a to determine N_a at equilibrium.

16.5 Solutions of problems of chapter 5

Problem 1. Free Fermi gas at thermodynamic limit:

One considers a free Fermi gas at the thermodynamic limit, of volume V.

a) One has (5.13):

$$N = (2S + 1) \int_0^\infty dE \rho(E) \frac{1}{e^{\beta(E-\mu)} + 1} \qquad (16.211)$$

and (5.15):

$$\overline{E} = (2S + 1) \int_0^\infty dE \rho(E) \frac{E}{e^{\beta(E-\mu)} + 1} \qquad (16.212)$$

For J, one uses (5.6):

$$J = -k_B T \sum_\lambda \ln[1 - f(E_\lambda, T, \mu)]$$

$$= -k_B T (2S+1) \int_0^\infty dE \rho(E) \ln\left(1 - \frac{1}{e^{\beta(E-\mu)} + 1}\right)$$

$$\tag{16.213}$$

Replacing $\rho(E) = AE^{1/2}$ [see (2.36)] then integrating by parts, one gets

$$J = -k_B T (2S+1) \int_0^\infty dE \rho(E) \ln\left(1 - \frac{1}{e^{\beta(E-\mu)} + 1}\right)$$

$$= -k_B T (2S+1) [\frac{2}{3} AE^{3/2} \ln\left(1 - \frac{1}{e^{\beta(E-\mu)} + 1}\right) \Big|_0^\infty$$

$$- \frac{2}{3} A \int_0^\infty E^{3/2} \frac{\frac{-\beta e^{\beta(E-\mu)}}{(e^{\beta(E-\mu)}+1)^2}}{(1 - \frac{1}{e^{\beta(E-\mu)}+1})} dE]$$

$$= -k_B T (2S+1) \left[0 + \beta \frac{2}{3} A \int_0^\infty E^{3/2} \frac{1}{e^{\beta(E-\mu)} + 1} dE\right]$$

$$= -(2S+1) \frac{2}{3} A \int_0^\infty E^{3/2} \frac{1}{e^{\beta(E-\mu)} + 1} dE$$

$$= -\frac{2}{3} \overline{E} \tag{16.214}$$

b) The grand-canonical entropy S is given by (5.7). It suffices to change the sums into integrals for a large system. One can also use $S = -\partial J / \partial T$ with J given above.

c) The grand-canonical pressure p is

$$p = -\frac{\partial J}{\partial V} = \frac{2}{3} \frac{\partial \overline{E}}{\partial V}$$

$$= \frac{2\overline{E}}{3V} \tag{16.215}$$

where in the last equality one has used the factor A of $\rho(E) = AE^{1/2}$ which is proportional to V [see (2.36)].

Problem 2. Fermi gas at low temperatures:

One considers the Fermi gas of the previous exercise at low temperatures.

For the grand potential J at low T, one can use the relation $J = -\frac{2}{3}\overline{E}$ where $\overline{E}$ has been obtained by the Sommerfeld's expansion in (5.29). One has then

$$J = -\frac{2}{3}\overline{E}_0 \left[1 + \frac{5\pi^2}{12}\left(\frac{k_B T}{E_F}\right)^2 + O(T^4)\right] \qquad (16.216)$$

The grand-canonical entropy $\mathcal{S}$ is

$$\mathcal{S} = -\frac{\partial J}{\partial T} = \overline{E}_0 \frac{5\pi^2}{9}\left(\frac{k_B}{E_F}\right)^2 T + ... \qquad (16.217)$$

The pressure p is given at low T by

$$p = \frac{2\overline{E}}{3V} = \frac{2\overline{E}_0}{3V}\left[1 + \frac{5\pi^2}{12}\left(\frac{k_B T}{E_F}\right)^2 + O(T^4)\right] \qquad (16.218)$$

These results are very different from those of a classical ideal gas.

Problem 3. Free Fermi gas in one and two dimensions:

Using the densities of states for one and two dimensions obtained in Problem 5 of chapter 2, one calculates in the following the Fermi energy and the average kinetic energy per electron at 0 Kelvin in a metal in one and two dimensions as functions of the electron density n:

- One dimension:

One has

$$N = \int_0^{E_F} \rho(E)dE = \sqrt{\frac{m}{2}}\frac{L}{\hbar\pi}\int_0^{E_F} E^{-1/2}dE$$

$$= \sqrt{\frac{m}{2}}\frac{2L}{\hbar\pi}E_F^{1/2} \qquad (16.219)$$

from which one gets

$$E_F = \frac{2}{m}\left(\frac{\hbar\pi N}{2L}\right)^2 = \frac{2}{m}\left(\frac{\hbar\pi n}{2}\right)^2 \qquad (16.220)$$

where $n = N/L$ is the electron density.

The kinetic energy is

$$\overline{E} = \int_0^{E_F} \rho(E)EdE = \sqrt{\frac{m}{2}}\frac{L}{\hbar\pi}\int_0^{E_F} E^{1/2}dE$$

$$= \sqrt{\frac{m}{2}}\frac{L}{\hbar\pi}\frac{2}{3}E_F^{3/2} \qquad (16.221)$$

- Two dimensions:
One has

$$N = \int_0^{E_F} \rho(E)dE = \frac{mL^2}{\hbar^2\pi} \int_0^{E_F} dE$$

$$= \frac{mL^2}{\hbar^2\pi} E_F \qquad (16.222)$$

from which one obtains

$$E_F = \frac{\hbar^2\pi}{m} n \qquad (16.223)$$

The kinetic energy is

$$\bar{E} = \int_0^{E_F} \rho(E)EdE = \frac{mL^2}{\hbar^2\pi} \int_0^{E_F} EdE$$

$$= \frac{mL^2}{\hbar^2\pi} \frac{1}{2} E_F^2 \qquad (16.224)$$

Problem 4. Free Fermi gas in two dimensions:

a) Using the result of the previous exercise, one establishes the following relation between the electron density and the Fermi wave vector:

$$E_F = \frac{\hbar^2\pi}{m} n = \frac{\hbar^2 k_F^2}{2m} \to k_F = \sqrt{2\pi n} \qquad (16.225)$$

b) One has at $T = 0$

$$N = \int_0^{\mu_0} \rho(E)dE = D \int_0^{\mu_0} dE = D\mu_0 \to \mu_0 = N/D \qquad (16.226)$$

At $T \neq 0$, if one uses the Sommerfeld's expansion, only the first term is not zero because the derivatives of D are zero for the next terms. One has thus

$$N = D \int_0^{\mu} dE = D\mu \qquad (16.227)$$

This shows that $\mu = \mu_0 \equiv E_F$ for all temperatures if one uses the Sommerfeld's expansion.

c) At low temperatures, one uses the Sommerfeld's expansion for $\overline{E}$. One obtains

$$\overline{E} = D \int_0^\infty \frac{E dE}{\exp[\beta(E-\mu)]+1}$$

$$\simeq D \left[\int_0^\mu E dE + \frac{\pi^2}{6}(k_B T)^2 + ... \right]$$

$$= D \left[\mu^2/2 + \frac{\pi^2}{6}(k_B T)^2 + ... \right]$$

$$= \frac{D\mu^2}{2} \left[1 + \frac{\pi^2}{3}\left(\frac{k_B T}{\mu}\right)^2 + ... \right] \tag{16.228}$$

Note that μ in this expression is equal to μ_0 as found above by the Sommerfeld's expansion. Thus, only the second term of $\overline{E}$ depends on T. The calorific capacity is thus

$$C_V = D\frac{\pi^2}{3}k_B^2 T \tag{16.229}$$

d) At high temperatures, $\exp[\beta(E-\mu)] \gg 1$ (this condition implies $\mu < 0$). One has

$$N = D \int_0^\infty \frac{dE}{\exp[\beta(E-\mu)]+1}$$

$$\simeq D \int_0^\infty \exp[-\beta(E-\mu)]dE$$

$$= Dk_B T \exp(\beta\mu) \tag{16.230}$$

$$N/D = \mu_0 = k_B T \exp(\beta\mu) \tag{16.231}$$

This shows that μ depends on T at high temperatures. The Sommerfeld's expansion for μ is not valid at high temperatures. For $\overline{E}$ at high temperatures, one has

$$\overline{E} = D \int_0^\infty \frac{E dE}{\exp[\beta(E-\mu)]+1}$$

$$\simeq D \int_0^\infty E \exp[-\beta(E-\mu)]dE$$

$$= D \exp(\beta\mu) \int_0^\infty E \exp(-\beta E)dE$$

$$= D(k_B T)^2 \exp(\beta\mu) \int_0^\infty x \exp(-x)dx$$

$$= D(k_B T)^2 \exp(\beta\mu)\Gamma(2)$$

$$= D(k_B T)^2 \exp(\beta\mu) = Nk_B T \tag{16.232}$$

where one has replaced in the last equality $N = Dk_BT \exp(\beta\mu)$ given by (16.230). This result is that of a bi-dimensional classical ideal gas.

e) One now calculates μ exactly. One has

$$N = D \int_0^\infty \frac{dE}{\exp[\beta(E-\mu)]+1}$$

$$= Dk_BT \int_{-\beta\mu}^\infty \frac{dx}{\exp(x)+1}$$

Putting $u = e^{-x} \to du = -e^{-x}dx = -udx$, one has

$$N = -Dk_BT \int_{\exp(\beta\mu)}^0 \frac{du}{u(u^{-1}+1)}$$

$$= -Dk_BT \int_{\exp(\beta\mu)}^0 \frac{du}{u+1}$$

$$= -Dk_BT \ln(1+u)|_{\exp(\beta\mu)}^0$$

$$\frac{N}{D} = k_BT \ln[1+\exp(\beta\mu)]$$

$$\mu_0 = k_BT \ln[1+\exp(\beta\mu)] \tag{16.233}$$

This formula connects μ and μ_0. The last equation can be rewritten as

$$\exp(\mu_0/k_BT) = 1 + \exp(\beta\mu)$$

$$\exp(\beta\mu) = \exp(\mu_0/k_BT) - 1$$

$$\mu = k_BT \ln[\exp(\mu_0/k_BT) - 1] \tag{16.234}$$

Problem 5. Pressure of a free Fermi gas:

One recalls the following expressions of a free Fermi gas:

$$N = \sum_i \frac{1}{e^{\beta(E_i-\mu)}+1},$$

$$J = -k_BT \ln \mathcal{Z}, \quad p = -\frac{\partial J}{\partial V}$$

so that [see (5.6) for J]

$$pV = -J = k_BT \sum_i \ln(1 + e^{-\beta(E_i-\mu)}).$$

At very high temperatures, these equations give the results of a classical ideal gas. One considers the moderate temperature regime where the condition $e^{-\beta(E-\mu)} < 1$ is fulfilled. One can then make the following expansion with the notation (fugacity) $\lambda = \exp(\beta\mu)$:

$$pV = k_B T \sum_i \ln(1 + e^{-\beta(E_i - \mu)})$$

$$\simeq k_B T \sum_i \sum_{n=1}^{\infty} (-1)^{n-1} \frac{\lambda^n}{n} e^{-n\beta E_i}$$

$$= k_B T \sum_{n=1}^{\infty} (-1)^{n-1} \frac{\lambda^n}{n} \sum_i e^{-n\beta E_i} \qquad (16.235)$$

$$N = \sum_i \frac{1}{e^{\beta(E_i - \mu)} + 1}$$

$$\simeq \sum_i \sum_{n=1}^{\infty} (-1)^{n-1} \lambda^n e^{-n\beta E_i}$$

$$= \sum_{n=1}^{\infty} (-1)^{n-1} \lambda^n \sum_i e^{-n\beta E_i} \qquad (16.236)$$

where $\lambda = \exp(\beta\mu)$. By comparing term by term with the same n, one sees that the sum on n in pV is stronger than that in N (attention at the alternate signs of the series). Thus,

$$pV = k_B T \sum_{n=1}^{\infty} (-1)^{n-1} \frac{\lambda^n}{n} \sum_i e^{-n\beta E_i}$$

$$> k_B T \sum_{n=1}^{\infty} (-1)^{n-1} \lambda^n \sum_i e^{-n\beta E_i} = k_B T N = p_c V$$

The pressure of a Fermi gas (left-hand side) is thus always stronger than that of a classical gas (p_c of the right-hand side).

Problem 6. Free Fermi gas with internal degrees of freedom:

a) One has

$$N = \sum_i \frac{1}{e^{\beta(E_i - \mu)} + 1} + \sum_i \frac{1}{e^{\beta(E_i + \epsilon - \mu)} + 1}$$

$$= \int_0^{\infty} \rho(E) dE \frac{1}{e^{\beta(E - \mu)} + 1}$$

$$+ \int_0^{\infty} \rho(E) dE \frac{1}{e^{\beta(E + \epsilon - \mu)} + 1} \qquad (16.237)$$

where $\rho(E) = (2s + 1)AE^{1/2}$ with A given in (2.36). One can rewrite this expression as

$$N = (2s + 1)A \int_0^\infty E^{1/2}dE \frac{1}{e^{\beta(E-\mu)} + 1}$$

$$+ (2s + 1)A \int_0^\infty E^{1/2}dE \frac{1}{e^{\beta(E+\epsilon-\mu)} + 1}$$

$$= (2s + 1)A \int_0^\infty E^{1/2}dE \frac{1}{e^{\beta(E-\mu)} + 1}$$

$$+ (2s + 1)A \int_\epsilon^\infty (E' - \epsilon)^{1/2}dE' \frac{1}{e^{\beta(E'-\mu)} + 1} \quad (16.238)$$

b) At $T = 0$, one has

$$N = (2s + 1)A \int_0^{\mu_0} E^{1/2}dE + (2s + 1)A \int_\epsilon^{\mu_0} (E' - \epsilon)^{1/2}dE'$$

$$= \frac{2}{3}(2s + 1)A \left[\mu_0^{3/2} + (\mu_0 - \epsilon)^{3/2} \right]$$

$$\simeq \frac{2}{3}(2s + 1)A\mu_0^{3/2} \left[1 + 1 - \frac{3}{2}(\frac{\epsilon}{\mu_0}) \right]$$

$$= \frac{4}{3}(2s + 1)A\mu_0^{3/2} \left[1 - \frac{3}{4}(\frac{\epsilon}{\mu_0}) \right] \quad (16.239)$$

where one has used the condition $\epsilon \ll \mu_0$. One solves this equation to determine μ_0:

$$\mu_0^{3/2} = C \left[1 + \frac{3}{4}(\frac{\epsilon}{\mu_0}) \right]$$

$$\mu_0 \simeq C^{2/3} \left[1 + \frac{2}{3} \times \frac{3}{4}(\frac{\epsilon}{\mu_0}) \right]$$

$$\mu_0^2 = C^{2/3} \left[\mu_0 + \frac{1}{2}\epsilon \right] \quad (16.240)$$

where $C = \frac{3N}{4(2s+1)A}$. This is an equation of second degree. The positive solution is

$$\mu_0 \simeq C^{2/3} + \frac{\epsilon}{2} \quad (16.241)$$

Replacing A in C by (5.20), one sees that the first term of the solution is E_F given by (5.21).

c) The energy at $T = 0$:

$$\overline{E}_0 = (2s + 1)A \int_0^{\mu_0} E^{3/2}dE + (2s + 1)A \int_\epsilon^{\mu_0} (E' - \epsilon)^{1/2}E'dE'$$

$$\simeq \frac{4}{5}(2s + 1)A\mu_0^{5/2} \left[1 - \frac{5}{12}\frac{\epsilon}{\mu_0} \right] \quad (16.242)$$

where details of the second integral have been omitted.

Problem 7. Ultra relativistic ideal gas:

One has $E = cp$ where p is the modulus of $\vec{p}$ and c the light velocity.

a) The sum on individual states can be decomposed as (see chapter 2, for example)

$$\sum_\lambda ... = \sum_{spin} \sum_p ... = (2S+1)\frac{V}{h^3}\int_0^\infty d\vec{p} ... \quad (S = 1/2, \text{ spin amplitude}).$$

b) • the number of particles N:

$$N = (2S+1)\frac{V}{h^3}\int_0^\infty d\vec{p}\frac{1}{e^{\beta(E-\mu)}+1}$$

$$= (2S+1)\frac{V}{h^3}4\pi\int_0^\infty p^2 dp\frac{1}{e^{\beta(E-\mu)}+1}$$

$$= \frac{8\pi V}{h^3 c^3}\int_0^\infty E^2 dE\frac{1}{e^{\beta(E-\mu)}+1} \quad (16.243)$$

• the average energy:

$\overline{E}$ is calculated in the same manner:

$$\overline{E} = (2S+1)\frac{V}{h^3}4\pi\int_0^\infty Ep^2 dp\frac{1}{e^{\beta(E-\mu)}+1}$$

$$= \frac{8\pi V}{h^3 c^3}\int_0^\infty E^3 dE\frac{1}{e^{\beta(E-\mu)}+1} \quad (16.244)$$

• the grand potential J:

$$J = -k_B T\sum_\lambda \ln\left(1 - \frac{1}{e^{\beta(E_\lambda-\mu)}+1}\right)$$

$$= -k_B T(2S+1)\frac{V}{h^3}4\pi\int_0^\infty p^2 dp\ln\left(1 - \frac{1}{e^{\beta(E-\mu)}+1}\right)$$

$$= -k_B T\frac{8\pi V}{h^3 c^3}\int_0^\infty E^2 dE\ln\left(1 - \frac{1}{e^{\beta(E-\mu)}+1}\right) \quad (16.245)$$

Integrating by parts yields

$$J = -k_B T\frac{8\pi V}{h^3 c^3}[\frac{1}{3}E^3\ln\left(1 - \frac{1}{e^{\beta(E-\mu)}+1}\right)\Big|_0^\infty$$

$$-\frac{1}{3}\int_0^\infty E^3\frac{\frac{-\beta e^{\beta(E-\mu)}}{(e^{\beta(E-\mu)}+1)^2}}{(1 - \frac{1}{e^{\beta(E-\mu)}+1})}dE]$$

$$= -k_B T\frac{8\pi V}{h^3 c^3}\left[0 + \beta\frac{1}{3}\int_0^\infty E^3\frac{1}{e^{\beta(E-\mu)}+1}dE\right]$$

$$= -\frac{8\pi V}{h^3 c^3}\frac{1}{3}\int_0^\infty E^3\frac{1}{e^{\beta(E-\mu)}+1}dE$$

$$= -\frac{1}{3}\overline{E} \quad (16.246)$$

from which one gets $a = -1/3$.

Remark: In the non relativistic case, one finds $a = -2/3$ (see Problem 1).

c) At $T = 0$, one replaces the Fermi-Dirac distribution in (16.243) and (16.244) by 1, and the upper integral bound by μ_0. One has

$$N = \frac{8\pi V}{h^3 c^3} \int_0^{\mu_0} E^2 dE = \frac{8\pi V}{3h^3 c^3} \mu_0^3$$

$$\mu_0 = hc \left(\frac{3n}{8\pi} \right)^{1/3} \tag{16.247}$$

$$\overline{E}_0 = \frac{8\pi V}{h^3 c^3} \int_0^{\mu_0} E^3 dE$$

$$= \frac{8\pi V}{4h^3 c^3} \mu_0^4 \tag{16.248}$$

The pressure is

$$P_0 = -\frac{\partial J_0}{\partial V} = \frac{\overline{E}_0}{3V} \tag{16.249}$$

d) At low temperatures, $|E - \mu| \ll k_B T$. One can use the Sommerfeld's expansion (Appendix B) for (16.243) and (16.244). One finds

$$\mu = \mu_0 \left[1 - \frac{1}{3} \left(\frac{\pi k_B T}{\mu_0} \right)^2 + \ldots \right]$$

$$\overline{E} = \overline{E}_0 \left[1 + \frac{2}{3} \left(\frac{\pi k_B T}{\mu_0} \right)^2 + \ldots \right] \tag{16.250}$$

The entropy of the gas at low temperatures is

$$S = -\frac{\partial J_0}{\partial T} = \frac{1}{3V} \frac{\partial \overline{E}_0}{\partial T}$$

$$= \frac{4}{9V} \overline{E}_0 \left(\frac{\pi k_B}{\mu_0} \right)^2 T \tag{16.251}$$

Problem 8. Electrons in Sodium:

a) The mass per volume unit of Na is $m_v = 0.97 \text{ gcm}^{-3}$ and the mass per mole is $m_m = 22.99 \text{ g/mole}$.

One has in three dimensions $E_F = \frac{\hbar^2}{2m}(3\pi^2 n)^{2/3}$. One calculates n as follows: the mass per atom is $m = \frac{m_m}{N_A} = \frac{22.99}{6.023 \times 10^{23}}$ g, so that the density of atoms is

$$n_a = \frac{m_v}{m} = \frac{0.97 \times 6.023 \times 10^{23}}{22.99} = 2.7 \times 10^{22} \text{ atoms per cm}^3.$$

Each atom Na gives one conducting electron so that the electron density n is equal to the density of atoms n_a. On has by replacing n in E_F: $E_F = 3.34$ eV.

The number of electrons occupying the energy interval $[E_F - 0.5eV, E_F]$ is

$$N_e = \int_{E_F - 0.5eV}^{E_F} \rho(E)dE$$

$$= \frac{1}{2\pi^2}\left(\frac{2m}{\hbar^2}\right)^{3/2}\int_{E_F-0.5eV}^{E_F} E^{1/2}dE$$

$$= 5.8 \times 10^{21} \text{ per cm}^3.$$

b) The Fermi temperature T_F :
One has $k_B T_F = E_F \rightarrow T_F = 3.34$ eV$/k_B$.

Problem 9. Pauli paramagnetism:
The Zeeman energy of an electron in an applied magnetic field $\vec{B}$ along the z axis: $E_\uparrow = E - \mu_B B$ and $E_\downarrow = E + \mu_B B$. The magnetic moment is $M = \mu_B(N_\uparrow - N_\downarrow)$ which is calculated as follows:

$$M = \mu_B \int_{\mu_B B}^{\infty} dE \frac{\rho(E)}{e^{\beta(E - \mu_B B - \mu)} + 1}$$

$$-\mu_B \int_{-\mu_B B}^{\infty} dE \frac{\rho(E)}{e^{\beta(E + \mu_B B - \mu)} + 1}$$

$$= \mu_B \int_0^{\infty} dE \left[\rho(E + \mu_B B) - \rho(E - \mu_B B)\right] f(E, T, \mu)$$

where one has changed the variables $E \rightarrow E \pm \mu_B B$. For weak fields, one can make the following expansion: $\rho(E \pm \mu_B B) \simeq \rho(E) \pm \mu_B B \left[\rho'(E)\right]_E$. One has

$$M \simeq 2\mu_B^2 B \int_0^{\infty} dE \rho'(E) f(E, T, \mu).$$

One finds the susceptibility

$$\chi = \frac{dM}{dB} \simeq 2\mu_B^2 \int_0^{\infty} dE \rho'(E) f(E, T, \mu) \qquad (16.252)$$

One can use the Sommerfeld's expansion (Appendix B) for this integral:

$$\chi \simeq 2\mu_B^2 \left[\int_0^\mu dE\rho'(E) + \frac{\pi^2}{6}\rho''(\mu)(k_BT)^2 \right]$$

$$= 2\mu_B^2 \left[\rho(\mu) - A\frac{\pi^2}{48\mu^{3/2}}(k_BT)^2 \right]$$

where one has used $\rho(E) = AE^{1/2}$ of (2.36). Using (5.28), one obtains

$$\chi = 2\mu_B^2\rho(E_F) \left[1 - \frac{\pi^2}{12}\left(\frac{k_BT}{E_F}\right)^2 \right] \tag{16.253}$$

The first term is independent of T as found in (5.44). The second term depends on T^2.

At high T, $f \simeq e^{-\beta(E-\mu)}$, (16.252) becomes

$$\chi \simeq 2\mu_B^2 \int_0^\infty dE\rho'(E)e^{-\beta(E-\mu)}$$

$$= 2\mu_B^2 \left[\rho'(E)e^{-\beta(E-\mu)}|_0^\infty + \beta \int_0^\infty dE\rho(E)e^{-\beta(E-\mu)} \right]$$

$$= 2\mu_B^2\beta\,[0 + N/2] = \frac{N\mu_B^2}{k_BT} \tag{16.254}$$

where N is the total number of electrons. This result is known as the "Curie law".

Remark: One has used (2.36), without the factor 2, for each kind of spin.

Problem 10. Electrons in Copper:

a) The velocity of electrons at the Fermi level as a function of electron density n is

$$\frac{1}{2}mv_F^2 = E_F = \frac{\hbar^2}{2m}(3\pi^2n)^{2/3}.$$

Numerical application:For Cu, $n = 8.45 \times 10^{22}$ cm^{-3}, $m = 9.1095 \times 10^{-28}$ g, $\hbar = 1.05459 \times 10^{-27}$ erg-s. One has $v_F = 1.57 \times 10^8$ cm/s.

b) The average kinetic energy per electron at 0 K in Rydberg (1 Ryd=$\frac{me^4}{2\hbar^2}$) is

$$E_c = 2.21/r_s^2 \text{ Rydberg}$$

where $r_s = \frac{r_0}{a_H}$ with $a_H = \frac{\hbar^2}{me^2}$ (Bohr radius) and r_0 is the radius of the sphere occupied by an electron in the gas [see the unit conversion at (10.22)].

16.6 Solutions of problems of chapter 6

Problem 1. Gas of photons:

The energy of a photon of frequency ν_i is $E_i = h\nu_i$. The energy of a gas of photons is written as $E(n_0, n_1, ...) = \sum_i n_i h\nu_i$ where n_i is the number of photons of frequency ν_i. Each microstate of the gas is defined by a set $(n_0, n_1, ...)$. The partition function of the gas is

$$Z = \sum_{n_0=0}^{\infty} \sum_{n_1=0}^{\infty} ...e^{-\beta n_0 h\nu_0}e^{-\beta n_1 h\nu_1}...$$

$$= \frac{1}{1 - e^{-\beta h\nu_0}} \frac{1}{1 - e^{-\beta h\nu_1}}... = \prod_i \frac{1}{1 - e^{-\beta h\nu_i}} \quad (16.255)$$

The average number of photon ν_i is

$$\bar{n}_i = \sum_{n_i=0}^{\infty} \frac{n_i e^{-n_i \beta h\nu_i}}{Z} = -\frac{1}{h\beta} \frac{\partial Z}{\partial \nu_i} \frac{1}{Z}$$

$$= -\frac{1}{h\beta} \frac{\partial \ln Z}{\partial \nu_i} = \frac{1}{e^{\beta h\nu_i} - 1} \quad (16.256)$$

This is the Bose-Einstein distribution with a zero chemical potential.

Problem 2. Bose-Einstein condensation:

The energy of a quantum particle in a box is given by (2.34):

$$E_k = E(n_x, n_y, n_z) = \frac{\hbar^2 k^2}{2m} = \frac{2\pi^2 \hbar^2}{mL^2}(n_x^2 + n_y^2 + n_z^2)$$

where $n_i = \pm 1, \pm 2, ...$ $(i = x, y, z)$.

a) The energy of the first excited state of a Helium-4 atom:
For $L = 1$ cm and $m \simeq 4 \times 1.67 \times 10^{-27}$ kg (He-4 mass), the distance between the first two levels is

$$\Delta E = \frac{\hbar^2}{2m} 3 \left(\frac{\pi}{L}\right)^2 \simeq 10^{-18} \text{eV} = 10^{-14} \text{K} \qquad (16.257)$$

b) For an ideal gas of He-4 of density 10^{22} per cm^3, the chemical potential at $T = 1$ mK is calculated by the formula (6.24). One has

$$\mu/k_B = -T/N = -10^{-3}/10^{22} = -10^{-25} \text{K} \qquad (16.258)$$

The number of atoms occupying the first level is (taking $\mu \simeq 0$):

$$n = \frac{1}{\exp[(E_1 - \mu)/k_B T] - 1} \simeq k_B T/E_1 \simeq 10^{11} \qquad (16.259)$$

The fraction of He-4 atoms of the first energy level is thus $n/N \simeq 10^{-11}$. If one uses the Boltzmann distribution, one has $n = N \exp(-E_1/k_B T) \rightarrow n/N \simeq 1 - E_1/k_B T \simeq 1 - 10^{-11} \simeq 1$. This is not true experimentally.

Problem 3. Pressure in a gas of bosons:

a) At high temperatures, the expression (6.49) gives the equation of state of a classical ideal gas: it suffices to replace $\overline{E}$ by its limit at high temperatures (see section 6.4). One has then $pV = Nk_B T$.

b) One recalls the following expressions for a gas of bosons:

$$N = \sum_i \frac{1}{e^{\beta(E_i - \mu)} - 1}, \quad J = -k_B T \ln \mathcal{Z},$$

$$p = -\frac{\partial J}{\partial V} \rightarrow pV = -J = -k_B T \sum_i \ln(1 - e^{-\beta(E_i - \mu)}) \quad [\text{see } (6.6)].$$

These equations yield at high temperatures the results of a classical ideal gas. Now, since $\mu \leq 0$, one has $e^{-\beta(E-\mu)} < 1$ at any T. Thus, one can make the following expansion, using the notation (fugacity) $\lambda = \exp(\beta\mu)$,

$$pV = -k_B T \sum_i \ln(1 - e^{-\beta(E_i - \mu)}) \simeq k_B T \sum_i \sum_{n=1}^{\infty} \frac{\lambda^n}{n} e^{-n\beta E_i}$$

$$= k_B T \sum_{n=1}^{\infty} \frac{\lambda^n}{n} \sum_i e^{-n\beta E_i} \tag{16.260}$$

$$N = \sum_i \frac{1}{e^{\beta(E_i - \mu)} - 1} \simeq \sum_i \sum_{n=1}^{\infty} \lambda^n e^{-n\beta E_i}$$

$$= \sum_{n=1}^{\infty} \lambda^n \sum_i e^{-n\beta E_i} \tag{16.261}$$

One sees that the sum on n in pV is smaller than that in N (by comparing terms of the same n in the sums). Thus

$$pV = k_B T \sum_{n=1}^{\infty} \frac{\lambda^n}{n} \sum_i e^{-n\beta E_i}$$

$$< k_B T \sum_{n=1}^{\infty} \lambda^n \sum_i e^{-n\beta E_i} = k_B T N = p_c V \tag{16.262}$$

The pressure in a gas of bosons (p of the left-hand side) is always weaker than the pressure in a classical ideal gas (p_c of the right-hand side).

Problem 4. Two-dimensional gas of bosons:
One has

$$N = \sum_i \frac{1}{e^{\beta(E_i - \mu)} - 1} = \sum_i \frac{e^{-\beta(E_i - \mu)}}{1 - e^{-\beta(E_i - \mu)}}$$

$$= \sum_i e^{-\beta(E_i - \mu)} \sum_{m=0}^{\infty} e^{-m\beta(E_i - \mu)} = \sum_i \sum_{m=0}^{\infty} e^{-(m+1)\beta(E_i - \mu)}$$

$$= \sum_i \sum_{n=1}^{\infty} e^{-n\beta(E_i - \mu)} = \sum_{n=1}^{\infty} e^{n\beta\mu} \sum_i e^{-n\beta E_i}$$

$$= \sum_{n=1}^{\infty} e^{n\beta\mu} \int_0^{\infty} D dE e^{-n\beta E} = D \sum_{n=1}^{\infty} e^{n\beta\mu} (-\frac{1}{n\beta}) e^{-n\beta E} \Big|_0^{\infty}$$

$$= D k_B T \sum_{n=1}^{\infty} \frac{e^{n\beta\mu}}{n} \tag{16.263}$$

where one has replaced the sum on i by an integral on E allowed by the thermodynamic limit. The density of states is a constant D in two dimensions. The constant A is thus $A = Dk_B$.

The Bose-Einstein condensation requires that μ is very small with the order of $1/N \simeq 0$ (see lecture). This is not the case with (16.263): one sees that if μ is zero, the series does not converge. On the other hand, one can always find a value of μ which satisfies this relation at any temperature. One concludes that there is no Bose-Einstein condensation in two dimensions.

Problem 5. Gas of bosons having internal degrees of freedom: Bose-Einstein condensation

One supposes that each particle, in addition to its kinetic energy, possesses an internal degree of freedom taking the values $\epsilon_0 = 0$ or ϵ_1 ($\epsilon_1 > 0$).

a) The number of particles at temperatures higher than the Bose temperature is given by

$$N_e = \sum_i \frac{1}{e^{\beta(E_i + \epsilon_0 - \mu)} - 1} + \sum_i \frac{1}{e^{\beta(E_i + \epsilon_1 - \mu)}}$$

$$= (2S + 1) \int_0^\infty dE\rho(E) \frac{1}{\phi^{-1}e^{\beta E} - 1}$$

$$+ (2S + 1) \int_0^\infty dE\rho(E) \frac{1}{\psi^{-1}e^{\beta E} - 1} \qquad (16.264)$$

where $\phi = \exp(\beta\mu)$ and $\psi = \exp[\beta(\mu - \epsilon_1)]$. The sums have been replaced by integrals. Following the method used after Eq. (6.32), one arrives at

$$N_e = (2S + 1)A(k_B T)^{3/2} \frac{\sqrt{\pi}}{2} [F(\phi) + F(\psi)] \qquad (16.265)$$

where

$$A = \frac{V}{4\pi^2} \left(\frac{2m}{\hbar^2}\right)^{3/2} \qquad (16.266)$$

and

$$F(x) = \sum_{m=1}^\infty \frac{x^m}{m^{3/2}} \qquad (16.267)$$

b) The Bose temperature T_B is defined as the temperature at which $N_e = N$ and $\mu \simeq 0$, namely $\phi = 1$ and $\psi = \psi_0 = \exp(-\epsilon_1/k_B T_B)$. One has

$$N = (2S+1)A(k_B T_B)^{3/2}\frac{\sqrt{\pi}}{2}[F(1) + F(\psi_0)] \tag{16.268}$$

Explicitly,

$$F(1) = \sum_{m=1}^{\infty}\frac{1}{m^{3/2}} \simeq 2.612 \tag{16.269}$$

$$F(\psi_0) = \sum_{m=1}^{\infty}\frac{\exp(-m\epsilon_1/k_B T_B)}{m^{3/2}} \tag{16.270}$$

One calculates T_B as a function of T_B^0: If one takes into account only the first term of the series (16.270), one has

$$N \simeq (2S+1)A(k_B T_B)^{3/2}\frac{\sqrt{\pi}}{2}F(1)\left[1 + \frac{\exp(-\epsilon_1/k_B T_B)}{F(1)}\right] \tag{16.271}$$

hence,

$$\left(\frac{T_B^0}{T_B}\right)^{3/2} = 1 + \frac{\exp(-\epsilon_1/k_B T_B)}{2.612} \tag{16.272}$$

One sees that $T_B < T_B^0$.

c) If $\epsilon_1 \gg k_B T_B$, one neglects $F[\exp(-\epsilon_1/k_B T_B)]$ in (16.268). One has from (16.271) and (16.272)

$$N = 2.612\frac{\sqrt{\pi}}{2}(2S+1)A(k_B T_B^0)^{3/2}$$

$$T_B = T_B^0$$

The first equation is the equation (6.21).

Problem 6. Einstein's model for the vibration of atoms on a lattice: One uses the results (16.83) and (16.85) of Problem 6 of chapter 3. At high temperatures, $e^{\beta\hbar\omega_E} \simeq 1 - \beta\hbar\omega_E$, one has

$$C_V \simeq 3Nk_B \tag{16.273}$$

This result is that obtained for N classical harmonic oscillators given by the theorem of energy equipartition.

At low temperatures, one has

$$C_V \simeq \frac{3N}{k_B T^2}(\hbar\omega_E)^2 e^{-\beta\hbar\omega_E} \tag{16.274}$$

Thus, C_V tends to 0 exponentially as T tends to 0, in disagreement with experiments which show that C_V tends to 0 as T^3.

Problem 7. Wien displacement law:

One demonstrates the formula (6.61):

To determine the wave length at which $u(\lambda, T)$ is maximum, one looks for the solution of $du(\lambda, T)/d\lambda = 0$. One has

$$\frac{du(\lambda, T)}{d\lambda} = \frac{du(\nu, T)}{d\nu}\frac{d\nu}{d\lambda}$$

$$= \frac{8\pi hV}{c^3}[3\nu^2(\exp(\beta h\nu) - 1) - \beta h \exp(\beta h\nu)\nu^3]$$

$$\times \frac{1}{[\exp(\beta h\nu) - 1]^2} \times (-\frac{c}{\lambda^2}) \qquad (16.275)$$

At high temperatures, $\exp(\beta h\nu) \gg 1$. One has thus

$$\frac{du(\lambda, T)}{d\lambda} = 0 \rightarrow 3\exp(\beta h\nu) - \beta h \exp(\beta h\nu)\nu = 0 \qquad (16.276)$$

from which

$$\nu_{max} = \frac{3\exp(\beta h\nu)}{\beta h \exp(\beta h\nu)} = \frac{3}{\beta h} \qquad (16.277)$$

One gets

$$\lambda_{max} = c/\nu_{max} = \frac{c\beta h}{3} \qquad (16.278)$$

One finds thus the Wien maximum displacement law:

$$\lambda_{max}T = \text{constante} = \frac{ch}{3k_B} \qquad (16.279)$$

where λ_{max} is the wave length giving the maximum emitted energy.

Problem 8. Equation of state of a gas of photons:

One demonstrates the formula (6.62): $p = \frac{1}{3}\frac{\overline{E}}{V}$.

The energy of a photon is $E_i = h\nu_i = hc/\lambda_i = \hbar ck_i$ where k_i is the modulus of the wave vector $\vec{k}_i$, c the light velocity and λ_i the wave length. The gas energy is $E = \sum_i E_i$. The sum on photon individual states i is performed over the photon polarization states and over the wave vectors (see chapter 2):

$$\sum_k ... = 2\frac{V}{(2\pi)^3}\int_0^\infty d\vec{k}...$$

The factor 2 is the two-fold degeneracy of the photon polarization.

The number of particles N is

$$
\begin{aligned}
N &= 2\frac{V}{(2\pi)^3}\int_0^\infty d\vec{k}\,\frac{1}{e^{\beta E}-1} \\
&= 2\frac{V}{(2\pi)^3}4\pi\int_0^\infty k^2 dk\,\frac{1}{e^{\beta E}-1} \\
&= \frac{8\pi V}{(2\pi\hbar c)^3}\int_0^\infty E^2 dE\,\frac{1}{e^{\beta E}-1}
\end{aligned}
\tag{16.280}
$$

The average energy $\overline{E}$ is

$$
\begin{aligned}
\overline{E} &= 2\frac{V}{(2\pi)^3}4\pi\int_0^\infty Ek^2 dk\,\frac{1}{e^{\beta E}-1} \\
&= \frac{8\pi V}{(2\pi\hbar c)^3}\int_0^\infty E^3 dE\,\frac{1}{e^{\beta E}-1}
\end{aligned}
\tag{16.281}
$$

The grand potential J is calculated from (6.5):

$$
\begin{aligned}
J &= -k_B T\sum_l \ln(1+\frac{1}{e^{\beta E_l}-1}) \\
&= -k_B T2\frac{V}{(2\pi)^3}4\pi\int_0^\infty k^2 dk\,\ln\left(1+\frac{1}{e^{\beta E}-1}\right) \\
&= -k_B T\frac{8\pi V}{(2\pi\hbar c)^3}\int_0^\infty E^2 dE\,\ln\left(1+\frac{1}{e^{\beta E}-1}\right)
\end{aligned}
\tag{16.282}
$$

Integrating by parts gives

$$
\begin{aligned}
J &= -k_B T\frac{8\pi V}{(2\pi\hbar c)^3}[\frac{1}{3}E^3\ln\left(1+\frac{1}{e^{\beta E}-1}\right)\Big|_0^\infty \\
&\quad -\frac{1}{3}\int_0^\infty E^3\,\frac{\frac{-\beta e^{\beta E}}{(e^{\beta E}-1)^2}}{(1+\frac{1}{e^{\beta E}-1})}dE] \\
&= -k_B T\frac{8\pi V}{(2\pi\hbar c)^3}[0+\beta\frac{1}{3}\int_0^\infty E^3\,\frac{1}{e^{\beta E}-1}dE] \\
&= -\frac{8\pi V}{(2\pi\hbar c)^3}\frac{1}{3}\int_0^\infty E^3\,\frac{1}{e^{\beta E}-1}dE \\
&= -\frac{1}{3}\overline{E}
\end{aligned}
\tag{16.283}
$$

Remark: The same expression has been demonstrated for the relativistic fermion case in Problem 7 of chapter 5.

16.7   Solutions of problems of chapter 7

Problem 1. Exercise of commutation and anticommutation relations:
One has

$$B^+_{\vec{k}\sigma} = b^+_{\vec{k}\sigma} b^+_{-\vec{k}-\sigma}$$
$$B_{\vec{k}\sigma} = b_{-\vec{k}-\sigma} b_{\vec{k}\sigma}$$

where $b^+_{\vec{k}\sigma}$ and $b_{\vec{k}\sigma}$ are fermion creation and annihilation operators of state $(\vec{k}, \sigma)$.

a) $B^+_{\vec{k}\sigma}$ and $B_{\vec{k}\sigma}$ are operators creating and annihilating a pair of fermions of states $(\vec{k}, \sigma)$ and $(-\vec{k}, -\sigma)$.

b) One has

$$\left[B_{\vec{k}\sigma}, B^+_{\vec{k}'\sigma}\right] = [b_{-\vec{k}-\sigma} b_{\vec{k}\sigma}, b^+_{\vec{k}'\sigma} b^+_{-\vec{k}'-\sigma}]$$

$$= b_{-\vec{k}-\sigma} b_{\vec{k}\sigma} b^+_{\vec{k}'\sigma} b^+_{-\vec{k}'-\sigma} - b^+_{\vec{k}'\sigma} b^+_{-\vec{k}'-\sigma} b_{-\vec{k}-\sigma} b_{\vec{k}\sigma}$$

$$= b_{-\vec{k}-\sigma}[\delta(\vec{k}, \vec{k}') - b^+_{\vec{k}'\sigma} b_{\vec{k}\sigma}]b^+_{-\vec{k}'-\sigma} - b^+_{\vec{k}'\sigma}[\delta(\vec{k}, \vec{k}') - b_{-\vec{k}-\sigma} b^+_{-\vec{k}'-\sigma}]b_{\vec{k}\sigma}$$

$$= b_{-\vec{k}-\sigma} b^+_{-\vec{k}'-\sigma} \delta(\vec{k}, \vec{k}') - b_{-\vec{k}-\sigma} b^+_{\vec{k}'\sigma} b_{\vec{k}\sigma} b^+_{-\vec{k}'-\sigma} - b^+_{\vec{k}'\sigma} b_{\vec{k}\sigma} \delta(\vec{k}, \vec{k}') + b^+_{\vec{k}'\sigma} b_{-\vec{k}-\sigma} b^+_{-\vec{k}'-\sigma} b_{\vec{k}\sigma}$$

$$= [\delta(\vec{k}, \vec{k}') - n_{-\vec{k}-\sigma}\delta(\vec{k}, \vec{k}')] - [-b^+_{\vec{k}'\sigma} b_{-\vec{k}-\sigma}]$$
$$\times [-b^+_{-\vec{k}'-\sigma} b_{\vec{k}\sigma}]$$
$$- n_{\vec{k}\sigma}\delta(\vec{k}, \vec{k}') + b^+_{\vec{k}'\sigma} b_{-\vec{k}-\sigma} b^+_{-\vec{k}'-\sigma} b_{\vec{k}\sigma}$$

$$= (1 - n_{\vec{k}\sigma} - n_{-\vec{k}-\sigma})\delta(\vec{k}, \vec{k}')$$

where $n_{\vec{k}\sigma} = b^+_{\vec{k}\sigma} b_{\vec{k}\sigma}$.
Next,

$$[B_{\vec{k}\sigma}, B_{\vec{k}'\sigma}] = b_{-\vec{k}-\sigma} b_{\vec{k}\sigma} b_{-\vec{k}'-\sigma} b_{\vec{k}'\sigma} - b_{-\vec{k}'-\sigma} b_{\vec{k}'\sigma} b_{-\vec{k}-\sigma} b_{\vec{k}\sigma}$$
$$= 0$$

where the last line is obtained by moving in the first term the third operator to the first position and the fourth operator to the second position: in doing so one sees that the two terms cancel out.
For the anticommutation relation, one writes

$$[B_{\vec{k}\sigma}, B_{\vec{k}'\sigma}]_+ = b_{-\vec{k}-\sigma} b_{\vec{k}\sigma} b_{-\vec{k}'-\sigma} b_{\vec{k}'\sigma} + b_{-\vec{k}'-\sigma} b_{\vec{k}'\sigma} b_{-\vec{k}-\sigma} b_{\vec{k}\sigma} \tag{16.284}$$

If $\vec{k} \neq \vec{k}'$, then moving in the second term the third and fourth operators to the first and second position one sees that the two terms are identical so that

$$\left[B_{\vec{k}\sigma}, B_{\vec{k}'\sigma}\right]_+ = 2b_{-\vec{k}-\sigma}b_{\vec{k}\sigma}b_{-\vec{k}'-\sigma}b_{\vec{k}'\sigma} = 2B_{\vec{k}\sigma}B_{\vec{k}'\sigma} \qquad (16.285)$$

If $\vec{k} = \vec{k}'$ then

$$\begin{aligned}
\left[B_{\vec{k}\sigma}, B_{\vec{k}'\sigma}\right]_+ &= b_{-\vec{k}-\sigma}b_{\vec{k}\sigma}b_{-\vec{k}-\sigma}b_{\vec{k}\sigma} + b_{-\vec{k}-\sigma}b_{\vec{k}\sigma}b_{-\vec{k}-\sigma}b_{\vec{k}\sigma} \\
&= b_{-\vec{k}-\sigma}[-b_{-\vec{k}-\sigma}b_{\vec{k}\sigma}]b_{\vec{k}\sigma} + b_{-\vec{k}-\sigma}[-b_{-\vec{k}-\sigma}b_{\vec{k}\sigma}]b_{\vec{k}\sigma} \\
&= 0 \qquad (16.286)
\end{aligned}$$

because $b_{\vec{k}\sigma}b_{\vec{k}\sigma}|\rangle = 0$ for any fermion state (one particle at most). Combining the two cases, one has

$$\left[B_{\vec{k}\sigma}, B_{\vec{k}'\sigma}\right]_+ = 2B_{\vec{k}\sigma}B_{\vec{k}'\sigma}\left[1 - \delta(\vec{k}, \vec{k}')\right] \qquad (16.287)$$

Problem 2. One shows that $[\hat{\mathcal{H}}, \hat{N}]=0$ where $\hat{N}$ is the occupation number field operator defined in (7.70) and $\hat{\mathcal{H}}$ the Hamiltonian in the second quantization (7.69):
One has

$$\begin{aligned}
\hat{\mathcal{H}} = &\sum_\sigma \int d\vec{r}\,\hat{\Psi}_\sigma^+(\vec{r})H(\vec{r})\hat{\Psi}_\sigma(\vec{r}) \\
&- \frac{1}{2}\sum_{\sigma\sigma'} \int \int d\vec{r}_1 d\vec{r}_2\,\hat{\Psi}_\sigma^+(\vec{r}_1)\hat{\Psi}_{\sigma'}^+(\vec{r}_2)V(\vec{r}_1, \vec{r}_2)\hat{\Psi}_\sigma(\vec{r}_1)\hat{\Psi}_{\sigma'}(\vec{r}_2)
\end{aligned}$$

and

$$\hat{N} = \sum_\beta \int d\vec{r}_3\,\hat{\Psi}_\beta^+(\vec{r}_3)\hat{\Psi}_\beta(\vec{r}_3) \qquad (16.288)$$

To calculate $[\hat{\mathcal{H}}, \hat{N}]$, one decomposes the operators in the commutators as follows

$$[AB, C] = A[B, C] - [C, A]B \qquad (16.289)$$
$$[AB, C] = A[B, C]_+ - [A, C]_+ B \qquad (16.290)$$

for bosons and fermions, respectively. For $[\hat{\mathcal{H}}, \hat{N}]$, one should decompose several times. For example, with the kinetic term of $\hat{\mathcal{H}}$, one puts

$A = \hat{\Psi}_\sigma^+(\vec{r})$, $B = H(\vec{r})\hat{\Psi}_\sigma(\vec{r})$, $C = \hat{\Psi}_\beta^+(\vec{r}_3)$, $D = \hat{\Psi}_\beta(\vec{r}_3)$. In the boson case, one has

$$[AB, CD] = A[B, CD] - [CD, A]B$$
$$= A[B, C]D - AC[D, B] - C[D, A]B + [A, C]DB$$

One uses now the commutation relations between field operators A, B, C, and D to get the desired results.

Problem 3. One shows that $\hat{\Psi}(\vec{r})\hat{N} = (\hat{N} + 1)\hat{\Psi}(\vec{r})$ for both boson and fermion cases:

$$\hat{\Psi}(\vec{r})\hat{N} = \hat{\Psi}(\vec{r}) \int d\vec{r}' \hat{\Psi}^+(\vec{r}')\hat{\Psi}(\vec{r}')$$

$$= \int d\vec{r}' [\delta(\vec{r} - \vec{r}') \pm \hat{\Psi}^+(\vec{r}')\hat{\Psi}(\vec{r})]\hat{\Psi}(\vec{r}')$$

$$= \hat{\Psi}(\vec{r}) \pm \int d\vec{r}' \hat{\Psi}^+(\vec{r}')\hat{\Psi}(\vec{r})\hat{\Psi}(\vec{r}')$$

$$= \hat{\Psi}(\vec{r}) + \int d\vec{r}' \hat{\Psi}^+(\vec{r}')\hat{\Psi}(\vec{r}')\hat{\Psi}(\vec{r})$$

$$= (1 + \hat{N})\hat{\Psi}(\vec{r}) \qquad (16.291)$$

where the signs $\pm$ correspond to the boson and fermion cases, respectively. The last line is valid for two cases.

Problem 4. One shows that $\hat{\Psi}^+(\vec{r})|\text{vac} >$ ("vac" stands for vacuum) is a state in which there is a particle localized at $\vec{r}$.

Let $\rho(\vec{r}) = \hat{\Psi}^+(\vec{r})\hat{\Psi}(\vec{r})$ be the density operator. We have

$$\rho(\vec{r})\hat{\Psi}^+(\vec{r}')|\text{vac} > = \int d\vec{r}'' \hat{\Psi}^+(\vec{r}'')\hat{\Psi}(\vec{r}'')\delta(\vec{r} - \vec{r}'')\hat{\Psi}^+(\vec{r}')|\text{vac} >$$

$$= \int d\vec{r}'' \hat{\Psi}^+(\vec{r}'')\delta(\vec{r} - \vec{r}'')[\delta(\vec{r}'' - \vec{r}')$$
$$- \hat{\Psi}^+(\vec{r}')\hat{\Psi}(\vec{r}'')]|\text{vac} >$$

$$= \int d\vec{r}'' \hat{\Psi}^+(\vec{r}'')\delta(\vec{r} - \vec{r}'')\delta(\vec{r}'' - \vec{r}')|\text{vac} >$$

$$= \delta(\vec{r} - \vec{r}')\hat{\Psi}^+(\vec{r}')|\text{vac} > \qquad (16.292)$$

The last equality means that $\delta(\vec{r} - \vec{r}')$ is the eigenvalue of $\rho(\vec{r})$ corresponding to the eigenvector $\Psi^+(\vec{r}')|\text{vac} >$. There is thus one particle at $\vec{r}$ in the state $\Psi^+(\vec{r})|\text{vac} >$.

Problem 5. Boson Hamiltonian:

The general Hamiltonian (7.38) is

$$\hat{\mathcal{H}} = \sum_{p,r}\langle r|H|p\rangle a_r^+ a_p + \frac{1}{2}\sum_{pqrs}\langle rs|V|pq\rangle a_r^+ a_s^+ a_p a_q \qquad (16.293)$$

where

$$\langle r|H|p\rangle = \int d\vec{r}\,\phi_r^*(\vec{r})\left[\frac{1}{2m}(-i\hbar\nabla)^2\right]\phi_p(\vec{r})$$

$$\langle rs|V|pq\rangle = \int\int d\vec{r}d\vec{r}'\,\phi_r^*(\vec{r})\phi_s^*(\vec{r}')V(\vec{r},\vec{r}')\phi_p(\vec{r})\phi_q(\vec{r}')$$

One calculates the kinetic term, using plane waves $\frac{1}{\sqrt{\Omega}}\exp(i\vec{k}\cdot\vec{r})$ (Ω: volume):

$$\langle\vec{k}_1|H|\vec{k}_2\rangle = \frac{1}{\Omega}\int d\vec{r}\,e^{-i(\vec{k}_1-\vec{k}_2)\cdot\vec{r}}\frac{\hbar^2 k_2^2}{2m}$$

$$= \frac{1}{\Omega}\delta(\vec{k}_1 - \vec{k}_2)\frac{\hbar^2 k_2^2}{2m} \qquad (16.294)$$

where $\delta(...)$ is the Kronecker symbol.

The potential term is, putting $\vec{r} = \vec{r}_1 - \vec{r}_2$ on the way,

$$\langle\vec{k}_1\vec{k}_2|V|\vec{k}_3\vec{k}_4\rangle = \frac{1}{\Omega^2}\int e^{-i(\vec{k}_1-\vec{k}_3)\cdot\vec{r}_1}V(\vec{r}_1 - \vec{r}_2)e^{i(\vec{k}_4-\vec{k}_2)\cdot\vec{r}_2}d\vec{r}_1 d\vec{r}_2$$

$$= \frac{1}{\Omega^2}\int e^{-i(\vec{k}_1-\vec{k}_3)\cdot\vec{r}}e^{-i(\vec{k}_1-\vec{k}_3)\cdot\vec{r}_2}V(\vec{r})e^{i(\vec{k}_4-\vec{k}_2)\cdot\vec{r}_2}$$
$$\times d\vec{r}d\vec{r}_2$$

$$= \frac{1}{\Omega^2}\int e^{-i(\vec{k}_1-\vec{k}_3)\cdot\vec{r}}V(\vec{r})e^{-i(\vec{k}_1+\vec{k}_2-\vec{k}_3-\vec{k}_4)\cdot\vec{r}_2}d\vec{r}d\vec{r}_2$$

$$= \frac{1}{\Omega^2}\Omega\delta(\vec{k}_1 + \vec{k}_2 - \vec{k}_3 - \vec{k}_4)\int V(\vec{r})e^{-i(\vec{k}_1-\vec{k}_3)\cdot\vec{r}}d\vec{r}$$

$$= \frac{1}{\Omega}\delta(\vec{k}_1 + \vec{k}_2 - \vec{k}_3 - \vec{k}_4)\mathcal{V}(\vec{k}_1 - \vec{k}_3) \qquad (16.295)$$

where $\mathcal{V}(\vec{k}_1 - \vec{k}_3)$ is the Fourier transform of $V(\vec{r})$. One obtains thus

$$\hat{\mathcal{H}} = \sum_{\vec{k}}\epsilon_{\vec{k}}a_{\vec{k}}^+ a_{\vec{k}}$$

$$+ \frac{1}{2\Omega}\sum_{\vec{k}_1,\vec{k}_2,\vec{k}_3,\vec{k}_4}\mathcal{V}(\vec{k}_1 - \vec{k}_3)a_{\vec{k}_1}^+ a_{\vec{k}_2}^+ a_{\vec{k}_3} a_{\vec{k}_4}\delta(\vec{k}_1 + \vec{k}_2 - \vec{k}_3 - \vec{k}_4)$$

where $\epsilon_{\vec{k}} = \frac{\hbar^2 k^2}{2m}$ is the kinetic energy of the state $\vec{k}$, a and a^+ are annihilation and creation operators, respectively. This Hamiltonian is used to study phonons excited in a gas of Helium-4 in chapter 9.

Problem 6. Gas of interacting bosons:

a) In the ground state of a boson system, all momenta are zero: $\vec{k} = 0$. The kinetic energy is thus zero. One calculates now the first contribution from the potential. One has

$$H_1 = \frac{1}{2} \sum_{\vec{k}_1, \vec{k}_2, \vec{k}_3, \vec{k}_4} V_{\vec{k}_1 \vec{k}_2, \vec{k}_3 \vec{k}_4} a^+_{\vec{k}_1} a^+_{\vec{k}_2} a_{\vec{k}_3} a_{\vec{k}_4} \qquad (16.296)$$

where, using the result of the previous exercise (Problem 5),

$$V_{\vec{k}_1 \vec{k}_2, \vec{k}_3 \vec{k}_4} = \langle \vec{k}_1 \vec{k}_2 | V(\vec{r}_1, \vec{r}_2) | \vec{k}_3 \vec{k}_4 \rangle$$

$$= \frac{1}{\Omega} \mathcal{V}(\vec{k}_1 - \vec{k}_3) \delta(\vec{k}_1 + \vec{k}_2 - \vec{k}_3 - \vec{k}_4)$$

The average value in the ground state $|f_0^N\rangle$ of N particles is

$$E_1 = \langle f_0^N | H_1 | f_0^N \rangle$$

$$= \frac{1}{2\Omega} \sum_{\vec{k}_1, \vec{k}_2, \vec{k}_3, \vec{k}_4} \mathcal{V}(\vec{k}_1 - \vec{k}_3) \delta(\vec{k}_1 + \vec{k}_2 - \vec{k}_3 - \vec{k}_4)$$

$$\times \langle f_0^N | a^+_{\vec{k}_1} a^+_{\vec{k}_2} a_{\vec{k}_3} a_{\vec{k}_4} | f_0^N \rangle$$

$$= \frac{1}{2\Omega} \sum_{\vec{k}_1, \vec{k}_2, \vec{k}_3, \vec{k}_4} \mathcal{V}(\vec{k}_1 - \vec{k}_3)$$

$$\times \langle f_0^N | a^+_{\vec{k}_1} [-\delta(\vec{k}_2, \vec{k}_3) + a_{\vec{k}_3} a^+_{\vec{k}_2}] a_{\vec{k}_4} | f_0^N \rangle$$

The $\langle f_0^N | ... | f_0^N \rangle$ is not zero if $\vec{k}_1 = \vec{k}_4$ and $\vec{k}_2 = \vec{k}_3$ Note that in the ground state all $\vec{k}_1 = ... = \vec{k}_4 = 0$, so that this condition is fulfilled. One has

$$\langle f_0^N | a^+_{\vec{k}_1} [-\delta(\vec{k}_2, \vec{k}_3) + a_{\vec{k}_3} a^+_{\vec{k}_2}] a_{\vec{k}_4} | f_0^N \rangle = -n_{\vec{k}_1} + n_{\vec{k}_1} n_{\vec{k}_2} \quad (16.297)$$

from which the sums on $\vec{k}_1, ..., \vec{k}_4$ give

$$E_1 = \langle H_1 \rangle = \frac{1}{2\Omega} \mathcal{V}(0)[-N + N^2]$$

$$\frac{E_1}{N} = \frac{1}{2\Omega} \mathcal{V}(0)(N - 1)$$

b) The next order of the energy is given by the perturbation theory

$$E_2 = \sum_{j \neq 0} \frac{|\langle j|H_1|f_0\rangle|^2}{E_0 - E_j} \tag{16.298}$$

where $\langle j|$ is the excited state of energy E_j and $|f_0\rangle$ is the ground state of energy E_0 $(= 0)$ (no interaction ground state). Let us examine $\langle j|H_1|f_0\rangle$:

$$\langle j|H_1|f_0\rangle = \frac{1}{2\Omega} \sum_{\vec{k}_1,\vec{k}_2,\vec{k}_3,\vec{k}_4} \mathcal{V}(\vec{k}_1 - \vec{k}_3)\delta(\vec{k}_1 + \vec{k}_2 - \vec{k}_3 - \vec{k}_4)$$

$$\times \langle j|a^+_{\vec{k}_1} a^+_{\vec{k}_2} a_{\vec{k}_3} a_{\vec{k}_4}|f_0\rangle$$

$$= \frac{1}{2\Omega} \sum_{\vec{k},\vec{p},\vec{q}\neq 0} \mathcal{V}(\vec{q})\langle j|a^+_{\vec{k}+\vec{q}} a^+_{\vec{k}-\vec{q}} a_{\vec{k}} a_{\vec{p}}|f_0\rangle$$

where one has put $\vec{k} = \vec{k}_3$, $\vec{p} = \vec{k}_4$, $\vec{k}_1 = \vec{k} + \vec{q}$, and $\vec{k}_2 = \vec{k} - \vec{q}$. The condition $\delta(\vec{k}_1 + \vec{k}_2 - \vec{k}_3 - \vec{k}_4)$ is thus fulfilled. Note that in the ground state all wave vectors are zero. Since $a_{\vec{p}}|f_0\rangle = \sqrt{N}|f_0^{N-1}\rangle$, one has

$$a_{\vec{k}} a_{\vec{p}}|f_0\rangle = \sqrt{N-1}\sqrt{N}|f_0^{N-2}\rangle$$

Operating $a^+_{\vec{k}+\vec{q}} a^+_{\vec{k}-\vec{q}}$ on $\sqrt{N-1}\sqrt{N}|f_0^{N-2}\rangle$ one has

$$a^+_{\vec{k}+\vec{q}} a^+_{\vec{k}-\vec{q}}\sqrt{N-1}\sqrt{N}|f_0^{N-2}\rangle = \sqrt{N-1}\sqrt{N}|-\vec{q},\vec{q},(N-2)\times 0\rangle$$

$$= \sqrt{N-1}\sqrt{N}|j'\rangle$$

where $\vec{q}$ should be non zero. In order that the above $\langle j|H_1|j'\rangle$ is not zero, one should take the excited state $|j\rangle$ equal to $|j'\rangle$. Such a choice leads to

$$\langle j'|a^+_{\vec{q}} a^+_{-\vec{q}} a_{\vec{k}} a_{\vec{p}}|j'\rangle = \sqrt{(N-1)N}$$

Since

$$E_0 - E_j = \epsilon_k + \epsilon_p - \epsilon_{p-q} - \epsilon_{k+q}$$

$$= 0 + 0 - \frac{\hbar^2(-q)^2}{2m} - \frac{\hbar^2 q^2}{2m} = -\frac{\hbar^2 q^2}{m}$$

one has

$$E_2 = \frac{1}{2} \sum_{\vec{q} \neq 0} \left[\frac{\mathcal{V}(\vec{q})}{\Omega} \sqrt{N(N-1)} \right]^2 \left[-\frac{\hbar^2 q^2}{m} \right]^{-1}$$

$$= -\frac{1}{2} \frac{\Omega}{(2\pi)^3} \left[\frac{1}{\Omega} \right]^2 N(N-1) \int \frac{|\mathcal{V}(\vec{q})|^2}{\hbar^2 q^2/m} d\vec{q}$$

$$\frac{E_2}{N} = -\frac{N-1}{2\Omega} \frac{1}{(2\pi)^3} \int \frac{|\mathcal{V}(\vec{q})|^2}{\hbar^2 q^2/m} d\vec{q}$$

Problem 7. Method of diagonalization of Hamiltonian in second quantization:

The Hamiltonian is

$$\hat{\mathcal{H}} = \sum_k H_k$$

$$H_k = [A_k(a_k^+ a_k + a_{-k}^+ a_{-k}) + B_k(a_k a_{-k} + a_{-k}^+ a_k^+)]$$

The transformation

$$c_k = u_k a_k - v_k a_{-k}^+$$
$$c_k^+ = u_k a_k^+ - v_k a_{-k}$$

a) One has

$$[c_k, c_{k'}^+] = [u_k a_k - v_k a_{-k}^+, u_{k'} a_{k'}^+ - v_{k'} a_{-k'}]$$
$$= u_k u_{k'}[a_k, a_{k'}^+] + v_k v_{k'}[a_{-k}^+, a_{-k'}]$$
$$\quad - u_k v_{k'}[a_k, a_{-k'}] - v_k u_{k'}[a_{-k'}^+, a_{k'}^+]$$
$$= u_k u_{k'} \delta_{k,k'} + v_k v_{k'}(-\delta_{k,k'}) - 0 - 0$$
$$= (u_k^2 - v_k^2)\delta_{k,k'} = \delta_{k,k'} \tag{16.299}$$

b)

$$[c_k^+, H_k] = [u_k a_k^+ - v_k a_{-k}, A_k(a_k^+ a_k + a_{-k}^+ a_{-k}) + B_k(a_k a_{-k} + a_{-k}^+ a_k^+)]$$

Using (16.289) in Problem 2 to decompose the chain of operators, then following the same procedure as in the previous question, one obtains

$$[c_k^+, H_k] = u_k[-A_k a_k^+] - B_k a_{-k} - v_k[A_k a_{-k} + B_k a_k^+)] \tag{16.300}$$

Using $[c_k^+, H_k] = -\lambda_k c_k^+$ one has

$$u_k[-A_k a_k^+ - B_k a_{-k}] - v_k[A_k a_{-k} + B_k a_k^+)] = -\lambda[u_k a_k^+ - v_k a_{-k}]$$
$$\tag{16.301}$$

Identifying term by term of the two sides of the above equation, one gets

$$A_k u_k + B_k v_k = \lambda_k u_k \rightarrow (A_k - \lambda_k) u_k + B_k v_k = 0$$
$$B_k u_k + A_k v_k = -\lambda_k v_k \rightarrow B_k u_k + (A_k + \lambda_k) v_k = 0$$

Non trivial solutions imply

$$(A_k - \lambda_k)(A_k + \lambda_k) - B_k^2 = 0$$

One finds

$$\lambda_k^2 = A_k^2 - B_k^2 \tag{16.302}$$

In summary, using a transformation to operators c_k and c_k^+, the original Hamiltonian becomes diagonal $H_k = \lambda_k c_k^+ c_k$ with λ_k is given by (16.302).

Solutions of Problems of Part 2

17.1 Solutions of problems of chapter 8

Problem 1. Reciprocal lattice of a triangular lattice:

The basis vector of the reciprocal lattice are $(\vec{b}_1, \vec{b}_2)$ where

$$\vec{b}_j \cdot \vec{a}_i = 2\pi\delta_{ij}$$

with

$$\vec{a}_1 = a\vec{e}_1, \quad \text{and} \quad \vec{a}_2 = \frac{a}{2}\vec{e}_1 + \frac{a\sqrt{3}}{2}\vec{e}_2.$$

One obtains $\vec{b}_1 \perp \vec{a}_2$, from which $\vec{b}_1 \cdot \vec{a}_1 = b_1 a \cos(\pi/6) = 2\pi$. One has thus $b_1 = \frac{4\pi}{a\sqrt{3}}$.

Similarly, $\vec{b}_2 \perp \vec{a}_1$ so that

$$\vec{b}_2 \cdot \vec{a}_2 = b_2 a_2 \cos(\pi/6) = 2\pi.$$

One has then $b_2 = \frac{4\pi}{a\sqrt{3}}$.

See Fig. 17.1 for a graphic presentation of the reciprocal lattice with the first Brillouin zone.

Problem 2. The honeycomb lattice is not a Bravais lattice because there are two types of site with different environments: sites A and sites B (see Fig. 17.2 for examples of non Bravais lattices). To construct the reciprocal lattice, one has to divide the lattice into two sublattice, one contains A sites and the other B sites. Each sublattice is a Bravais lattice of triangular structure with lattice constant $a\sqrt{3}$, distance between two nearest A (or B) sites. The reciprocal lattice of each triangular sublattice is constructed as described in the preceding problem.

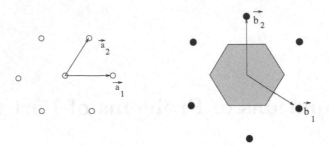

Fig. 17.1   Reciprocal lattice (right) of the triangular lattice (left). The gray area indicates the first Brillouin zone.

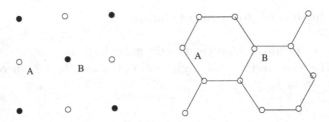

Fig. 17.2   Examples of non Bravais lattices. Left: square lattice with two types of atom. Right: honeycomb lattice with A-sites and B-sites.

Problem 3. Face-centered cubic lattice — Centered cubic lattice:

(1) Reciprocal lattice of a face-centered cubic lattice:

The face-centered cubic lattice has the basis vectors given by (8.5)-(8.7). Using (8.14)-(8.16), one obtains the basis vectors of the reciprocal lattice:

$\vec{b}_1 = \frac{2\pi}{a}(\vec{e}_1 + \vec{e}_2 - \vec{e}_3)$

$\vec{b}_2 = \frac{2\pi}{a}(-\vec{e}_1 + \vec{e}_2 + \vec{e}_3)$

$\vec{b}_3 = \frac{2\pi}{a}(\vec{e}_1 - \vec{e}_2 + \vec{e}_3)$

One recognizes that these vectors generate a body-centered cubic lattice by comparison with the definition (8.2)-(8.4).

(2) Reciprocal lattice of a body-centered cubic lattice:

The body-centered cubic lattice has the basis vectors given by (8.2)-(8.4). Using (8.14)-(8.16), one obtains the following basis vectors of the reciprocal lattice:

$\vec{b}_1 = \frac{2\pi}{a}(\vec{e}_1 + \vec{e}_2)$

$\vec{b}_2 = \frac{2\pi}{a}(\vec{e}_2 + \vec{e}_3)$

$\vec{b}_3 = \frac{2\pi}{a}(\vec{e}_3 + \vec{e}_1)$

By comparison with (8.5)-(8.7) one sees that the above vectors generate a face-centered cubic lattice.

Problem 4. Chain of two types of atom:

A chain of two types of atom is not a Bravais lattice. One divides it into two sublattices, one with A atoms, the other with B atoms. Each sublattice has a lattice constant $2a$. The reciprocal lattice is a chain of constant $2\pi/(2a) = \pi/a$, twice shorter than the constant of the reciprocal lattice of a mono-atomic chain. The first Brillouin zone is limited at $\pm\frac{\pi}{2a}$. The periodic boundary condition applies for a chain (sublattice) of $N/2$ atoms, so that there are $N/2$ states in the first Brillouin zone.

Problem 5. Periodic potential:

See solution of Problem 2 of chapter 11.

Problem 6. Fourier transform of the Coulomb potential $\frac{1}{r}$:

$$V(\vec{r}) = \int_{ZB} d\vec{k}\, e^{-i(\vec{k})\cdot\vec{r}} A(\vec{k})$$

where $A(\vec{k})$ is the Fourier component of $V(\vec{r})$:

$$A(\vec{k}) = \int_{\Omega\to\infty} d\vec{r}\, e^{i(\vec{k})\cdot\vec{r}} V(\vec{r})$$

$$= \int_{\Omega\to\infty} \frac{d\vec{r}\, e^{i\vec{k}\cdot\vec{r}}}{|\vec{r}|}$$

Using the spherical coordinates, one writes

$$A(\vec{k}) = 2\pi \int_0^\infty r^2 dr \int_0^\pi \sin\theta d\theta \frac{e^{ikr\cos\theta}}{r}$$

$$= 2\pi\lim_{\mu\to 0} \int_0^\infty e^{-\mu r} r\, dr \int_1^{-1} d(-\cos\theta) e^{ikr\cos\theta}$$

$$= \frac{2\pi}{ik}\lim_{\mu\to 0} \int_0^\infty dr[e^{(-\mu+ik)r} - e^{(-\mu-ik)r}]$$

$$= \frac{2\pi}{ik}\lim_{\mu\to 0}\left[\frac{-1}{-\mu+ik} + \frac{-1}{\mu+ik}\right]$$

$$= \lim_{\mu\to 0}\frac{4\pi}{k^2+\mu^2}$$

$$= \frac{4\pi}{k^2} \tag{17.1}$$

Problem 7. Fourier analysis:

One has

$$q_r = \frac{1}{N^{1/2}} \sum_k Q_k e^{ikr}$$

where $k = \frac{2\pi n}{Na}$. One writes

$$\sum_r q_r e^{-ik'r} = \frac{1}{N^{1/2}} \sum_r \sum_k Q_k e^{i(k-k')r}$$

$$= \frac{1}{N^{1/2}} \sum_k Q_k \sum_r e^{i(k-k')r} = \frac{1}{N^{1/2}} \sum_k Q_k N \delta_{k,k'}$$

$$= N^{1/2} Q_{k'} \tag{17.2}$$

from which

$$Q_k = \frac{1}{N^{1/2}} \sum_r q_r e^{-ikr}.$$

Problem 8. Fourier analysis of function in continuous real space:

One has

$$q(x) = \frac{1}{L^{1/2}} \sum_k Q_k e^{ikx}$$

where $k = \frac{2\pi n}{L}$. One writes

$$\int_{L/2}^{L/2} dx q(x) e^{-ik'x} = \frac{1}{L^{1/2}} \sum_k Q_k \int_{L/2}^{L/2} dx e^{i(k-k')x}$$

$$= \frac{1}{L^{1/2}} \sum_k Q_k \int_{L/2}^{L/2} dx e^{i(k-k')x}$$

$$= \frac{1}{L^{1/2}} \sum_k Q_k \frac{1}{i(k-k')} \left[2\sin(k-k')L/2\right]$$

$$= \frac{1}{L^{1/2}} \sum_k Q_k \frac{1}{i(k-k')} \left[2\sin\pi(n-n')\right]$$

$$= 0 \quad \text{if } n \neq n' \tag{17.3}$$

If $k = k'$, then

$$\int_{L/2}^{L/2} dx q(x) e^{-ik'x} = \frac{1}{L^{1/2}} \sum_k Q_k \int_{L/2}^{L/2} dx e^{i(k-k')x} = \frac{1}{L^{1/2}} Q_{k'} L \tag{17.4}$$

from which one obtains

$$Q_k = \frac{1}{L^{1/2}} \int_{L/2}^{L/2} dx q(x) e^{-ikx}.$$

Problem 9. Structure factors:

One has

$$\gamma_{\vec{k}} = \frac{1}{Z} \sum_{\vec{R}} e^{i\vec{k}\cdot\vec{R}}$$

where

- Simple cubic lattice $Z = 6$: $\vec{R} = \pm a\vec{e}_1, \pm a\vec{e}_2, \pm a\vec{e}_3$, from which one gets

$$\gamma_{\vec{k}} = \frac{1}{3}\left[\cos k_x a + \cos k_y a + \cos k_z a\right]$$

- Body-centered cubic lattice: see (12.36)
- Face-centered cubic lattice: see (12.37)
- Triangular lattice $Z = 6$:
  $\vec{R} = \pm a\vec{e}_1, (\pm a\vec{e}_1 \pm a\frac{\sqrt{3}}{2}\vec{e}_2)$, from which,

$$\gamma_{\vec{k}} = \frac{1}{3}\left[\cos k_x a + 2\cos k_x a \cos(k_y a\sqrt{3}/2)\right].$$

17.2 Solutions of problems of chapter 9

Problem 1. Demonstration of (9.36):

One has

$$u_i(t) = \frac{1}{\sqrt{N}} \sum_k X_k(t) e^{ikR_i}$$

$$p_i(t) = \frac{1}{\sqrt{N}} \sum_k P_k(t) e^{ikR_i}$$

and

$$X_k(t) = \frac{1}{\sqrt{N}} \sum_i u_i(t) e^{-ikR_i}$$

$$P_k(t) = \frac{1}{\sqrt{N}} \sum_i p_i(t) e^{-ikR_i}$$

One writes

$$\sum_i p_i^2 = \frac{1}{N} \sum_i \sum_k \sum_{k'} P_{k'}^* P_k e^{-i(k'-k)R_i}$$

$$= \frac{1}{N} \sum_k \sum_{k'} P_{k'}^* P_k \sum_i e^{-i(k'-k)R_i}$$

$$= \frac{1}{N} \sum_k \sum_{k'} P_{k'}^* P_k N \delta_{k,k'} = \sum_k P_k^* P_k$$

and

$$\sum_{i,j}(u_i - u_j)^2 = \sum_{i,j}[u_i^2 + u_j^2 - 2u_iu_j]$$

$$= \frac{1}{N}\sum_{i,j}[\sum_k\sum_{k'}X_{k'}^*X_ke^{-i(k'-k)R_i}$$

$$+\sum_k\sum_{k'}X_{k'}^*X_ke^{-i(k'-k)R_j}$$

$$-2\sum_k\sum_{k'}X_{k'}^*X_ke^{-i(k'R_i-kR_j)}]$$

$$= 2\sum_k\sum_k{}'X_{k'}^*X_k\delta_{k,k'}\sum_{j\in NN}1$$

$$-\frac{2}{N}\sum_k\sum_{k'}\sum_{i,j}X_{k'}^*X_ke^{-i(k'-k)R_i}e^{ik(R_j-R_i)}$$

$$= 4\sum_kX_k^*X_k - \frac{2}{N}\sum_k\sum_{k'}\sum_iX_{k'}^*X_ke^{-i(k'-k)R_i}$$

$$\times\sum_{i-j}e^{ik(R_j-R_i)}$$

$$= 4\sum_kX_k^*X_k - \frac{2}{N}\sum_k\sum_{k'}X_{k'}^*X_kN\delta_{k,k'}$$

$$\times[e^{ika} + e^{-ika}]$$

$$= 4\sum_kX_k^*X_k - 4\sum_kX_k^*X_k\cos ka$$

$$= 4\sum_kX_k^*X_k[1 - \cos ka]$$

$$= 4\sum_kX_k^*X_k\sin^2(ka/2) \tag{17.5}$$

where one has used the sum rules in Appendix 8, NN stands for nearest neighbors of a site. One obtains by replacing $\sin^2(ka/2)$ as a function of de $\omega(k)^2$ (dispersion relation),

$$E = \sum_k\frac{P_k^+P_k}{2m} + \sum_k\frac{m\omega(k)^2}{2}X_k^+X_k \tag{17.6}$$

$$= \sum_kE_k \tag{17.7}$$

where

$$E_k = \frac{P_k^*P_k}{2m} + \frac{m\omega(k)^2}{2}X_k^*X_k \tag{17.8}$$

Problem 2. Chain of two types of atoms:

a) The first Brillouin zone is limited by $\pm \frac{\pi}{2a}$ (see Appendix 8)

b) The atoms of mass m occupy the even sites and those of mass M the odd sites of the chain. One has the equations of motion

$$m\ddot{u}_{2i} = -K(2u_{2i} - u_{2i+1} - u_{2i-1})$$
$$M\ddot{u}_{2i+1} = -K(2u_{2i+1} - u_{2i+2} - u_{2i})$$

The solutions are

$$u_{2i} = Ae^{i(k2ra - \omega t)}$$
$$u_{2i+1} = Be^{i[k(2r+1)a - \omega t]}$$

Replacing these into the above equations of motion one has

$$(2K - m\omega^2)A - 2K\cos ka B = 0$$
$$-2K\cos ka A + (2K - M\omega^2)B = 0$$

The non trivial solutions correspond to

$$\omega_k^2 = K(\frac{1}{m} + \frac{1}{M}) \pm K\left[\frac{1}{m} + \frac{1}{M} - \frac{4\sin^2 ka}{mM}\right]^{1/2} \tag{17.9}$$

The periodic boundary condition $e^{ik2ra} = e^{ik2(r+N)a}$ yields $k = \frac{2\pi n}{2Na} = \frac{\pi n}{Na}$.

As seen above, ω_k has two solutions: branch $-$ is an acoustic one because $\omega_k \to 0$ when $k \to 0$ and branch $+$ is an optical branch, as shown in Fig. 17.3. One observes the existence of a forbidden band at the borders $\pm\frac{\pi}{2a}$ of the Brillouin zone.

Replacing ω_k in the equations of motion, one finds

$$\frac{B}{A} = \frac{2K - m\omega_k^2}{2K\cos ka} = \frac{2K\cos ka}{2K - M\omega_k^2}.$$

c) Particular vibration modes:

- For the acoustic branch:
  - $k \to 0$, $\frac{B}{A} = 1$: uniform mode, sound velocity: $v_s = (\frac{2Ka^2}{M+m})^{1/2}$.
  - $k \to \pi/(2a)$, $\frac{B}{A} \to \infty$: light atoms do not move, heavy atoms vibrate in opposite phases (optical mode).

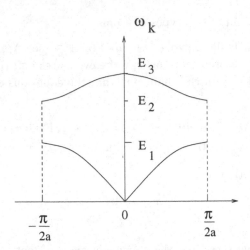

Fig. 17.3   Dispersion relation for a chain of two types of atoms with $m < M$: upper branch corresponds to optical modes, lower branch to acoustic ones. $E_1 = (\frac{2K}{M})^{1/2}$, $E_2 = (\frac{2K}{m})^{1/2}$ and $E_3 = [2K(\frac{1}{m} + \frac{1}{M})]^{1/2}$. There is a forbidden band between E_1 et E_2.

- For the optical branch:
  - $k \rightarrow 0$, $\omega_k = E_3$, $\frac{B}{A} = -\frac{m}{M} < 0$: neighboring atoms vibrate in opposite phases.
  - $k \rightarrow \pi/(2a)$, $\omega_k = E_2$, $\frac{B}{A} \rightarrow 0$: heavy atoms do not move, light atoms vibrate in optical mode.

Problem 3. Interaction between next nearest neighbors in a chain:

a) Interaction between next nearest neighbors does not change the periodicity of the lattice. Therefore, the reciprocal lattice does not change.

b) One writes the equation of motion for an atom:

$$M\ddot{u}_n = -K_1(2u_n - u_{n+1} - u_{n-1}) + -K_2(2u_n - u_{n+2} - u_{n-2}) \tag{17.10}$$

Taking a solution of the form

$$u_n = Ae^{i(kna - \omega t)} \tag{17.11}$$

one obtains

$$\omega^2 = \frac{2K_1}{M}[1 - \cos(ka)] + \frac{2K_2}{M}[1 - \cos(2ka)] \tag{17.12}$$

c) The dispersion relation is shown in Fig. 17.4. The effect of interaction between next nearest neighbors is strong near the Brillouin zone borders [see Fig. 9.2 of (9.8) for comparison]. The frequency maximum is not at the zone border.

d) When $k \to 0$: $\omega \propto k$, the sound velocity is $v_s = v_s^0[1 + 4\frac{K_2}{K_1}]^{1/2}$ where v_s^0 is the sound velocity in the absence of K_2 given by (9.12). The sound velocity is increased under the effect of K_2.

e) When $K_2/K_1 \to \infty$, the chain is decoupled into two "independent" interpenetrating chains of lattice constant $2a$. The second term in (17.12) dominates, making the frequency maximum at $k = \pi/(2a)$ which is the new border of the Brillouin zone of the chain of lattice constant $2a$. Note that the curve will go down to zero at π/a.

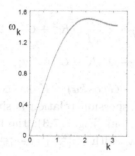

Fig. 17.4 Dispersion relation ω_k versus $k \in [0, \pi/a]$ (half of the Brillouin zone) for a chain with interactions K_1 (between nearest neighbors) and K_2 (between next nearest neighbors). Curve is plotted with $K_2/K_1 = 0.5$ and $a = 1$.

Problem 4. Chain of two types distances:

a) The periodicity of the lattice is $a = a_1 + a_2$. The unit cell contains two atoms. It is convenient to divide the lattice into two sublattices: the interatomic distance within each sublattice is a. The lattice constant of the reciprocal lattice is thus $2\pi/a$ and the first Brillouin zone is limited at $\pm\pi/a$.

b) Equations of motion: let $u_1(n)$ be the position of the atom of sublattice 1 at the n-th cell, and $u_2(n)$ that of the atom of sublattice 2 of the same cell. One has

$$M\ddot{u}_1(n) = -K[u_1(n) - u_2(n)] - G[u_1(n) - u_2(n-1)]$$

$$M\ddot{u}_2(n) = -K[u_2(n) - u_1(n)] - G[u_2(n) - u_1(n+1)]$$

where the interaction between two neighbors separated by a_1 is K and that between neighbors separated by a_2 is G. The solutions are

$$u_1(n) = Ae^{i(kna-\omega t)}$$

$$u_2(n) = Be^{i(kna-\omega t)}$$

Replacing in the above equations of motion one gets

$$(K + G - M\omega^2)A - (K + Ge^{-ika})B = 0$$
$$-(K + Ge^{ika})A + (K + G - M\omega^2)B = 0$$

c) The non trivial solutions correspond to

$$\omega_k^2 = \frac{K+G}{M} \pm \frac{1}{M}) \left[K^2 + G^2 + 2KG\cos ka\right]^{1/2}$$

$$= \frac{K+G}{M} \pm |K + G\cos ka| \qquad (17.13)$$

and $A/B = \pm(-K - G\cos ka)/(|K + G\cos ka|)$.

The figure of the dispersion relation is similar to that of a chain with two types of atom (Fig. 17.3): the forbidden band is at the zone borders with a width of $[\frac{2K}{M}]^{1/2} - [\frac{2G}{M}]^{1/2}$.

d) Particular modes:

- When $k \to 0$: on has an optical mode with

$$\omega_+ \simeq [\frac{K+G}{M}]^{1/2}, \quad \frac{A}{B} = -1,$$

and an acoustic mode with

$$\omega_- \simeq k, \quad \frac{A}{B} = 1.$$

- When $k \to \pi/a$: one has an optical mode with

$$\omega_+ \simeq [\frac{2K}{M}]^{1/2}, \quad \frac{A}{B} = -1,$$

and an acoustic mode with

$$\omega_+ \simeq [\frac{2G}{M}]^{1/2}, \quad \frac{A}{B} = 1.$$

Problem 5. Phonons in a rectangular lattice:

Guide: If the restoring force between two atoms is on the axis connecting their positions at equilibrium, then the equations of motion of u_x and u_y are independent from each other. Assuming the same constant K, one has

$$M\ddot{u}_x(i,j) = -K[2u_x(i,j) - u_x(i+1,j) - u_x(i-1,j)]$$

$$M\ddot{u}_y(i,j) = -K[2u_y(i,j) - u_y(i,j+1) - u_y(i,j-1)].$$

The solutions are

$$u_x(i,j) = Ae^{i(\vec{k}\cdot\vec{R}(i,j)-\omega t)}$$

$$u_y(i,j) = Be^{i(\vec{k}\cdot\vec{R}(i,j)-\omega t)}.$$

Replacing in the equations of motion, one gets

$$\omega^2 = 2\left(\frac{K}{M}\right)^{1/2}\left[1 - \cos(\vec{k}\cdot\vec{a}_1)\right]$$

and

$$\omega^2 = 2\left(\frac{K}{M}\right)^{1/2}\left[1 - \cos(\vec{k}\cdot\vec{a}_2)\right].$$

where $\vec{a}_1$ ($\parallel \vec{Ox}$) and $\vec{a}_2$ ($\parallel \vec{Oy}$) are the basis vectors of the rectangular lattice. These are frequencies of vibration modes propagating along $\vec{Ox}$ and $\vec{Oy}$, respectively.

Problem 6. Phonons in a simple cubic lattice:

One considers a simple cubic lattice of constant a where interactions are between nearest neighbors and between next nearest neighbors, K_1 and K_2, respectively.

One uses (9.21) to calculate the force which acts on atom i sitting at the origin of a Cartesian system of coordinates. One puts $x_{ij} = u_i^x - u_j^x$, $y_{ij} = u_i^y - u_j^y$, $z_{ij} = u_i^z - u_j^z$, $X_{ij} = R_i^x - R_j^x$, $Y_{ij} = R_i^y - R_j^y$, and $Z_{ij} = R_i^z - R_j^z$. From (9.22), one obtains

$$F_{ij}^x = -\frac{\partial U}{\partial x_{ij}}$$

$$= -\frac{K(\vec{R}_i - \vec{R}_j)}{(|\vec{R}_i - \vec{R}_j|)^2}[x_{ij}X_{ij} + y_{ij}Y_{ij} + z_{ij}Z_{ij}]X_{ij} \quad (17.14)$$

Similarly, one obtains the other components of $\vec{F}_{ij}$.

One observes that the 6 nearest neighbors are at positions $\vec{\rho}_{ij} = a(\pm 1, 0, 0), a(0, \pm 1, 0), a(0, 0, \pm 1)$, and the 12 next nearest neighbors are at $\vec{\rho}_{ij} = a(\pm 1, \pm 1, 0), a(0, \pm 1, \pm 1), a(\pm 1, 0, \pm 1)$. Using (17.14) for each pair of atoms one obtains the x component of the force on atom i:

$$
\begin{aligned}
F_i^x = &-K_1[(u_{000}^x(t) - u_{100}^x(t) + u_{000}^x(t) - u_{-100}^x(t)] \\
&-K_2[(u_{000}^x(t) - u_{110}^x(t) + u_{000}^y(t) - u_{110}^y(t)] \\
&-K_2[(u_{000}^x(t) - u_{1-10}^x(t) - u_{000}^y(t) + u_{1-10}^y(t)] \\
&-K_2[(u_{000}^x(t) - u_{-110}^x(t) - u_{000}^y(t) + u_{-110}^y(t)] \\
&-K_2[(u_{000}^x(t) - u_{-1-10}^x(t) + u_{000}^y(t) - u_{-1-10}^y(t)] \\
&-K_2[(u_{000}^x(t) - u_{101}^x(t) + u_{000}^z(t) - u_{101}^z(t)] \\
&-K_2[(u_{000}^x(t) - u_{10-1}^x(t) - u_{000}^z(t) + u_{10-1}^z(t)] \\
&-K_2[(u_{000}^x(t) - u_{-101}^x(t) - u_{000}^z(t) + u_{-101}^z(t)] \\
&-K_2[(u_{000}^x(t) - u_{-10-1}^x(t) + u_{000}^z(t) - u_{-10-1}^z(t)]
\end{aligned}
$$

where u_{-100}^x is the x component of $\vec{u}$ of the neighbor at position $a(-1, 0, 0)$ etc.

Taking a solution of the form

$$
\vec{u}_i = (u_i^x \hat{i} + u_i^y \hat{j} + u_i^z \hat{k}) e^{-i\vec{k}\cdot\vec{R}_i},
$$

one obtains

$$
\begin{aligned}
F_i^x = &-2K_1[1 - \cos(k_x a)]u_{000}^x \\
&- 4K_2[2 - \cos(k_x a)\cos(k_y a) - \cos(k_x a)\cos(k_z a)]u_{000}^x \\
&- 4K_2 \sin(k_x a)\sin(k_y a)u_{000}^y - 4K_2 \sin(k_x a)\sin(k_z a)u_{000}^z
\end{aligned}
$$

Using

$$
F_i^x = m\frac{d^2 u_{000}^x}{dt^2} = -m\omega^2 u_{000}^x
$$

one writes the equation of motion for the x component:

$$
\begin{aligned}
\omega^2 u_{000}^x = \{&\frac{2K_1}{m}[1 - \cos(k_x a)] + \frac{4K_2}{m}[2 - \cos(k_x a)\cos(k_y a) \\
&- \cos(k_x a)\cos(k_z a)]\}u_{000}^x \\
&+ \frac{4K_2}{m}\sin(k_x a)\sin(k_y a)u_{000}^y \\
&+ \frac{4K_2}{m}\sin(k_x a)\sin(k_z a)u_{000}^z
\end{aligned}
$$

One proceeds in the same manner to the other components of $\vec{F}_i$. The dynamic matrix D is thus

$$D = \begin{pmatrix} D_{xx} & D_{xy} & D_{xz} \\ D_{yx} & D_{yy} & D_{yz} \\ D_{zx} & D_{zy} & D_{zz} \end{pmatrix}$$

where

$$D_{xx} = \frac{2K_1}{m}[1 - \cos(k_x a)] + \frac{4K_2}{m}[2 - \cos(k_x a)\cos(k_y a)$$
$$- \cos(k_x a)\cos(k_z a)]$$

$$D_{xy} = \frac{4K_2}{m}\sin(k_x a)\sin(k_y a)$$

The other elements are obtained by permutation of indices. The diagonalization of this matrix for a given $\vec{k}$ gives the dispersion relations. For example, if $k_x = k$, $k_y = k_z = 0$, only diagonal elements of matrix D are non zero:

$$D_{xx} = \frac{2K_1}{m}[1 - \cos(ka)] + \frac{8K_2}{m}[1 - \cos(ka)]$$

$$D_{yy} = \frac{4K_2}{m}[1 - \cos(ka)]$$

$$D_{zz} = \frac{4K_2}{m}[1 - \cos(ka)]$$

Therefore, the dispersion relations are

$$\omega_1^2 = \frac{2K_1 + 8K_2}{m}[1 - \cos(ka)]$$

$$\omega_2^2 = \omega_3^2 = \frac{4K_2}{m}[1 - \cos(ka)]$$

ω_1 is the longitudinal mode, ω_2 and ω_3 are transversal degenerate modes.

Problem 7. Density of modes in a square lattice:

For a square lattice, one uses the result of the previous exercise putting $|\vec{a}_1| = |\vec{a}_2| \equiv a$:

$$\omega^2 = 2\left(\frac{K}{M}\right)^{1/2}\left[1 - \cos(\vec{k} \cdot \vec{a})\right]$$

For the density of modes, one follows (9.73) to write down for two dimensions:

$$g(\omega)d\omega = G(\vec{k})d\vec{k} = \frac{1}{\frac{(2\pi)^2}{L^2}}d\vec{k} = \frac{L^2}{(2\pi)^2}2\pi k dk$$

One finds

$$g(\omega) = \frac{L^2}{(2\pi)^2}\frac{2\pi k}{|\vec{\nabla}_{\vec{k}}\omega|}.$$

There is no difficulty to calculate $|\vec{\nabla}_{\vec{k}}\omega|$ using the dispersion relation given above. However, it is complicated to replace k as a function of ω. One takes the limit of small k for simplicity:

$$\omega = 2\left(\frac{K}{M}\right)^{1/2}|\sin(\vec{k}.\vec{a})/2| \simeq \left(\frac{K}{M}\right)^{1/2}ak \equiv v_s k$$

where v_s is the sound velocity. One has thus $k = \omega/v_s$. Replacing these relations in $g(\omega)$ one obtains

$$g(\omega) = \frac{L^2}{(2\pi)^2}\frac{2\pi\omega}{v_s^2} = \frac{L^2}{2\pi v_s^2}\omega.$$

Problem 8. Energy and specific heat at finite temperatures:
One has

$$\omega^2(\vec{k}) = 2C^2(3 - \cos k_x a - \cos k_y a - \cos k_z a) \qquad (17.15)$$

a) This is an acoustic branch because $w(\vec{k}) \to 0$ when $\vec{k} \to 0$:

$$\omega^2(\vec{k}) \to 2C^2\left[(k_x a)^2 + (k_y a)^2 + (k_z a)^2\right]/2 = C^2 a^2 k^2.$$

One sees that $w(\vec{k})$ does not depend on the direction at small k. The sound velocity is thus $v_s = \frac{d\omega}{dk} = Ca$.
b) See lecture.
c) For a branch in three dimensions, one has

$$g(\omega) = \frac{L^3}{(2\pi)^3}\frac{4\pi k^2}{|\vec{\nabla}_{\vec{k}}\omega|}.$$

In the Debye approximation, one replaces the real dispersion relation $w(\vec{k})$ by $\omega(\vec{k}) = v_s k$ as if k is always small. One has then

$$g(\omega) = g_D(\omega) = \frac{L^3}{(2\pi)^3}\frac{4\pi\omega^2}{v_s^3} = \frac{L^3}{2\pi^2}\frac{\omega^2}{v_s^3}.$$

One recovers (9.75). The remaining part of this problem is similar to what follows this formula in the lecture.

Problem 9. Phonons in a chain with long-range interaction:

a) Guide: One writes the equation of motion taking into account interactions with all neighbors: two neighbors at the symmetric positions $\pm ma$ give rise to a term $1 - \cos kma = 2\sin^2(kma/2)$ so that

$$\omega(k) = 2\left[\sum_{m>0}(K_m/M)\sin^2(mka/2)\right]^{1/2}.$$

b) At small k, one has $\sin(mka/2) \simeq mka/2$, hence

$$\omega(k) = ka\left[\sum_{m>0}K_m m^2/M\right]^{1/2}.$$

If $\sum_{m>0}K_m m^2$ converges, the sound velocity is

$$v_s = a\left[\sum_{m>0}K_m m^2/M\right]^{1/2}.$$

c) This is a series which converges to the required result.
d) A straightforward calculation.
    For the last two questions, the interaction is limited to the nearest neighbors: this was done in the lecture. For the integral $\int_0^\infty g(\omega)d\omega$, one replaces the upper bound by the highest frequency ω_{max}, the integral is equal to N, the number of atoms in the chain.

Problem 10. Phonons and melting:

a) In classical mechanics, the energy of a harmonic oscillator is given by $E = E_c + E_p$ where $E_c = \frac{v^2}{2M}$ (kinetic energy) and $E_p = \frac{M\omega^2 x^2}{2}$ (potential energy). The average value of each term is the same and equal to $\frac{k_B T}{2}$ for each degree of freedom, as demonstrated by the theorem of energy equipartition in section 3.6.4.
    In one dimension, $< E > = k_B T$ or $< E > = M\omega^2 < x^2 >$. When $\omega \to 0$, $< x^2 >$ diverges: this means that the sinusoidal motion around the equilibrium position is destroyed, giving rise to an instability of the solid phase. The consequence is the melting of the solid.

b) See section 9.4.2 for the commutation relations of phonon operators which are bosons. See also the relation (7.43) for $< a_k^+ a_k >=< n_k >$. Note that $< n_k >$ is the average value of the number of bosons (here phonons) excited at T which is given by (6.1). To prove $< a^+ a^+ >= 0$, one writes $< a^+ a^+ >=< n|a^+ a^+|n >\propto< n|n + 2 >= 0$ (orthogonality of state vectors). Similar demonstration for $< aa >= 0$.

c) One has to make the sum on ω_k to get the total energy.

d) One studies below longitudinal vibrations in a mono-atomic chain of constant d.

One uses (9.38):

$$\hat{X}_k = \frac{1}{\sqrt{N}} \sum_m \hat{u}_m e^{-ikR_m}$$

$$\hat{u}_m = \frac{1}{\sqrt{N}} \sum_k \hat{X}_k e^{ikR_m}.$$

Replacing $\hat{X}_k = \frac{1}{2} A_k^{-1}[a_k + a_{-k}^+]$ given by (9.54), one obtains the demanded result (with $R_m = md$):

$$u_m = \sum_{k \in BZ} \left(\frac{\hbar}{2NM\omega_k} \right)^{1/2} [e^{ikmd} a_k + e^{-ikmd} a_k^+].$$

One takes the square of this relation and calculates its average value:

$$< u_m^2 > = \sum_{k \in BZ} \sum_{k' \in BZ} \frac{\hbar}{2NM\omega_k \omega_{k'}'} [e^{i(k-k')md} < a_k a_{k'} >$$
$$+ e^{-i(k-k')md} < a_k^+ a_{k'}^+ > + ...]$$
$$= \sum_{k \in BZ} \frac{< E(k) >}{NM\omega_k^2}$$

where one has used the sum rules and $< a^+ a^+ >= 0$ etc.

e) One replaces the sum on k by $\int g(\omega)d\omega....$ In one dimension, $g(\omega) = $ constant [see (9.76)]. One has therefore

$$< u_m^2 > \propto \int \frac{< E(k) >}{\omega^2} d\omega = \int \frac{\hbar \omega}{\omega^2 (e^{\beta \hbar \omega} - 1)} d\omega.$$

This integral diverges at the lower bound ($\omega = 0$): it means that the chain cannot be stabilized at a non zero temperature.

f) In two dimensions, the factor of the integral changes, the density of modes $g(\omega) \propto \omega$ with $\omega \propto k$ for small k (see Problem 7). The denominator of $< u_m^2 >$ remains unchanged. One has

$$< u_m^2 > \propto \int \frac{\hbar \omega}{\omega^2 (e^{\beta \hbar \omega} - 1)} \omega d\omega.$$

The above integral diverges at the lower bound as $\frac{1}{\omega}$ when $T \neq 0$: one concludes that the crystalline ordering in two dimensions does not survive at non zero temperatures.

17.3 Solutions of problems of chapter 10

Problem 1. System of two electrons — Fermi hole:

a) The Slater determinant wave function $\Psi(\vec{r}_1, \sigma_1, \vec{r}_2, \sigma_2)$ for a two-electron system is written as

$$\begin{vmatrix} \psi_{f_1}(q_1) & \psi_{f_1}(q_2) \\ \psi_{f_2}(q_1) & \psi_{f_2}(q_2) \end{vmatrix}$$

where $\psi_{f_i}(q_i) = \psi_{1s,\sigma_i}(\vec{r}_i, \zeta_i) = \varphi_{1s}(\vec{r}_i) S_{\sigma_i}(\zeta_i)$ [see Eq. (7.5)].
For two electrons in a He atom in the ground state for example, one has $\sigma_1 = +$ and $\sigma_2 = -$:
$S_{\sigma_1}(\uparrow) = 1, S_{\sigma_1}(\downarrow) = 0, S_{\sigma_2}(\uparrow) = 0, S_{\sigma_2}(\downarrow) = 1$.
One writes $\Psi(\vec{r}_1, \sigma_1, \vec{r}_2, \sigma_2) = \varphi_{1s}(\vec{r}_1) S_{\sigma_1}(\uparrow) \varphi_{1s}(\vec{r}_2) S_{\sigma_2}(\downarrow) = \varphi_{1s}(\vec{r}_1) \varphi_{1s}(\vec{r}_2)$
The Hamiltonian is

$$\mathcal{H} = \frac{p_1^2}{2m} + \frac{p_2^2}{2m} + \frac{e^2}{|\vec{r}_1 - \vec{r}_2|} \tag{17.16}$$

The system energy is thus

$$E = \langle \varphi_{1s}(\vec{r}_1) \varphi_{1s}(\vec{r}_2) | \frac{p_1^2}{2m} + \frac{p_2^2}{2m} + \frac{e^2}{|\vec{r}_1 - \vec{r}_2|} | \varphi_{1s}(\vec{r}_1) \varphi_{1s}(\vec{r}_2) \rangle$$

$$= E_1 + E_2 + \langle \varphi_{1s}(\vec{r}_1) \varphi_{1s}(\vec{r}_2) | \frac{e^2}{|\vec{r}_1 - \vec{r}_2|} | \varphi_{1s}(\vec{r}_1) \varphi_{1s}(\vec{r}_2) \rangle$$

where $E_1 = E_2 =$ kinetic energy of each electron. The integral represents the direct interaction energy. The exchange interaction does not exist because the spins are antiparallel.

b) One takes the Slater determinant with

$$\psi_{f_i}(q_j) = \frac{1}{\sqrt{2}} \frac{1}{\sqrt{\Omega}} e^{i\vec{k}_i \cdot \vec{r}_j} S_{\sigma_i}(\zeta_j)$$

where the normalization factor has been introduced.

One writes for the case of two antiparallel spins: $\sigma_1 = +, \sigma_2 = -$. Hence $S_{\sigma_1}(\uparrow) = 1$, $S_{\sigma_1}(\downarrow) = 0$, $S_{\sigma_2}(\downarrow) = 1$, $S_{\sigma_2}(\uparrow) = 0$. The corresponding wave function is

$$\Psi_{\vec{k}_1,\vec{k}_2}(\vec{r}_1, \uparrow, \vec{r}_2, \downarrow) = \frac{1}{2} \frac{1}{\Omega} e^{i\vec{k}_1 \cdot \vec{r}_1} e^{i\vec{k}_2 \cdot \vec{r}_2}$$

The probability of this state is

$$P(\uparrow, \downarrow) = \sum_{\vec{k}_1,\vec{k}_2} |\Psi_{\vec{k}_1,\vec{k}_2}(\vec{r}_1, \uparrow, \vec{r}_2, \downarrow)|^2 = \frac{1}{4} \frac{1}{\Omega^2} \sum_{\vec{k}_1,\vec{k}_2} 1 = \frac{1}{4} \frac{N^2}{\Omega^2}$$

$$(17.17)$$

One sees that it is independent of the distance. The same thing is obtained for $P(\downarrow, \uparrow)$. The probability to find two antiparallel spins is thus $2 \times \frac{1}{4} \frac{N^2}{\Omega^2}$

In the case of two parallel spins, one has $\sigma_1 = +, \sigma_2 = +$, thus

$$\Psi_{\vec{k}_1,\vec{k}_2}(\vec{r}_1, \uparrow, \vec{r}_2, \uparrow) = \frac{1}{2} \frac{1}{\Omega} \left[e^{i\vec{k}_1 \cdot \vec{r}_1} e^{i\vec{k}_2 \cdot \vec{r}_2} - e^{i\vec{k}_2 \cdot \vec{r}_1} e^{i\vec{k}_1 \cdot \vec{r}_2} \right] \quad (17.18)$$

One has then

$$P(\uparrow, \uparrow) = \sum_{\vec{k}_1,\vec{k}_2} |\Psi_{\vec{k}_1,\vec{k}_2}(\vec{r}_1, \uparrow, \vec{r}_2, \uparrow)|^2$$

$$\doteq \frac{1}{4} \frac{1}{\Omega^2} \sum_{\vec{k}_1,\vec{k}_2} \left(e^{-i\vec{k}_1 \cdot \vec{r}_1} e^{-i\vec{k}_2 \cdot \vec{r}_2} - e^{-i\vec{k}_2 \cdot \vec{r}_1} e^{-i\vec{k}_1 \cdot \vec{r}_2} \right)$$

$$\times \left(e^{i\vec{k}_1 \cdot \vec{r}_1} e^{i\vec{k}_2 \cdot \vec{r}_2} - e^{i\vec{k}_2 \cdot \vec{r}_1} e^{i\vec{k}_1 \cdot \vec{r}_2} \right)$$

$$= \frac{1}{4} \frac{1}{\Omega^2} \sum_{\vec{k}_1,\vec{k}_2} 2 - \frac{1}{4} \frac{1}{\Omega^2} \sum_{\vec{k}_1,\vec{k}_2} [e^{i\vec{k}_1 \cdot (\vec{r}_2 - \vec{r}_1)} e^{-i\vec{k}_2 \cdot (\vec{r}_2 - \vec{r}_1)}$$

$$+ e^{-i\vec{k}_1 \cdot (\vec{r}_2 - \vec{r}_1)} e^{i\vec{k}_2 \cdot (\vec{r}_2 - \vec{r}_1)}]$$

$$= 2 \times \frac{1}{4} \frac{N^2}{\Omega^2} - \frac{1}{4} \frac{1}{\Omega^2} \times 2I^2 \quad (17.19)$$

where

$$I \equiv \sum_{\vec{k}} e^{-i\vec{k}\cdot(\vec{r}_2 - \vec{r}_1)}$$

$$= \frac{\Omega}{(2\pi)^3} \int d\vec{k} e^{-ikr\cos\theta} = \frac{\Omega}{(2\pi)^3} 2\pi \int_0^{k_F} k^2 dk$$

$$\times \int_{-1}^{1} d(\cos\theta) e^{-ikr\cos\theta}$$

$$= \frac{\Omega}{(2\pi)^3} 2\pi \int_0^{k_F} k^2 dk [e^{ikr} - e^{-ikr}] \frac{1}{ikr}$$

$$= \frac{\Omega}{(2\pi)^3} \frac{4\pi}{r} \int_0^{k_F} k dk \sin kr = \frac{\Omega}{(2\pi)^3} \frac{4\pi}{r} [-k_F \cos k_F r$$

$$+ \int_0^{k_F} dk \cos kr]$$

$$= \frac{\Omega}{(2\pi)^3} 4\pi \left[\frac{-k_F \cos k_F r}{r} + \frac{\sin k_F r}{r^2} \right] \tag{17.20}$$

Since $N = \sum_{\vec{k}} = \frac{\Omega}{(2\pi)^3} \int_0^{k_F} 4\pi k^2 dk = \frac{\Omega}{(2\pi)^3} [\frac{4\pi k_F^3}{3}]$, Eq. (17.19) becomes

$$P(\uparrow, \uparrow) = 2 \times \frac{1}{4} \frac{N^2}{\Omega^2} \left[1 - \frac{1}{N^2} (4\pi)^2 \left(\frac{-k_F \cos k_F r}{r} + \frac{\sin k_F r}{r^2} \right) \right]$$

$$= \frac{1}{2} \frac{1}{(2\pi)^6} \left(\frac{4\pi k_F^3}{3} \right)^2$$

$$\times \left[1 - \left(\frac{3}{4\pi} \right)^2 (4\pi)^2 \left(\frac{-k_F \cos k_F r + \sin k_F r}{k_F^3 r^3} \right)^2 \right].$$

When $r \to 0$,

$$3 \left(\frac{-k_F \cos k_F r + \sin k_F r}{k_F^3 r^3} \right) \to 1 - \frac{1}{10} k_F^3 r^2$$

(expansion at the order of $(k_F r)^5$ of $\sin k_F r$ and of $\cos k_F r$). One sees that $P(\uparrow, \uparrow) \to 0$ when $r \to 0$. This region of small r has a deficit of parallel spins. It explains the Pauli exclusion principle. It is called the Fermi hole. One shows in Fig. 17.5 the function

$$\left[1 - 9 \left(\frac{-k_F \cos k_F r + \sin k_F r}{k_F^3 r^3} \right)^2 \right].$$

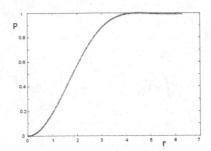

Fig. 17.5   Probability to find two parallel spins versus their relative distance, with $k_F = 1$.

Problem 2. Screened Coulomb potential, Thomas-Fermi approximation:

a) The Poisson equation is

$$\nabla^2 \varphi(\vec{r}) = \frac{e}{\epsilon_0}[n(\vec{r}) - n_0] \tag{17.21}$$

where $n(\vec{r})$ is the density of electrons at $\vec{r}$, and n_0 the non perturbed electron density. One has in the non perturbed region

$$\epsilon_F^0 = \frac{\hbar^2}{2m}(3\pi^2 n_0)^{2/3} \quad \text{(Fermi level at } T = 0) \tag{17.22}$$

b) In the perturbed region

$$\mu = \epsilon_F(\vec{r}) - e\varphi(\vec{r})$$

$$\simeq \frac{\hbar^2}{2m}[3\pi^2 n(\vec{r})]^{2/3} - e\varphi(\vec{r})$$

One supposes the perturbation is weak, one has $\mu \simeq \epsilon_F^0$ so that

$$\epsilon_F(\vec{r}) - \epsilon_F^0 \simeq e\varphi(\vec{r}) \tag{17.23}$$

One expands $\epsilon_F(\vec{r})$ around ϵ_F^0:

$$\epsilon_F(\vec{r}) \simeq \epsilon_F^0 + (n - n_0)\left(\frac{d\epsilon_F(\vec{r})}{dn}\right)_{n=n_0}$$

$$\epsilon_F(\vec{r}) - \epsilon_F^0 = (n - n_0)\left(\frac{d\epsilon_F(\vec{r})}{dn}\right)_{n=n_0}$$

One has thus

$$(n - n_0)\left(\frac{d\epsilon_F(\vec{r})}{dn}\right)_{n=n_0} = e\varphi(\vec{r}) \tag{17.24}$$

One calculates

$$\left(\frac{d\epsilon_F(\vec{r})}{dn}\right)_{n=n_0} = \frac{2}{3}\frac{\epsilon_F^0}{n_0} \tag{17.25}$$

From (17.24), one has

$$(n - n_0) = \frac{3}{2} \frac{n_0}{\epsilon_F^0} e\varphi(\vec{r}) \tag{17.26}$$

Equation (17.21) becomes

$$\nabla^2 \varphi(\vec{r}) = \frac{3n_0 e^2}{2\epsilon_0 \epsilon_F^0} \varphi(\vec{r}) \equiv \lambda^2 \varphi(\vec{r}) \tag{17.27}$$

If $\varphi(\vec{r}) = \varphi(r)$ (spherical symmetry), then

$$\nabla^2 \varphi(r) = \left(\frac{d^2}{dr^2} + \frac{2}{r} \frac{d}{dr} \right) \varphi(r) = \lambda^2 \varphi(r) \tag{17.28}$$

One finds a solution of the form $\varphi(r) = q\frac{e^{-\lambda r}}{r}$ by substituting this into the above equation. The larger λ is, the more strongly $\varphi(r)$ decreases with increasing r.

Remark: λ is proportional to $n_0^{1/2}$ so that the denser the gas is, the stronger the screening becomes. The above approximation is called "Thomas-Fermi approximation".

Problem 3. Hartree-Fock approximation:

a) One shows that the density of states $\rho(\epsilon_i)$ of an electron calculated using the Hartree-Fock result given by (10.17) is equal to 0 at $k_i = k_F$. The density of states [see (2.27)] can be written as

$$\rho(\epsilon) = \frac{d\phi(\epsilon)}{d\epsilon} = \frac{d\phi(\epsilon)}{d\vec{k}} \cdot \frac{d\vec{k}}{d\epsilon} \tag{17.29}$$

Using (10.17), one calculates $\frac{d\epsilon}{dk_i}$. One will see that this derivative diverges at $k_i = k_F$ due to the logarithmic term. The density of states is thus zero at k_F. This contradicts the definition of k_F which is the last *occupied* level.

One calculates the electron effective mass m^* defined by (10.34):

$$\frac{1}{m^*_{\alpha\beta}} = \frac{1}{\hbar^2} \frac{\partial^2 \epsilon_{\vec{k}_i}}{\partial k_\alpha \partial k_\beta} \tag{17.30}$$

where $\alpha, \beta = x, y, z$. Again here, the second derivative diverges at k_F using (10.17). So the effective mass is 0. This is not physically correct.

b) One introduces the screening factor $e^{-\mu r}$ to the Coulomb potential where μ is a screening constant [see Eq. (10.58)]. Using (10.13) without taking the limit $\mu \to 0$, one can calculate in the same manner as for (10.17) and one obtains

$$\epsilon_i = \frac{\hbar^2 k_i^2}{2m} - \frac{e^2}{2\pi} [\frac{k_F^2 + \mu^2 - k_i^2}{k_i} \ln \frac{(k_F + k_i)^2 + \mu^2}{(k_F - k_i)^2 + \mu^2} + 2k_F$$

$$+ 2\mu(\arctan(\frac{k_i - k_F}{\mu}) - \arctan(\frac{k_i + k_F}{\mu}))] \qquad (17.31)$$

The effective mass and the density of states do not diverge anymore at $k_i = k_F$. One has

$$\frac{1}{m^*} = \frac{1}{m} + \frac{1}{2\pi \hbar^2 k_F} \left[\left(1 + \frac{\mu^2}{2k_F^2} \right) \ln \left(1 + \frac{4k_F^2}{\mu^2} \right) - 2 \right] \qquad (17.32)$$

where m is the free electron mass (from the kinetic term). The second term is positive, therefore $m^* < m$.

Problem 4. Paramagnetic-ferromagnetic transition in an electron gas: One shows in the following that the ferromagnetic phase is more stable than the paramagnetic when the electron density is smaller than a critical value.

One has calculated the total energy of an electron gas in a phase where all spin orientations are allowed: one has used k_F by supposing that each energy level is occupied by two antiparallel spins (spin degeneracy=2). One calculates now the total energy in the ferromagnetic phase where all spins are parallel. The same calculation as it has been done for (10.19) without spin degeneracy leads to

$$\tilde{k}_F = \left(6\pi^2 \frac{N}{\Omega} \right)^{1/3} \qquad (17.33)$$

where factor 6 replaces factor 3 in the parentheses found for the paramagnetic case. The relation between k_F and $\tilde{k}_F$ is $\tilde{k}_F = 2^{1/3} k_F$. Thus, the ferromagnetic kinetic energy is related to the paramagnetic one is

$$E'_c = \frac{3}{5} N \frac{\hbar^2 \tilde{k}_F^2}{2m} = 2^{2/3} E_c \qquad (17.34)$$

The exchange energies of the two phases are connected by

$$E'_{ex} = -\frac{e^2 \Omega}{(2\pi)^3} \tilde{k}_F^4 = 2^{1/3} E_{ex} \qquad (17.35)$$

where one has taken off the factor 2 due to the spin degeneracy in (10.23) and replaced k_F by $\tilde{k}_F$ in (10.31).

The difference of the total energy of the two phases is thus

$$\Delta E = E'_{ex} - E_{ex} + E'_c - E_c = (2^{1/3} - 1)E_{ex} + (2^{2/3} - 1)E_c$$

$$= -\frac{0.916}{r_s}(2^{1/3} - 1) + \frac{2.21}{r_s^2}(2^{2/3} - 1) \qquad (17.36)$$

where one has used (10.22) and (10.32).

The ferromagnetic phase is favorable if $\Delta E < 0$, namely if

$$-\frac{0.916}{r_s}(2^{1/3}-1)+\frac{2.21}{r_s^2}(2^{2/3}-1) < 0 \quad \Rightarrow \quad r_s > 5.46 = r_s^c \quad (17.37)$$

One concludes that when the electron density is smaller than that corresponding to r_s^c the electron gas undergoes a transition to the ferromagnetic state.

Problem 5. Gas of interacting fermions in second quantization:

a) Replacing the potential in the second term one has

$$\frac{g}{2} \sum_{\sigma,\sigma'} \int d\vec{r}_1 \Psi_\sigma^+(\vec{r}_1) \Psi_{\sigma'}^+(\vec{r}_1) \Psi_{\sigma'}(\vec{r}_1) \Psi_\sigma(\vec{r}_1)$$

The simple argument to show that this term is zero is that if $\sigma = \sigma'$, two identical spins cannot occupy the same site (Pauli principle).

b) Following the procedure outlined in section 7.7 with the use of the decomposition of operator chains (7.75) one obtains, with spin $\beta \neq \sigma$,

$$i\hbar\frac{d\Psi_\sigma(\vec{r},t)}{dt} = \frac{p^2}{2m}\Psi_\sigma(\vec{r},t) + g\Psi_\beta^+(\vec{r},t)\Psi_\beta(\vec{r},t)\Psi_\sigma(\vec{r},t) \quad (17.38)$$

$$i\hbar\frac{d\Psi_\sigma^+(\vec{r},t)}{dt} = -\frac{1}{2m}\Psi_\sigma^+(\vec{r},t)p^2 - g\Psi_\sigma^+(\vec{r},t)\Psi_\beta^+(\vec{r},t)\Psi_\beta(\vec{r},t)$$

$$(17.39)$$

Using

$$\Psi_\sigma(\vec{r},t) = \sum_{\vec{k}} b_{\vec{k}\sigma} e^{-i\omega_{\vec{k}}t}\varphi_{\vec{k}}(\vec{r}) \qquad (17.40)$$

$$\Psi_\sigma^+(\vec{r},t) = \sum_{\vec{k}} b_{\vec{k}\sigma}^+ e^{i\omega_{\vec{k}}t}\varphi_{\vec{k}}^+(\vec{r}) \qquad (17.41)$$

one has from (17.38)

$$\sum_{\vec{k}} \hbar\omega_k b_{\vec{k},\sigma} e^{-\omega_k t} \varphi_{\vec{k}}(\vec{r}) = \sum_{\vec{k}} \frac{\hbar^2 k^2}{2m} b_{\vec{k},\sigma} e^{-\omega_k t} \varphi_{\vec{k}}(\vec{r})$$

$$+g < \Psi_\beta^+(\vec{r},t)\Psi_\beta(\vec{r},t) > \quad \sum_{\vec{k}} b_{\vec{k},\sigma} e^{-\omega_k t} \varphi_{\vec{k}}(\vec{r})$$

By simplifying the two sides, one has

$$\hbar\omega_k \varphi_{\vec{k},\sigma}(\vec{r}) = \left[\frac{\hbar^2 k^2}{2m} + g \left(< \sum_{\vec{k}'} n_{\vec{k}',\beta} \varphi_{\vec{k}',\beta}^*(\vec{r}) \varphi_{\vec{k}',\beta}(\vec{r}) > \right) \right] \varphi_{\vec{k},\sigma}(\vec{r})$$

$$(17.42)$$

Note that the factor behind g in the bracket is the average number of particles of spin β at $\vec{r}$ in contact interaction with the particle of spin σ at $\vec{r}$. The above equation is a Schrödinger equation with an interaction term averaged over all states. This is equivalent to the Hartree-Fock approximation.

c) One uses the Hamiltonian to calculate the kinetic and interaction terms:

-kinetic term:

$$E_c = \sum_\sigma \int d\vec{r} \Psi_\sigma^+(\vec{r}) \frac{p^2}{2m} \Psi_\sigma(\vec{r})$$

$$= \frac{1}{\Omega} \sum_\sigma \sum_{\vec{k},\vec{k}'} \frac{\hbar^2 k'^2}{2m} b_{\vec{k},\sigma}^+ b_{\vec{k}',\sigma} \int e^{i(\vec{k}'-\vec{k})\cdot\vec{r}} d\vec{r}$$

$$= \frac{1}{\Omega} \sum_\sigma \sum_{\vec{k},\vec{k}'} \frac{\hbar^2 k'^2}{2m} b_{\vec{k},\sigma}^+ b_{\vec{k}',\sigma} \Omega\delta(\vec{k}-\vec{k}')$$

$$= \sum_\sigma \sum_{\vec{k}} \frac{\hbar^2 k^2}{2m} b_{\vec{k},\sigma}^+ b_{\vec{k},\sigma} \quad (17.43)$$

-interaction term:

$$E_i = \frac{g}{2\Omega^2} \sum_\sigma \sum_{\beta \neq \sigma} \sum_{\vec{k}_1,...,\vec{k}_4} b^+_{\vec{k}_1,\sigma} b^+_{\vec{k}_2,\beta} b_{\vec{k}_3,\beta} b_{\vec{k}_4,\sigma}$$

$$\times \int e^{i(\vec{k}_3 + \vec{k}_4 - \vec{k}_1 - \vec{k}_2)\cdot\vec{r}} d\vec{r}$$

$$= \frac{g}{2\Omega^2} \sum_\sigma \sum_{\beta \neq \sigma} \sum_{\vec{k}_1,...,\vec{k}_4} b^+_{\vec{k}_1,\sigma} b^+_{\vec{k}_2,\beta} b_{\vec{k}_3,\beta} b_{\vec{k}_4,\sigma}$$

$$\times \Omega \delta(\vec{k}_3 + \vec{k}_4 - \vec{k}_1 - \vec{k}_2)$$

$$= \frac{g}{2\Omega^2} \sum_\sigma \sum_{\beta \neq \sigma} \sum_{\vec{k}_3,\vec{k}_4,\vec{q}} b^+_{\vec{k}_3+\vec{q},\sigma} b^+_{\vec{k}_4-\vec{q},\beta} b_{\vec{k}_3,\beta} b_{\vec{k}_4,\sigma} \quad (17.44)$$

where one has put $\vec{k}_3 - \vec{k}_1 \equiv -\vec{q}$, thus $\vec{k}_4 - \vec{k}_2 = \vec{q}$.

d) The average energy is written as

$$\overline{E} = \langle f | \hat{\mathcal{H}} | f \rangle = \overline{E}_c + \overline{E}_i$$

where $|f\rangle$ is the ground state. One has

$$\overline{E}_c = \sum_\sigma \sum_{\vec{k}} \frac{\hbar^2 k^2}{2m} \langle f | b^+_{\vec{k},\sigma} b_{\vec{k},\sigma} | f \rangle = \sum_\sigma \sum_{\vec{k}} \frac{\hbar^2 k^2}{2m}$$

$$= \frac{\Omega}{(2\pi)^3} \left[\int_0^{k_F^+} \frac{\hbar^2 k^2}{2m} 4\pi k^2 dk + \int_0^{k_F^-} \frac{\hbar^2 k^2}{2m} 4\pi k^2 dk \right]$$

$$= \frac{\Omega}{2\pi^2} \frac{\hbar^2}{2m} \frac{1}{5} \left[(k_F^+)^5 + (k_F^-)^5 \right]$$

$$= \frac{3}{5} \frac{\hbar^2}{2m} \left(\frac{6\pi^2}{\Omega} \right)^{2/3} \left(N_+^{5/3} + N_-^{5/3} \right) \quad (17.45)$$

where the two terms correspond to the sum on $\sigma = \pm$, and where one has used the results of (10.18) and (10.19) for each kind of spin.

Using $N_\pm = N(1 \pm \varsigma)/2$ one has

$$\overline{E}_c = \frac{3}{5} \frac{\hbar^2}{2m} \left(\frac{6\pi^2}{\Omega} \right)^{2/3} \left(\frac{N}{2} \right)^{5/3} \left[(1+\varsigma)^{5/3} + (1-\varsigma)^{5/3} \right]$$

$$= \frac{3}{5} \frac{\hbar^2}{2m} \left(\frac{6\pi^2}{\Omega} \right)^{2/3} \left(\frac{N}{2} \right)^{2/3} \frac{N}{2} \left[(1+\varsigma)^{5/3} + (1-\varsigma)^{5/3} \right]$$

$$= \frac{3}{5} \frac{\hbar^2}{2m} \left(\frac{3\pi^2 N}{\Omega} \right)^{2/3} \frac{N}{2} \left[(1+\varsigma)^{5/3} + (1-\varsigma)^{5/3} \right]$$

$$= \frac{3}{5} E_F \frac{N}{2} \left[(1+\varsigma)^{5/3} + (1-\varsigma)^{5/3} \right]$$

where E_F is the Fermi energy in the paramagnetic case with the degeneracy of up and down spin states [see (5.21) with $S = 1/2$]. As for the interaction term, one has

$$\overline{E}_i = \frac{g}{2\Omega^2} \sum_\sigma \sum_{\beta \neq \sigma} \sum_{\vec{k}', \vec{k}, \vec{q}} \langle f | b^+_{\vec{k}'+\vec{q}, \sigma} b^+_{\vec{k}-\vec{q}, \beta} b_{\vec{k}', \beta} b_{\vec{k}, \sigma} | f \rangle$$

$$= \frac{g}{2\Omega^2} \sum_\sigma \sum_{\beta \neq \sigma} \sum_{\vec{k}', \vec{k}, \vec{q}} \langle f | b^+_{\vec{k}'+\vec{q}, \sigma} b_{\vec{k}, \sigma} b^+_{\vec{k}-\vec{q}, \beta} b_{\vec{k}', \beta} | f \rangle$$

where moving twice the last operator to the second position does not change the sign of the second line. In order to have non zero $\langle f | ... | f \rangle$, one should choose $\vec{k}' + \vec{q} = \vec{k}$ so that

$$\overline{E}_i = \frac{g}{2\Omega^2} \sum_\sigma \sum_{\beta \neq \sigma} \sum_{\vec{k}', \vec{k}} n_{\vec{k}\sigma} n_{\vec{k}'\beta}$$

$$= \frac{g}{2\Omega^2} 2(N_+)(N_-) = \frac{g}{\Omega^2} N_+ N_- = \frac{g}{4\Omega^2} N^2(1 - \zeta^2)$$

One finally obtains

$$\overline{E} = \frac{3}{5} E_F \frac{N}{2} \left[(1+\zeta)^{5/3} + (1-\zeta)^{5/3} \right] + \frac{g}{4\Omega^2} N^2(1 - \zeta^2) \quad (17.46)$$

Minimizing with respect to ζ one has

$$\frac{1}{N} \frac{\partial \overline{E}}{\partial \zeta} = \frac{E_F}{2} \left[(1+\zeta)^{2/3} - (1-\zeta)^{2/3} \right] - \frac{gN}{4\Omega} 2\zeta \quad (17.47)$$

$$\simeq \frac{E_F}{2} \left[1 + \frac{2}{3}\zeta - 1 + \frac{2}{3}\zeta \right] - \frac{gN}{2\Omega} \zeta \quad (17.48)$$

where one has supposed $\zeta << 1$. If $\overline{E}$ tends to a minimum as ζ increases, one should have $\partial \overline{E}/\partial \zeta < 0$, namely

$$\frac{E_F}{2} \left[1 + \frac{2}{3}\zeta - 1 + \frac{2}{3}\zeta \right] - \frac{gN}{2\Omega} \zeta < 0$$

$$\zeta \left[\frac{2E_F}{3} - \frac{gN}{2\Omega} \right] < 0$$

$$\frac{gN}{\Omega E_F} > \frac{4}{3}$$

$$\frac{gN}{\Omega \epsilon_c} > \frac{20}{9} \quad (17.49)$$

where one has used (5.24): $\overline{E}_c/N = \frac{3}{5} E_F$, namely $E_F = \frac{5}{3}\epsilon_c$. The above result shows that as ζ increases from zero, the energy decreases, namely the partially magnetized state ($\zeta \neq 0$) is more stable than the paramagnetic state ($\zeta = 0$).

Let us consider the situation when $\zeta = 1 - \eta$ where $\eta \ll 1$. Expanding (17.47) one has

$$\frac{1}{N}\frac{\partial \overline{E}}{\partial \zeta} = \frac{E_F}{2}\left[(2-\eta)^{2/3} - \eta^{2/3}\right] - \frac{gN}{2\Omega}(1-\eta)$$

$$= \frac{E_F}{2}\left[2^{2/3}(1-\eta/2)^{2/3} - \eta^{2/3}\right] - \frac{gN}{2\Omega}(1-\eta)$$

$$\simeq \frac{E_F}{2}2^{2/3} - \frac{gN}{2\Omega} \tag{17.50}$$

One sees here that if

$$\frac{1}{N}\frac{\partial \overline{E}}{\partial \zeta} > 0$$

$$\text{then} \Rightarrow \frac{gN}{\Omega} < E_F 2^{2/3} \Rightarrow \frac{gN}{\Omega E_F} < 2^{2/3}$$

$$\Rightarrow \frac{gN}{\Omega \epsilon_c} < \frac{5}{3}2^{2/3} \simeq 2.64 \tag{17.51}$$

Thus, $\zeta \to 1$, $\overline{E}$ increases if the above condition is satisfied. The situation is shown in Fig. 17.6. One concludes that between $\zeta = 0$ and $\zeta = 1$, $\overline{E}$ goes through a minimum somewhere in the between: this minimum corresponds to a stable state which is partially magnetized. This happens if

$$\frac{20}{9} < \frac{gN}{\Omega \epsilon_c} < 2.64. \tag{17.52}$$

Problem 6. Fermion gas as a function of density:

One has

$$\hat{\mathcal{H}} = \sum_{\vec{k}\sigma, \vec{k}'\sigma'} \langle \vec{k}|T|\vec{k}'\rangle b^{+}_{\vec{k}\sigma} b_{\vec{k}'\sigma'}$$

$$+ \frac{1}{2}\sum_{\sigma,\sigma'}\sum_{\vec{k}_1,\vec{k}_2,\vec{k}_3,\vec{k}_4} \langle \vec{k}_1\sigma\vec{k}_2\sigma'|V(\vec{r}_1 - \vec{r}_2)|\vec{k}_3\sigma\vec{k}_4\sigma'\rangle$$

$$\times b^{+}_{\vec{k}_1\sigma} b^{+}_{\vec{k}_2\sigma'} b_{\vec{k}_4\sigma'} b_{\vec{k}_3\sigma}$$

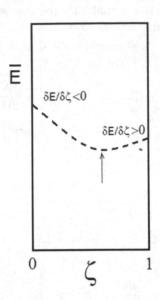

Fig. 17.6   The energy $\overline{E}$ versus the parameter ζ: its derivative is negative near $\zeta = 0$ and positive near $\zeta = 1$ if the condition (17.52) is satisfied. $\overline{E}$ should go through a minimum (indicated by an arrow) between these two limits as schematically shown by the broken line. The system is partially magnetized.

a) Let us calculate $[b^+_{\vec{k}'\sigma'} b_{\vec{k}''\sigma''}, b^+_{\vec{k}\sigma} b_{\vec{k}\sigma}]$. To simplify the writing, one puts $i = (\vec{k}'\sigma')$, $j = (\vec{k}''\sigma'')$ and $m = (\vec{k}\sigma)$. One has

$$[b^+_i b_j, b^+_m b_m] = b^+_i b_j b^+_m b_m b^+_m b_m - b^+_m b_m b^+_i b_j$$
$$= b^+_i [\delta(j,m) - b^+_m b_j] b_m - b^+_m [\delta(i,m) - b^+_i b_m] b_j$$
$$= b^+_i b_j - b^+_i b^+_m b_j b_m - b^+_i b_j + b^+_m b^+_i b_m b_j = 0$$

In the same manner, one has

$$[b^+_i b^+_j b_l b_k, b^+_m b_m] = [b^+_i [\delta(j,l) - b_l b^+_j] b_k, b^+_m b_m]$$
$$= [b^+_i b_k \delta(j,l), b^+_m b_m] - [b^+_i b_l b^+_j b_k, b^+_m b_m]$$

Using the decomposition $[AB, C] = A[B, C] - [A, C]B$ (see Problem 2 of chapter 7 for details) one can show that the above commutation relation is zero. One concludes that $[\hat{\mathcal{H}}, \hat{N}] = 0$.

b) The system energy of the ground state is composed of three terms: the kinetic term is given by (10.18), the direct interaction term by the N^2 term of (10.42), and the exchange term by (10.31). Using

$\epsilon_F = \hbar^2 k_F^2/(2m)$ with $k_F = \left(3\pi^2 N/\Omega\right)^{1/3}$ [see (10.19)-(10.20)], one has

$$\frac{E}{N} = \frac{\hbar^2}{2m}\frac{3}{5}\left(3\pi^2\right)^{2/3}\rho^{2/3} + g\frac{2\pi e^2}{\mu^2}\rho - g\frac{3e^2}{4\pi}\left(3\pi^2\right)^{1/3}\rho^{1/3} \quad (17.53)$$

where $\rho = N/\Omega$ (particle density), e the particle charge, $\mu > 0$ and $g \gtrless 0$ are constants given in

$$V(\vec{r}_1 - \vec{r}_2) = g\frac{e^2 e^{-\mu|\vec{r}_1 - \vec{r}_2|}}{|\vec{r}_1 - \vec{r}_2|}$$

c) If $g < 0$, the system energy becomes negative at high density as seen by examining the equation (17.53) rewritten as

$$\frac{E}{N} = A\rho^{2/3} - B\rho + C\rho^{1/3} = Ax^2 - Bx^3 + Cx = x\left(Ax - Bx^2 + C\right) \quad (17.54)$$

where A, B and C are positive constants defined in (17.53) and one has put $x = \rho^{1/3} > 0$. The quantity in the parentheses of the last equality is a second-order polynomial which changes its sign from positive at small x to negative at a value of x (see Fig. 17.7). It means that an attractive interaction between fermions of charge e can give rise to bound states.

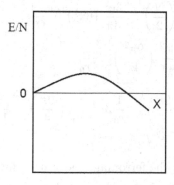

Fig. 17.7   Energy E/N given by (17.54) versus $x = \rho^{1/3}$: it becomes negative at a large value of x (high density).

d) One assumes $g = 1$ (electron gas). Following the same method as in Problem 5 above and using (17.53), one obtains

$$E(\zeta) = A \left(\frac{N}{2}\right)^{5/3} \left[(1+\zeta)^{5/3} + (1-\zeta)^{5/3}\right]$$

$$-C \left(\frac{N}{2}\right)^{4/3} \left[(1+\zeta)^{4/3} + (1-\zeta)^{4/3}\right]$$

$$+B \left(\frac{N}{2}\right)^{2} \left[(1+\zeta)^{2} + (1-\zeta)^{2} + 2(1-\zeta^2)\right] \quad (17.55)$$

where the first, second and third lines correspond to the kinetic, exchange and direct terms, respectively. A, B and C are positive defined by comparison with (17.53) where $g = 1$. Note that in the last line one has used

$$N^2 = (N^+ + N^-)^2 = (N^+)^2 + (N^-)^2 + 2N^+ N^-$$

$$= \frac{N^2}{4}[(1+\zeta)^2 + (1-\zeta)^2 + 2(1-\zeta^2)]$$

To compare the energies of the paramagnetic and ferromagnetic states, one calculates

$$\Delta E = E(\zeta = 0) - E(\zeta = 1) = A'\left(2 - 2^{5/3}\right) - C'\left(2 - 2^{4/3}\right)$$

where $A' = \dfrac{N}{2} \dfrac{\hbar^2}{2m} \dfrac{3}{5} (6\pi^2)^{2/3} \left(\dfrac{N}{2\Omega}\right)^{2/3}$

$$= \frac{N}{2} \frac{3e^2}{10} a_H \left(\frac{9\pi}{4}\right)^{2/3} \frac{1}{r_0^2}$$

and $C' = \dfrac{N}{2} \left(\dfrac{N}{2}\right)^{1/3} \dfrac{3e^2}{4\pi} (6\pi^2)^{1/3} \dfrac{1}{\Omega^{1/3}}$

$$= \frac{N}{2} \frac{3e^2}{4\pi} \left(\frac{9\pi^2}{4\pi}\right)^{1/3} \frac{1}{r_0}$$

One has

$$\frac{\Delta E}{N/2} = \left(\frac{9\pi}{4}\right)^{1/3} \frac{1}{r_0} 3e^2 [\frac{a_H}{r_0} \left(\frac{9\pi}{4}\right)^{1/3} \frac{1}{10}\left(2 - 2^{5/3}\right)$$

$$-\frac{1}{4\pi}\left(2 - 2^{4/3}\right)]$$

$\Delta E \geq 0$, namely the ferromagnetic state is favorable, if

$$\frac{a_H}{r_0} \geq \frac{1}{4\pi} \frac{2 - 2^{4/3}}{2 - 2^{5/3}} 10 \left(\frac{4}{9\pi}\right)^{1/3}$$

or $r_s = \dfrac{r_0}{a_H} \geq \dfrac{2\pi}{5} \left(\dfrac{9\pi}{4}\right)^{1/3} \left(1 + 2^{1/3}\right) \simeq 5.45$

One recovers the result of Problem 4.

Some curves of $E(\zeta)$ given by (17.55) are shown in Fig. 17.8.

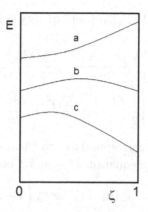

Fig. 17.8 Energy $E(\zeta)$ given by (17.55) versus ζ for $r_s < 5.45$ (a), $r_s = 5.45$ (b) and $r_s > 5.45$ (c).

17.4 Solutions of problems of chapter 11

Problem 1. Perturbation near the boundary of a Brillouin zone:

One considers the state at the boundary of the first Brillouin zone: $k_0 = K_1/2 = \pi/a$ (one dimension) (a: lattice constant). This state is degenerate with the state at the opposite boundary $k_0' = k_0 - K_1 = -K_1/2$. For $k = k_0 - \epsilon$ ($\epsilon \ll \pi/a$), one retains in the central equation only two coefficients $C(k)$ and $C(k - K_1)$ as one did for (11.57)-(11.58):

$$\left[\frac{\hbar^2 k^2}{2m} - E\right] C(k) + C(k - K_1)V_{k-K_1} = 0 \quad (17.56)$$

$$\left[\frac{\hbar^2 (k - K_1)^2}{2m} - E\right] C(k - K_1) + C(k)V_k = 0 \quad (17.57)$$

The secular equation for a non trivial solution is

$$E^2 - E\left[\frac{\hbar^2(k-K_1)^2}{2m} + \frac{\hbar^2 k^2}{2m}\right] + \frac{\hbar^2(k-K_1)^2}{2m}\frac{\hbar^2 k^2}{2m} - V_1^2 = 0$$

$$(17.58)$$

where $V_1 \simeq V_k = V_{k-K_1}$ (exact equality if $k = k_0$). One gets

$$E = \frac{1}{2}\left[\frac{\hbar^2(k-K_1)^2}{2m} + \frac{\hbar^2 k^2}{2m}\right]$$

$$\pm \left[\frac{1}{2}\left(\frac{\hbar^2(k-K_1)^2}{2m} + \frac{\hbar^2 k^2}{2m}\right)^2 + V_1^2\right]^{1/2} \quad (17.59)$$

where the signs $\pm$ correspond to two "bands". One expands various terms of the above equation around k_0: one has for example

$$k^2 = (k_0 - \epsilon)^2 \simeq k_0^2\left(1 - 2\frac{\epsilon}{k_0}\right).$$

With all terms expanded, one obtains

$$E^+ \simeq E_+(k_0) + \frac{\hbar^2 \epsilon^2}{2m}\left(1 + \frac{\hbar^2 k_0^2}{2mV_1}\right)$$

$$E^- \simeq E_-(k_0) + \frac{\hbar^2 \epsilon^2}{2m}\left(1 - \frac{\hbar^2 k_0^2}{2mV_1}\right)$$

where $E_\pm$ is defined by (11.59). The corrections of the second order show that the energy of each band is parabolic around $k_0 = \pi/a$.

Problem 2. Electrons in a square lattice:

a) See lecture

b) One writes in general

$$V(\vec{r}) = \sum_{\vec{k}} W_{\vec{k}} e^{i\vec{k}\cdot\vec{r}}$$

where $W_{\vec{k}}$ is to be determined. Since $V(\vec{r} + \vec{R}) = V(\vec{r})$ (periodic function), where $\vec{R}$ is a vector connecting two sites of the lattice, one has

$$\sum_{\vec{k}} W_{\vec{k}} e^{i\vec{k}\cdot(\vec{r}+\vec{R})} = \sum_{\vec{k}} W_{\vec{k}} e^{i\vec{k}\cdot\vec{r}},$$

from which one has

$$e^{i\vec{k}.\vec{R}} = 1, \quad \text{namely} \quad \vec{k}.\vec{R} = 2\pi m (m : \text{integer}).$$

One recognizes that $\vec{k}$ which verifies this relation is a vector of the reciprocal lattice $\vec{K}$ [see (8.27)]. Therefore,

$$V(\vec{r}) = \sum_{\vec{K}} W_{\vec{K}} e^{i\vec{K}.\vec{r}}.$$

The inverse transform gives

$$\int_S d\vec{r} e^{-i\vec{K}'.\vec{r}} V(\vec{r}) = \sum_{\vec{K}} W_{\vec{K}} \int d\vec{r} e^{i(\vec{K}-\vec{K}').\vec{r}}$$

$$= \sum_{\vec{K}} W_{\vec{K}} S \delta_{\vec{K},\vec{K}'} \qquad (17.60)$$

where S is the system surface [see (8.26)]. Thus,

$$W_{\vec{K}'} = \frac{1}{S} \int d\vec{r} e^{-i\vec{K}'.\vec{r}} V(\vec{r}).$$

Since $V(\vec{r})$ is real and symmetric, $V^*(\vec{r}) = V(\vec{r})$ and $V(\vec{r}) = V(-\vec{r})$, one has $W^*_{-\vec{K}} = W_{\vec{K}}$ and $W_{\vec{K}} = W_{-\vec{K}}$. One obtains therefore $W^*_{-\vec{K}} = W_{-\vec{K}}$ or $W^*_{\vec{K}} = W_{\vec{K}}$. This shows that $W_{\vec{K}}$ is real.

c) The energy correction at first order:

$$|\vec{k}\rangle_0 = \phi_{\vec{k}}(\vec{r}) = \frac{1}{S} e^{-i\vec{k}.\vec{r}},$$

$$\epsilon_1(\vec{k}) = \int d\vec{r} \phi^*_{\vec{k}}(\vec{r}) V(\vec{r}) \phi_{\vec{k}}(\vec{r}) = \frac{1}{S} \int d\vec{r} e^{-i\vec{k}.\vec{r}} V(\vec{r}) e^{i\vec{k}.\vec{r}} = W_0.$$

d) The energy correction at second order:

$$\epsilon_2(\vec{k}) = \sum_{\vec{k}' \neq \vec{k}} \frac{\left| \int d\vec{r} \phi^*_{\vec{k}'}(\vec{r}) V(\vec{r}) \phi_{\vec{k}}(\vec{r}) \right|^2}{\epsilon_0(\vec{k}) - \epsilon_0(\vec{k}')}$$

$$= \frac{1}{S} \sum_{\vec{k}' \neq \vec{k}} \frac{I^2_{\vec{k},\vec{k}'}}{\epsilon_0(\vec{k}) - \epsilon_0(\vec{k}')}$$

where

$$I_{\vec{k},\vec{k}'} = \int d\vec{r} e^{-i\vec{k}'.\vec{r}} V(\vec{r}) e^{i\vec{k}.\vec{r}} = \int d\vec{r} e^{-i\vec{k}'.\vec{r}} \sum_{\vec{K}} W_{\vec{K}} e^{i\vec{K}.\vec{r}} e^{i\vec{k}.\vec{r}}$$

$$= \sum_{\vec{K}} W_{\vec{K}} S \delta_{\vec{k}',\vec{k}+\vec{K}} = W_{\vec{k}'-\vec{k}} S \qquad (17.61)$$

One obtains

$$\epsilon_2(\vec{k}) = \sum_{\vec{k}' \neq \vec{k}} \frac{W_{\vec{k}' - \vec{k}}}{\epsilon_0(\vec{k}) - \epsilon_0(\vec{k}')}$$

$$= \sum_{\vec{G} \neq 0} \frac{W_{\vec{G}}^2}{\epsilon_0(\vec{k}) - \epsilon_0(\vec{k} + \vec{G})} \qquad (17.62)$$

where one has used $\vec{k}' = \vec{k} + \vec{G}$.

e) If $\vec{k}$ is on the boundary of the Brillouin zone, $\vec{k} + \vec{G}$ is degenerate with $\vec{k}$, the denominator is thus zero. Therefore, $\epsilon_2(\vec{k})$ cannot be calculated by the above relation.

f) One takes $\vec{K}_1 = (\pi/a, \pi/a)$. The degenerate states are given by $\vec{K}_1^2 = (\vec{K}_1 + \vec{G})^2$. These states are situated at the other three corners of the first Brillouin zone, namely at $(-\pi/a, \pi/a), (-\pi/a, -\pi/a), (\pi/a, -\pi/a)$. One defines $\vec{G}_1 = (-2\pi/a, 0)$, $\vec{G}_2 = (0, -2\pi/a)$ and $\vec{G}_3 = (-2\pi/a, -2\pi/a)$. One sees that $\vec{G}_1 \perp \vec{G}_2$ and $\vec{G}_3 = \vec{G}_1 + \vec{G}_2$.

One has

$$W_{\vec{G}_1} = \frac{1}{S} \int dx \int dy V(x, y) e^{i2\pi x/a}$$

$$= \frac{1}{S} \int dx \int dy V(x, y) e^{i2\pi y/a} = W_{\vec{G}_2}$$

where one has interchanged the dummy variables to obtain the last equality.

$$W_{\vec{G}_3} = \frac{1}{S} \int dx \int dy V(x, y) e^{i2\pi x/a + i2\pi y/a}$$

$$= \frac{1}{S} \int dx \int dy V(x, -y) e^{i2\pi x/a - i2\pi y/a}$$

$$= \frac{1}{S} \int dx \int dy V(x, y) e^{i2\pi x/a - i2\pi y/a} = W_{\vec{G}_1 - \vec{G}_2}$$

g) One takes the wave function as a linear combination of the degenerate states:

$$\Psi_{\vec{K}_1} = \alpha_0 |\vec{K}_1\rangle_0 + \sum_{j=1}^{3} \alpha_j |\vec{G}_j\rangle_0.$$

One calculates

$$H\Psi_{\vec{K}_1} = (H_0 + V)\Psi_{\vec{K}_1} = \epsilon \Psi_{\vec{K}_1}.$$

Multiplying from the left of the two sides of the equation by $\phi^*_{\vec{K}_1}$ and integrating with respect to $\vec{r}$, one obtains

$$\epsilon_0(\vec{K}_1)\alpha_0 + W_{-\vec{G}_1}\alpha_1 + W_{-\vec{G}_2}\alpha_2 + W_{-\vec{G}_3}\alpha_3 = \epsilon\alpha_0$$

where one has taken $W_0 = 0$.

One proceeds in the same manner for $\phi^*_{\vec{K}_1+\vec{G}_1}$, $\phi^*_{\vec{K}_1+\vec{G}_2}$ and $\phi^*_{\vec{K}_1+\vec{G}_3}$. One obtains 4 equations with the unknowns $\alpha_i, i = 0, 1, 2, 3$ which can be put under a matrix form:

$$\begin{pmatrix} \epsilon_0(\vec{K}_1) - \epsilon & W & W & W' \\ W & \epsilon_0(\vec{K}_1) - \epsilon & W' & W \\ W & W' & \epsilon_0(\vec{K}_1) - \epsilon & W \\ W' & W & W & \epsilon_0(\vec{K}_1) - \epsilon \end{pmatrix} \begin{pmatrix} \alpha_0 \\ \alpha_1 \\ \alpha_2 \\ \alpha_3 \end{pmatrix} = 0$$

where $W = W_{-\vec{G}_1} = _{-\vec{G}_2}$ and $W' = W_{-\vec{G}_3}$. For a non trivial solution, one has to satisfy the secular equation which is a 4×4 determinant. To simplify this determinant, one can construct the equivalent determinant by

First step:
- new first column= old first column - old fourth column
- new second column= old second column - old third column
- new third column= old second column + old third column
- new fourth column= old first column + old fourth column

Second step: with the new determinant, one constructs the final one by subtract the first line by the fourth, the second line by the third, add the second and third line, and add the first and fourth line. One has

$$2 \begin{vmatrix} \epsilon_0(\vec{K}_1) - \epsilon - W' & 0 & 0 & 0 \\ 0 & \epsilon_0(\vec{K}_1) - \epsilon - W' & 0 & 0 \\ 0 & 0 & \epsilon_0(\vec{K}_1) - \epsilon + W' & 2W \\ 0 & 0 & 2W & \epsilon_0(\vec{K}_1) - \epsilon + W' \end{vmatrix} = 0$$

Solving this equation, one gets

$$\left(\epsilon_0(\vec{K}_1) - \epsilon - W'\right)^2 \left[\left(\epsilon_0(\vec{K}_1) - \epsilon + W'\right)^2 - 4W^2\right] = 0 \quad (17.63)$$

from which the four solutions are
$\epsilon = \epsilon_0(\vec{K}_1) - W'$: doubly degenerate
$\epsilon = \epsilon_0(\vec{K}_1) + W' \pm 2W$.

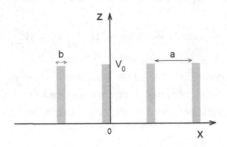

Fig. 17.9   Kronig-Penney model.

The degeneracy is partially removed: the first solution is still two-fold degenerate.

Problem 3. Electrons in a rectangular lattice:

Guide: The reciprocal lattice is given by translation vectors $\vec{K} = (\frac{2\pi m}{Na}, \frac{2\pi n}{Nb})$: it is a rectangular lattice. The first Brillouin zone is a rectangle limited by $(\pm\frac{\pi}{a}, \pm\frac{\pi n}{b})$. Since $b > a$, the state $\vec{K}_1 = (\frac{\pi}{a}, 0)$ has only one degenerate state at the opposite boundary $\vec{K}_1' = (-\frac{\pi}{a}, 0)$ when $V = 0$. One follows the same method as in the preceding problem: one finds the secular equation which is a 2×2 determinant. One obtains thus two solutions for ϵ of the form $\epsilon_0(\vec{K}_1) \pm W$.

Problem 4. Energy band in the Kronig-Penney potential model:

One considers a one-dimensional lattice of N sites, of constant $c = a + b$ with $a \gg b$ as shown in Fig. 17.9. The potential in this lattice is defined in the first cell by

$$V(x) = \begin{cases} 0 & \text{if } 0 \le x < a \\ V_0 & \text{if } a \le x < a + b \end{cases} \tag{17.64}$$

This potential is periodic, namely $V(x + c) = V(x)$. V_0 is finite but can go to infinity.

a) The Schrödinger equations for the wave function $\varphi(x)$ in the well

$(0 \leq x \leq a)$ and for $\phi(x)$ inside the barrier $(a \leq x \leq a+b)$ are

$$-\frac{\hbar^2}{2m}\frac{\partial^2\varphi(x)}{\partial x^2} = E\varphi(x)$$

$$\left[-\frac{\hbar^2}{2m}\frac{\partial^2}{\partial x^2} + V_0\right]\phi(x) = E\phi(x)$$

The respective solutions are

$$\varphi(x) = \alpha e^{ikx} + \beta e^{-ikx} \tag{17.65}$$

$$\phi(x) = \alpha' e^{qx} + \beta' e^{-qx} \tag{17.66}$$

Replacing these into the Schrödinger equations, one has

$$k = \sqrt{\frac{2mE}{\hbar^2}} \quad \text{and} \quad q = \sqrt{\frac{2m(V_0 - E)}{\hbar^2}}$$

The energy conservation at the boundary yields

$$k^2 = -q^2 + 2mV_0 \tag{17.67}$$

b) At the well-barrier boundary $x = 0$, one has

$$\varphi(0) = \alpha + \beta$$

$$\frac{\partial\varphi(0)}{\partial x} = (i\alpha k - i\beta k)$$

from which

$$\alpha = \frac{1}{2}\left[\frac{1}{ik}\frac{\partial\varphi(0)}{\partial x} + \varphi(0)\right]$$

$$\beta = \frac{1}{2}\left[\varphi(0) - \frac{1}{ik}\frac{\partial\varphi(0)}{\partial x}\right]$$

Using these for (17.65) one has at $x = a$:

$$\varphi(a) = \alpha e^{ika} + \beta e^{-ika}$$

$$= \frac{1}{2}\left[\frac{1}{ik}\frac{\partial\varphi(0)}{\partial x} + \varphi(0)\right]e^{ika}$$

$$+ \frac{1}{2}\left[\varphi(0) - \frac{1}{ik}\frac{\partial\varphi(0)}{\partial x}\right]e^{-ika}$$

$$= \cos(ka)\varphi(0) - \frac{1}{k}\sin(ka)\frac{\partial\varphi(0)}{\partial x}$$

Similarly,

$$\frac{\partial\varphi(0)}{\partial x} = -k\sin(ka)\varphi(0) + \cos(ka)\frac{\partial\varphi(0)}{\partial x}$$

Combining these two equations, one writes

$$\begin{pmatrix} \varphi(a) \\ \frac{\partial\varphi(a)}{\partial x} \end{pmatrix} = \begin{pmatrix} \cos(ka) & \sin(ka)/k \\ k\sin(ka) & \cos(ka) \end{pmatrix} \begin{pmatrix} \varphi(0) \\ \frac{\partial\varphi(0)}{\partial x} \end{pmatrix} \equiv T_a \begin{pmatrix} \varphi(0) \\ \frac{\partial\varphi(0)}{\partial x} \end{pmatrix}$$

$$\tag{17.68}$$

c) For x inside a barrier, using the wave function $\phi(x)$ at $x = a$ one has

$$\phi(a) = \alpha' e^{qa} + \beta' e^{-qa}$$

$$\frac{\partial \phi(a)}{\partial x} = q(\alpha' e^{qa} - \beta' e^{-qa})$$

Putting $P_1 = \alpha' e^{qa}$, $P_2 = \beta' e^{-qa}$, one has at $x = a + b$:

$$\phi(a + b) = P_1 e^{qb} + P_2 e^{-qb} \tag{17.69}$$

$$\frac{\partial \phi(a + b)}{\partial x} = q(P_1 e^{qb} - P_2 e^{-qb}) \tag{17.70}$$

As for α and β above, one obtains

$$P_1 = \frac{1}{2}\left[\phi(a) + \frac{1}{q}\frac{\partial \phi(a)}{\partial x}\right]$$

$$P_2 = \frac{1}{2}\left[\phi(a) - \frac{1}{q}\frac{\partial \phi(a)}{\partial x}\right]$$

Replacing these into (17.69) and (17.70) and after simplifying, one has

$$\phi(a + b) = \cosh(qb)\phi(a) + \frac{\sinh(qb)}{q}\frac{\partial \phi(a)}{\partial x} \tag{17.71}$$

$$\frac{\partial \phi(a + b)}{\partial x} = q\sinh(qb)\phi(a) + \cosh(qb)\frac{\partial \phi(a)}{\partial x} \tag{17.72}$$

Or

$$\begin{pmatrix} \phi(a + b) \\ \frac{\partial \phi(a+b)}{\partial x} \end{pmatrix} = \begin{pmatrix} \cosh(qb) & \sinh(qb)/q \\ q\sinh(qb) & \cosh(qb) \end{pmatrix} \begin{pmatrix} \phi(a) \\ \frac{\partial \phi(a)}{\partial x} \end{pmatrix} \equiv T_b \begin{pmatrix} \phi(a) \\ \frac{\partial \phi(a)}{\partial x} \end{pmatrix} \tag{17.73}$$

By the continuity conditions of the wave function and its derivative at boundaries, one has

$$\phi(a + b) = \varphi(a + b), \quad \phi(a) = \varphi(a)$$

$$\frac{\partial \phi(a + b)}{\partial x} = \frac{\partial \varphi(a + b)}{\partial x}, \quad \frac{\partial \phi(a + b)}{\partial x} = \frac{\partial \varphi(a + b)}{\partial x}$$

so that

$$T_b \times T_a \begin{pmatrix} \varphi(0) \\ \frac{\partial \varphi(0)}{\partial x} \end{pmatrix} = T_b \begin{pmatrix} \varphi(a) \\ \frac{\partial \varphi(a)}{\partial x} \end{pmatrix} = \begin{pmatrix} \varphi(a + b) \\ \frac{\partial \varphi(a+b)}{\partial x} \end{pmatrix} \tag{17.74}$$

Thus, in the translation $0 \rightarrow a + b$, the wave function and its derivative are transformed by the transfer matrix $T = T_a \times T_b$ where T_a and T_b are the 2×2 matrices given by (17.68) and (17.73).

d) The periodic boundary conditions are given by

$$\varphi(x) = \varphi(x + Nc), \quad \frac{\partial\varphi(x)}{\partial x} = \frac{\partial\varphi(x + Nc)}{\partial x}$$

so that

$$T^N \begin{pmatrix} \varphi(x) \\ \frac{\partial\varphi(x)}{\partial x} \end{pmatrix} = \begin{pmatrix} \varphi(x + Nc) \\ \frac{\partial\varphi(x+Nc)}{\partial x} \end{pmatrix} = \begin{pmatrix} \varphi(x) \\ \frac{\partial\varphi(x)}{\partial x} \end{pmatrix} \qquad (17.75)$$

Since, by definition, the eigenvalue of T is given by

$$T \begin{pmatrix} \varphi(x) \\ \frac{\partial\varphi(x)}{\partial x} \end{pmatrix} = \lambda \begin{pmatrix} \varphi(x) \\ \frac{\partial\varphi(x)}{\partial x} \end{pmatrix} \Rightarrow T^N \begin{pmatrix} \varphi(x) \\ \frac{\partial\varphi(x)}{\partial x} \end{pmatrix} = \lambda^N \begin{pmatrix} \varphi(x) \\ \frac{\partial\varphi(x)}{\partial x} \end{pmatrix}$$
$$(17.76)$$

Comparing this relation to (17.75) one gets

$$\lambda^N = 1 \Rightarrow \lambda^N = e^{2\pi n} \Rightarrow \lambda = e^{2\pi n/N} \qquad (17.77)$$

where n is an integer.

e) One has $T = T_a \times T_b$. Putting

$$T = \begin{pmatrix} A & B \\ C & D \end{pmatrix}$$

and using the matrices T_a and T_b, one has

$$A = \cos(ka)\cosh(qb) + \frac{q}{k}\sin(ka)\sinh(qb),$$

$$B = \cos(ka)\frac{\sinh(qb)}{q} + \frac{\sin(ka)}{k}\cosh(qb),$$

$$C = k\sin(ka)\cosh(qb) + q\cos(ka)\sinh(qb),$$

$$D = \frac{k}{q}\sinh(qb) + \cos(ka)\cosh(qb).$$

One diagonalizes T now. Writing

$$T = \begin{pmatrix} A & B \\ C & D \end{pmatrix} \begin{pmatrix} M_1 \\ M_2 \end{pmatrix} = \lambda \begin{pmatrix} M_1 \\ M_2 \end{pmatrix}$$

one sees that to find the non trivial matrix elements (M_1, M_2), one has to solve the secular equation

$$(A - \lambda)(D - \lambda) - CB = 0 \qquad (17.78)$$

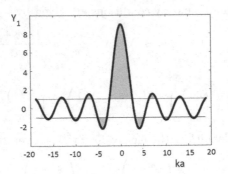

Fig. 17.10   Plot of Eq. (17.80) using $mV_0ab = 8$: Gray areas indicate forbidden energy bands in the Kronig-Penney model.

This gives

$$\lambda_\pm = \frac{A + D \pm \sqrt{(A+D)^2 - 4(AD - CB)}}{2}$$

Since $\lambda_+ + \lambda_- = A + D$ and knowing that $\lambda_+ + \lambda_- = e^{i2\pi n/N} + e^{-i2\pi n/N} = 2\cos(2\pi n/N)$, one obtains

$$2\cos(ka)\cosh(qb) + \left(\frac{k}{q} + \frac{q}{k}\right)\sin(ka)\sinh(qb) = 2\cos\left(\frac{2\pi n}{N}\right) \tag{17.79}$$

f) When $b \to 0$ and $V_0 \to \infty$ but qb remains finite, taking $\cosh(qb) \simeq 1$, one can rewrite (17.79) as

$$\cos(ka) + \frac{1}{2}(k^2 + q^2)ab\frac{\sin(ka)}{ka}\frac{\sinh(qb)}{qb} = \cos\left(\frac{2\pi n}{N}\right)$$

$$\cos(ka) + mV_0ab\frac{\sin(ka)}{ka} = \cos\left(\frac{2\pi n}{N}\right) \tag{17.80}$$

where one has used (17.67) and $\frac{\sinh(qb)}{qb} \to 1$. One plots in Fig. 17.10 the function of the left-hand side, Y_1, versus ka. Solutions of (17.80) require that $|Y_1| = |\cos\left(\frac{2\pi n}{N}\right)| \le 1$: the gray zones indicate regions where there are no solutions (forbidden bands). They start at $0, \pm\pi, \pm2\pi, \ldots$ and get narrower and narrower.

Problem 5. Tight-binding approximation:

This exercise reformulates the lecture in paragraph 11.4.1 under the form of questions. The reader finds the answers to these questions using that paragraph.

Problem 6. Tight-binding approximation in a square lattice:

a) The first Brillouin zone is a square limited by $k_x = \pm\pi/a$ and $k_y = \pm\pi/a$.

Using (11.96), one writes the energy for the case of a square lattice:

$$E_{\vec{k}} = E^{(0)} + E_{\vec{k}}^{(0)} + 2E^{(1)}[\cos(k_x a) + \cos(k_y a)] \qquad (17.81)$$

where

$$E_{\vec{k}}^{(0)} = \int_{\text{cell}} \phi^*(\vec{r})[U(\vec{r}) - V(\vec{r})]\phi(\vec{r})d\vec{r} \qquad (17.82)$$

$$E^{(1)} = \int_{\text{cell}} \phi^*(\vec{r} + \vec{R}_l)[U(\vec{r}) - V(\vec{r})]\phi(\vec{r})d\vec{r} \qquad (17.83)$$

$\vec{R}_l$ being the vectors connecting a site with its four nearest neighbors.

b) When $k \to 0$, one has

$$E_{\vec{k}} \simeq E^{(0)} + E_{\vec{k}}^{(0)} + 2E^{(1)}[1 - (k_x a)^2/2 + 1 - (k_y a)^2/2]$$

$$= E^{(0)} + E_{\vec{k}}^{(0)} + 4E^{(1)}[1 - (ka)^2/4] \equiv C + \frac{\hbar k^2}{2m^*}$$

where C is a constant and m^* is the effective mass.

The density of states $\rho(E)$ is calculated with the general definition: [see remarks after Eq. (2.36)]

$$g(\vec{k})d\vec{k} \equiv \rho(E)dE \qquad (17.84)$$

where $g(\vec{k})$ is the density of states in the $\vec{k}$ space. One has

$$g(\vec{k})d\vec{k} = \frac{\Omega}{(2\pi)^3}d\vec{k} \qquad (17.85)$$

(Ω: volume). One writes

$$\rho(E) = g(\vec{k})\frac{dS}{|\vec{\nabla}E_{\vec{k}}|} = \frac{\Omega}{(2\pi)^3}4\pi k^2\frac{1}{|\vec{\nabla}E_{\vec{k}}|} \qquad (17.86)$$

where dS is the equi-energetic surface where $E_{\vec{k}} = E$. One uses (17.81) to calculate $|\vec{\nabla} E_{\vec{k}}|$: one obtains $|\vec{\nabla} E_{\vec{k}}| = \frac{\hbar k}{m^*}$ from which one has

$$\rho(E) = \frac{\Omega}{4\pi^2}\left(\frac{2m^*}{\hbar^2}\right)^{3/2}[E_{\vec{k}} - C]^{1/2} \propto k \qquad (17.87)$$

Problem 7. Tight-binding approximation in a face-centered cubic lattice:

a) The reciprocal lattice of a face-centered cubic lattice is a body-centered cubic lattice: see Problem 3 of chapter 8. The first Brillouin zone can be constructed. It is shown in Fig. 17.11 for 1/8 of the first zone.

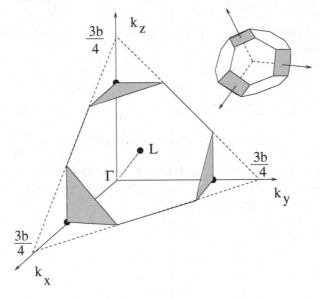

Fig. 17.11   1/8 of the first Brillouin zone of the reciprocal lattice of centered cubic structure (left). The whole Brillouin zone is shown on the right: it is a truncated octahedron. Γ is the point at the origin and L is the central point of the hexagon perpendicular to the diagonal direction of the cube. b is the lattice constant of the reciprocal lattice.

b) Using the structure factors (Problem 9 of chapter 8), one has

$$E_{\vec{k}} = E^{(0)} + \sum_l e^{i\vec{k}\cdot\vec{R}_l} \int_{\text{cell}} \phi^*(\vec{r}+\vec{R}_l)[U(\vec{r}) - V(\vec{r})]\phi(\vec{r})d\vec{r}$$

$$= E^{(0)} + E_{\vec{k}}^{(0)} + 4V[\cos(\frac{k_x a}{2})\cos(\frac{k_y a}{2}) + \cos(\frac{k_y a}{2})\cos(\frac{k_z a}{2})$$

$$+ \cos(\frac{k_z a}{2})\cos(\frac{k_x a}{2})] \tag{17.88}$$

where

$$\int_{\text{cell}} \phi^*(\vec{r}+\vec{R}_l)[U(\vec{r}) - V(\vec{r})]\phi(\vec{r})d\vec{r} = V \quad \text{or} \quad E_{\vec{k}}^{(0)}$$

if $|\vec{R}_l| = a$ or 0.

c) The energy is extremal when the quantity in the parentheses is equal to -3 or +3. The largest band width is thus $24|V|$. The points corresponding to these values are:
- For the maximum of the energy $(V < 0)$: $(k_x a = k_y a = k_z a = \pm\pi)$ (8 points L, see Fig. 17.11).
- For the minimum of the energy $(V < 0)$: $(k_x a = k_y a = k_z a = 0)$ (point Γ, see Fig. 17.11).

d) Guide: For small k near Γ, an expansion of the cosinus in the expression of the energy gives $E_{\vec{k}} \propto k^2$. The proportionality coefficient is identified with $\frac{\hbar^2}{2m^*}$, from which one has the effective mass m^*. The density of states in this region is calculated as in the preceding exercise.

17.5 Solutions of problems of chapter 12

Problem 1. Ground-state spin configuration of a chain of Ising spins:
Chain of Ising spins: Let J_1 and J_2 be the interactions between nearest neighbors and between next nearest neighbors, respectively. One supposes $J_1 > 0$ (ferromagnetic). When $J_2 = 0$, the ground state is ferromagnetic. If $J_2 > 0$, the ground state remains ferromagnetic. Now, if $J_2 < 0$, and if $|J_2| \to \infty$, the ground state is obviously dominated by J_2: the ground state is composed of two independent, interpenetrating antiferromagnetic sublattices such as ... ↑⇑↓⇓↑⇑ ... where thin arrows are spins of sublattice 1 and double arrows are spins of sublattice 2. One calculates now the

critical value of J_2 when the ground state change from ferromagnetic to the two-independent-sublattice structure. To that end, one compares the energies of the two states:

Energy of a spin in the ferromagnetic state: $E_F = -2J_1 + 2|J_2|$

Energy of a spin in the $\uparrow\uparrow\downarrow\downarrow$ state: $E_A = -2|J_2|$

The $\uparrow\uparrow\downarrow\downarrow$ state is more stable when $E_A < E_F$, namely when $|J_2|/J_1 > 1/2$ with $J_2 < 0$.

Problem 2. Chain of Heisenberg spins:

a) If one supposes a ferromagnetic state when J_2 is introduced, then one has $\omega = 2J_1 SZ[1 - \cos(ka)] + 2J_2 SZ[1 - \cos(2ka)]$ ($Z=2$: the numbers of nearest neighbors and next nearest neighbors). The first term is from (12.66) and the second term is obtained similarly by summing the couplings to next nearest neighbors at distance $2a$.

Let us examine the spin-wave "stiffness" D defined as $\omega = Dk^2$ at small k. As long as D is positive, the spin waves can be excited. One has

$$\omega \simeq_{k\to 0} 2J_1 SZ \left[\frac{(ka)^2}{2}\right]^2 + 2J_2 SZ \left[\frac{(2ka)^2}{2}\right]^2$$

$$= 2J_1 SZ \frac{(ka)^2}{2} \left[1 + \frac{4J_2}{J_1}\right]$$

$$= Dk^2 \qquad\qquad (17.89)$$

where the spin-wave stiffness is

$$D = 2J_1 SZ \frac{a^2}{2} \left[1 + \frac{4J_2}{J_1}\right]$$

If J_2 is positive (ferromagnetic interaction), then the spin-wave stiffness D is always positive. The stability of the ferromagnetic state is ensured.

b) If $J_2 < 0$, one has $\omega = 2J_1 SZ[1 - \cos(ka)] - 2|J_2|SZ[1 - \cos(2ka)]$. One sees that ω is strongly affected by J_2. D is

$$D = 2J_1 SZ \frac{a^2}{2} \left[1 - \frac{4|J_2|}{J_1}\right]$$

The stiffness D becomes zero at $1 - \frac{4|J_2|}{J_1} = 0$, namely $|J_2| = J_1/4$. From this "critical" value, the ferromagnetic state is therefore unstable. For $|J_2| > J_1/4$ or $J_2 < -J_1/4$. , the helical state takes place [see paragraph (12.3.3)].

One can find the critical value of J_2 by minimizing ω with respect to the "angle" (or "phase") ka between two neighboring spins: one has

$$\frac{d\omega}{dk} = 0$$

$$0 = 2J_1 SZa \sin(ka) - 4a|J_2|SZ \sin(2ka)$$

$$0 = 2SZa[J_1 \sin(ka) - 4|J_2| \sin(ka) \cos(ka)]$$

$$0 = 2SZa \sin(ka)[J_1 - 4|J_2| \cos(ka)]$$

There are two solutions: $\sin(ka) = 0$ (corresponding to the maximum of ω) and $\cos(ka) = \frac{J_1}{4|J_2|} = -\frac{J_1}{4J_2}$ ($J_2 < 0$). The second solution corresponds to the minimum of ω, namely a stable solution. The angle ka given by this solution generates the helical spin configuration as seen in Fig. 12.5.

Problem 3. Commutation relations:

One shows that the operators a^+ and a defined in the Holstein-Primakoff approximation, Eqs. (12.49) and (12.50), respect rigorously the commutation relations between the spin operators:

$$[S_l^+, S_m^-] = 2S_l^z \delta_{lm} \quad \text{and} \quad [S_l^z, S_m^\pm] = \pm S_l^\pm \delta_{lm}$$

Demonstration: Replacing the spin operators by the Holdstein-Primakoff operators, one has

$$[S_l^+, S_m^-] = 2S[f_l(S)a_l, a_m^+ f_m(S)]$$
$$= 2S[f_l(S)a_l a_m^+ f_m(S) - a_m^+ f_m(S)f_l(S)a_l]$$

If $l = m$, one has

$$[S_l^+, S_l^-] = 2S\left[f(S)aa^+ f(S) - a^+ f(S)f(S)a\right]$$
$$= 2S[(1 - \frac{a^+a}{2S})^{1/2}(a^+a + 1)(1 - \frac{a^+a}{2S})^{1/2}$$
$$- a^+(1 - \frac{a^+a}{2S})a]$$
$$= 2S\left[(1 - \frac{a^+a}{2S})(a^+a + 1) - a^+a + \frac{a^+a^+aa}{2S}\right]$$
$$= 2S[1 - \frac{a^+a}{S}]$$
$$= 2[S - a^+a] = 2S_l^z \tag{17.90}$$

If $l \neq m$, one obtains in the same manner $[S_l^+, S_m^-] = 0$. For the second relation, when $l = m$ one has

$$
\begin{aligned}
[S_l^z, S_l^+] &= \sqrt{2S}[(S - a^+a)fa - fa(S - a^+a)] \\
&= -\sqrt{2S}[faS - faa^+a - Sfa + a^+afa] \\
&= -\sqrt{2S}[-faa^+a + fa^+aa] \\
&= -\sqrt{2S}[-faa^+a + f(aa^+ - 1)a] \\
&= \sqrt{2S}fa = S_l^+
\end{aligned}
\tag{17.91}
$$

If $l \neq m$, one obtains $[S_l^z, S_m^{\pm}] = 0$. Similarly, one has $[S_l^z, S_m^-] = -S_l^- \delta_{lm}$.

Remark: One has used $a^+af = fa^+a$ in the demonstration of (17.90) because

$$
a^+af = a^+a\left(1 - \frac{a^+a}{2S}\right)^{1/2} = \left(a^+aa^+a - \frac{a^+aa^+aa^+a}{2S}\right)^{1/2}
$$

$$
= \left(1 - \frac{a^+a}{2S}\right)^{1/2} a^+a = fa^+a
$$

Problem 4. Heisenberg model in two dimensions:

a) One has from (12.66) or (12.29):

$$
\omega = 2JSZ(1 - \gamma_k)
$$

where $Z = 4$ (number of nearest neighbors in a square lattice), $\gamma_k = [\cos(k_x a) + \cos(k_y a)]/2$. $\omega \to 2JS(ka)^2$ when $\vec{k} \to 0$.

b) One has

$$
< S^z > = 1/2 - A \int_{ZB} \frac{2\pi k dk}{\exp(\beta\omega) - 1}
$$

where A is a constant (see paragraph 12.2.3). The most important contribution to the above integral comes from the small k modes where $\omega \to 2JS(ka)^2$. One has

$$
< S^z > \simeq 1/2 - A \int_{ZB} \frac{2\pi k dk}{1 + \beta JS(ka)^2 - 1} \simeq 1/2 - A \int_{ZB} \frac{2\pi k dk}{\beta JS(ka)^2}.
$$

This integral diverges at the lower bound $k = 0$: $< S^z >$ is not therefore defined when $T \neq 0$. The long-range order does not survive at finite temperatures in two dimensions (theorem by Mermin and Wagner).

Note that in three dimensions, one replaces in the integral $2\pi k dk$ by $4\pi k^2 dk$. The integral does not diverge at $k = 0$: the long-range order therefore exists at $T \neq 0$ in three dimensions.

Problem 5. Magnon-phonon interaction:

The diagonalization of a Hamiltonian in the second quantization is very important since when it is put under the diagonal form, namely $\mathcal{H} = \sum_k \lambda_k a_k^+ a_k$, one obtains directly the dispersion relation given by the coefficient λ_k (see also Problem 7 of chapter 7).

One considers the following Hamiltonian describing the interaction between magnon and phonon:

$$\mathcal{H} = \sum_{\vec{k}} \left[\omega_{\vec{k}}^m a_{\vec{k}}^+ a_{\vec{k}} + \omega_{\vec{k}}^p b_{\vec{k}}^+ b_{\vec{k}} + V_{\vec{k}}(a_{\vec{k}} b_{\vec{k}}^+ + a_{\vec{k}}^+ b_{\vec{k}}) \right] \qquad (17.92)$$

where $V_{\vec{k}}$ is the coupling constant, $\omega_{\vec{k}}^m$ and $\omega_{\vec{k}}^p$ are eigenfrequencies of magnon and phonon, respectively, a and a^+ denote annihilation and creation operators of magnon, while b and b^+ denote those of phonon.

a) One has

$$a_{\vec{k}} = \cos \theta_{\vec{k}} c_{\vec{k}} + \sin \theta_{\vec{k}} d_{\vec{k}} \qquad (17.93)$$
$$b_{\vec{k}} = \cos \theta_{\vec{k}} d_{\vec{k}} - \sin \theta_{\vec{k}} c_{\vec{k}} \qquad (17.94)$$

Multiplying the first equation by $\cos \theta_{\vec{k}}$ and the second by $\sin \theta_{\vec{k}}$, then taking the difference side by side, one has

$$a_{\vec{k}} \cos \theta_{\vec{k}} - b_{\vec{k}} \sin \theta_{\vec{k}} = c_{\vec{k}} \qquad (17.95)$$

where one used $\cos^2 \theta_{\vec{k}} + \sin^2 \theta_{\vec{k}} = 1$. Multiplying now (17.93) by $\sin \theta_{\vec{k}}$ and (17.94) by $\cos \theta_{\vec{k}}$ and adding two equations side by side one obtains

$$a_{\vec{k}} \sin \theta_{\vec{k}} + b_{\vec{k}} \cos \theta_{\vec{k}} = d_{\vec{k}} \qquad (17.96)$$

One demonstrates now the commutation relation $[c_{\vec{k}}, c_{\vec{k}'}^+] = \delta_{k,k'}$:

$$[c_{\vec{k}}, c_{\vec{k}'}^+] = [a_{\vec{k}}\cos\theta_{\vec{k}} - b_{\vec{k}}\sin\theta_{\vec{k}}, a_{\vec{k}'}^+\cos\theta_{\vec{k}'} - b_{\vec{k}'}^+\sin\theta_{\vec{k}'}]$$

$$= [a_{\vec{k}}\cos\theta_{\vec{k}}, a_{\vec{k}'}^+\cos\theta_{\vec{k}'}] - [a_{\vec{k}}\cos\theta_{\vec{k}}, b_{\vec{k}'}^+\sin\theta_{\vec{k}'}]$$

$$- [b_{\vec{k}}\sin\theta_{\vec{k}}, a_{\vec{k}'}^+\cos\theta_{\vec{k}'}] + [b_{\vec{k}}\sin\theta_{\vec{k}}, b_{\vec{k}'}^+\sin\theta_{\vec{k}'}]$$

$$= \cos\theta_{\vec{k}}\cos\theta_{\vec{k}'}[a_{\vec{k}}, a_{\vec{k}'}^+] - \cos\theta_{\vec{k}}\sin\theta_{\vec{k}'}[a_{\vec{k}}, b_{\vec{k}'}^+]$$

$$- \sin\theta_{\vec{k}}\cos\theta_{\vec{k}'}[b_{\vec{k}}, a_{\vec{k}'}^+] + \sin\theta_{\vec{k}}\sin\theta_{\vec{k}'}[b_{\vec{k}}, b_{\vec{k}'}^+]$$

$$= \cos^2\theta_{\vec{k}}\delta_{k,k'} - 0 - 0 + \sin^2\theta_{\vec{k}}\delta_{k,k'}$$

$$= \delta_{k,k'} \tag{17.97}$$

The other relations can be proved in the same manner.

b) Replacing the operators a and b in terms of c and d, one has

$$\mathcal{H} = \sum_{\vec{k}}\{\omega_{\vec{k}}^m[c_{\vec{k}}^+c_{\vec{k}}\cos^2\theta_{\vec{k}} + d_{\vec{k}}^+d_{\vec{k}}\sin^2\theta_{\vec{k}}$$

$$+ (c_{\vec{k}}^+d_{\vec{k}} + d_{\vec{k}}^+c_{\vec{k}})\cos\theta_{\vec{k}}\sin\theta_{\vec{k}}]$$

$$+ \omega_{\vec{k}}^p[d_{\vec{k}}^+d_{\vec{k}}\cos^2\theta_{\vec{k}} + c_{\vec{k}}^+c_{\vec{k}}\sin^2\theta_{\vec{k}}$$

$$- (d_{\vec{k}}^+c_{\vec{k}} + c_{\vec{k}}^+d_{\vec{k}})\cos\theta_{\vec{k}}\sin\theta_{\vec{k}}]$$

$$+ V_{\vec{k}}[c_{\vec{k}}d_{\vec{k}}^+\cos^2\theta_{\vec{k}} - d_{\vec{k}}c_{\vec{k}}^+\sin^2\theta_{\vec{k}}$$

$$+ (d_{\vec{k}}d_{\vec{k}}^+ - c_{\vec{k}}c_{\vec{k}}^+)\cos\theta_{\vec{k}}\sin\theta_{\vec{k}}$$

$$+ c_{\vec{k}}^+d_{\vec{k}}\cos^2\theta_{\vec{k}} - d_{\vec{k}}^+c_{\vec{k}}\sin^2\theta_{\vec{k}}$$

$$- (c_{\vec{k}}^+c_{\vec{k}} - d_{\vec{k}}^+d_{\vec{k}})\cos\theta_{\vec{k}}\sin\theta_{\vec{k}}]\} \tag{17.98}$$

Collecting non diagonal terms one has

$$\mathcal{H}_{\text{nd}} = \sum_{\vec{k}}[(\omega_{\vec{k}}^m - \omega_{\vec{k}}^p)(c_{\vec{k}}^+d_{\vec{k}} + d_{\vec{k}}^+c_{\vec{k}})\cos\theta_{\vec{k}}\sin\theta_{\vec{k}}$$

$$+ V_{\vec{k}}(c_{\vec{k}}d_{\vec{k}}^+ + c_{\vec{k}}^+d_{\vec{k}})\cos^2\theta_{\vec{k}}$$

$$- V_{\vec{k}}(d_{\vec{k}}c_{\vec{k}}^+ + d_{\vec{k}}^+c_{\vec{k}})\sin^2\theta_{\vec{k}}]$$

$$= \sum_{\vec{k}}[c_{\vec{k}}^+d_{\vec{k}} + d_{\vec{k}}^+c_{\vec{k}}]\{(\omega_{\vec{k}}^m - \omega_{\vec{k}}^p)\cos\theta_{\vec{k}}\sin\theta_{\vec{k}}$$

$$+ V_{\vec{k}}[\cos^2\theta_{\vec{k}} - \sin^2\theta_{\vec{k}}]\}$$

$$= \sum_{\vec{k}}\{(\omega_{\vec{k}}^m - \omega_{\vec{k}}^p)\sin(2\theta_{\vec{k}})/2 + V_{\vec{k}}\cos(2\theta_{\vec{k}})\}[c_{\vec{k}}^+d_{\vec{k}} + d_{\vec{k}}^+c_{\vec{k}}]$$

The non diagonal term is zero if the coefficient in the curly brackets is zero:

$$\{(\omega_{\vec{k}}^m - \omega_{\vec{k}}^p)\sin(2\theta_{\vec{k}})/2 + V_{\vec{k}}\cos(2\theta_{\vec{k}})\} = 0 \tag{17.99}$$

namely,

$$\tan(2\theta_{\vec{k}}) = \frac{2V_{\vec{k}}}{\omega_{\vec{k}}^p - \omega_{\vec{k}}^m} \tag{17.100}$$

One collects the diagonal terms of (17.98), putting $\omega_{\vec{k}}^m = \omega_{\vec{k}}^p = \omega$:

$$\begin{aligned}
\mathcal{H}_d &= \sum_{\vec{k}} \{\omega[c_{\vec{k}}^+ c_{\vec{k}} + d_{\vec{k}}^+ d_{\vec{k}}] - V_{\vec{k}}[c_{\vec{k}}^+ c_{\vec{k}} + c_{\vec{k}} c_{\vec{k}}^+ \\
&\quad - (d_{\vec{k}}^+ d_{\vec{k}} + d_{\vec{k}} d_{\vec{k}}^+)] \cos\theta_{\vec{k}} \sin\theta_{\vec{k}}\} \\
&= \sum_{\vec{k}} \left\{\omega[c_{\vec{k}}^+ c_{\vec{k}} + d_{\vec{k}}^+ d_{\vec{k}}] - V_{\vec{k}}[2c_{\vec{k}}^+ c_{\vec{k}} + 1 - (2d_{\vec{k}}^+ d_{\vec{k}} + 1)]\frac{1}{2}\right\}
\end{aligned}$$

where one has used the fact that when $\omega_{\vec{k}}^m = \omega_{\vec{k}}^p$ one has $\tan(2\theta_{\vec{k}}) = \infty$, namely $2\theta_{\vec{k}} = \pi/2$ or $\theta_{\vec{k}} = \pi/4$, or $\cos\theta_{\vec{k}} \sin\theta_{\vec{k}} = 1/2$. One gets

$$\mathcal{H} = \sum_{\vec{k}} \left[(\omega - V_{\vec{k}})c_{\vec{k}}^+ c_{\vec{k}} + (\omega + V_{\vec{k}})d_{\vec{k}}^+ d_{\vec{k}}\right] \tag{17.101}$$

The magnon dispersion curve crosses the phonon dispersion curve when $\omega_{\vec{k}}^m = \omega_{\vec{k}}^p$. Due to their interaction, the degeneracy at $\omega \equiv \omega_{\vec{k}}^m = \omega_{\vec{k}}^p$ is removed: ω is split into two levels $\omega \pm V_{\vec{k}}$ as seen in the above equation.

Problem 6. Magnons in antiferromagnets:

The operators a, a^+, b and b^+ obey the commutation relations (Problem 3 above). Replacing operators $S^\pm$ and S^z in (12.88) by these operators, one gets

$$\begin{aligned}
\mathcal{H} = J \sum_{(l,m)} [&-S^2 + Sf_l(S)a_l f_m(S)b_m + Sa_l^+ f_l(S)b_m^+ f_m(S) + \\
&Sa_l^+ a_l + Sb_m^+ b_m - a_l^+ a_l b_m^+ b_m] - g\mu_B H[\sum_l (S - a_l^+ a_l) \\
&- \sum_m (-S + b_m^+ b_m)]
\end{aligned} \tag{17.102}$$

In a first approximation , one supposes that the number of excited spin waves n is small with respect to $2S$, namely $a_l^+ a_l \ll 2S$ and $b_m^+ b_m \ll 2S$, so that an expansion is possible for $f_l(S)$ and $f_m(S)$. One has then

$$f_l(S) \simeq 1 - \frac{a_l^+ a_l}{4S} + \dots \tag{17.103}$$

$$f_m(S) \simeq 1 - \frac{b_m^+ b_m}{4S} + \dots \tag{17.104}$$

Equation (17.102) becomes at the quadratic order

$$\mathcal{H} \simeq -\frac{ZJNS^2}{2} + JS \sum_{(l,m)} \left(a_l^+ a_l + b_m^+ b_m + a_l^+ b_m^+ + a_l b_m\right)$$

$$+ g\mu_B H \left(\sum_l a_l^+ a_l - \sum_m b_m^+ b_m\right) \tag{17.105}$$

where Z is the coordination number and the following relation has been used

$$\sum_{(l,m)} 1 = \sum_l \sum_{\vec{R}} 1 = Z \sum_l 1 = Z\frac{N}{2} \tag{17.106}$$

$\vec{R}$ is a vector connecting the spin at l to a nearest neighbor belonging to the other sublattice, $N/2$ the total number of spins in a sublattice.

The first term of (17.105) is the classical ground-state energy where neighboring spins are antiparallel (Néel ground state). One introduces now the following Fourier transformations

$$a_l^+ = \sqrt{\frac{2}{N}} \sum_{\vec{k}} e^{i\vec{k}\cdot\vec{l}} a_{\vec{k}}^+ \tag{17.107}$$

$$a_l = \sqrt{\frac{2}{N}} \sum_{\vec{k}} e^{-i\vec{k}\cdot\vec{l}} a_{\vec{k}} \tag{17.108}$$

$$b_m^+ = \sqrt{\frac{2}{N}} \sum_{\vec{k}} e^{-i\vec{k}\cdot\vec{m}} b_{\vec{k}}^+ \tag{17.109}$$

$$b_m = \sqrt{\frac{2}{N}} \sum_{\vec{k}} e^{i\vec{k}\cdot\vec{m}} b_{\vec{k}} \tag{17.110}$$

One can show that the Fourier components $a_{\vec{k}}$, $a_{\vec{k}}^+$, $b_{\vec{k}}$ and $b_{\vec{k}}^+$ obey the boson commutation relations. Putting these into (17.105), one gets

$$\mathcal{H} = -\frac{ZJNS^2}{2} + ZJS \sum_{\vec{k}} [(1+h)a_{\vec{k}}^+ a_{\vec{k}} + (1-h)b_{\vec{k}}^+ b_{\vec{k}} +$$

$$\gamma_{\vec{k}}(a_{\vec{k}} b_{\vec{k}} + a_{\vec{k}}^+ b_{\vec{k}}^+)] \tag{17.111}$$

where

$$\gamma_{\vec{k}} = \frac{1}{Z} \sum_{\vec{R}} e^{i\vec{k}\cdot\vec{R}} \tag{17.112}$$

and

$$h = \frac{g\mu_B H}{ZJS} \qquad (17.113)$$

One sees that $\mathcal{H}$ of Eq. (17.111) does not have a diagonal form of the "harmonic oscillator", namely $a_{\vec{k}}^+ a_{\vec{k}}$ and $b_{\vec{k}}^+ b_{\vec{k}}$. This is because of the existence of the term $a_{\vec{k}} b_{\vec{k}} + a_{\vec{k}}^+ b_{\vec{k}}^+$. One can diagonalize $\mathcal{H}$ with the following transformation

$$\alpha_{\vec{k}} = a_{\vec{k}} \cosh \theta_k + b_{\vec{k}}^+ \sinh \theta_k \qquad (17.114)$$

$$\alpha_{\vec{k}}^+ = a_{\vec{k}}^+ \cosh \theta_k + b_{\vec{k}} \sinh \theta_k \qquad (17.115)$$

$$\beta_{\vec{k}} = a_{\vec{k}}^+ \sinh \theta_k + b_{\vec{k}} \cosh \theta_k \qquad (17.116)$$

$$\beta_{\vec{k}}^+ = a_{\vec{k}} \sinh \theta_k + b_{\vec{k}}^+ \cosh \theta_k \qquad (17.117)$$

where θ_k is a variable to be determined as follows. The inverse transformation gives

$$a_{\vec{k}}^+ = \alpha_{\vec{k}}^+ \cosh \theta_k - \beta_{\vec{k}} \sinh \theta_k \qquad (17.118)$$

$$a_{\vec{k}} = \alpha_{\vec{k}} \cosh \theta_k - \beta_{\vec{k}}^+ \sinh \theta_k \qquad (17.119)$$

$$b_{\vec{k}}^+ = -\alpha_{\vec{k}} \sinh \theta_k + \beta_{\vec{k}}^+ \cosh \theta_k \qquad (17.120)$$

$$b_{\vec{k}} = -\alpha_{\vec{k}}^+ \sinh \theta_k + \beta_{\vec{k}} \cosh \theta_k \qquad (17.121)$$

One can verify that the new operators also obey the commutation relations:

$$\begin{aligned}
[\alpha_{\vec{k}}, \alpha_{\vec{k}'}^+] &= [a_{\vec{k}} \cosh \theta_k + b_{\vec{k}}^+ \sinh \theta_k, a_{\vec{k}'}^+ \cosh \theta_k' + b_{\vec{k}'} \sinh \theta_k'] \\
&= \cosh \theta_k \cosh \theta_k' [a_{\vec{k}}, a_{\vec{k}'}^+] + \cosh \theta_k \sinh \theta_k' [a_{\vec{k}}, b_{\vec{k}'}^+] \\
&\quad + \sinh \theta_k \cosh \theta_k' [b_{\vec{k}}^+, a_{\vec{k}'}^+] + \sinh \theta_k \sinh \theta_k' [b_{\vec{k}}^+, b_{\vec{k}'}] \\
&= \cosh \theta_k \cosh \theta_k' \delta(k, k') + 0 + 0 - \sinh \theta_k \sinh \theta_k' \delta(k, k') \\
&= [\cosh^2 \theta_k - \sinh^2 \theta_k] \delta(k, k') = \delta(k, k') \qquad (17.122)
\end{aligned}$$

The same demonstration is done for the other relations. Replacing now (17.118)-(17.121) in (17.111) one gets

$$\begin{aligned}
\mathcal{H} = &-\frac{ZJNS^2}{2} + ZJS \sum_{\vec{k}} \{\cosh(2\theta_k) - 1 - \gamma_{\vec{k}} \sinh(2\theta_k) \\
&+ [\cosh(2\theta_k) - \gamma_{\vec{k}} \sinh(2\theta_k) + h] \alpha_{\vec{k}}^+ \alpha_{\vec{k}} \\
&+ [\cosh(2\theta_k) - \gamma_{\vec{k}} \sinh(2\theta_k) - h] \beta_{\vec{k}}^+ \beta_{\vec{k}} \\
&- [\sinh(2\theta_k) - \gamma_{\vec{k}} \cosh(2\theta_k)] (\alpha_{\vec{k}} \beta_{\vec{k}} + \alpha_{\vec{k}}^+ \beta_{\vec{k}}^+) \} \qquad (17.123)
\end{aligned}$$

In order that $\mathcal{H}$ is diagonal, the coefficient before the term $\alpha_{\vec{k}}\beta_{\vec{k}} + \alpha_{\vec{k}}^+\beta_{\vec{k}}^+$ should be zero. This requirement allows us to determine the variable θ_k. One has

$$\tanh(2\theta_k) = \gamma_{\vec{k}} \qquad (17.124)$$

Replacing $\sinh(2\theta_k)$ and $\cosh(2\theta_k)$ as functions of $\tanh(2\theta_k) = \gamma_{\vec{k}}$ one obtains

$$\mathcal{H} = -\frac{ZJNS^2}{2} + ZJS\sum_{\vec{k}}\left[\sqrt{1 - \gamma_{\vec{k}}^2} - 1\right]$$

$$+ZJS\sum_{\vec{k}}\left\{\left[\sqrt{1 - \gamma_{\vec{k}}^2} + h\right]\alpha_{\vec{k}}^+\alpha_{\vec{k}} + \left[\sqrt{1 - \gamma_{\vec{k}}^2} - h\right]\beta_{\vec{k}}^+\beta_{\vec{k}}\right\}$$

$$(17.125)$$

One recognizes that for a given wave vector $\vec{k}$, there are two modes corresponding to

$$\epsilon_{\vec{k}}^{\pm} = ZJS\left[\sqrt{1 - \gamma_{\vec{k}}^2} \pm h\right] \qquad (17.126)$$

Without an applied field, these modes are degenerate. It is important to note that for small k, $\gamma_{\vec{k}}^2 \simeq (1 + ak^2 + ...)$ which leads to

$$\epsilon_{\vec{k}}^{\pm} \propto k \qquad (17.127)$$

This result for antiferromagnets is different from $\epsilon_{\vec{k}}^{\pm} \propto k^2$ obtained for ferromagnets. One expects therefore that thermodynamic properties are different for the two cases in particular at low temperatures where small k modes dominate. This will be indeed seen below.

Problem 7. Properties at low temperatures of antiferromagnets:

If one knows the dispersion relation $\epsilon_{\vec{k}}$ one can in principle use formulas of statistical mechanics to study properties of a system as a function of the temperature (see chapter 3). One writes the partition function as follows [see Eq. (3.11) with a change of the notation to avoid a confusion with Z the coordination number used above]:

$$\Xi = \text{Tr} e^{-\beta\mathcal{H}}$$

$$= \sum_{n_k=0}^{\infty}\sum_{n_k'=0}^{\infty}\exp\left\{-\beta\left[E_0 + \sum_{\vec{k}}(n_k\epsilon_{\vec{k}}^+ + n_k'\epsilon_{\vec{k}}^-)\right]\right\}$$

$$= e^{-\beta E_0}\prod_{\vec{k}}\left[\frac{1}{1 - e^{-\beta\epsilon_{\vec{k}}^+}}\frac{1}{1 - e^{-\beta\epsilon_{\vec{k}}^-}}\right] \qquad (17.128)$$

where one has used

$$n_k = \alpha_{\vec{k}}^+ \alpha_{\vec{k}} \tag{17.129}$$

$$n_k' = \beta_{\vec{k}}^+ \beta_{\vec{k}} \tag{17.130}$$

$$E_0 = -\frac{ZJNS^2}{2} + ZJS \sum_{\vec{k}} \left[\sqrt{1 - \gamma_{\vec{k}}^2} - 1 \right] \tag{17.131}$$

The free energy is written as

$$F = -k_B T \ln \Xi = E_0 + k_B T \sum_{\vec{k}} \left[\ln(1 - \epsilon_{\vec{k}}^+) + \ln(1 - \epsilon_{\vec{k}}^-) \right]$$

$$= E_0 + 2k_B T \sum_{\vec{k}} \ln(1 - \epsilon_{\vec{k}}) \quad \text{if } h = 0 \tag{17.132}$$

To calculate various thermodynamic properties, one uses the above expression of F as seen below.

For $h = 0$, one can calculate the energy as follows

$$\mathcal{H} = -\frac{ZJNS^2}{2} + ZJS\frac{N}{2} + \sum_{\vec{k}} \epsilon_{\vec{k}}(\alpha_{\vec{k}}^+ \alpha_{\vec{k}} + \beta_{\vec{k}}^+ \beta_{\vec{k}} + 1)$$

$$= -\frac{ZJNS(S+1)}{2} + 2\sum_{\vec{k}} \epsilon_{\vec{k}}(n_k + \frac{1}{2}) \tag{17.133}$$

where one has used $\sum_{\vec{k}} 1 = \frac{N}{2}$ = number of microscopic states in the first Brillouin zone which is equal the number of spins in each sublattice. At $T = 0$, $n_k = n_k' = 0$ one obtains

$$E(T = 0) = -\frac{ZJNS(S+1)}{2} + \sum_{\vec{k}} \epsilon_{\vec{k}} \tag{17.134}$$

The second term is a correction to the classical ground-state energy $-\frac{ZJNS^2}{2}$ (Néel state). This correction is due to quantum fluctuations in analogy with the zero-point phonon energy.

At low temperatures, one calculates the magnon energy by the use of a low-temperature expansion. One gets

$$< E >= -T^2 \frac{\partial}{\partial T}(\frac{F}{T}) \simeq E(T = 0) + aT^4 \tag{17.135}$$

where a is a coefficient proportional to ZJS. One notes that the power of T is different from that of the ferromagnetic case [Eq. (12.77)]. This is a consequence of $\epsilon_{\vec{k}} \propto k$ for small k.

To calculate the sublattice magnetization, one writes for the $\uparrow$ sublattice

$$
\begin{aligned}
M &= \sum_l (S - < a_l^+ a_l >) \\
&= \frac{N}{2} S - \frac{2}{N} \sum_{\vec{k}} \sum_{\vec{k}'} < a_{\vec{k}}^+ a_{\vec{k}'} > \sum_{\vec{l}} e^{i(\vec{k} - \vec{k}') \cdot \vec{l}} \\
&= \frac{N}{2} S - \frac{2}{N} \sum_{\vec{k}} \sum_{\vec{k}'} < a_{\vec{k}}^+ a_{\vec{k}'} > \frac{N}{2} \delta_{\vec{k}, \vec{k}'} \\
&= \frac{N}{2} S - \sum_{\vec{k}} < a_{\vec{k}}^+ a_{\vec{k}} > \\
&= \frac{N}{2} S - \sum_{\vec{k}} < \cosh^2 \theta_k a_{\vec{k}}^+ a_{\vec{k}} + \sinh^2 \theta_k \beta_{\vec{k}}^+ \beta_{\vec{k}} + \sinh^2 \theta_k >
\end{aligned}
$$

$$(17.136)$$

where one has used successively the Fourier transformation and relations (17.118)-(17.121). One expresses now $\cosh^2 \theta_k$ and $\sinh^2 \theta_k$ in terms of $\gamma_{\vec{k}}$ using (17.124), then one uses

$$
< \alpha_{\vec{k}}^+ \alpha_{\vec{k}} > = < \beta_{\vec{k}}^+ \beta_{\vec{k}} > = \frac{1}{e^{\beta \epsilon_{\vec{k}}} - 1} \qquad (17.137)
$$

to obtain M. At low temperatures, using an expansion of $\frac{1}{e^{\beta \epsilon_{\vec{k}}} - 1}$ for small k with $\epsilon_{\vec{k}} \propto k$, one gets

$$
M \simeq \frac{N}{2} (S - \Delta S - AT^2) \qquad (17.138)
$$

where $\Delta S = \frac{2}{N} \sum_{\vec{k}} \sinh^2 \theta_k$ is independent of T, and A a coefficient.

One sees that at $T = 0$, the magnetization is $S - \Delta S$ which is smaller than the spin magnitude S. ΔS is called the zero-point spin contraction. ΔS depends on the lattice: $\Delta S \simeq 0.197$ for an antiferrromagnetic square lattice, $\Delta S \simeq 0.078$ for a cubic antiferromagnet of NaCl type.

Note that the sublattice magnetization of an antiferromagnet depends on T^2 while the ferromagnetic magnetization depends on $T^{3/2}$ [Eq. (12.74)].

Problem 8. Magnons in helimagnets:

Let $(\vec{\xi}_i, \vec{\eta}_i, \vec{\zeta}_i)$ be the unit vectors making a direct trihedron at the site i, namely $\vec{\eta}_i$ is parallel to the y axis as shown in Fig. 17.12.

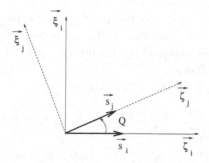

Fig. 17.12 Local coordinates defined for two spins $\vec{S}_i$ and $\vec{S}_j$. The axis $\vec{\eta}$ is common for the two spins.

One supposes in addition that the quantization axis of the spin $\vec{S}_i$ coincides with the local axis $\vec{\zeta}_i$.

One uses now the following transformation in the local coordinates associated with $\vec{S}_i$ and $\vec{S}_j$

$$\vec{\eta}_j = \vec{\eta}_i \tag{17.139}$$

$$\vec{\zeta}_j = \cos Q \vec{\zeta}_i + \sin Q \vec{\xi}_i \tag{17.140}$$

$$\vec{\xi}_j = -\sin Q \vec{\zeta}_i + \cos Q \vec{\xi}_i \tag{17.141}$$

One writes

$$\vec{S}_i = S_i^x \vec{\xi}_i + S_i^y \vec{\eta}_i + S_i^z \vec{\zeta}_i \tag{17.142}$$

$$\vec{S}_j = S_j^x \vec{\xi}_j + S_j^y \vec{\eta}_j + S_j^z \vec{\zeta}_j \tag{17.143}$$

Their scalar product becomes

$$
\begin{aligned}
\vec{S}_i \cdot \vec{S}_j &= [S_i^x \vec{\xi}_i + S_i^y \vec{\eta}_i + S_i^z \vec{\zeta}_i] \cdot [S_j^x(-\sin Q \vec{\zeta}_i + \cos Q \vec{\xi}_i) \\
&\quad + S_j^y \vec{\eta}_i + S_j^z(\cos Q \vec{\zeta}_i + \sin Q \vec{\xi}_i)] \\
&= S_i^z(-\sin Q S_j^x + \cos Q S_j^z) + S_i^y S_j^y \\
&\quad + S_i^x(\cos Q S_j^x + \sin Q S_j^z) \\
&= \cos Q S_i^z S_j^z - \sin Q S_i^z \left[\frac{S_j^+ + S_j^-}{2}\right] - \frac{(S_i^+ - S_i^-)(S_j^+ - S_j^-)}{4} \\
&\quad + \cos Q \frac{(S_i^+ + S_i^-)(S_j^+ + S_j^-)}{4} + \sin Q \left[\frac{S_i^+ + S_i^-}{2}\right] S_j^z
\end{aligned}
$$

To be general, the angle Q should depend on positions of $\vec{S}_i$ and $\vec{S}_j$. One defines $\cos Q = \cos(\vec{Q} \cdot \vec{R}_{ij})$ where $\vec{Q}$ is the vector of modulus Q, perpendicular to the plane of the angle Q, namely plane $(\vec{\zeta}, \vec{\xi})$, and $\vec{R}_{ij}$ the vector connecting the positions of $\vec{S}_i$ and $\vec{S}_j$. One shall keep in the following $J(\vec{R}_{ij})$ as interaction between $\vec{S}_i$ and $\vec{S}_j$ which will be replaced by J_1 and J_2 depending on $\vec{R}_{ij}$ at the end of the calculation.

Equation (12.82) is rewritten as

$$
\begin{aligned}
\mathcal{H} = -\frac{1}{4}\sum_{(i,j)} J(\vec{R}_{ij})\{&(S_i^+ S_j^- + S_i^- S_j^+)[1 + \cos(\vec{Q}\cdot\vec{R}_{ij})] \\
&-(S_i^+ S_j^+ + S_i^- S_j^-)[1 - \cos(\vec{Q}\cdot\vec{R}_{ij})] + 4S_i^z S_j^z \cos(\vec{Q}\cdot\vec{R}_{ij}) \\
&+2[(S_i^+ + S_i^-)S_j^z - S_i^z(S_j^+ + S_j^-)]\sin(\vec{Q}\cdot\vec{R}_{ij})\} \\
&-\frac{D}{4}\sum_i (S_i^+ - S_i^-)^2
\end{aligned}
\tag{17.144}
$$

With this Hamiltonian, one can choose an appropriate method to calculate the magnon dispersion relation. One can use the Holstein-Primakoff method by replacing the operators $S^\pm$ and S^z by (12.52)-(12.55): one obtains the magnon dispersion relation

$$
\begin{aligned}
\mathcal{H} = -NSJ(\vec{Q}) + \frac{S}{2}\sum_{\vec{k}}[&A(\vec{k},\vec{Q})(a_{\vec{k}} a_{\vec{k}}^+ + a_{\vec{k}}^+ a_{\vec{k}}) \\
&+B(\vec{k},\vec{Q})(a_{\vec{k}} a_{-\vec{k}} + a_{\vec{k}}^+ a_{-\vec{k}}^+)]
\end{aligned}
\tag{17.145}
$$

where

$$
J(\vec{k}) = \sum_{\vec{R}_{ij}} J(\vec{R}_{ij})\exp(\vec{k}\cdot\vec{R}_{ij})
\tag{17.146}
$$

$$
A(\vec{k},Q) = \left[2J(\vec{Q}) - J(\vec{k}) - \frac{1}{2}[J(\vec{k}+\vec{Q}) + J(\vec{k}-\vec{Q})] + D\right]
\tag{17.147}
$$

$$
B(\vec{k},\vec{Q}) = \left[J(\vec{k}) - \frac{1}{2}[J(\vec{k}+\vec{Q}) + J(\vec{k}-\vec{Q})] - D\right]
\tag{17.148}
$$

The Hamiltonian (17.145) can be diagonalized by introducing the new operators $\alpha_{\vec{k}}$ and $\alpha_{\vec{k}}^+$ just as in the antiferromagnetic case studied above

$$
a_{\vec{k}} = \alpha_{\vec{k}}\cosh\theta_k - \alpha_{-\vec{k}}^+\sinh\theta_k
\tag{17.149}
$$

$$
a_{\vec{k}}^+ = \alpha_{\vec{k}}^+\cosh\theta_k - \alpha_{-\vec{k}}\sinh\theta_k
\tag{17.150}
$$

where $\alpha_{\vec{k}}$ and $\alpha_{\vec{k}}^+$ obey the boson commutation relations [see similar transformation in Eqs. (17.114)-(17.117)]. Hamiltonian (17.145) is diagonal if one takes

$$\tanh(2\theta_k) = \frac{B(\vec{k}, \vec{Q})}{A(\vec{k}, \vec{Q})} \tag{17.151}$$

One then has

$$\mathcal{H} = \frac{S}{2} \sum_{\vec{k}} \hbar\omega_{\vec{k}}[\alpha_{\vec{k}}^+ \alpha_{\vec{k}} + \alpha_{\vec{k}} \alpha_{\vec{k}}^+] \tag{17.152}$$

where the energy of the magnon mode $\vec{k}$ is

$$\hbar\omega_{\vec{k}} = \sqrt{A(\vec{k}, \vec{Q})^2 - B(\vec{k}, \vec{Q})^2} \tag{17.153}$$

In the case of the body-centered cubic lattice, one has

$$J(\vec{k}) = 8J_1 \cos(k_x a/2)\cos(k_y a/2)\cos(k_z c/2) + 2J_2 \cos(k_z c)$$
$$= 2J_2\left[-4\cos Q \cos(k_x a/2)\cos(k_y a/2)\cos(k_z c/2) + \cos(k_z c)\right]$$

where $\epsilon_c = 1$ has been used, a and c being the lattice constants (one uses c for the helical axis). Figure 12.6 shows the magnon spectrum for J_2/J_1 corresponding to $Q = \pi/3$.

Problem 9. Triangular antiferromagnet:

The energy of the three interacting spins on a triangle is written as

$$E = J(\vec{S}_1.\vec{S}_2 + \vec{S}_2.\vec{S}_3 + \vec{S}_3.\vec{S}_1) \tag{17.154}$$

where J is positive (antiferromagnetic). One writes E under the following equivalent form

$$E = \frac{1}{2}J\left(\sum_{i=1}^{3} \vec{S}_i\right)^2 - \frac{3}{2}JS^2 \tag{17.155}$$

where one has used $S_i^2 = S^2$ ($i = 1, 2, 3$). Since the last term is a constant, one sees that the minimum of E is the minimum of $\left(\sum_{i=1}^{3} \vec{S}_i\right)^2$, namely,

$$0 = \left(\sum_{i=1}^{3} \vec{S}_i\right)^2 = (\vec{S}_1 + \vec{S}_2 + \vec{S}_3)^2$$
$$\text{or } 0 = (\vec{S}_1 + \vec{S}_2 + \vec{S}_3)$$

The sum of the three vectors is equal zero, it means that they form a closed path. Since they have the same modulus, the closed path is an equilateral triangle: the angle between two spins is thus $+2\pi/3$ or $-2\pi/3$, going from $\vec{S}_1$ to $\vec{S}_3$. These two values indicate two degenerate $\pm$ chiralities. The $+$ chirality is shown on the left of Fig. 12.7.

The above result is valid for vector spins, namely XY and Heisenberg spins.

Problem 10. Ground state of Villain's model:

The energy of a plaquette of the 2D Villain's model with XY spins defined in Fig. 12.7 with S_1 and S_2 linked by the antiferromagnetic interaction η, is written as

$$H_p = \eta S_1 \cdot S_2 - S_2 \cdot S_3 - S_3 \cdot S_4 - S_4 \cdot S_1 \qquad (17.156)$$

where $(S_i)^2 = 1$. The variational method gives

$$\delta \left[H_p - \frac{1}{2} \sum_{i=1}^{4} \lambda_i (S_i)^2 \right] = 0 \qquad (17.157)$$

By symmetry, $\lambda_1 = \lambda_2 \equiv \lambda$, $\lambda_3 = \lambda_4 \equiv \mu$. We have

$$\lambda S_1 - \eta S_2 + S_4 = 0 \qquad (17.158)$$
$$-\eta S_1 + \lambda S_2 + S_3 = 0 \qquad (17.159)$$
$$S_2 + \mu S_3 + S_4 = 0 \qquad (17.160)$$
$$S_1 + S_3 + \mu S_4 = 0 \qquad (17.161)$$

Hence,

$$(\lambda - \mu)(S_1 + S_2) + (S_3 + S_4) = 0 \qquad (17.162)$$
$$(S_1 + S_2) + (\mu + 1)(S_3 + S_4) = 0 \qquad (17.163)$$

We deduce

$$\mu = -\left[\frac{1 + \eta}{\eta} \right]^{1/2} \qquad (17.164)$$

$$\lambda = \eta\mu = -[\eta(1 + \eta)]^{1/2} \qquad (17.165)$$

To calculate the angle between two spins, for instance S_1 and S_4, we write

$$(\lambda S_1 + S_4)^2 = (-\eta S_2)^2$$

Hence

$$S_1 \cdot S_4 = \cos\theta_{14} = \frac{1}{2\lambda}(\eta^2 - \lambda^2 - 1) = \frac{1}{2}\left(\frac{\eta+1}{\eta}\right)^{1/2}$$

We find in the same manner,

$$\cos\theta_{23} = \cos\theta_{34} = \cos\theta_{41} = \frac{1}{2}\left(\frac{\eta+1}{\eta}\right)^{1/2}$$

We have

$$\theta_{14} = \theta_{23} = \theta_{34} = \theta_{41} \equiv \theta.$$

Note that $|\theta_{12}| = 3|\theta|$. These solutions exist if $|\cos\theta| \le 1$, namely $\eta > \eta_c = 1/3$. When $\eta = 1$, we have $\theta = \pi/4$, $\theta_{12} = 3\pi/4$ (see Fig. 12.7).

17.6 Solutions of problems of chapter 13

Problem 1. Ising spin model in the mean-field approximation:

The energy of the spin σ_i in the mean field of its neighbors is

$$E_i = -JC < \sigma > \sigma_i \qquad (17.166)$$

where C is the coordination number and $< \sigma >$ the average value of a neighboring spin. The partition function of the spin σ_i is

$$Z_i = \sum_{\sigma_i = \pm 1} \exp(\beta JC < \sigma > \sigma_i) = \cosh(\beta JC < \sigma >) \qquad (17.167)$$

The average value of the spin σ_i is thus

$$< \sigma_i > = \sum_{\sigma_i = \pm 1} \sigma_i \exp(\beta JC < \sigma > \sigma_i)/Z_i = \tanh(\beta JC < \sigma >)$$

$$(17.168)$$

In a homogeneous system, $< \sigma_i >$ should be identical with the average value of its neighbors, namely $< \sigma >$. Therefore, replacing the left-hand side $< \sigma_i >$ by $< \sigma >$, one obtains the mean-field equation $< \sigma > = \tanh(\beta JC < \sigma >)$.

Problem 2. Interaction between next-nearest neighbors in a centered cubic lattice:

a) All spins are parallel at $T = 0$. All interactions are fully satisfied.

b) The hypothesis of the mean-field theory: all neighboring spins of a spin are replaced by an average value which is used to calculate the value of the spin under consideration.

c) The energy of a spin at $T = 0$ is $E = -Z_1 J_1 - Z_2 J_2$ where Z_1 and Z_2 are the numbers of nearest neighbors and of next-nearest neighbors, respectively. For a body-centered cubic lattice, $Z_1 = 8$, $Z_2 = 6$. E is the energy which maintains the spin ordering: the lower it is, the higher the temperature is needed to destroy the ordering. Thus, the stronger J_2 is, the higher the transition temperature becomes.

d) The same calculation as that in the chapter by replacing CJ with $Z_1 J_1 + Z_2 J_2$ in Eqs. (13.19)-(13.21) and in the following equations to obtain the final mean-field equation.

e) The critical temperature is obtained by replacing CJ in Eq. (13.31) by $Z_1 J_1 + Z_2 J_2$.

f) Now one supposes $J_2 < 0$. When $|J_2| >> J_1$, it is obvious that the J_2 interaction imposes the antiferromagnetic ordering to make the overall energy negative. The spins on the cube corners form an antiferromagnetic sublattice, the centered spins form another antiferromagnetic sublattice, independent of the first one. Since each spin has 4 up neighbors and 4 down neighbors (make a figure to convince yourself), its interaction energy with nearest neighbors is zero, independent of J_1. The energy of such a spin configuration is thus $E_{Antif} = Z_2 J_2 = -Z_2 |J_2|$.

If $|J_2| << J_1$ then the ferromagnetic state is more favorable. Its energy is $E_{Ferro} = -Z_1 J_1 - Z_2 J_2 = -Z_1 J_1 + Z_2 |J_2|$.

The critical value of $|J_2|$ below which the ferromagnetic state is stable is determined by solving $E_{Ferro} < E_{Antif}$. One has $|J_2^c| = \frac{Z_1 J_1}{2 Z_2}$ or $J_2^c = -\frac{Z_1 J_1}{2 Z_2}$. When $|J_2| < |J_2^c|$ (or $J_2 > J_2^c$) one has the ferromagnetic ordering. Otherwise, one has the antiferromagnetic one.

Problem 3. Interaction between next-nearest neighbors in a square lattice:

Guide: One follows the same method as in the previous exercises.

Problem 4. System of two spins:

a) One writes

$$\vec{S}_1 \cdot \vec{S}_2 = S_1^z S_2^z + (S_1^+ S_2^- + S_1^- S_2^+)/2$$

The states of two spins $1/2$ are

$$\phi_1 = |1/2, 1/2\rangle, \quad \phi_2 = |1/2, -1/2\rangle,$$

$$\phi_3 = |-1/2, 1/2\rangle, \quad \phi_4 = |-1/2, -1/2\rangle.$$

To calculate

$$[-2J[S_1^z S_2^z + (S_1^+ S_2^- + S_1^- S_2^+)/2] - D[(S_1^z)^2 + (S_2^z)^2] - B(S_1^z + S_2^z)]|\phi_i\rangle$$

one uses

$$S^{\pm}|jm\rangle = [j(j+1) - m(m \pm 1)]^{1/2}\hbar|j, m \pm 1\rangle \quad (17.169)$$
$$(j = 1/2, m = \pm 1/2)$$
$$S^z|m\rangle = \hbar m|m\rangle \quad (17.170)$$

One obtains a matrix 4×4. A simple diagonalization gives the following eigenvalues for the two-spin cluster:
$\epsilon_1 = -J/2 - D/2 - B$ ($\uparrow \uparrow$), $\epsilon_2 = 3J/2 - D/2$ ($\uparrow \downarrow - \downarrow \uparrow$),
$\epsilon_3 = -J/2 - D/2$ ($\uparrow \downarrow + \downarrow \uparrow$), $\epsilon_4 = -J/2 - D/2 + B$ ($\downarrow \downarrow$) (one has taken $\hbar = 1$).

Problem 5. Improvement of the mean-field approximation:

a) One puts the cluster of two spins $\vec{S}_i$ and $\vec{S}_j$ found in the preceding exercise in a lattice: it has $(Z - 1)$ neighbors. One treats the interaction of the 4 cluster configurations in zero field ($B = 0$) with these neighbors by the mean-field theory. The energy of the cluster in the crystal depends on the embedded cluster spin configuration, they are in increasing energies:
$\phi_1 = (\uparrow \uparrow) \rightarrow E_1 = -J/2 - 2J(Z - 1) < S^z >$
$(\phi_2 + \phi_3)/2 = (\uparrow \downarrow + \downarrow \uparrow) \rightarrow E_2 = -J/2$
$\phi_4 = (\downarrow \downarrow) \rightarrow E_3 = -J/2 + 2J(Z - 1) < S^z >$
$(\phi_2 - \phi_3)/2 = (\uparrow \downarrow - \downarrow \uparrow) \rightarrow E_4 = 3J/2$

One considers the cluster of two spins as a superspin with the z component $S^z = (S^z_i + S^z_j)/2$. One has

$$< S^z >= Tr\frac{1}{2}(S^z_i + S^z_j)\exp(-\beta E)/Tr\exp(-\beta E)$$

where

$$Tr\exp(-\beta E) = \exp(\beta J/2)\exp(\beta X) + \exp(\beta J/2)$$
$$+ \exp(\beta J/2)\exp(-\beta X)$$
$$+ \exp(-\beta 3J/2)$$
$$(X \equiv 2J(Z-1) < S^z >)$$

$$Tr\frac{1}{2}(S^z_i + S^z_j)\exp(-\beta E) = \frac{1}{2}\exp(\beta J/2)\exp(\beta X) + 0$$
$$- \frac{1}{2}\exp(\beta J/2)\exp(-\beta X) + 0$$
$$= \exp(\beta J/2)\sinh\beta X$$

Hence,

$$< S^z >= \frac{\sinh\beta X}{2\left[\cosh\beta X + \exp(-\beta J)\cosh\beta J\right]} \qquad (17.171)$$

One sees that $< S^z >= 0$ is a solution of this equation. An expansion around $< S^z >= 0$ gives

$$2 < S^z > \left[\frac{-3 + 2\beta(Z-1)J - e^{-2\beta J}}{2}\right] = \beta^2 4(Z-1)^2 J^2 < S^z >^3 \qquad (17.172)$$

The solution $< S^z >\neq 0$ is possible if

$$-3 + 2\beta(Z-1)J - e^{-2\beta J} > 0.$$

T_c is obtained by solving $-3 + 2\beta_c(Z-1)J - e^{-2\beta_c J} = 0$ where $\beta_c = (k_B T_c)^{-1}$. One obtains

$$e^{-2J/k_B T_c} + 3 - 2(Z-1)J/k_B T_c = 0 \qquad (17.173)$$

Problem 6. Chain of Ising spins by micro-canonical method:

a) In the ground state all spins are parallel, the system energy is thus $E = -JN$. The ground-state is two-fold degenerate: all spins are $\uparrow$ or $\downarrow$.

b) The first excited state: one spin is reversed ... ↑↑↑↓↑↑↑ ... or a block of spins is reversed ... ↑↑↑↓↓↓↑↑↑ .... The energies of the two states are identical with two "unsatisfied bonds" of energy $+J$. The energy of the system in the first excited state is thus $E(2) = -J(N-2) + 2J = -J(N - 2 \times 2)$ where 2 is the number of unsatisfied bonds. The degeneracy of the first excited state is equal to the number of ways to choose two unsatisfied bonds among N bonds of the system. This is given by $g(2) = 2C_N^2$ where the factor 2 results from the global reversal of the whole system.

One deduces that the energy $E(2n)$ of a state in which there are $2n$ unsatisfied bonds is $E(2n) = -J(N - 2n) + 2nJ = -J(N - 2 \times 2n)$. The degeneracy is $g(2n) = 2C_N^{2n}$.

The energy of the system is maximum when all bonds are unsatisfied, namely $2n = N$ so that $E_{max} = 0$. The degeneracy of this state is $g = 2C_N^N = 2$.

c) For a given energy $E(2n)$, the entropy is

$$S = k_B \ln g(2n) = k_B[\ln 2 + \ln N! - \ln(2n)! - ln(N - 2n)!]$$
$$\simeq k_B[\ln 2 + N \ln N - 2n \ln 2n - (N - 2n) \ln(N - 2n)]$$

where the Stirling formula (A.12) has been used. The micro-canonical temperature is

$$T^{-1} = \frac{\partial S}{\partial E} = \frac{\partial S}{\partial n}\frac{\partial n}{\partial E}$$
$$= \frac{k_B}{2J} \ln \frac{N - 2n}{2n} \qquad (17.174)$$

from which,

$$\frac{N - 2n}{2n} = \exp(2J/k_B T) = \frac{N}{2n} - 1 = 1/x - 1 \qquad (17.175)$$

where x is the percentage of unsatisfied bonds $x = \frac{2n}{N}$. One obtains

$$x = \frac{1}{1 + \exp(2J/k_B T)} \qquad (17.176)$$

At low temperatures, $x \to 0$: all spins are parallel (the system is ferromagnetic). At high temperatures, $x \to 1/2$: half of the bonds are unsatisfied, the system is disordered (paramagnetic).

Problem 7. Chain of Ising spins by canonical method:

One considers again the system defined in Problem 6 but in the canonical situation: it is maintained at the temperature T.

a) The partition function is given by

$$Z = \sum_{n=0}^{N} g(2n) \exp(-\beta E(2n)) = 2 \sum_{n=0}^{N} C_N^{2n} \exp[\beta J(N - 4n)]$$

$$= 2 \exp(\beta JN) \sum_{n=0}^{N} C_N^{2n} \exp(-\beta 4Jn) \tag{17.177}$$

In combining the two following Newton relations, one has

$$(1 + u)^N + (1 - u)^N = \sum_{n=0}^{N} C_N^n u^n [1 + (-1)^n]$$

$$= 2 \sum_{n'=0}^{N/2} C_N^{2n'} u^{2n'} \tag{17.178}$$

since all odd terms in n are canceled out. Putting $u = \exp(-\beta 2J)$, and using (17.178), one rewrites (17.177) as

$$Z = 2 \exp(\beta JN) \sum_{n=0}^{N} C_N^{2n} \exp(-\beta 4Jn)$$

$$= \exp(\beta JN) \left[[1 + \exp(-\beta 2J)]^N + [1 - \exp(-\beta 2J)]^N \right]$$

$$= \left[[\exp(\beta J) + \exp(-\beta J)]^N + [\exp(\beta J) - \exp(-\beta J)]^N \right]$$

$$= 2^N \left[\cosh^N(\beta J) + \sinh^N(\beta J) \right] \tag{17.179}$$

This relation of Z is exact (see Problem 10 below).

b) The system energy is thus

$$\overline{E} = -\frac{\partial \ln E}{\partial \beta} = -\frac{\partial}{\partial \beta} \ln 2^N [\cosh^N(\beta J) + \sinh^N(\beta J)]$$

$$= -JN \left[\frac{\cosh^{N-1}(\beta J) \sinh(\beta J) + \sinh^{N-1}(\beta J) \cosh(\beta J)}{\cosh^N(\beta J) + \sinh^N(\beta J)} \right]$$

c) The average percentage $\overline{x}$ of unsatisfied bonds is

$$
\overline{x} = \frac{2\overline{n}}{N} = \frac{1}{2} + \frac{\overline{E}}{JN}
$$

$$
= \frac{1}{2}\left[1 - \frac{\cosh^{N-1}(\beta J)\sinh(\beta J) + \sinh^{N-1}(\beta J)\cosh(\beta J)}{\cosh^{N}(\beta J) + \sinh^{N}(\beta J)}\right]
$$

$$
= \frac{e^{-\beta J}}{2}\left[\frac{\cosh^{N-1}(\beta J) - \sinh^{N-1}(\beta J)}{\cosh^{N}(\beta J) + \sinh^{N}(\beta J)}\right]
$$

At low temperatures, $\overline{x} \to 0$. At high temperatures, $\overline{x} \to 1/2$. One finds again here the results using the micro-canonical method found in Problem 6.

The curves $\overline{E}$ and the calorific capacity C are plotted in Problem 10 below (Fig. 17.16).

Problem 8. Mean-field approximation for antiferromagnets:

a) Ground-state spin configuration:
In zero applied field, the neighboring spins are antiparallel, except in geometrically frustrated systems. A few antiferromagnetic systems are displayed in Fig. 17.13.

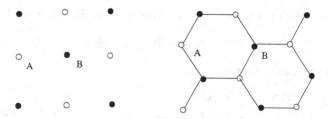

Fig. 17.13 Antiferromagnetic ordering: black and white circles denote ↑ and ↓ spins, respectively.

b) One defines two sublattices: one contains the up spins and the other the down spins, indicated by indices l and m respectively. The mean-field theory applied above to ferromagnets can be applied in the same manner to antiferromagnets. write two coupled mean-field equations for two sublattices. One has the following mean-field energies of l and m spins

$$H_l = CJ < S_-^z > S_l^z + [CJ < \Delta S_- > -g\mu_B H_0] S_l^z$$
$$(17.180)$$

$$H_m = CJ < S_+^z > S_m^z + [CJ < \Delta S_+ > -g\mu_B H_0] S_m^z$$
$$(17.181)$$

where C is the coordination number, $< S_l^z >=< S_+^z > + < \Delta S_+ >$ denotes the average value of S_l^z, and $< \Delta S_+ >$ the spin variation induced by the applied field. The amplitude of $\vec{H}_0$ is supposed to be very small hereafter.

With H_l, one calculates $< S_l^z >$ as follows

$$< S_l^z > = < S_+^z > + < \Delta S_+ > = \frac{\mathrm{Tr} S_l^z e^{-\beta H_l}}{\mathrm{Tr} e^{-\beta H_l}}$$
$$= S B_S(x) \tag{17.182}$$

where $B_S(x)$ is the Brillouin function given by

$$B_S(x) = \frac{2S+1}{2S} \coth \frac{(2S+1)x}{2S} - \frac{1}{2S} \coth \frac{x}{2S} \tag{17.183}$$

with

$$x = \beta[-CJS(< S_-^z > + < \Delta S_- >) + g\mu_B S H_0] \tag{17.184}$$

For weak fields, one expands the function $B_S(x)$ around

$$x_0 = -\beta CJS < S_-^z > .$$

One then obtains

$$< S_+^z > + < \Delta S_+ >$$
$$\simeq S B_S(-\beta CJS < S_-^z >) - \beta[CJS^2 < \Delta S_- >$$
$$-g\mu_B S^2 H_0] B_S'(x_0)$$

therefore

$$< S_+^z > \simeq S B_S(-\beta CJS < S_-^z >) \tag{17.185}$$
$$< \Delta S_+ > \simeq -\beta \left[CJS^2 < \Delta S_- > -g\mu_B S^2 H_0 \right] B_S'(x_0) \tag{17.186}$$

$B_S'(x_0)$ being the derivative of $B_S(x_0)$ with respect to x taken at x_0.

In the same manner, one obtains for the down spin $< S_m^z >$:

$$< S_-^z > \simeq S B_S(-\beta C J S < S_+^z >) \tag{17.187}$$

$$< \Delta S_- > \simeq -\beta \left[C J S^2 < \Delta S_+ > -g\mu_B S^2 H_0 \right] B_S'(x_0^-) \tag{17.188}$$

with $x_0^- = -\beta C J S < S_+^z >$.

c) If the two sublattices are symmetric, namely

$$< S_+^z >= - < S_-^z > \equiv < S^z >$$

Equations (17.185) and (17.187) are equivalent because the Brillouin function is an odd function. One then has only one implicit equation for $< S^z >$ to solve:

$$< S^z >= S B_S(\beta C J S < S^z >) \tag{17.189}$$

This mean-field equation for a sublattice spin is the same as the mean-field equation for ferromagnets. One has thus the same result on the temperature dependence and on the critical temperature. Note that the critical temperature for antiferromagnets is called "Néel temperature" and denoted by T_N:

$$\frac{k_B T_N}{J} = \frac{C S(S+1)}{3} \tag{17.190}$$

Note that one did not use the factor 2 in the Hamiltonian (13.124). So, this result is the same as (13.31) for ferromagnets.

d) Susceptibility: For $< \Delta S_\pm >$, one has

$$< \Delta S_+ >=< \Delta S_- >\equiv< \Delta S >.$$

$B_S'(x)$ is an even function of x, therefore

$$< \Delta S >= -\beta \left[C J S^2 < \Delta S > -g\mu_B S^2 H_0 \right] B_S'(\beta C J S < S^z >) \tag{17.191}$$

The susceptibility is given by

$$\chi_\| = (\frac{\partial M}{\partial H_0})_{H_0=0} = \frac{N g\mu_B < \Delta S >}{H_0}$$

$$= \frac{N(g\mu_B S)^2 B_S'(\beta C J S < S^z >)}{k_B T + C J S^2 B_S'(\beta C J S < S^z >)} \tag{17.192}$$

When $T \to 0$, $B_S'(...)$ tends to 0 faster than T. One deduces that $\chi_\| = 0$. On the contrary, for $T \geq T_N$, $B_S'(...) \simeq \frac{S+1}{3S}$, one gets

$$\chi_\| = \frac{N(g\mu_B)^2 S(S+1)}{3k_B(T + T_N)} \tag{17.193}$$

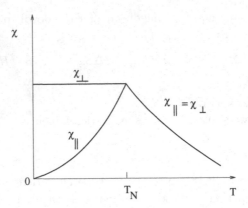

Fig. 17.14   Susceptibility $\chi_\parallel$ and $\chi_\perp$ of an antiferromagnet versus T.

where one notices the $+$ sign in front of T_N, in contrast to the ferromagnetic case. There is thus no divergence of the susceptibility at the phase transition for an antiferromagnet.

In the case where the applied field is also weak but perpendicular to the z axis, for example $\vec{H}_0 \parallel \vec{Ox}$, one modifies (17.180) and (17.181) to obtain

$$\chi_\perp(T \geq T_N) = \frac{N(g\mu_B)^2 S(S+1)}{3k_B(T+T_N)} = \chi_\parallel(T \geq T_N) \qquad (17.194)$$

and

$$\chi_\perp(T \leq T_N) = \frac{N(g\mu_B)^2}{4CJ} = \text{constant}. \qquad (17.195)$$

Figure 17.14 shows $\chi_\parallel$ and $\chi_\perp$ versus T.

Comment: In materials which have magnetic domains or in powdered systems, experimental susceptibility at $T \leq T_N$ is an average with spatial weight coefficients $1/3$ and $2/3$:

$$\chi(T \leq T_N) = \frac{1}{3}\chi_\parallel + \frac{2}{3}\chi_\perp \qquad (17.196)$$

Problem 9. Ferrimagnets by mean-field theory:

Consider a system of Heisenberg spins which is composed of two sublattices, sublattice A containing $\uparrow$ spins of amplitude S_A and sublattice B containing $\downarrow$ spins of amplitude S_B. The Hamiltonian is written as

$$\mathcal{H} = J_1 \sum_{(l,m)} \vec{S}_l \cdot \vec{S}_m \qquad (17.197)$$

where l and m indicate the sites of A and B, respectively. One can start with equations (17.185) and (17.187) for two sublattices in zero applied field:

$$< S_A^z > = S_A B_{S_A}(-\beta C J_1 S_A < S_B^z >) \qquad (17.198)$$
$$< S_B^z > = S_B B_{S_B}(-\beta C J_1 S_B < S_A^z >) \qquad (17.199)$$

where C is the coordination number. Since the sublattices are not equivalent because $S_A \neq S_B$, one has to solve these two coupled equations by iteration. One puts $M_A =< S_A^z >$ and $M_B =< S_B^z >$. At $T = 0$, the expansion of the functions $B_{S_A}(...)$ and $B_{S_B}(...)$ gives $M_A = S_A$ and $M_B = S_B$ (see section 13.3). At low temperatures, one can obtain the solution for M_A and M_B by solving graphically Eqs. (17.198)-(17.199). However, it is more complicated to calculate the critical temperature. The high-temperature expansion similar to (13.29) gives two equations containing M_A and M_B of the form $M_A = a(S_A, T)M_B + b(S_A, T)M_B^3 + ...$ and $M_B = c(S_B, T)M_A + d(S_B, T)M_A^3 + ...$ where $a(S_A, T)$, $b(S_A, T)$, $c(S_B, T)$ and $d(S_B, T)$ are coefficients depending on S_A, S_B and T. An explicit expression of the critical temperature T_N can be obtained:

$$k_B T_N = \frac{C J_1}{3} \sqrt{S_A(S_A + 1)S_B(S_B + 1)} \qquad (17.200)$$

This result is equivalent to (17.190) for antiferromagnets if $S_A = S_B$. Let us give a qualitative argument. One supposes that $S_A > S_B$, when M_B becomes very small M_A is still large. It induces a local field on its B neighbors, keeping them from going to zero. As long as M_A is not zero, M_B is maintained at a non zero value. However, fluctuations of M_B affect in turn M_A. Therefore, the critical temperature is somewhere between the two temperatures where the respective sublattices become disordered when each occupies the entire lattice, namely

$$\frac{k_B T_A}{J_1} = \frac{C S_A(S_A + 1)}{3} > \frac{k_B T_N}{J_1} > \frac{k_B T_B}{J_1} = \frac{C S_B(S_B + 1)}{3}.$$

For $S_A = 2$, $S_B = 1$ and $C = 8$ (body-centered cubic lattice), one has $k_B T_A/J_1 = 16$, $k_B T_B/J_1 = 16/3 \simeq 5.33$, and $k_B T_N/J_1 = 8\sqrt{12}/3 \simeq 9.2376$. The numerical solution of (17.198) and (17.199) for the above values of S_A, S_B and C is shown as a function of T in Fig. 17.15.

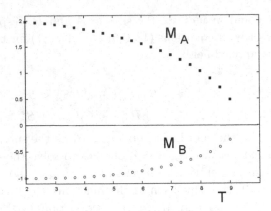

Fig. 17.15   Numerical solution of (17.198) and (17.199) is shown as a function of T for $S_A = 2$, $S_B = 1$, $C = 8$ and $k_B/J_1 = 1$. See text for comments.

Problem 10. Chain of Ising spins by exact method:
One writes the Hamiltonian as

$$\mathcal{H} = -J \sum_{n=1}^{N} \sigma_n \sigma_{n+1} \tag{17.201}$$

with $\sigma_{N+1} = \sigma_1$. The partition function is written as

$$Z = \sum_{\sigma_1 = \pm 1} \sum_{\sigma_2 = \pm 1} \cdots \sum_{\sigma_N = \pm 1} e^{\beta \sum_{n=1}^{N} \sigma_n \sigma_{n+1}}$$

$$= \sum_{\sigma_1 = \pm 1} \sum_{\sigma_2 = \pm 1} \cdots \sum_{\sigma_N = \pm 1} \prod_{n=1}^{N} e^{\beta \sigma_n \sigma_{n+1}} \tag{17.202}$$

where $\beta = J/k_B T$. Since $\sigma_n \sigma_{n+1} = \pm 1$, one has the following identity (by verification)

$$e^{\beta \sigma_n \sigma_{n+1}} = \cosh \beta + \sigma_n \sigma_{n+1} \sinh \beta \tag{17.203}$$

Equation (17.202) becomes

$$Z = \sum_{\sigma_1 = \pm 1} \sum_{\sigma_2 = \pm 1} \cdots \sum_{\sigma_N = \pm 1} [\cosh \beta + \sigma_1 \sigma_2 \sinh \beta]$$
$$\times [\cosh \beta + \sigma_2 \sigma_3 \sinh \beta] \ldots$$
$$= \sum_{\sigma_1 = \pm 1} \sum_{\sigma_2 = \pm 1} \cdots \sum_{\sigma_N = \pm 1} [(\cosh \beta)^N + (\cosh \beta)^{N-1} \sinh \beta (\sigma_2 \sigma_3)$$
$$+ \ldots + \cosh \beta (\sinh \beta)^{N-1} (\sigma_3 \sigma_4)(\sigma_4 \sigma_5) \ldots$$
$$\ldots (\sigma_{N+1} \sigma_1) + (\sinh \beta)^N (\sigma_1 \sigma_2)(\sigma_2 \sigma_3) \ldots (\sigma_{N+1} \sigma_1)] \tag{17.204}$$

Except the first and the last terms, all other terms of the sum in the square brackets [...] are zero because in each term there is one σ which appears once in the factor giving rise, when summed up, two terms of opposite signs. The first term in (17.204) does not depend on σ, it gives $2^N (\cosh \beta)^N$. The last term yields $2^N (\sinh \beta)^N$ because each σ appears twice in its factor. One has

$$Z = 2^N \left[(\cosh \beta)^N + (\sinh \beta)^N \right] \tag{17.205}$$

For $T \neq 0$ ($\beta \neq \infty$), one has $\cosh \beta > \sinh \beta$. With $N \gg 1$, one can neglect $(\sinh \beta)^N$ compared to $(\cosh \beta)^N$. Thus,

$$Z = 2^N (\cosh \beta)^N \left[1 + \left(\frac{\sinh \beta}{\cosh \beta} \right)^N \right] = 2^N (\cosh \beta)^N \tag{17.206}$$

The free energy is $F = -k_B T \ln Z = -N k_B T \ln[2 \cosh(J/k_B T)]$. The average energy is [see (3.17)]:

$$\overline{E} = -\frac{\partial \ln Z}{\partial \beta} = -N J \tanh(\beta J) \tag{17.207}$$

The heat capacity is

$$C = dE/dT = N k_B \left[\frac{k_B T}{J} \cosh \frac{J}{k_B T} \right]^{-2} \tag{17.208}$$

The energy and the heat capacity per spin are shown in Fig. 17.16. One sees that C has a maximum but no divergence. Thus, there is no phase transition at finite T.

Remark: The expressions of $\overline{E}$ and C are similar to those obtained for a two-level system given in chapters 2 and 3).

17.7 Solutions of problems of chapter 14

Problem 1. Demonstration of the Ginzburg-Landau equations (14.13] and (14.14):

f_s given by (14.12) is a function of Ψ and $\vec{\nabla}\Psi = (\partial_1 \Psi, \partial_2 \Psi, \partial_3 \Psi)$. One minimizes f_s with boundary conditions at Ω: $\Psi|_\Omega = 0$ and

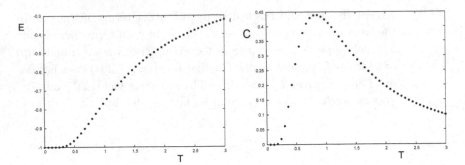

Fig. 17.16   Energy E (left) and heat capacity C (right) per spin versus temperature T [from Eqs. (17.207) and (17.208)]. $J/k_B = 1$ has been used.

$\vec{\nabla}\Psi|_\Omega = 0$. One uses the Euler-Lagrange equation:

$$\frac{\partial f_s}{\partial \Psi} - \sum_i \partial_i \frac{\partial f_s}{\partial(\partial_i \Psi)} = 0 \qquad (17.209)$$

$$\frac{\partial f_s}{\partial \Psi^*} - \sum_i \partial_i \frac{\partial f_s}{\partial(\partial_i \Psi^*)} = 0 \qquad (17.210)$$

Using (14.12) , one has

$$\frac{\partial f_s}{\partial \Psi^*} = \alpha\Psi + \beta|\Psi|^2\Psi - \frac{e^*\vec{A}}{2m^*} \cdot (\frac{\hbar}{i}\vec{\nabla} - e^*\vec{A})\Psi \qquad (17.211)$$

$$\sum_i \partial_i \frac{\partial f_s}{\partial(\partial_i \Psi^*)} = -\frac{1}{2m^*}\frac{\hbar}{i}\sum_i \partial_i \left[\frac{\hbar}{i}\partial_i - e^*\vec{A}\right]\Psi$$

$$= -\frac{1}{2m^*}\left[(\frac{\hbar}{i})^2\vec{\nabla}^2\Psi - \frac{\hbar}{i}e^*\vec{A}\cdot\vec{\nabla}\Psi\right]\Psi \qquad (17.212)$$

Replacing (17.211) and (17.212) in (17.210), one has

$$\alpha\Psi + \beta|\Psi|^2\Psi - \frac{e^*\vec{A}}{2m^*}\cdot(\frac{\hbar}{i}\vec{\nabla} - e^*\vec{A})\Psi +$$

$$\frac{1}{2m^*}\left[(\frac{\hbar}{i})^2\vec{\nabla}^2\Psi - \frac{\hbar}{i}e^*\vec{A}\cdot\vec{\nabla}\Psi\right]\Psi = 0$$

$$\alpha\Psi + \beta|\Psi|^2\Psi + \frac{1}{2m^*}\left[(\frac{\hbar}{i})^2\vec{\nabla}^2\Psi - \frac{\hbar}{i}e^*\vec{A}\cdot\vec{\nabla}\Psi + e^{*2}\vec{A}^2\Psi\right] = 0$$

$$\alpha\Psi + \beta|\Psi|^2\Psi + \frac{1}{2m^*}\left[\frac{\hbar}{i}\vec{\nabla} - e^*\vec{A}\right]^2\Psi = 0 \qquad (17.213)$$

The last line is the first GL equation (14.13).

For the second GL equation, the variational parameters are $\vec{A}$ and $\vec{\nabla} \wedge \vec{A} = \vec{B}$ with boundary conditions at Ω: $\vec{A}|_\Omega = 0$ and $\vec{\nabla} \wedge \vec{A}|_\Omega = 0$ (namely $\partial_i A_j|_\Omega = 0$). One has similarly

$$\frac{\partial f_s}{\partial A_j} - \sum_i \partial_i \frac{\partial f_s}{\partial(\partial_i A_j)} = 0 \tag{17.214}$$

$$\frac{\partial f_s}{\partial A_j} = -\frac{e^*}{2m^*}\left[\Psi^*\left(\frac{\hbar}{i}\vec{\nabla} - e^*\vec{A}\right)_j \Psi + \Psi\left(-\frac{\hbar}{i}\vec{\nabla} - e^*\vec{A}\right)_j \Psi^*\right]$$

$$= -\frac{e^*}{2m^*}\left(\frac{\hbar}{i}\right)\left(\Psi^*\vec{\nabla}\Psi - \Psi\vec{\nabla}\Psi^*\right)_j + \frac{e^{*2}A_j}{m^*}|\Psi|^2 \tag{17.215}$$

and

$$(\vec{\nabla} \wedge \vec{A})^2 = \sum_{lmnqr} \epsilon_{lmn}\epsilon_{lqr}(\partial_m A_n)(\partial_q A_r)$$

$$\sum_i \partial_i \frac{\partial f_s}{\partial(\partial_i A_j)} = \frac{1}{2\mu_0}\sum_i \sum_{lmnqr} \epsilon_{lmn}\epsilon_{lqr}\partial_i \frac{\partial(\partial_m A_n)(\partial_q A_r)}{\partial(\partial_i A_j)}$$

$$= \frac{1}{2\mu_0}\sum_i \sum_{lmnqr} \epsilon_{lmn}\epsilon_{lqr}\partial_i[\partial_{mi}\partial_{nj}(\partial_q A_r) +$$

$$\partial_{qi}\partial_{rj}(\partial_m A_n)]$$

$$= \frac{1}{\mu_0}\sum_i \sum_{lmn} \epsilon_{lmn}\epsilon_{lij}\partial_i(\partial_m A_n)$$

$$= \frac{1}{\mu_0}\sum_i \epsilon_{ijl}\partial_i(\vec{\nabla} \wedge \vec{A})_l$$

$$= -\frac{1}{\mu_0}(\vec{\nabla} \wedge \vec{\nabla} \wedge \vec{A})_j$$

$$= -\frac{1}{\mu_0}(\vec{\nabla} \wedge \vec{B})_j$$

$$= -(\vec{J})_j \tag{17.216}$$

Replacing (17.215) and (17.216) in (17.214), one obtains

$$-\frac{e^*}{2m^*}\left(\frac{\hbar}{i}\right)\left(\Psi^*\vec{\nabla}\Psi - \Psi\vec{\nabla}\Psi^*\right)_j + \frac{e^{*2}A_j}{m^*}|\Psi|^2 + (\vec{J})_j = 0 \tag{17.217}$$

This is the second GL equation (14.14).

Problem 2. Current density $\vec{J}$: gauge-invariance

One shows below that $\vec{J}$ is gauge-invariant.

a) Equation (14.14) is

$$\vec{J} = \frac{e^*\hbar}{2m^*i}\left(\Psi^*\vec{\nabla}\Psi - \Psi\vec{\nabla}\Psi^*\right) - \frac{e^{*2}\vec{A}}{m^*}|\Psi|^2 \qquad (17.218)$$

Assuming that $\Psi = |\Psi|e^{i\varphi}$, one obtains

$$\Psi^*\vec{\nabla}\Psi - \Psi\vec{\nabla}\Psi^* = 2i|\Psi|^2\vec{\nabla}\varphi.$$

Replacing this into (17.218), one gets (14.15):

$$\vec{J} = \frac{e^*}{m^*}\left(\hbar\vec{\nabla}\varphi - e^*\vec{A}\right)|\Psi|^2 \qquad (17.219)$$

b) One sees in (14.11) that $|\Psi|^2 = n_s$ (density of superconducting electrons). If one writes

$$\vec{J} = n_s e^*\vec{v}_s$$

where $\vec{v}_s$ is the electron velocity in the superconducting regime, then by comparison to (17.219), one has

$$m^*\vec{v}_s = \left(\hbar\vec{\nabla}\varphi - e^*\vec{A}\right).$$

c) Gauge invariance: Under the gauge transformation, the wave function is transformed as

$$\Psi'(\vec{r}) = e^{ie^*\chi/\hbar}\Psi(\vec{r})$$

This implies

$$\varphi' = \varphi + \frac{e^*\chi}{\hbar}.$$

Since $\vec{A}' = \vec{A} + \vec{\nabla}\chi$, one has

$$\begin{aligned}
\vec{v}_s' &= \frac{1}{m^*}\left(\hbar\vec{\nabla}\varphi' - e^*\vec{A}'\right) \\
&= \frac{1}{m^*}\left[\hbar\vec{\nabla}\left(\varphi + \frac{e^*\chi}{\hbar}\right) - e^*(\vec{A} + \vec{\nabla}\chi)\right] \\
&= \frac{1}{m^*}\left(\hbar\vec{\nabla}\varphi - e^*\vec{A}\right) = \vec{v}_s
\end{aligned}$$

$\vec{v}_s$ is thus gauge-invariant. Since $\vec{J} = n_s e^*\vec{v}_s$, one concludes that $\vec{J}$ is also gauge-invariant.

Problem 3. Theory of Gorter-Casimir [70]:

a) The free energy is given by

$$F(x,T) = \sqrt{x}f_n(T) + (1-x)f_s(T) = -\sqrt{x}\frac{\gamma}{2}T^2 - (1-x)\beta \quad (17.220)$$

$F(x,T)$ is minimum when $\partial F(x,T)/\partial x = 0$. The solution is

$$x = \frac{1}{16}\frac{\gamma^2}{\beta^2}T^4 \quad (17.221)$$

When $x = 1$, the superconducting component disappears. This corresponds to T_c given by

$$1 = \frac{1}{16}\frac{\gamma^2}{\beta^2}T_c^4 \Rightarrow T_c^2 = \frac{4\beta}{\gamma} \quad (17.222)$$

Equation (17.221) becomes

$$x = \left(\frac{T}{T_c}\right)^4 \quad (17.223)$$

b) One has

$$F_s(T) = -\beta[1 + (T/T_c)^4]$$

and

$$F_n(T) = -\frac{\gamma}{2}T^2\sqrt{x} = -\frac{\gamma}{2}T^2\left(\frac{T}{T_c}\right)^2 = -\frac{\gamma}{2}T_c^2\left(\frac{T}{T_c}\right)^2$$

$$= -\frac{\gamma}{2}\frac{4\beta}{\gamma}\left(\frac{T}{T_c}\right)^2 = -2\beta\left(\frac{T}{T_c}\right)^2 \quad (17.224)$$

Using Eq. (14.29), one has

$$\frac{\mu_0 H_c^2}{2} = -(F_s - F_n) = \beta\left[1 + \left(\frac{T}{T_c}\right)^4 - 2\left(\frac{T}{T_c}\right)^2\right]$$

$$= \beta\left[1 - \left(\frac{T}{T_c}\right)^2\right]^2 \quad (17.225)$$

so that $H_c(T) = H_c(0)(1 - t^2)$ where $t = T/T_c$ and $H_c(0) = \sqrt{2\beta/\mu_0}$.

Note: In spite of a phenomenological character in its demonstration, this relation is experimentally verified within a few percents.

Problem 4. Energy of a vortex:

a) The London's equation is written as

$$\vec{\nabla}^2 \vec{B} - \frac{1}{\lambda^2}\vec{B} = -\frac{\phi_0}{\lambda^2}\delta(\vec{r}) \qquad (17.226)$$

where the right-hand side represents the vortex core, ϕ_0 being the flux given by (14.36). Writing this equation in cylindrical coordinates (r, θ, z), one has the following radial part

$$\frac{1}{r}\frac{\partial}{\partial r}\left(r\frac{\partial B}{\partial r}\right) - \frac{B}{\lambda^2} = -\frac{\phi_0}{\lambda^2}\delta(r) \qquad (17.227)$$

where r is in the xy plane perpendicular to the applied field axis. The solution of this equation is a Bessel function

$$B(r) = B_0 K_0(\frac{r}{\lambda}) \quad [B(r) \text{ is the } z \text{ component of } \vec{B}] \qquad (17.228)$$

The Bessel function has the following limits

$$K_0(\frac{r}{\lambda} \to 0) \to \ln(\frac{\lambda}{r}), \quad K_0(\frac{r}{\lambda} \to \infty) \to \exp(-\frac{r}{\lambda})$$

Of course, the lower bound diverges when $r \to 0$ but, as said earlier, the London's solution is valid for $r \gg \xi$, namely not close to the normal-superconducting boundary. Now, the field gradient induces the current

$$\mu_0\vec{J} = \vec{\nabla} \wedge \vec{B}$$

so that

$$J(r) = J_0 K_1(\frac{r}{\lambda}) \quad [J(r) \text{ lies on the } \theta \text{ axis}] \qquad (17.229)$$

The Bessel function $K_1(u)$ is connected to $K_0(u)$ by

$$K_1(u) = -\frac{dK_0(u)}{du}$$

where $u \equiv \frac{r}{\lambda}$. It has the following limits

$$K_1(\frac{r}{\lambda} \to 0) \to \frac{\lambda}{r}, \quad K_1(\frac{r}{\lambda} \to \infty) \to \exp(-\frac{r}{\lambda})$$

Vortices, or circular supercurrents, occur thus at distance $\xi < r < \lambda$. For $r \gg \lambda$ the field and thus current are screened because they exponentially decay.

Let us determine the parameter B_0: since the flux is quantized, one has

$$\int_0^\infty B_0 K_0(\frac{r}{\lambda})2\pi r dr = \phi_0$$

Using the tabulated integral

$$\int_0^\infty u K_0(u) du = 1$$

one obtains

$$B_0 = \frac{\phi_0}{2\pi\lambda^2}.$$

This gives

$$J_0 = \frac{\phi_0}{2\pi\mu_0\lambda^3}.$$

One is now ready for calculating the energy of a single vortex. London has proposed to include in the energy the magnetic energy and the kinetic energy of circulating electrons in the supercurrent. One has

$$\epsilon_1 = \int_0^\infty \left[\frac{B^2}{2\mu_0} + \frac{1}{2} n_s m^* v_s^2 \right] 2\pi r dr$$

where the second term can be expressed in term of J_s (supercurrent) using $J_s = n_s e^* v_s$:

$$\frac{1}{2} n_s m^* v_s^2 = \frac{1}{2}\lambda^2 \mu_0 J_s^2.$$

One has

$$\epsilon_1 = \frac{1}{2\mu_0} \int_0^\infty \left[B^2 + \lambda^2 \mu_0^2 J_s^2 \right] 2\pi r dr$$

$$= \frac{1}{2\mu_0} \left(\frac{\phi_0}{2\pi\lambda^2}\right)^2 2\pi \int_\xi^\lambda \left\{ r \left[\ln(\lambda/r)\right]^2 + r(\lambda/r)^2 \right\} dr$$

$$= \frac{\phi_0^2}{4\pi\mu_0\lambda^2} \int_{\xi/\lambda}^1 \left\{ x (\ln x)^2 + \frac{1}{x} \right\} dx$$

$$= \frac{\phi_0^2}{4\pi\mu_0\lambda^2} \left\{ \frac{x^2}{2} \left[\frac{1}{2} - \ln x + (\ln x)^2 \right] + \ln x \right\}_{\xi/\lambda}^1$$

$$\simeq \frac{\phi_0^2}{4\pi\mu_0\lambda^2} \ln(\frac{\lambda}{\xi}) \tag{17.230}$$

b) One admits the following interaction energy between two vortices (see Refs. [103, 159]):

$$E_{int} = A[K_0(d/\lambda) - K_0(\sqrt{2}d/\xi)]$$

where A is a positive constant. The two terms compete with each other. The first term (> 0) is a repulsive force similar to the force between two wires with currents in opposite directions. In our case, the current is caused by the electrons rotating around the vortex. Two vortices placed side by side will have currents running in the opposite directions and will be repelled. We can see from Eq. (17.229) that the current is in the θ direction around the vortex. The second term, with a minus sign, is an attractive force caused by the superconducting state which prefers to have no vortices: the attraction will merge two nearby vortices, and will finally make disappear all vortices.

c) When the first term is larger, E_{int} is positive, the vortices repel. When the second term is larger, the vortices merge into each other. To see when one case is favored with respect to the other, let us compare the arguments of the Bessel function. $K_0(x)$ is a positive, decreasing function of x. The vortices repel, i. e. E_{int} is positive, when

$$\frac{d}{\lambda} < \frac{d\sqrt{2}}{\xi}.$$

Simplifying this relation and writing it using the definition of $\kappa = \lambda/\xi$, one has the vortices repel when

$$\kappa > \frac{1}{\sqrt{2}}.$$

This case corresponds to type II superconductors: vortices are separated and form a triangular lattice [1].

The other case, namely when

$$\kappa < \frac{1}{\sqrt{2}},$$

corresponds to the attraction of vortices: vortices will collapse to give rise to a pure superconducting state of type I superconductors.

Problem 5. Electron gas in a strong magnetic field: Landau's levels, Landau diamagnetism

We show below that a system of electrons of effective mass m^* under a strong magnetic field gives rise to a negative susceptibility. This phenomenon is called "Landau diamagnetism" which has the same mechanism as the vortices leading to the Meissner effect.

When the applied magnetic field is strong, we cannot use the perturbation theory. We have to incorporate the action of the field in the Hamiltonian via the vector potential $\vec{A}(\vec{r})$.

We suppose that the field $\vec{B}$ is applied along the z axis. We solve the problem in the following.

a) Landau's levels:

The Schrödinger equation for an electron of effective mass m^* under the applied field $\vec{B}$ is written as

$$\frac{1}{2m^*}\left[\frac{\hbar}{i}\vec{\nabla} + e\vec{A}(\vec{r})\right]^2 \Psi(\vec{r}_i) = E\Psi(\vec{r}_i) \qquad (17.231)$$

where $\vec{B} = \text{rot}\vec{A}$. For simplicity, we choose $\vec{A} = (0, xB, 0)$. We have

$$\frac{\partial^2\Psi}{\partial x^2} + \left(\frac{\partial}{\partial y} - \frac{ieB}{\hbar}x\right)^2\Psi + \frac{\partial^2\Psi}{\partial z^2} + \frac{2m^*E}{\hbar^2}\Psi = 0 \qquad (17.232)$$

The structure of this equation suggests a solution of the form

$$\Psi(x, y, z) = u(x)\exp[i(k_y y + k_z z)] \qquad (17.233)$$

Equation (17.232) becomes

$$\frac{\partial^2 u(x)}{\partial x^2} + \left[\frac{2m^*E'}{\hbar^2} - (k_y - \frac{eB}{\hbar}x)^2\right]u(x) = 0 \qquad (17.234)$$

where

$$E' = E - \frac{\hbar^2 k_z^2}{2m^*} \qquad (17.235)$$

Thus, the motion of the electron in the z direction is that of a free electron. For the motion in the xy plane, we have to solve Eq. (17.234). We rewrite it as

$$-\frac{\hbar^2}{2m^*}\frac{\partial^2 u(x)}{\partial x^2} + \frac{1}{2}m^*\left(\frac{eB}{m^*}x - \frac{\hbar k_y}{m^*}\right)^2 u(x) = E'u(x) \qquad (17.236)$$

We recognize that the above equation is the Schrödinger equation of a harmonic oscillator of pulsation

$$\omega_c = \frac{eB}{m^*} \qquad (17.237)$$

centered at

$$x_0 = \frac{1}{\omega_c}\frac{\hbar k_y}{m^*} \qquad (17.238)$$

The energy of this oscillator is therefore

$$E' = (n + \frac{1}{2})\hbar\omega_c \qquad (17.239)$$

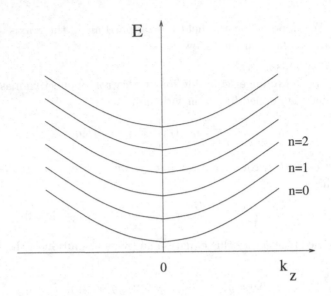

Fig. 17.17   Landau's levels given by Eq. (17.240).

where n is an integer ≥ 0. With Eq. (17.235), we obtain the total electron energy

$$E = E_n = (n + \frac{1}{2})\hbar\omega_c + \frac{\hbar^2 k_z^2}{2m^*} \qquad (17.240)$$

where ω_c is called "cyclotron pulsation" and the different energy levels corresponding to different values of n are called "Landau's levels". These levels are shown in Fig. 17.17.

b) Degeneracy of Landau's levels:

The electron energy E is doubly quantized: the allowed values of k_z are given by the periodic condition in the z direction, and index n quantizes the energy of the harmonic motion in the xy plane. However, the quantization by n is subject to a double condition:

- the harmonic motion in the xy plane takes place only if the center of the oscillator x_0 lies inside the xy plane of the material, namely

$$0 < x_0 < L_x \qquad (17.241)$$

where L_x is the length of the system in the x direction. Replacing x_0 by Eq. (17.238), the condition (17.241) becomes

$$0 < k_y < \frac{m^*\omega_c}{\hbar}L_x = \frac{eB}{\hbar}L_x \qquad (17.242)$$

- the quantization of k_y by the periodic condition in the y direction, namely $k_y = 2\pi n_y/L_y$. The distance between two successive levels of k_y is $2\pi/L_y$. Therefore, the number of values of k_y inside the limits given by (17.242) is

$$d = \frac{\frac{eBL_x}{\hbar}}{\frac{2\pi}{L_y}} = \frac{eB}{2\pi\hbar}L_xL_y \qquad (17.243)$$

d by definition is the degeneracy of the level E_n. We show now that this degeneracy of Landau's levels is the same as that in zero field. Without $\vec{B}$, the components of the wave vector $\vec{k}$ are quantized uniquely by the periodic conditions in three directions. The distances between successive values of k_x and k_y are $2\pi/L_x$ and $2\pi/L_y$. the number of states in a circular surface δA in the space (k_x, k_y) is

$$d' = \frac{\delta A}{\frac{(2\pi)^2}{L_xL_y}} = \frac{L_xL_y}{(2\pi)^2}2\pi k_\parallel dk_\parallel \qquad (17.244)$$

where $k_\parallel$ is the modulus of the wave vector in the (k_x, k_y) plane. Energy $E_\parallel$ of the state (k_x, k_y) is $E_\parallel = \frac{\hbar^2 k_\parallel^2}{2m^*}$. We have

$$d' = \frac{L_xL_y}{(2\pi)^2}\frac{2\pi m^*}{\hbar^2}dE_\parallel \qquad (17.245)$$

To compare d' to d, we take $dE_\parallel = \hbar\omega_c$ which is the separation between two successive Landau's levels. We then have

$$d' = \frac{L_xL_y}{(2\pi)^2}\frac{2\pi m^*}{\hbar^2}\hbar\omega_c = \frac{eB}{2\pi\hbar}L_xL_y \qquad (17.246)$$

We see that $d = d'$.

c) Quantization of electron orbit:

Using the Bohr quantization relation

$$\oint \vec{p}\cdot d\vec{r} = (m+\gamma)2\pi\hbar \qquad (17.247)$$

where m is an integer and $\gamma = 1/2$ a phase correction, we show that the projection of an electron trajectory in the k space on the (k_x, k_y) plane is given by

$$\mathcal{S}_k = 2\pi\hbar^{-1}eB(m+\gamma) \qquad (17.248)$$

Demonstration: We have $\vec{p} = \hbar\vec{k} + e\vec{A}$. We write

$$\oint \vec{p} \cdot d\vec{r} = \oint \hbar\vec{k} \cdot d\vec{r} + e \oint \vec{A} \cdot d\vec{r}$$

$$= e \oint \vec{r} \wedge \vec{B} \cdot d\vec{r} + e \int \text{rot}\vec{A} \cdot d\vec{S}$$

$$= -e\vec{B} \cdot \oint \vec{r} \wedge d\vec{r} + e \int \vec{B} \cdot d\vec{S} \qquad (17.249)$$

where we have transformed the circular integral on the trajectory into a surface integral by the Stokes's theorem and used a property of the mixed product between vectors. Since $\oint \vec{r} \wedge d\vec{r}$ is equal twice the surface S limited by the closed trajectory in the real space and $\int \vec{B} \cdot d\vec{S} = \vec{B} \cdot \int d\vec{S}$, we can write

$$\oint \vec{p} \cdot d\vec{r} = -2eBS + eBS = -e\phi \qquad (17.250)$$

where $\phi = BS$ is the magnetic flux passing through the surface limited by the electron trajectory. Comparison of Eq. (17.250) to Eq. (17.247) gives

$$-e\phi = (m + \gamma)2\pi\hbar$$

$$-eBS = (m + \gamma)2\pi\hbar$$

$$S = -\frac{1}{eB}(m + \gamma)2\pi\hbar \qquad (17.251)$$

Since the real space is connected to the reciprocal space by $\hbar\vec{k} = m^*\vec{v} = e\vec{B} \wedge \vec{r}$, we have

$$\hbar\Delta\vec{k} = e\vec{B} \wedge \Delta\vec{r}$$

$$\hbar^2(\Delta k)^2 = (eB)^2(\Delta r)^2$$

$$(\Delta r)^2 = \frac{\hbar^2}{(eB)^2}(\Delta k)^2$$

$$S = = \frac{\hbar^2}{(eB)^2}S_k \qquad (17.252)$$

The surface S of the trajectory in real space is connected to the surface S_k of the trajectory in reciprocal space by the last equality. Replacing this relation in Eq. (17.251), one obtains Eq. (17.248). We have seen that each Landau's level is d-fold degenerate. Replacing L_xL_y by the surface of the material L^2, we can express d of Eq. (17.243) as

$$d = \frac{BL^2}{\frac{2\pi\hbar}{e}} \qquad (17.253)$$

The numerator is the magnetic flux passing through the surface of the sample and the denominator is the flux quantum. The Landau's degeneracy is thus the number of magnetic flux quanta crossing the surface of the material.

Putting $\zeta \equiv \frac{L^2}{\frac{2\pi\hbar}{e}}$, we write $d = \zeta B$.

d) Diamagnetic susceptibility:

We show in the following that the susceptibility of electrons on Landau's levels is negative. The simplest way to do is to use the partition function for these levels

$$Z = \frac{z^N}{N!} \tag{17.254}$$

where z is the partition function for an electron. $N!$ expresses the indiscernibility of electrons. Taking into account the degeneracy of each level n, we have

$$z = \sum_n d \exp(-\beta E_n) = \sum_n d \exp[-\beta(n + 1/2)\hbar\omega_c)]$$

$$= \sum_{n=0}^{\infty} d \exp[-\beta(2n+1)\mu_B B] = \frac{\exp(-\beta\mu_B B)}{1 - \exp(-2\beta\mu_B B)}$$

$$= \frac{1}{2\sinh(\beta\mu_B B)} \tag{17.255}$$

where we have replaced ω_c by eB/m^* and $\mu_B = \frac{e\hbar}{2m^*}$. The formula of the geometric series has been used on the second line. The average magnetic moment M of the system is written as

$$\overline{M} = \frac{1}{Z}\sum_l M_l e^{-\beta E_l} = \frac{1}{Z}\sum_l M_l e^{\beta B M_l}$$

$$= \frac{1}{\beta}\frac{1}{Z}\frac{\partial}{\partial B}\sum_l e^{\beta B M_l} = \frac{1}{\beta}\frac{1}{Z}\frac{\partial Z}{\partial B} = \frac{1}{\beta}\frac{\partial \ln Z}{\partial B}$$

$$= -\frac{\partial F}{\partial B} \tag{17.256}$$

from which we have

$$\overline{M} = -\frac{\partial F}{\partial B} = -k_B T\frac{\partial \ln Z}{\partial B}$$

$$= -N k_B T\frac{\partial \ln z}{\partial B}$$

$$= -N\mu_B \mathcal{L}(x) \tag{17.257}$$

where

$$\mathcal{L}(x) = \coth(x) - \frac{1}{x} \tag{17.258}$$

with $x = \beta \mu_B B$.

At high T, one has $x \ll 1$, $\mathcal{L}(x) \simeq x/3$. This gives

$$\overline{M} \simeq -\frac{N\mu_B^2 B}{3k_B T} \tag{17.259}$$

so that

$$\chi = -\frac{N\mu_B^2}{3k_B T} \tag{17.260}$$

The negative sign of χ shows the diamagnetic character of the electrons considered here. This diamagnetism is called "Landau's diamagnetism".

Remark: If we include the energy associated with k_z in E_n [cf. Eq. (17.240)], we obtain the following factor for the partition function z: $\int_{-\infty}^{\infty} \exp(-\beta \frac{\hbar^2 k_z^2}{2m^*}) dk_z = \sqrt{\frac{2\pi m^*}{\beta \hbar^2}}$. This factor does not depend on B.

17.8   Solutions of problems of chapter 15

Problem 1. Effect of magnetic field:

Demonstration of Eq. (15.24):

When we apply a magnetic field $\vec{B}$, the force becomes $\vec{F} = -e[\vec{\varepsilon} + \vec{v} \wedge \vec{B}]$. We cannot therefore replace f by f_0 in $\frac{\vec{F}}{\hbar} \cdot \vec{\nabla}_{\vec{k}} f$ because the field $\vec{B}$ will not appear in the final equation. To see the effect of $\vec{B}$, we have to go to the second order: we have to replace f by $f_0 + \varphi$, not by f_0:

$$\begin{aligned}
\frac{\vec{F}}{\hbar} \cdot \vec{\nabla}_{\vec{k}} f &= \frac{\vec{F}}{\hbar} \cdot \vec{\nabla}_{\vec{k}} (f_0 + \varphi) \\
&= \frac{\vec{F}}{\hbar} \cdot \left[\frac{\partial f_0}{\partial E} \vec{\nabla}_{\vec{k}} E + \vec{\nabla}_{\vec{k}} \varphi \right] \\
&= -e\vec{\varepsilon} \cdot \left[\frac{\partial f_0}{\partial E} \hbar \vec{v} + \frac{1}{\hbar} \vec{\nabla}_{\vec{k}} \varphi \right] - e(\vec{v} \wedge \vec{B}) \cdot \frac{\partial f_0}{\partial E} \vec{v} \\
&\quad - \frac{e}{\hbar} (\vec{v} \wedge \vec{B}) \cdot \vec{\nabla}_{\vec{k}} \varphi \tag{17.261}
\end{aligned}$$

We see here that for the effect of the electric field we can neglect the second term in [...], namely $\vec{\nabla}_{\vec{k}}\varphi$. However, for the magnetic field the first term $(\vec{v} \wedge \vec{B}) \cdot \frac{\partial f_0}{\partial E}\vec{v}$ is zero because of the mixed product. We have to retain the last term of the above equation which depends on $\vec{B}$. We rewrite this term as

$$-\frac{e}{\hbar}(\vec{v} \wedge \vec{B}) \cdot \vec{\nabla}_{\vec{k}}\varphi = -\frac{e}{\hbar}\vec{B} \cdot \left[\vec{\nabla}_{\vec{k}}\varphi \wedge \vec{v}\right]$$

$$= \frac{e}{\hbar}\vec{B} \cdot \left[\vec{\nabla}_{\vec{k}}\phi\frac{\partial f_0}{\partial E} \wedge \frac{\vec{\nabla}_{\vec{k}}E}{\hbar}\right]$$

$$= \frac{e}{\hbar^2}\vec{B} \cdot \left[\vec{\nabla}_{\vec{k}}\phi \wedge \vec{\nabla}_{\vec{k}}E\right]\frac{\partial f_0}{\partial E} \quad (17.262)$$

The solution (15.23) becomes

$$\phi = \tau\vec{A} \cdot \vec{v} + \tau\frac{e}{\hbar^2}\vec{B} \cdot \left[\vec{\nabla}_{\vec{k}}\phi \wedge \vec{\nabla}_{\vec{k}}E\right] \quad (17.263)$$

This is a differential equation.

Problem 2. Electrons in a strong electric field: an approximation

In a metal, when the applied electric field $\vec{\varepsilon}$ is weak, the current $\vec{j}$ is proportional to $\vec{\varepsilon}$ (Ohm's law). For strong fields, a deviation from the Ohm's law is observed.

One considers a gas of conducting electrons of effective mass m^*. In weak fields, one has $\vec{j} = \sigma\vec{\varepsilon} \equiv -ne < \vec{v}_c >$ where $< \vec{v}_c >$ is the average velocity due to the field. One obtains then

$$< \vec{v}_c >= -\frac{\sigma\vec{\varepsilon}}{ne} = -\frac{e^2n\tau}{m^*ne}\vec{\varepsilon} = -\frac{e\tau}{m^*}\vec{\varepsilon} \quad (17.264)$$

where τ, the relaxation time between two collisions, is defined as l/v, l being the mean free path and v the average velocity of the electron.

In strong fields, one shows that σ is no more a constant of the material but it depends on $\vec{\varepsilon}$. One supposes that the loss of energy of an electron due to a collision with a phonon is proportional to its kinetic energy. In the stationary regime the energy balance is

$$\left(\frac{\partial E}{\partial t}\right)_{field} + \left(\frac{\partial E}{\partial t}\right)_{collision} = 0 \quad (17.265)$$

where

$$\left(\frac{\partial E}{\partial t}\right)_{\text{field}} = -e\varepsilon < v_F > = \frac{e^2 l}{m^* v}\varepsilon^2 \quad (17.266)$$

$$\left(\frac{\partial E}{\partial t}\right)_{\text{collision}} = -\frac{\alpha^2 m^* v^2}{\tau} = -\frac{\alpha^2 m^* v^3}{l} \quad (17.267)$$

One obtains from (17.265)

$$v = \sqrt{\frac{el}{\alpha m^*}}\sqrt{\varepsilon} \quad (17.268)$$

If one writes $|j| = \sigma\varepsilon \equiv nev$, then σ is given by

$$\sigma = \frac{nev}{\varepsilon} = en\sqrt{\frac{el}{\alpha m^*\varepsilon}} \quad (17.269)$$

from which one gets

$$|j| = en\sqrt{\frac{el}{\alpha m^*}}\sqrt{\varepsilon} \quad (17.270)$$

Electrons obeying this relation are called "hot electrons".
In the region of intermediate fields, one has

$$|j| = \sigma_0\varepsilon(1 + a\varepsilon^2) \quad (17.271)$$

where $\sigma_0 = \frac{e^2 n\tau(\varepsilon=0)}{m^*}$ is the conductivity of the Ohm's law ($|j| = \sigma_0\varepsilon$) and a denotes a constant. The above relation can be demonstrated by expanding τ as follows

$$\tau(\varepsilon) \simeq \tau(\varepsilon = 0) + \frac{d\tau}{dE}\left(\frac{\partial E}{\partial t}\right)_{\text{field}} = \tau(\varepsilon = 0)(1 + a\varepsilon^2) \quad (17.272)$$

where one has used $(\frac{\partial E}{\partial t})_{\text{field}} \propto \varepsilon^2$ [see Eq. (17.266)]. Replacing $\tau(\varepsilon)$ in $\vec{j} = -\frac{ne^2\tau(\varepsilon)}{m^*}\vec{\varepsilon}$ one obtains (17.271). Figure 17.18 shows the different regimes of j as a function of the applied electric field.

Problem 3. Semiconductors: effect of temperature on conductivity
   The current density in a semiconductor without doping is written as $\vec{j} = -ne\vec{v}_n + ep\vec{v}_p$ where n is the number of electrons of velocity $\vec{v}_n$ and p the number of holes of velocity $\vec{v}_p$. In an intrinsic semiconductor one has $n = p = n_i$, so that

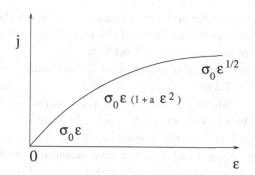

Fig. 17.18   Current j versus applied electric field ε.

$$j = n_i e(\mu_n + \mu_p)\varepsilon = \sigma\varepsilon$$

where μ_n and μ_p are electron and hole mobilities. In principle, μ_n, $\mu_p \propto T^{-3/2}$. This neutralizes the dependence on T in N_c and N_p (see paragraph 11.3.2). One has

$$\sigma \propto n_i = C \exp(-\frac{E_g}{2k_B T})$$

from which,

$$\ln \sigma = \ln C - \frac{E_g}{2k_B T}.$$

For $\sigma' = 2\sigma$, one has

$$\ln \sigma' = \ln C - \frac{E_g}{2k_B T'}$$

or

$$\ln \sigma + \ln 2 = \ln C - \frac{E_g}{2k_B T'}.$$

One then gets

$$\ln 2 = -\frac{E_g}{2k_B T'} + \frac{E_g}{2k_B T} \simeq \frac{E_g(T' - T)}{2k_B T^2}.$$

At $T = 300$ K: to increase twice the conductivity one needs $\Delta T = T' - T = 15$ K for $E_g = 0.6$ eV (Ge) and $\Delta T = T' - T = 9$ K for $E_g = 1$ eV (Si).

Problem 4. Semiconductor: effect of magnetic field on the gap

One calculates the variation of the forbidden energy gap in a semi-conductor under an applied field $\vec{B}$.

In a strong applied magnetic field $\vec{B}$, the Landau's levels are given by Eq. (17.240): $E_n = (n + 1/2)\hbar\omega_c + \hbar^2 k_z^2/2m^*$ where $\omega_c = eB/m^*$. This is for electrons in the conduction band (CB) and valence band (VB) with $m^* = m_c^*$ (CB) and $m^* = m_v^*$ (VB). The zero-field CB and VB become multiple bands (Landau's levels) and the initial band gap becomes wider under the effect of the applied field. Its variation is given by the distance between the first Landau's levels of CB and VB, namely $\Delta E_g = \hbar eB/2m_c^* + \hbar eB/2|m_v^*|$.

Problem 5. Doped semiconductors:

The impurity energy level E_D is located in the forbidden band at a distance distance Δ from the bottom of the conduction band. The number of electrons which occupy this level, taking into account the spin degeneracy, is written by [see (4.51)]

$$n_d = \frac{N_d}{\frac{1}{2}e^{\beta(E_D - E_F)} + 1}$$

$$= \frac{N_d}{\frac{1}{2}e^{\beta(E_g - \Delta - E_F)} + 1}$$

The number of conducting electrons provided by N_d impurities is then

$$n = N_d - n_d = \frac{N_d}{2e^{-\beta(E_g - \Delta - E_F)} + 1} \tag{17.273}$$

One sets this number equal to the number of electrons of the conduction band given by (11.67), one has

$$x^2 + \frac{g}{2d}x - \frac{1}{2}\frac{N_d}{N_c}\frac{g^2}{2} = 0$$

where

$$x \equiv e^{\beta E_F}, \quad g \equiv e^{\beta E_g}, \quad d \equiv e^{\beta\Delta}$$

The solution is

$$x = \frac{g}{4d}\left[-1 + \left(1 + 8d\frac{N_d}{N_c}\right)^{1/2}\right]$$

One examines the following limits:

If $8d\frac{N_d}{N_c} \ll 1$, then $x \simeq g\frac{N_d}{N_c}$ or $E_F \simeq E_g + k_BT \ln \frac{N_d}{N_c}$. Comparing to (11.67), one has $n = N_d$.

If $8d\frac{N_d}{N_c} \gg 1$ then $x \simeq \frac{g}{d^{1/2}}(\frac{N_d}{2N_c})^{1/2}$ or $E_F \simeq E_g - \frac{\Delta}{2} + \frac{k_BT}{2} \ln \frac{N_d}{2N_c}$. Comparing to (11.67), one has $n = (\frac{N_cN_d}{2})^{1/2}e^{-\beta\Delta/2}$.

At $T = 300$ K with $N_d = 10^{14}$ cm^{-3}, one has $E_F = 0.79$ eV. The number of electrons is $n = 10^{14}$ cm^{-3} and the number of holes is $p = \frac{n_ip_i}{n} = 2 \times 10^5$ cm^{-3} negligible with respect to n.

At $T = 150$ K, one has $N_c = 7.5 \times 10^{17}$ cm^{-3}. With $N_d = 10^{17}$ cm^{-3}, one has $E_F = 0.973$ eV. The number of electrons is $n = 8.7 \times 10^{16}$ cm^{-3} and the number of holes is $p = \frac{n_ip_i}{n} << n$.

The above result shows that when N_d increases, n increases. Since np is constant, p decreases.

The p-type doping is treated in a similar manner: one finds the inverse tendency.

Problem 6. Swallow impurity states in semiconductors: an approximation.

Consider a crystal in which the first conduction band ($j = 0$) has a minimum at $\vec{k} = 0$. In the absence of impurities, the Hamiltonian of an electron in the crystal is

$$\mathcal{H}_0 = \frac{p^2}{2m} + V(\vec{r})$$

where $V(\vec{r})$ is the periodic potential in the real lattice. The eigenfunctions of $\mathcal{H}_0$ are Bloch functions $|\vec{k}, j >$ of energy $E_0(\vec{k}, j)$; $\vec{k}$: wave vector and j: band index.

One replaces an atom of the crystal by an impurity composed of an ion of charge $+e$ and an electron to ensure the electric neutrality of the system. The presence of this impurity adds the following potential to $\mathcal{H}_0$: $U(\vec{r}) = -e^2/\epsilon r$, ϵ: dielectric constant.

The method introduced by W. Kohn [Nuovo Cimento, vol. VII, p. 713 (1958)] consists of using a combination of non perturbed Bloch functions. For the answer to this exercise the reader can skip the following steps of the original Kohn's paper, and go directly to the last question. One gives however the main steps of the calculation.

a) One looks for the eigenfunctions of $\mathcal{H}$ by using the following linear combination

$$|\Psi\rangle = \sum_{\vec{k},j} A(\vec{k},j)|\vec{k},j\rangle.$$

One can show that the coefficients $A(\vec{k},j)$ and the corresponding energy E obey the following equation

$$[E_0(k,j) - E]A(k,j) + \sum_{\vec{k}',j'} \langle \vec{k},j|U(\vec{r})|\vec{k}',j'\rangle A(\vec{k}',j') = 0 \quad (17.274)$$

b) One denotes by $u_{\vec{k},j}$ the periodic factor of $|\vec{k},j\rangle$. One can explicit this factorization by writing the function $u^*_{\vec{k},j}(\vec{r})u_{\vec{k}',j'}(\vec{r})$ as

$$u^*_{\vec{k},j}(\vec{r})u_{\vec{k}',j'}(\vec{r}) = \sum_{\vec{b}} C_{\vec{b}}(\vec{k},j;\vec{k}',j')e^{i\vec{b}\cdot\vec{r}}.$$

where the sum should be performed over the reciprocal lattice vectors (see chapter 11).

c) One then calculates $\langle \vec{k},j|U|\vec{k}',j'\rangle$ as a function of coefficients $C_{\vec{b}}(\vec{k},j;\vec{k}',j')$.

For a weak interaction between an electron and the impurity one shall use the following approximation

$$A(\vec{k},j) = A(\vec{k})\delta_{j,0}.$$

d) One evaluates the coefficients $C_{\vec{b}}(\vec{k},0;\vec{k}',0)$ (guide: taking into account the orthogonality relations of Bloch functions).

e) One supposes, in addition, that an approximation using small wave vectors can be used so that the dispersion relation $E_0(\vec{k},0)$ is that of a free electron of effective mass m^*. One then shows that the equation (17.274) is equivalent to the Schrödinger equation for a hydrogen-like atom:

$$\left[-\frac{\hbar^2\nabla^2}{2m^*} - \frac{e^2}{\epsilon r}\right]\Psi(\vec{r}) = E\Psi(\vec{r}) \quad (17.275)$$

The solution of this is that of the hydrogen atom with an effective Bohr radius

$$a^* = \frac{\hbar^2\epsilon}{m^*e^2} = a_0\frac{\epsilon}{m^*/m_0}$$

where a_0 is the Bohr radius and m_0 the electron mass at rest. The energy is hydrogen-like and is given by

$$E_n = -\frac{(e^2/\epsilon)^2}{2\hbar^2 m^* n^2} = -\frac{1}{n^2 \epsilon^2 (m^*/m_0)} \text{Rydberg} \qquad (17.276)$$

where $n = 1, 2, 3, ...$ and 1 Rydberg$= m_0 e^2/(2\hbar) = 13.6$ eV. These energy levels lie in the gap just below the bottom of the conduction band ($E = 0$): the lowest level ($n = 1$) is at the distance $-\frac{1}{\epsilon^2(m^*/m_0)} = -0.1$ Rydberg, using $m^* \simeq 0.1 m_0$, $\epsilon \simeq 10$. This level is close to the bottom of the conduction band: it is called swallow impurity level. All impurity levels ($n = 1, 2, 3, 4, ..., \infty$) form a pseudo energy band between the lowest level and the bottom of the conduction band.

In the case of an acceptor impurity, the same calculation can be done. The acceptor impurity levels lie closer to the top of the valence band (see Fig. 17.19).

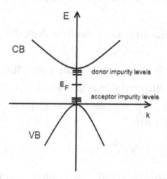

Fig. 17.19 Impurity energy levels lie below the conduction band (CB) are from donor impurities. Those from acceptor impurities lie closer to the top of the valence band (VB).

Problem 7. Recombinations in semiconductors:

(1) Direct recombination rate in a doped semiconductor:

For direct recombinations, the electron recombination rate is equal to the hole one:

$$r = r_p = r_n = knp$$

where k is a constant of the material. Note that these are total rates. One should extract the generation due to thermal excitation, namely $g_T = kn_i^2$. One then has $r = k(np - n_i^2)$. Replacing $n = n_0 + \Delta n$ and $p = p_0 + \Delta p$ into this, one has

$$
\begin{aligned}
r &= k[(n_0 + \Delta n)(p_0 + \Delta p) - n_i^2] \\
&= k[n_0 \Delta p + p_0 \Delta n + \Delta n \Delta p + n_0 p_0 - n_i^2] \\
&= k[n_0 + p_0 + \Delta n]\Delta n = \frac{\Delta n}{\tau(\Delta n)}
\end{aligned}
$$

where one has used $\Delta n = \Delta p$, $n_0 p_0 = n_i^2$ and

$$\tau(\Delta n) \equiv \frac{1}{k(n_0 + p_0 + \Delta n)} \tag{17.277}$$

$\tau(\Delta n)$ is called the life-time of excited carriers in semiconductors. If $n_0 >> p_0$ then $\tau(\Delta n) = 1/kn_0$, then $r = kn_0 \Delta n$.

(2) Indirect recombination rate in a doped semiconductor:

One calculates the recombination rate due to impurity centers. Let E_R be the impurity energy level (in the gap) and N_R the impurity density. The numbers of occupied impurity sites N_R^0 and of unoccupied N_R^u are

$$N_R^0 = N_R f_R \quad \text{and} \quad N_R^u = N_R(1 - f_R)$$

where f_R is the occupation probability given by the Fermi-Dirac distribution

$$f_R = \frac{1}{\exp[\beta(E_R - E_F)] + 1}.$$

The recombination rates are

$$r_n = N_R[C_n n(1 - f_R) - E_n f_R] \tag{17.278}$$

$$r_p = N_R[C_p p f_R - E_p(1 - f_R)] \tag{17.279}$$

where C_n, C_p, E_n and E_p are capture and emission coefficients of electrons and holes at the impurity level (see Fig. 17.20). These

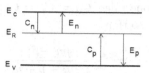

Fig. 17.20　Indirect recombination by impurity centers: E_R is the impurity energy level. The coefficients C_n, C_p, E_n and E_p are defined in the text.

coefficients are related as seen here: at equilibrium $n = n_0$, $p = p_0$, $r_n = r_p = 0$. Therefore

$$E_n = C_n n_0 \frac{1 - f_R}{f_R}$$

$$E_p = C_p p_0 \frac{f_R}{1 - f_R}$$

Replacing f_R and using (15.120) and (15.121), one has

$$E_n = C_n n_i e^{\beta(E_R - E_F)}$$

$$E_p = C_p n_i e^{-\beta(E_R - E_F)}$$

Replacing these into (17.278) and (17.279), one gets

$$r_n = C_n N_R \left[n(1 - f_R) - f_R n_i e^{\beta(E_R - E_F)} \right] \qquad (17.280)$$

$$r_p = C_p N_R \left[p f_R - (1 - f_R) n_i e^{-\beta(E_R - E_F)} \right] \qquad (17.281)$$

An impurity is a recombination center if $r_p = r_n$. With this condition, one obtains

$$f_R = \frac{C_n n + C_p n_i e^{-\beta(E_R - E_F)}}{C_n (n + n_i e^{\beta(E_R - E_F)}) + C_p (p + n_i e^{-\beta(E_R - E_F)})} \qquad (17.282)$$

One then has

$$r = r_n = r_p$$

$$= C_n C_p N_R \frac{pn - n_i^2}{C_n (n + n_i e^{\beta(E_R - E_F)}) + C_p (p + n_i e^{-\beta(E_R - E_F)})}$$

Putting

$$\tau_{p0} = \frac{1}{C_p N_R} \quad \text{and} \quad \tau_{n0} = \frac{1}{C_n N_R}$$

one finally has

$$r = \frac{pn - n_i^2}{\tau_{p0}(n + n_i e^{\beta(E_R - E_F)}) + \tau_{n0}(p + n_i e^{-\beta(E_R - E_F)})} \qquad (17.283)$$

In practical cases, one takes $C_n = C_p = C$, $E_R \simeq E_F$ and $\tau_{p0} = \tau_{n0} = \tau_m = 1/CN_R$ so that

$$r = \frac{1}{\tau_m} \frac{pn - n_i^2}{2n_i + p + n} \tag{17.284}$$

Note: if $C_n >> C_p$, the impurity is an electron trap (see Problem 5 of chapter 4). In the reverse case, $C_p >> C_n$, the impurity is a hole capturer. No recombination occurs at such impurities.

Problem 8. Dielectric relaxation:

In a semiconductor the charge neutrality is observed at equilibrium. Consider an n-type semiconductor. One has $n_0 >> n_i >> p_0$. A charge excess ρ is artificially created: $\rho = -e\Delta n$ where Δn is the number of electrons in excess. One has

$$\frac{\partial n}{\partial t} = \frac{1}{e} \vec{\nabla} \cdot \vec{j}_n \quad \text{or} \quad \frac{\partial \rho}{\partial t} = -\vec{\nabla} \cdot \vec{j}_n \tag{17.285}$$

One has

$$\vec{j}_n = \sigma \vec{\epsilon} + eD_n \vec{\nabla} n \quad \text{where} \quad \sigma = n_0 e \mu_n$$
$$\Rightarrow \vec{\nabla} \cdot \vec{j}_n = \sigma \vec{\nabla} \cdot \vec{\epsilon} + eD_n \nabla^2 n$$
$$= \frac{\sigma}{\epsilon_s} \rho - D_n \nabla^2 \rho$$

where one has used the Gauss theorem $\vec{\nabla} \cdot \vec{\epsilon} = \rho/\epsilon_s$, ϵ_s being the semiconductor dielectric constant. Replacing this into (17.285) one obtains

$$\frac{\partial \rho}{\partial t} = -\frac{\sigma}{\epsilon_s} \rho + D_n \nabla^2 \rho \tag{17.286}$$

If ρ is uniform, namely $\nabla^2 \rho = 0$, one has

$$\frac{d\rho}{\rho} = -\frac{dt}{\tau} \quad \Rightarrow \quad \rho(t) = \rho_0 \exp(-t/\tau) \tag{17.287}$$

where $\tau_d = \epsilon_s/\sigma$ (dielectric relaxation time). ρ tends exponentially to zero with a dielectric relaxation time τ_d.

Problem 9. : Polarized $p - n$ junction: direct current

In section 15.8.6 a non polarized $p - n$ junction has been studied. Consider in this exercise a pn junction under an applied forward-bias voltage V_0. One supposes:

(i) the hole concentration in the n side (minority carriers) is very small, namely $\delta p(x_d) << n_0$ (very weak injection, n_0 being electron concentration at equilibrium)

(ii) the recombinations in the depletion zone are negligible due to a short time needed to cross this narrow zone ($X_D \sim 0.4 \ \mu m$) compared to a long diffusion length of minority carriers ($L_D \sim 50 \ \mu m$)

(iii) the quasi Fermi level of each kind of majority carriers is conserved until the entrance of the other majority zone, namely $E_{Fp}(x) = E_{Fp0}$ for $x \leq x_d$, and $E_{Fn}(x) = E_{Fn0}$ for $x \geq -x_a$.

One calculates the direct current which flows across the junction (see Fig. 15.7). One has for electrons

$$n_n(x_d) = N_c e^{-\beta(E_{cn}-E_{Fn0})} \quad (n \text{ side})$$
$$n_p(-x_a) = N_c e^{-\beta(E_{cp}-E_{Fn0})} \quad (p \text{ side})$$

where E_{cn} and E_{cp} are the energies of the bottom of the conduction band in the n and p zones, respectively. One deduces

$$\frac{n_n(x_d)}{n_p(-x_a)} = \exp\left(\frac{E_{cp} - E_{cn}}{k_B T}\right) = \exp\left[\frac{e(V_D - V_0)}{k_B T}\right]$$
$$= \frac{n_{n0}}{n_{p0}} \exp\left(\frac{-eV_0}{k_B T}\right)$$

where one has used (15.136) in the last equality. Using the hypothesis $n_n(x_d) \simeq n_{n0}$ and simplifying the above relation one has

$$n_p(-x_a) = n_{p0} \exp\left(\frac{eV_0}{k_B T}\right) \tag{17.288}$$

This relation yields the following "additional" electron concentration (or concentration in excess) which enters the p zone as minority carriers, under the effect of V_0:

$$\delta n_p(-x_a) = n_p(-x_a) - n_{p0} = n_{p0}\left[\exp\left(\frac{eV_0}{k_B T}\right) - 1\right] \tag{17.289}$$

In the same manner, one can calculate the excess of the hole concentration which enters the n zone under the effect of V_0:

$$\delta p_n(x_d) = p_n(x_d) - p_{n0} = p_{n0}\left[\exp\left(\frac{eV_0}{k_B T}\right) - 1\right] \tag{17.290}$$

Since the transport of the minority carriers (holes) into the n zone is entirely diffusive (the same for electrons in the p zone), one solves the Fick diffusion equation in the stationary regime for $x_d \leq x < \infty$:

$$D_p \frac{\partial^2 (p_n(x) - p_{n0})}{\partial x^2} - \frac{\partial (p_n(x) - p_{n0})}{\tau_{p0}} = 0$$

$$\delta p_n(x) = p_n(x) - p_{n0} = \delta p_n(x_d) \exp\left(-\frac{x - x_d}{L_p}\right)$$

$$p_n(x) = p_{n0} \left\{1 + \left[\exp\left(\frac{eV_0}{k_BT}\right) - 1\right] \exp\left(-\frac{x - x_d}{L_p}\right)\right\}$$

Note that the second term in the first equality is the recombination rate. In the above equations, D_p is the diffusion coefficient of minority holes in the n zone and one has used (17.289). The diffusion length L_p of minority holes is defined by

$$L_p = \sqrt{D_p \tau_{p0}}, \quad (\tau_{p0}: \text{relaxation time of minority holes}).$$

Similarly, the excess of minority electrons in the p zone is given by

$$n_p(x) = n_{p0} \left\{1 + \left[\exp\left(\frac{eV_0}{k_BT}\right) - 1\right] \exp\left(-\frac{x + x_a}{L_n}\right)\right\}$$

with the diffusion length $L_n = \sqrt{D_n \tau_{n0}}$. The corresponding current densities of minority carriers are thus

$$I_p(\text{holes in } n \text{ region}) = -eD_p\left(\frac{\partial p_n(x)}{\partial x}\right)$$

$$= \frac{eD_p}{L_p} p_{n0} \left[\exp\left(\frac{eV_0}{k_BT}\right) - 1\right]$$

$$\times \exp\left(-\frac{x - x_d}{L_p}\right)$$

$$I_n(\text{electrons in } p \text{ region}) = eD_n\left(\frac{\partial n_p(x)}{\partial x}\right)$$

$$= \frac{eD_n}{L_n} n_{p0} \left[\exp\left(\frac{eV_0}{k_BT}\right) - 1\right]$$

$$\times \exp\left(\frac{x + x_a}{L_n}\right)$$

The above relations allow us to calculate the current densities of majority carriers using the condition of charge neutrality in each zones (n or p).

One can use now the assumption that there is no recombination in the depletion zone to calculate the current flowing across the junction. One has

$$I = I_n(-x_a) + I_p(x_d) = \left(\frac{eD_n}{L_n}n_{p0} + \frac{eD_p}{L_p}p_{n0}\right)\left[\exp\left(\frac{eV_0}{k_BT}\right) - 1\right]$$

$$= I_S\left[\exp\left(\frac{eV_0}{k_BT}\right) - 1\right] \qquad (17.291)$$

where I_S is called "saturation current" defined by

$$I_S = \left(\frac{eD_n}{L_n}n_{p0} + \frac{eD_p}{L_p}p_{n0}\right)$$

Using $n_i^2 = n_{n0}p_{n0} = n_{p0}p_{p0}$ [see (11.73)] and noticing that $n_{n0} \simeq N_d$ and $p_{p0} \simeq N_a$, one obtains

$$I_S = en_i^2\left(\frac{D_n}{L_nN_a}n_{p0} + \frac{D_p}{L_pN_d}\right) \qquad (17.292)$$

Note that I_S depends on T via $n_i^2 \propto \exp(-E_g/k_BT)$ [see (11.73)]. I_s is strongly increased with T (see Problem 3 above): it is doubled every ~ 10 K.

The above calculation was made for a forward-bias voltage. For the reverse-bias case, it suffices to replace V_0 by $-V_0$ in the above formulas.

Numerical values:
$X_D \sim 0.5~\mu m$, $L_D = \sqrt{D\tau} \sim 0.15$ mm, $\mu \sim 1000$ cm^{2}V^{-1}s^{-1}, $D = \mu k_BT/e \sim 25$ cm^{2}s^{-1}, $\tau \sim 10^{-5}$ s.
With $N_a = N_d = 10^{16}$ cm^{-3}, $D_n/L_n = D_p/L_p \sim 15~10^2$ cms^{-1}:
i) for Ge ($n_i \sim 10^{13}$ cm^{-3}), one has $I_s \sim 4.8~10^{-6}$ Acm^{-2},
ii) for Si ($n_i \sim 10^{10}$ cm^{-3}), one has $I_s \sim 4.8~10^{-9}$ Acm^{-2}.

Problem 10. Transport in a superlattice:
One has for $n = 1$:

$$E(k_x) = E_0 + t\cos(k_x b) \qquad (17.293)$$

where $E_0 = \frac{\hbar^2}{2m}\frac{\pi^2}{a^2}$ and $t \ll E_0$. The second term in (17.293) results from a weak coupling between neighboring wells if b is not so small.

a) The electron velocity is

$$v_x = \frac{1}{\hbar}\frac{\partial E}{\partial k_x} = -\frac{tb}{\hbar}\sin(k_x b).$$

The effective mass m^* is defined as (see chapter 11):

$$\frac{1}{m^*} = \frac{1}{\hbar^2} \frac{\partial^2 E}{\partial k_x^2} = -\frac{tb^2}{\hbar^2} \cos(k_x b).$$

For $k_x \to 0$, $(m^*)^{-1} = -\frac{tb^2}{\hbar^2}$. At this limit,

$$E(k_x) = E_0 + t \cos(k_x b) \to E_0 + t \left(1 - \frac{k_x^2 b^2}{2}\right) = \text{constant} - \frac{\hbar^2 k_x^2}{2m^*}.$$

Therefore, the density of state is calculated as for a free electron of mass m^* in one dimension. The result is given in Problem 5 of chapter 2, namely Eq. (16.36) with m replaced by m^*.

b) One applies an electric field ε along the x direction. The Boltzmann's equation is

$$\left(\frac{\partial f}{\partial t}\right)_F = \frac{\vec{F}}{\hbar} \cdot \vec{\nabla}_{k_x} f = -\frac{e\varepsilon}{\hbar} \frac{\partial f}{\partial E} \frac{\partial E}{\partial k_x}$$

$$= -\left(\frac{\partial f}{\partial t}\right)_C = \frac{f - f_0}{\tau} = \frac{g}{\tau} \qquad (17.294)$$

Linearizing $\frac{\partial f}{\partial E} \simeq \frac{\partial f_0}{\partial E}$, one has

$$-e\varepsilon \frac{\partial f_0}{\partial E} \left(-\frac{tb}{\hbar} \sin(k_x b)\right) = \frac{g}{\tau}$$

from which

$$g = tb \frac{e\varepsilon\tau}{\hbar} \sin(k_x b) \frac{\partial f_0}{\partial E}.$$

The current density is

$$j = -e \int g v_x dk_x = -tb \frac{e^2 \varepsilon \tau}{\hbar} \int \sin(k_x b) \left(-\frac{tb}{\hbar} \sin(k_x b)\right) \frac{\partial f_0}{\partial E} dk_x$$

$$= t^2 b^2 \frac{e^2 \varepsilon \tau}{\hbar^2} \int \sin^2(k_x b) \frac{\partial f_0}{\partial E} dk_x$$

$$\equiv \sigma \varepsilon \qquad (17.295)$$

One sees that the linearized Boltzmann's equation yields the Ohm's law. The constant $C = \int \sin^2(k_x b) \frac{\partial f_0}{\partial E} dk_x \neq 0$ because the integrand is an even function of k_x in an integration with opposite bounds (Brillouin zone boundaries).

c) One takes into account non linear terms. One has

$$\frac{\partial f}{\partial k_x} = \frac{\partial f_0}{\partial k_x} + \frac{\partial g}{\partial k_x}$$

from which

$$\left(\frac{\partial f}{\partial t}\right)_F = -\left(\frac{\partial f}{\partial t}\right)_C$$

$$-\frac{e\varepsilon}{\hbar}\left(\frac{\partial f_0}{\partial E}\frac{\partial E}{\partial k_x} + \frac{\partial g}{\partial k_x}\right) = \frac{g}{\tau}$$

One can put the above equation under the following form

$$\frac{\partial g}{\partial k_x} + \frac{\hbar}{e\varepsilon\tau}g = tb\left(-\frac{\partial f_0}{\partial E}\right)\sin(k_x b)$$

$$\frac{\partial g}{\partial k_x} + Ag = B\sin(k_x b) \tag{17.296}$$

where

$$A = \frac{\hbar}{e\varepsilon\tau},$$

$$B = tb\left(-\frac{\partial f_0}{\partial E}\right).$$

One supposes that $-\frac{\partial f_0}{\partial E}$ is independent of k_x (relaxation-time approximation). B is therefore a constant hereafter. Equation (17.296) is a differential equation with a second member. The solution without second member is

$$g = g_0 e^{-k_x/A} = g_0 e^{-\frac{\hbar k_x}{e\varepsilon\tau}} \tag{17.297}$$

Making vary the constant g_0 one has, by comparison with (17.296),

$$\frac{\partial g_0}{\partial k_x} = B\sin(k_x b)e^{\frac{\hbar k_x}{e\varepsilon\tau}}$$

$$\frac{\partial g_0}{\partial k_x} = B\frac{e^{ik_x b} - e^{-ik_x b}}{2i}e^{\frac{\hbar k_x}{e\varepsilon\tau}}$$

$$= B\frac{e^{\left(ib + \frac{\hbar}{e\varepsilon\tau}\right)k_x} - e^{\left(-ib + \frac{\hbar}{e\varepsilon\tau}\right)k_x}}{2i}$$

$$g_0 = B\frac{e^{\left(ib + \frac{\hbar}{e\varepsilon\tau}\right)k_x}}{2i\left(ib + \frac{\hbar}{e\varepsilon\tau}\right)} - B\frac{e^{\left(-ib + \frac{\hbar}{e\varepsilon\tau}\right)k_x}}{2i\left(-ib + \frac{\hbar}{e\varepsilon\tau}\right)} + D$$

where D is an integration constant. One can rewrite the above equation as

$$g_0 = B\frac{\frac{\hbar}{e\varepsilon\tau}\sin(k_x b) - b\cos(k_x b)}{b^2 + \frac{\hbar^2}{e^2\varepsilon^2\tau^2}}e^{\frac{\hbar k_x}{e\varepsilon\tau}} + D \tag{17.298}$$

Replacing this into (17.297) one obtains

$$g = B\frac{\frac{\hbar}{e\varepsilon\tau}\sin(k_xb) - b\cos(k_xb)}{b^2 + \frac{\hbar^2}{e^2\varepsilon^2\tau^2}} + De^{-\frac{\hbar k_x}{e\varepsilon\tau}} \tag{17.299}$$

where the last term is to be neglected. One rewrites g as

$$g = B\frac{\frac{e\varepsilon\tau}{\hbar}\sin(k_xb) - b\frac{e^2\varepsilon^2\tau^2}{\hbar^2}\cos(k_xb)}{1 + \frac{b^2e^2\varepsilon^2\tau^2}{\hbar^2}} \tag{17.300}$$

The current density is thus

$$j = -e\int v_x g \, dk_x$$

$$= \int \frac{etb}{\hbar}\sin(k_xb)\frac{B\left(\frac{e\varepsilon\tau}{\hbar}\sin(k_xb) - b\frac{e^2\varepsilon^2\tau^2}{\hbar^2}\cos(k_xb)\right)}{1 + \frac{b^2e^2\varepsilon^2\tau^2}{\hbar^2}} dk_x$$

$$= \frac{e^2tbB\varepsilon\tau}{\hbar^2}\int\frac{\sin^2(k_xb)}{1 + \frac{b^2e^2\varepsilon^2\tau^2}{\hbar^2}}dk_x$$

$$- \frac{etb^2B}{\hbar}\frac{e^2\varepsilon^2\tau^2}{\hbar^2}\int\frac{\sin(k_xb)\cos(k_xb)}{1 + \frac{b^2e^2\varepsilon^2\tau^2}{\hbar^2}}dk_x$$

Note that the second integral is zero because its integrand is an odd function of k_x in an integral with opposite bounds (first Brillouin zone boundaries). Thus,

$$j = \frac{\sigma\varepsilon}{1 + \gamma\varepsilon^2} \tag{17.301}$$

where

$$\sigma = \frac{e^2tbB\tau}{\hbar^2}\int \sin^2(k_xb)dk_x \tag{17.302}$$

$$\gamma = \frac{b^2e^2\tau^2}{\hbar^2} \tag{17.303}$$

The current density j obeys the Ohm's law if ε is small (the second term in the denominator can be neglected). For strong fields, the function j given by (17.301) has a maximum ε_m determined by $\frac{dj}{d\varepsilon} = 0$: one has $\varepsilon_m = 1/\sqrt{\gamma} = \hbar/(e\tau b)$. The current density is shown in Fig. 17.21. For $\varepsilon > \varepsilon_m$, the differential resistance $\frac{dj}{d\varepsilon}$ is negative.

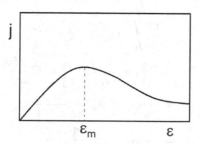

Fig. 17.21   Current density j versus applied electric field ε, see (17.301).

Problem 11. Hall effect - Magnetoresistance:

a) Moderate fields:

i) The equations of motion of an electron in the x and y directions are given by (15.80) and (15.81). For an electron, we rewrite

$$\frac{dv_x}{dt} = -\frac{e}{m_e}\epsilon_x - \omega_c v_y \qquad (17.304)$$

$$\frac{dv_y}{dt} = -\frac{e}{m_e}\epsilon_y + \omega_c v_x \qquad (17.305)$$

Putting $Z = v_x + iv_y$, we have, using the above equations,

$$\frac{dZ}{dt} = -\frac{e}{m_e}(\epsilon_x + i\epsilon_y) + i\omega_c Z.$$

The solution without the right-hand side is $Z = Ae^{i\omega_c t}$, and the stationary solution is

$$dZ/dt = 0 \rightarrow Z = \frac{e}{m_e}\frac{(\epsilon_x + i\epsilon_y)}{i\omega_c}.$$

The general solution is thus

$$Z = Z_0 + \left[\frac{1}{i\omega_c}\frac{e}{m_e}(\epsilon_x + i\epsilon_y)\right](1 - e^{i\omega_c t})$$

$$= Z_0 + C(1 - e^{i\omega_c t})$$

where C is the time-independent quantity in the square brackets and $Z_0 = v_x^0 + iv_y^0 = 0$ for simplicity.

ii) The time-averaged Z is given by

$$\overline{Z} = \frac{1}{\tau} \int_0^\infty Z e^{-t/\tau} dt$$

$$= \frac{C}{\tau} \int_0^\infty (1 - e^{i\omega_c t}) e^{-t/\tau}$$

$$= C \left[1 + \frac{1}{i\omega_c\tau - 1} \right] = C \frac{i\omega_c\tau}{i\omega_c\tau - 1}$$

$$= \frac{e}{m_e}(\epsilon_x + i\epsilon_y) \frac{\tau}{i\omega_c\tau - 1}$$

$$= \frac{e}{m_e}(\epsilon_x + i\epsilon_y) \frac{\tau(i\omega_c\tau + 1)}{\omega_c^2\tau^2 + 1} \tag{17.306}$$

Identifying the real part with v_x and the imaginary part with v_y, we get

$$j_x = -nev_x = \frac{ne^2}{m_e} \left[\frac{\tau}{1 + \omega_c^2\tau^2}\epsilon_x - \frac{\omega_c\tau^2}{1 + \omega_c^2\tau^2}\epsilon_y \right] \tag{17.307}$$

$$j_y = -nev_y = \frac{ne^2}{m_e} \left[\frac{\tau}{1 + \omega_c^2\tau^2}\epsilon_y + \frac{\omega_c\tau^2}{1 + \omega_c^2\tau^2}\epsilon_x \right] \tag{17.308}$$

iii) If there is one type of carriers, Eq. (17.308) when $j_y = 0$ yields $\epsilon_y/\epsilon_x = -\omega_c\tau = \tan\theta_e$ as before. We have the same Hall coefficient, namely $R = -1/ne$. When both electrons and holes participate in the conduction, the Hall coefficient is different as seen in the following. Writing (17.307) and (17.308) for each kind of carriers, we have

$$j_{xe} = C_{1e}\epsilon_x - C_{2e}\epsilon_y$$

$$j_{ye} = C_{2e}\epsilon_x + C_{1e}\epsilon_y$$

$$j_{xh} = C_{1h}\epsilon_x + C_{2h}\epsilon_y$$

$$j_{yh} = -C_{2h}\epsilon_x + C_{1h}\epsilon_y$$

where the coefficients C_{1e} etc. are defined from the coefficients in (17.307) and (17.308)

$$C_{1e} = \frac{ne^2}{m_e} \frac{\tau_e}{1 + \omega_{ce}^2\tau_e^2} \tag{17.309}$$

$$C_{2e} = \frac{ne^2}{m_e} \frac{\omega_{ce}\tau_e^2}{1 + \omega_{ce}^2\tau_e^2} \tag{17.310}$$

$$C_{1h} = \frac{pe^2}{m_h} \frac{\tau_h}{1 + \omega_{ch}^2\tau_h^2} \tag{17.311}$$

$$C_{2h} = \frac{pe^2}{m_h} \frac{\omega_{ch}\tau_h^2}{1 + \omega_{ch}^2\tau_h^2} \tag{17.312}$$

We have

$$j_x = j_{xe} + j_{xh} = (C_{1e} + C1h)\epsilon_x - (C_{2e} - C_{2h})\epsilon_y$$
$$j_y = j_{ye} + j_{yh} = (C_{2e} - C2h)\epsilon_x + (C_{1e} + C_{1h})\epsilon_y$$

Setting $j_y = 0$, we have

$$\epsilon_x = -\frac{(C_{1e} + C_{1h})}{(C_{2e} - C_{2h})}\epsilon_y,$$

hence

$$j_x = -\frac{(C_{1e} + C_{1h})^2 + (C_{2e} - C_{2h})^2}{(C_{2e} - C_{2h})}\epsilon_y.$$

We obtain

$$R = \frac{\epsilon_y}{\frac{j_x B}{c}} = -\frac{C_{2e} - C_{2h}}{(C_{1e} + C_{1h})^2 + (C_{2e} - C_{2h})^2}\frac{1}{\frac{B}{c}} \tag{17.313}$$

Replacing the coefficients, we obtain

$$R = \frac{\sigma_e^2 R_e + \sigma_h^2 R_h + \sigma_e^2\sigma_h^2 R_e R_h (R_e + R_h)B^2}{(\sigma_e + \sigma_h)^2 + \sigma_e^2\sigma_h^2(R_e + R_h)^2 B^2} \tag{17.314}$$

where $R_e = -1/ne$, $R_h = 1/pe$, $\sigma_e = ne\mu_e$, $\sigma_h = pe\mu_h$.

Comments:

- For small B, neglecting B^2 terms, we find the result of weak fields.
- If there is one kind of carriers the B terms in (17.314) are zero: R does not depend on B.
- If the material is intrinsic, namely $p = n$ and $R_e = -R_h$, then R does not depend on B, neither.
- For strong fields, (17.314) gives

$$R = \frac{R_e R_h}{R_e + R_h} = \frac{1}{(p - n)e}.$$

iv) The above results show that the presence of a magnetic field has no effect on the resistance when the conduction is due to only one type of carriers (n or p) and when the relaxation time is constant and when the effective mass is isotropic (spherical iso-energy surfaces). If one of these three conditions is not fulfilled, there is a correction to the initial resistivity ρ_0. We suppose that both electrons and holes participate to the conduction, keeping isotropic effective mass and constant τ, using R of (17.313) and $\sigma = j_x/\epsilon_x$ with j_x given by above, we write

$$\sigma = -\frac{(C_{1e} + C_{1h})^2 + (C_{2e} - C_{2h})^2}{C_{1e} + C_{1h}}$$

$$\sigma R = -\frac{C_{2e} - C_{2h}}{C_{1e} + C_{1h}} \frac{1}{\frac{B}{c}}$$

Replacing the coefficients and using R of (17.314) we obtain

$$\sigma = \frac{(\sigma_e + \sigma_h)^2 + \sigma_e^2 \sigma_h^2 (B/c)^2 (R_e + R_h)^2}{(\sigma_e + \sigma_h) + \sigma_e \sigma_h (B/c)^2 (\sigma_e R_e^2 + \sigma_h R_h^2)} \tag{17.315}$$

Putting $\sigma_0 \equiv \sigma_e + \sigma_h$ (at $B = 0$), and expanding the denominator with respect to B, we have

$$\frac{\Delta\sigma}{\sigma_0} = \frac{\sigma - \sigma_0}{\sigma_0}$$

$$= -\frac{np\mu_e\mu_h(\mu_e + \mu_h)^2 (B/c)^2}{(n\mu_e + p\mu_h)^2}$$

$$\equiv -\xi R_0^2 \sigma_0^2 (B/c)^2 \tag{17.316}$$

where the coefficient of the transverse magneto-resistance ξ is

$$\xi = \frac{np\mu_e\mu_h(\mu_e + \mu_h)^2}{(p\mu_h^2 - n\mu_e^2)^2}$$

and R_0 given by the weak-field approximation [Eq. (15.86)]. Note that

$$\frac{\Delta\rho}{\rho_0} = -\frac{\Delta\sigma}{\sigma_0}$$

$$= \xi R_0^2 \sigma_0^2 (B/c)^2 \tag{17.317}$$

We see that if there is one kind of carriers, this correction is zero (n or p is zero).

Numerical application: The maximum of $\frac{\Delta\rho}{\rho_0}$ is at $n\mu_e = p\mu_h$. In order to have $\frac{\Delta\rho}{\rho_0} \simeq 0.1$ in an applied field of 10^3 Gauss,

$$\frac{\Delta\rho}{\rho_0} = \frac{(\mu_e + \mu_h)^2}{4}(B/c)^2 \simeq 0.1 \rightarrow (\mu_e + \mu_h) \simeq 10^2 \times 630 \text{ cm}^2/\text{V-sec.}$$

b) Effects of collisions:

One supposes again that the effective mass m_e is isotropic and one type of carriers (n), but the relaxation time τ depends on the electron energy under the form $\tau = aE^{-s}$. This form represents several types of collision. The results for j_x and j_x in the weak-field approximation shown above are still valid provided that the coefficients $\frac{\tau}{1+\omega_c^2\tau^2}$ and $\frac{\omega_c\tau^2}{1+\omega_c^2\tau^2}$ are replaced by their values averaged over all energies.

i) Hall coefficient:

Using (17.313) for R but with averaged values for the coefficients of (17.309)-(17.312), we have

$$R = -\frac{K}{ne} \quad \text{where } K \text{ is given by}$$

$$K = \frac{<\frac{\tau^2}{1+\omega_c^2\tau^2}>}{<\frac{\tau}{1+\omega_c^2\tau^2}>^2 + \omega_c^2 <\frac{\tau^2}{1+\omega_c^2\tau^2}>^2}$$

$$\simeq \frac{<\tau^2>}{<\tau>^2} \tag{17.318}$$

where we supposed $\omega_c^2\tau^2 \ll 1$ (weak fields).

ii) We calculate $<\tau^2>$ and $<\tau>$:

$$<\tau> = \frac{\int_0^{E_{max}} \tau(E)e^{-E/k_BT}E^{3/2}dE}{\int_0^{E_{max}} e^{-E/k_BT}E^{3/2}dE}$$

$$= \frac{A\int_0^{E_{max}} E^{-s}e^{-E/k_BT}E^{3/2}dE}{\int_0^{E_{max}} e^{-E/k_BT}E^{3/2}dE}$$

$$\simeq a(k_BT)^{-s}\frac{\int_0^\infty u^{-s}e^{-u}u^{3/2}du}{\int_0^\infty e^{-u}u^{3/2}du}$$

$$= a(k_BT)^{-s}\frac{\Gamma(5/2 - s)}{\Gamma(5/2)} \tag{17.319}$$

where we used the Maxwell-Boltzmann statistics and put $u = E/k_BT$. The coefficient a is the conversion coefficient including A. The same calculation for $<\tau^2>$ leads to a similar result with $<\tau^2> = a^2(k_BT)^{-2s}\frac{\Gamma(5/2-2s)}{\Gamma(5/2)}$. Using the Γ integrals for

- $s = 1/2$ (phonon scattering), we have $K = 3\pi/8 \simeq 1.18$,
- $s = -3/2$ (scattering by charged impurities), we have $K = 315\pi/512 \simeq 1.93$.

We see that K has a strong deviation from the value 1 of the no collision case.

iii) One supposes there is only one kind of carriers, in the same manner we can show that

$$\frac{\Delta\rho}{\rho_0} = \frac{e^2 B^2}{(m_e)^2} \left[\frac{<\tau^3><\tau> - <\tau^2>^2}{<\tau>^2} \right]$$

where $<\tau^3>$ can be calculated as above.

PART 4
Appendices

Appendix A

Mathematical Complements and Table of Constants

A.1 Volume of a sphere in n dimensions

One considers a sphere in the space defined by n degrees of freedom $x_1, X_2, ..., x_n$. The radius r of the sphere is given by

$$r^2 = x_1^2 + x_2^2 + ... + x_n^2 \tag{A.1}$$

The volume of the sphere is $V_n(r) = A_n r^n$ where A_n is a coefficient to be determined. One considers the following integral

$$C = \int_{-\infty}^{\infty} dx_1 \int_{-\infty}^{\infty} dx_2 ... \int_{-\infty}^{\infty} dx_n \exp[-(x_1^2 + x_2^2 + ... + x_n^2)] = [\sqrt{\pi}]^n \tag{A.2}$$

where one has used n times the integral I_0 of section A.3 below. If one rewrites the integral C in spherical coordinates, one has

$$C = \int_0^{\infty} dr n A_n r^{n-1} \exp(-r^2) \tag{A.3}$$

where one has used $dV_n = n A_n r^{n-1} dr$. Putting $t = r^2$, one has

$$C = \frac{1}{2} \int_0^{\infty} dt n A_n t^{n/2-1} \exp(-t) = \frac{1}{2} A_n n \Gamma(n/2) = \frac{1}{2} A_n \Gamma(n/2 + 1) \tag{A.4}$$

where one has used the definition of the function Γ (see section A.4). Comparing (A.2) and (A.4), one obtains

$$A_n = \frac{\pi^{n/2}}{\Gamma(n/2 + 1)} \tag{A.5}$$

The volume of a sphere of radius r in n dimensions is thus

$$V_n(r) = \frac{\pi^{n/2}}{\Gamma(n/2 + 1)} r^n \tag{A.6}$$

573

A.2   Stirling formula

One shows below the Stirling formula which is frequently used in statistical physics.

One considers the $\Gamma(n+1)$ function

$$\Gamma(n+1) = \int_0^\infty x^n \exp(-x)dx = n! \tag{A.7}$$

Using the equality $u = \exp(\ln u)$ one can rewrite it under the form

$$n! = \int_0^\infty x^n \exp(-x)dx = \int_0^\infty \exp(-x + n\ln x)dx \tag{A.8}$$

Putting $y(x) \equiv n\ln x - x$, one has $y' = \frac{n}{x} - 1$, $y'' = -\frac{n}{x^2}$, etc. One sees that y is maximum when $y' = 0$, namely when $x = n$. One expands y around n:

$$y(x) = y(n) + y'(n)(x-n) + y''(n)\frac{(x-n)^2}{2} + ... \simeq n\ln n - n - \frac{(x-n)^2}{2n} \tag{A.9}$$

Replacing $y(x)$ in (A.8) and putting $u = x - n$, one has

$$n! = \int_{-n}^\infty \exp(-n + n\ln n)\exp[-u^2/(2n)]du \tag{A.10}$$

If n is very large, one can replace the lower bound of the integral by $-\infty$, then one uses the integral I_0 of A.3 below, one gets

$$n! = \exp(-n + n\ln n)\sqrt{2\pi n} \tag{A.11}$$

or

$$\ln n! = n\ln n - n + \frac{1}{2}\ln 2\pi n \tag{A.12}$$

This formula is called " Stirling formula". Note that when n is very large, the last term is negligible with respect to the other terms.

A.3   Gaussian integrals

$$I_n = \int_0^\infty x^n \exp(-ax^2)dx \tag{A.13}$$

$$I_0 = \frac{1}{2}\sqrt{\frac{\pi}{a}} \ , \ I_1 = \frac{1}{2a} \ , \ I_2 = \frac{1}{4a}\sqrt{\frac{\pi}{a}}$$

$$I_3 = \frac{1}{2a^2} \ , \ I_4 = \frac{3}{8a^2}\sqrt{\frac{\pi}{a}} \ , \ I_5 = \frac{1}{a^3} \tag{A.14}$$

A.4 Γ function

$$\Gamma(n+1) = \int_0^\infty t^n \exp(-t)dt = n! \tag{A.15}$$

$$\Gamma(n+1) = n\Gamma(n) \quad , \quad \Gamma(1/2) = \sqrt{\pi}$$

$$\Gamma(3/2) = \frac{1}{2}\sqrt{\pi} \quad , \quad \Gamma(5/2) = \frac{3}{4}\sqrt{\pi} \tag{A.16}$$

A.5 ζ series or Riemann's series

$$\zeta(x) = \sum_{n=1}^\infty \frac{1}{n^x} \, , \; \zeta(2) = \frac{\pi^2}{6} \, , \; \zeta(4) = \frac{\pi^4}{90} \, , \; \zeta(3/2) \simeq 2.612 \, , \; \zeta(5/2) \simeq 1.341 \tag{A.17}$$

Bose integral:

$$\int_0^\infty \frac{x^y}{e^x - 1}dx = \Gamma(y+1)\zeta(y+1) \tag{A.18}$$

$$\int_0^\infty \frac{x^3}{e^x - 1}dx = \frac{\pi^4}{15} \tag{A.19}$$

A.6 Other formulas

- Geometric series of N terms of ratio q

$$S = a + aq + aq^2 + ... + aq^{N-1} = a\frac{1-q^N}{1-q} \tag{A.20}$$

- Infinite series of terms with $x < 1$

$$S = 1 + x + x^2 + x^3 + ... = \sum_{n=0}^\infty x^n = \frac{1}{1-x} \tag{A.21}$$

$$S = 1 - x + x^2 - x^3 + ... = \sum_{n=0}^\infty (-1)^n x^n = \frac{1}{1+x} \tag{A.22}$$

- Expansion of an exponential

$$\exp(x) = \sum_{n=0}^\infty \frac{x^n}{n!} \quad , \quad \exp(-x) = \sum_{n=0}^\infty (-1)^n \frac{x^n}{n!} \tag{A.23}$$

A.7   Universal constants

- Boltzmann constant $k_B = 1.380664 \times 10^{-23}$ J/K (SI)
- Planck constant $h = 6.62618 \times 10^{-34}$ Js (SI), $\hbar = \frac{h}{2\pi} = 1.054590 \times 10^{-34}$ Js
- Electron charge $e = -1.60219 \times 10^{-19}$ C
- Electron mass $m = 9.10954 \times 10^{-31}$ kg
- Neutron mass $m_n = 1.67495 \times 10^{-27}$ kg
- Proton mass $m_p = 1.67265 \times 10^{-27}$ kg
- Gyromagnetic factor of electron (Landé factor) $g = 2.002319315$
- Magnetic moment of electron $\mu_e = 9.284832 \times 10^{-24}$ J/T
- Bohr magneton $\mu_B = \frac{e\hbar}{2m} = -9.27408 \times 10^{-24}$ J/T
- Avogadro number $N_A = 6.02205 \times 10^{23}$
- Constant of an ideal gas $R = N_A k_B = 8.31441$ J/K/mol
- Light velocity in vacuum $c = 2.99792458 \times 10^8$ m/s

Appendix B

Sommerfeld's Expansion at Low Temperatures

Consider the following integral

$$I = \int_{-\infty}^{\infty} h(E)f(E)dE \qquad (B.1)$$

where $h(E)$ is a function derivable at any order with respect to T and zero when $E \to -\infty$, and $f(E)$ the Fermi-Dirac distribution:

$$f(E) = \frac{1}{e^{\beta(E-\mu)} + 1} \qquad (B.2)$$

μ being the chemical potential.

At low temperatures, one shows that I can be expanded in powers of T as follows:

$$I \simeq \int_{-\infty}^{\mu} h(E)dE + \frac{\pi^2}{6}(k_B T)^2 \left[h^{(1)}(E)\right]_{E=\mu} + \frac{7\pi^4}{360}(k_B T)^4 \left[h^{(3)}(E)\right]_{E=\mu} + ... \qquad (B.3)$$

where $\left[h^{(n)}(E)\right]_{E=\mu}$ is the n-th derivative of $h(E)$ with respect to E taken at $E = \mu$.

Demonstration:
One defines $g(E) = \int_{-\infty}^{E} dE h(E)$. The integration by parts gives

$$I = \int_{-\infty}^{\infty} h(E)f(E)dE = [g(E)f(E)]_{-\infty}^{\infty} - \int_{-\infty}^{\infty} dE g(E)\frac{\partial f(E)}{\partial E}$$

$$= 0 + \int_{-\infty}^{\infty} dE g(E)(-\frac{\partial f(E)}{\partial E}) \qquad (B.4)$$

where one has used the limits $f \to 0$ when $E \to \infty$, and $g(E) \to 0$ when $E \to -\infty$.

At low T the function $-\frac{\partial f(E)}{\partial E}$ is significant only around μ. It is therefore justified to make an expansion of $g(E)$ around μ. One has

$$g(E) = g(\mu) + \sum_{n=1}^{\infty} \left[\frac{(E-\mu)^n}{n!}\right] \left[\frac{d^n g(E)}{dE^n}\right]_{E=\mu} \tag{B.5}$$

Replacing this expansion in (B.4), one has

$$I = g(\mu) + \sum_{n=1}^{\infty} \int_{-\infty}^{\infty} dE \frac{(E-\mu)^n}{n!} \left[\frac{d^n g(E)}{dE^n}\right]_{E=\mu} \left(-\frac{\partial f(E)}{\partial E}\right)$$

$$= g(\mu) + \sum_{n=1}^{\infty} \int_{-\infty}^{\infty} dE \frac{(E-\mu)^{2n}}{(2n)!} \left[\frac{d^{2n-1} h(E)}{dE^{2n-1}}\right]_{E=\mu} \left(-\frac{\partial f(E)}{\partial E}\right)$$

Only terms of even order are not zero while terms of odd order are zero because they have integrands of odd powers with opposite integral bounds [$\frac{\partial f(E)}{\partial E}$ is an even function with respect to $(E-\mu)$]. Putting $x = \beta(E-\mu)$, one obtains

$$I = g(\mu) + \sum_{n=1}^{\infty} c_{2n}(k_B T)^{2n} \left[\frac{d^{2n-1} h(E)}{dE^{2n-1}}\right]_{E=\mu} \tag{B.6}$$

where

$$c_{2n} = \frac{1}{(2n)!} \int_{-\infty}^{\infty} x^{2n} \left[-\frac{d}{dx}\frac{1}{e^x+1}\right] dx \tag{B.7}$$

Integrating by parts yields

$$c_{2n} = \frac{1}{(2n)!} \left[-[x^{2n}\frac{1}{e^x+1}]_{-\infty}^{\infty} + 2n \int_{-\infty}^{\infty} x^{2n-1}\frac{1}{e^x+1} dx\right]$$

$$= \frac{1}{(2n)!} \left[0 + 4n(1 - 2^{1-2n})\Gamma(2n)\zeta(2n)\right] \tag{B.8}$$

where one has used the formula

$$\int_0^{\infty} x^{n-1}\frac{1}{e^x+1} dx = (1 - 2^{1-n})\Gamma(n)\zeta(n) \tag{B.9}$$

where $\Gamma(n) =$ and $\zeta(n) =$ are given in A.4 and A.5.

Rarely one goes beyond $2n = 4$ in the Sommerfeld's expansion.

Appendix C

Origin of the Heisenberg Model

In this appendix, we show the origin of the magnetic interaction which leads to the Heisenberg spin model given in (12.1). We suppose that the reader has read chapter 7 on the second quantization and chapter 10 on the Hartree-Fock approximation.

We consider the Coulomb interaction between two electrons written in the second quantization [see (7.69)]

$$\hat{\mathcal{H}} = -\frac{1}{2} \sum_{\sigma;\sigma'} \int \int \hat{\psi}_\sigma^+(\vec{r}_1)\hat{\psi}_{\sigma'}^+(\vec{r}_2) \frac{e^2}{|\vec{r}_1 - \vec{r}_2|} \hat{\psi}_\sigma(\vec{r}_1)\hat{\psi}_{\sigma'}(\vec{r}_2) d\vec{r}_1 d\vec{r}_2 \qquad (C.1)$$

where $\hat{\psi}_\sigma$ and $\hat{\psi}_\sigma^+$ are field operators defined by

$$\hat{\psi}_\sigma(\vec{r}) = \sum_{m,n} b_{mn\sigma}\varphi_{nm}(\vec{r}) \qquad (C.2)$$

$$\hat{\psi}_\sigma^+(\vec{r}) = \sum_{m,n} b_{mn\sigma}^+\varphi_{nm}^+(\vec{r}) \qquad (C.3)$$

where $\varphi_{nm}(\vec{r})$ and $\varphi_{nm}^+(\vec{r})$ are wave functions of orbital m at the site n of the crystal, b and b^+ fermion annihilation and creation operators. The wave functions φ_{nm} constitute an orthogonal set. Equation (C.1) becomes

$$\hat{\mathcal{H}} = -\frac{1}{2} \sum \langle \varphi_{n_1 m_1 \sigma_1}(\vec{r}_1)\varphi_{n_2 m_2 \sigma_2}(\vec{r}_2)| \frac{e^2}{|\vec{r}_1 - \vec{r}_2|} |\varphi_{n_3 m_3 \sigma_3}(\vec{r}_1)\varphi_{n_4 m_4 \sigma_4}(\vec{r}_2)\rangle$$

$$\times b_{n_1 m_1 \sigma_1}^+ b_{n_2 m_2 \sigma_2}^+ b_{n_3 m_3 \sigma_3} b_{n_4 m_4 \sigma_4} \qquad (C.4)$$

where the sum runs over $(n_1, m_1, \sigma_1, \cdots, n_4, m_4, \sigma_4)$.

If $n_1 = n_2 = n_3 = n_4$, the interactions are between electrons of the same site. Equation (C.4) is the origin of the Hund's empirical rules. For simplicity, we suppose one electron per site and one orbital per electron in the following.

If $n_1 = n_3$ and $n_2 = n_4$, the Coulomb term is given by

$$\hat{\mathcal{H}}_c = -\frac{1}{2} \sum_{n_1, n_2} \langle n_1 n_2 | \frac{e^2}{|r_{12}|} | n_2 n_1 \rangle \sum_{\sigma_1, \sigma_2} b^+_{n_1 \sigma_1} b^+_{n_2 \sigma_2} b_{n_1 \sigma_1} b_{n_2 \sigma_2}.$$

If $n_1 = n_4$ and $n_2 = n_3$ (by consequence, $\sigma_1 = \sigma_2$), the exchange term becomes

$$\hat{\mathcal{H}}_{ex} = -\frac{1}{2} \sum_{n_1, n_2, \sigma_1 = \sigma_2} \langle n_1 n_2 | \frac{e^2}{|r_{12}|} | n_2 n_1 \rangle b^+_{n_1 \sigma_1} b^+_{n_2 \sigma_2} b_{n_1 \sigma_1} b_{n_2 \sigma_2}$$

$$= -\frac{1}{2} \sum_{n_1, n_2, \sigma_1 = \sigma_2} \langle n_1 n_2 | \frac{e^2}{|r_{12}|} | n_2 n_1 \rangle b^+_{n_1 \sigma_1} b_{n_1 \sigma_1} b^+_{n_2 \sigma_2} b_{n_2 \sigma_2}$$

$$= -\frac{1}{2} \sum_{n_1, n_2} J_{n_1 n_2} \sum_{\sigma_1, \sigma_2, \sigma_1 = \sigma_2} b^+_{n_1 \sigma_1} b_{n_1 \sigma_1} b^+_{n_2 \sigma_2} b_{n_2 \sigma_2} \qquad (C.5)$$

where

$$J_{n_1 n_2} \equiv \langle n_1 n_2 | \frac{e^2}{|r_{12}|} | n_2 n_1 \rangle. \qquad (C.6)$$

$$\sum_{\sigma_1, \sigma_2, \sigma_1 = \sigma_2} b^+_{n_1 \sigma_1} b_{n_1 \sigma_1} b^+_{n_2 \sigma_2} b_{n_2 \sigma_2} = b^+_{n_1 \uparrow} b_{n_1 \uparrow} b^+_{n_2 \uparrow} b_{n_2 \uparrow} + b^+_{n_1 \downarrow} b_{n_1 \downarrow} b^+_{n_2 \downarrow} b_{n_2 \downarrow}$$

$$= \frac{1}{2} \left(b^+_{n_1 \uparrow} b_{n_1 \uparrow} + b^+_{n_1 \downarrow} b_{n_1 \downarrow} \right)$$

$$\times \left(b^+_{n_2 \uparrow} b_{n_2 \uparrow} + b^+_{n_2 \downarrow} b_{n_2 \downarrow} \right)$$

$$+ \frac{1}{2} \left(b^+_{n_1 \uparrow} b_{n_1 \uparrow} - b^+_{n_1 \downarrow} b_{n_1 \downarrow} \right)$$

$$\times \left(b^+_{n_2 \uparrow} b_{n_2 \uparrow} - b^+_{n_2 \downarrow} b_{n_2 \downarrow} \right)$$

$$+ b^+_{n_1 \uparrow} b_{n_1 \downarrow} b^+_{n_2 \downarrow} b_{n_2 \uparrow}$$

$$+ b^+_{n_1 \downarrow} b_{n_1 \uparrow} b^+_{n_2 \uparrow} b_{n_2 \downarrow} \qquad (C.7)$$

where the last two terms have been added. Note that these terms do not affect the result because their averages are zero in the diagonal representation: $\langle \hat{\psi} | b^+_{n_1 \uparrow} b_{n_1 \downarrow} b^+_{n_2 \downarrow} b_{n_2 \uparrow} | \hat{\psi} \rangle = \langle \hat{\psi} | b^+_{n_1 \downarrow} b_{n_1 \uparrow} b^+_{n_2 \uparrow} b_{n_2 \downarrow} | \hat{\psi} \rangle = 0$.

We define next the following spin operators

$$S^z_n = \frac{1}{2} (b^+_{n \uparrow} b_{n \uparrow} - b^+_{n \downarrow} b_{n \downarrow}) \qquad (C.8)$$

$$S^+_n \equiv S^x_n + i S^y_n = b^+_{n \uparrow} b_{n \downarrow} \qquad (C.9)$$

$$S^-_n \equiv S^x_n - i S^y_n = b^+_{n \downarrow} b_{n \uparrow} \qquad (C.10)$$

As we suppose one electron per site, we have

$$\left(b_{n_1\uparrow}^+ b_{n_1\uparrow} + b_{n_1\downarrow}^+ b_{n_1\downarrow}\right) = 1$$

$$\left(b_{n_2\uparrow}^+ b_{n_2\uparrow} + b_{n_2\downarrow}^+ b_{n_2\downarrow}\right) = 1$$

The right-hand side of (C.7) becomes

$$\frac{1}{2} + 2\vec{S}_{n_1}.\vec{S}_{n_2} \tag{C.11}$$

By using

$$\vec{S}_{n_1}.\vec{S}_{n_2} = S_{n_1}^z S_{n_2}^z + S_{n_1}^y S_{n_2}^y + S_{n_1}^x S_{n_2}^x$$
$$= S_{n_1}^z S_{n_2}^z + \frac{1}{2}\left(S_{n_1}^+ S_{n_2}^- + S_{n_1}^- S_{n_2}^+\right).$$

we rewrite (C.5) as

$$\hat{\mathcal{H}}_{ex} = -\frac{1}{2}\sum_{n_1,n_2} J_{n_1 n_2}\left(\frac{1}{2} + 2\vec{S}_{n_1}.\vec{S}_{n_2}\right)$$

As the double sum $\sum_{n_1,n_2}$ is performed over the sites n_1 et n_2, we added a factor $\frac{1}{2}$ to remove the double counting of each pair (n_1, n_2). We have $\sum_{(n_1,n_2)}$ instead of $\frac{1}{2}\sum_{n_1,n_2}$ where (n_1, n_2) indicates *the pair* $(n_1 n_2)$.
Finally,

$$\hat{\mathcal{H}}_{ex} = -\sum_{(n_1 n_2)} J_{n_1 n_2}\left(\frac{1}{2} + 2\vec{S}_{n_1}.\vec{S}_{n_2}\right) \tag{C.12}$$

The first term does not depend on spins, while the second term does. This term is the Heisenberg model which shall be used later in (12.1).

Hamiltonian (C.12) is thus the origin of ferromagnetism observed in crystals such as CrO_2, $CrBr_3$, $\cdots$. In general, $J_{n_1 n_2}$ is positive:

$$J_{n_1 n_2} = \int\int \varphi_{n_1}^*(\vec{r}_1)\varphi_{n_2}^*(\vec{r}_2)\frac{e^2}{|\vec{r}_1 - \vec{r}_2|}\varphi_{n_1}(\vec{r}_1)\varphi_{n_2}(\vec{r}_2)d\vec{r}_1 d\vec{r}_2 \tag{C.13}$$

Using

$$\frac{e^2}{|\vec{r}_1 - \vec{r}_2|} = \frac{1}{\Omega}\sum_{\vec{k}}\frac{4\pi e^2}{k^2}e^{i\vec{k}\cdot(\vec{r}_1 - \vec{r}_2)} \tag{C.14}$$

we get

$$J_{n_1 n_2} = \frac{1}{\Omega} \sum_k \frac{4\pi e^2}{k^2} \int \varphi_{n_1}^*(\vec{r}_1)\varphi_{n_2}(\vec{r}_1)e^{-i\vec{k}\cdot\vec{r}_1}d\vec{r}_1 \int \varphi_{n_2}^*(\vec{r}_2)\varphi_{n_1}(\vec{r}_2)e^{i\vec{k}\cdot\vec{r}_2}d\vec{r}_2$$

$$= \frac{1}{\Omega} \sum_{\vec{k}} \frac{4\pi e^2}{k^2} I^2 \qquad\qquad\qquad\qquad (C.15)$$

where

$$I = \int_\Omega \varphi_{n_1}^*(\vec{r}_1)\varphi_{n_2}(\vec{r}_1)e^{-i\vec{k}\cdot\vec{r}_1}d\vec{r}_1 = \int_\Omega \varphi_{n_2}^*(\vec{r}_2)\varphi_{n_1}(\vec{r}_2)e^{-i\vec{k}\cdot\vec{r}_2}d\vec{r}_2$$

Note that these two integrals are identical because the indices and variables are dummy. $J_{n_1 n_2}$ is thus positive. From (C.12), we see that if $\vec{S}_{n_1}$ and $\vec{S}_{n_2}$ are parallel, then the energy is lower. The wave functions $\varphi_n(\vec{r})$ have been supposed to be orthogonal. They are Wannier wave functions constructed from linear combinations of Bloch wave functions.

Appendix D

Hubbard Model: Superexchange

The Hubbard model has been introduced by Hubbard in 1963 [87] and almost at the same time by Gutzwiller [75] and Kanamori [92]. The Hubbard model is interesting because by changing a few parameters it can describe insulators, conductors and various magnetic states. In one dimension, it can be exactly solved [114]: at half-filling it gives an insulating state for $U > 0$, U being the on-site repulsive interaction defined below. Away from the half-filling, it yields a conducting state. The general Hubbard model cannot be solved without approximations. The reader is referred to reviews and books such as Refs. [56, 67, 131, 148]. In the following, we give some basic properties of the Hubbard model.

We have seen that the direct exchange interaction has an origin in a Coulomb interaction. In the Hubbard model, we suppose that the Coulomb interaction U is between the electrons of opposite spins which belong to the same ion. In addition, we introduce an interaction between the electrons of neighboring ions. The Hubbard Hamiltonian is written as

$$\hat{\mathcal{H}} = \sum_{n,n',\sigma} V_{nn'} b_{n'\sigma}^+ b_{n\sigma} + U \sum_n b_{n\uparrow}^+ b_{n\uparrow} b_{n\downarrow}^+ b_{n\downarrow} \tag{D.1}$$

where

$$V_{nn'} \equiv \int \varphi_{n'}^*(\vec{r}) \mathcal{H}_{cryst} \varphi_n(\vec{r}) d\vec{r} \qquad (\mathcal{H}_{cryst} = \text{crystalline field}) \tag{D.2}$$

The Hubbard model is often used for d electrons which are not completely localized: they are more or less *itinerant* because of the interaction $V_{nn'}$ between electrons of neighboring ions. In general, U is positive [demonstrations analogous to that for (C.15)]. We deduce

- If $U \gg V$: the crystal is an insulator because electrons prefer to stay

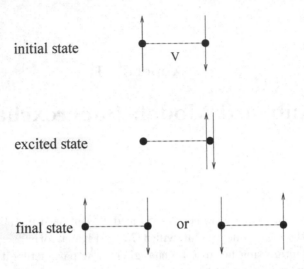

Fig. D.1   Spin exchange due to the second-order perturbation is schematically shown.

on different sites to reduce their interaction energy; the second term in
(D.1) is then equal to zero.

- If $U \ll V$: there is a possibility that electrons move from site to site
  under the action of V, the crystal is then a conductor.

We are interested in the first case where V is considered as a perturbation to U. We show below that the second-order perturbation in V gives
rise to a magnetically ordered state. The Hamiltonian of the second order
in V is written as

$$\mathcal{H}_2 = \sum_{n,n',\sigma,\sigma'} \frac{|V_{nn'}|^2}{E_0 - E_n} b_{n\sigma'}^+ b_{n'\sigma'} b_{n'\sigma}^+ b_{n\sigma} \tag{D.3}$$

where E_0 is the energy of the initial state in which each spin occupies a site
and E_n the energy of the excited state where two spins occupy the same
site. We thus have $E_0 = V$ and $E_n = U$.

It follows that

$$\mathcal{H}_2 \cong \sum_{n,n',\sigma,\sigma'} \frac{|V_{nn'}|^2}{U} b_{n\sigma'}^+ b_{n'\sigma'} b_{n'\sigma}^+ b_{n\sigma} \quad \text{with } V \ll U \tag{D.4}$$

The spin exchange due to the interaction (D.4) is schematically displayed in Fig. D.1.

Obviously, the exchange mechanism is possible only if the spins of the sites n and n' are initially antiparallel. Otherwise, the intermediate state is forbidden by the Pauli's principle. The sum on the spins in (C.15) clearly shows that fact:

$$\sum_{\sigma,\sigma'} b^+_{n\sigma'} b_{n'\sigma'} b^+_{n'\sigma} b_{n\sigma} = b^+_{n\uparrow} b_{n'\uparrow} b^+_{n'\uparrow} b_{n\uparrow} + b^+_{n\downarrow} b_{n'\downarrow} b^+_{n'\downarrow} b_{n\downarrow}$$

$$+ b^+_{n\uparrow} b_{n'\uparrow} b^+_{n'\downarrow} b_{n\downarrow} + b^+_{n\downarrow} b_{n'\downarrow} b^+_{n'\uparrow} b_{n\uparrow}$$

$$= b^+_{n\uparrow}(1 - b^+_{n'\uparrow} b_{n'\uparrow}) b_{n\uparrow} + b^+_{n\downarrow}(1 - b^+_{n'\downarrow} b_{n'\downarrow}) b_{n\downarrow}$$

$$- b^+_{n\uparrow} b^+_{n'\downarrow} b_{n'\uparrow} b_{n\downarrow} - b^+_{n\downarrow} b^+_{n'\uparrow} b_{n'\downarrow} b_{n\uparrow}$$

$$= (b^+_{n\uparrow} b_{n\uparrow} + b^+_{n\downarrow} b_{n\downarrow})$$

$$- \frac{1}{2}(b^+_{n\uparrow} b_{n\uparrow} + b^+_{n\downarrow} b_{n\downarrow})(b^+_{n'\uparrow} b_{n'\uparrow} + b^+_{n'\downarrow} b_{n'\downarrow})$$

$$- \frac{1}{2}(b^+_{n\uparrow} b_{n\uparrow} - b^+_{n\downarrow} b_{n\downarrow})(b^+_{n'\uparrow} b_{n'\uparrow} - b^+_{n'\downarrow} b_{n'\downarrow})$$

$$- b^+_{n\uparrow} b_{n\downarrow} b^+_{n'\downarrow} b_{n'\uparrow} - b^+_{n\downarrow} b_{n\uparrow} b^+_{n'\uparrow} b_{n'\downarrow} \tag{D.5}$$

Since there is only one electron per site, one has

$$b^+_{n\uparrow} b_{n\uparrow} + b^+_{n\downarrow} b_{n\downarrow} = b^+_{n'\uparrow} b_{n'\uparrow} + b^+_{n'\downarrow} b_{n'\downarrow} = 1,$$

Equation (D.5) becomes, using (C.8)-(C.12),

$$\sum_{\sigma,\sigma'} b^+_{n\sigma'} b_{n'\sigma'} b^+_{n'\sigma} b_{n\sigma} = \frac{1}{2} - 2\vec{S}_n.\vec{S}_{n'} \tag{D.6}$$

The Hamiltonian (D.4) is finally rewritten as

$$\mathcal{H}_2 = -\sum_{n,n'} \frac{|V_{nn'}|^2}{U}\left(\frac{1}{2} - 2\vec{S}_n.\vec{S}_{n'}\right) \tag{D.7}$$

We see here that if $\vec{S}_n$ and $\vec{S}_{n'}$ are parallel, namely $\vec{S}_n.\vec{S}_{n'} = \frac{1}{4}$, the energy given by (D.7) is indeed zero. If $\vec{S}_n$ and $\vec{S}_{n'}$ are antiparallel, this energy is minimum. Therefore, the stable state is an antiferromagnetic state. The Hamiltonian (D.7) is called "kinetic exchange interaction".

In the same manner, we can show that the interaction of the fourth order in V is given by

$$\mathcal{H}_4 \rightarrow -\frac{|V_{n'n}|^4}{U^3}\left(\vec{S}_n.\vec{S}_{n'}\right)^2 \tag{D.8}$$

This interaction is called "biquadratic exchange interaction" which is at the origin of the ferromagnetic state observed in some materials such as MnO.

In the case where there are many electrons per site, we can show that the direct and kinetic exchange interactions are given by (see demonstration below)

$$\mathcal{H}_{ex} = -2 \sum_{n_1, n_2} J_{n_1 n_2}^{eff} \vec{S}_{n_1} . \vec{S}_{n_2} \tag{D.9}$$

where the sum is performed on the different pairs (n_1, n_2), and $J_{n_1 n_2}^{eff}$ is defined by

$$J_{n_1 n_2}^{eff} = \frac{2}{(2S^2)} \sum_{m, m'} \left[-\frac{|V_{n_1 n_2}^{mm'}|^2}{U} + J_{n_1 n_2}^{mm'} \right] \tag{D.10}$$

with

$$V_{n_1 n_2}^{mm'} \equiv \int \varphi_m^*(\vec{r}, n_1) \mathcal{H}_{cryst} \varphi_{m'}(\vec{r}, n_2) d\vec{r} \tag{D.11}$$

$$J_{n_1 n_2}^{mm'} \equiv \int \int \varphi_m^*(\vec{r}_1, n_1) \varphi_{m'}^*(\vec{r}_2, n_2) \frac{e^2}{|\vec{r}_1 - \vec{r}_2|} \varphi_m(\vec{r}_2, n_2) \varphi_{m'}(\vec{r}_1, n_1) d\vec{r}_1 d\vec{r}_2 \tag{D.12}$$

where $S = |\vec{S}|$ is the total spin per site. $J_{n_1 n_2}^{eff}$ is called "effective exchange integral". The sums on m and m' are made on different pairs (m, m') of occupied orbitals in the case of a less-than-half-filled shell. In the case of a more-than-half-filled shell, the sums are made on non occupied orbitals.

Application: Let us consider a simplest case: the one-site Hubbard model defined by the Hamiltonian:

$$\hat{\mathcal{H}} = -t \sum_{i, i', \sigma} b_{i'\sigma}^+ b_{i\sigma} + U \sum_i b_{i\uparrow}^+ b_{i\uparrow} b_{i\downarrow}^+ b_{i\downarrow} - \mu \sum_i \left(b_{i\uparrow}^+ b_{i\uparrow} + b_{i\downarrow}^+ b_{i\downarrow} \right) \tag{D.13}$$

where t is the hopping term between lattice sites i and i', U the on-site Coulomb interaction. One has added here the chemical potential term.

In the case where $t = 0$, there is no hopping. The system is a collection of independent sites. For one site, Hamiltonian (D.13) becomes

$$\hat{\mathcal{H}} = U b_\uparrow^+ b_\uparrow b_\downarrow^+ b_\downarrow - \mu \left(b_\uparrow^+ b_\uparrow + b_\downarrow^+ b_\downarrow \right) \tag{D.14}$$

There are four eigen-states $|0>$, $|\uparrow>$, $|\downarrow>$ and $|\uparrow\downarrow>$ with eigen-energies 0, $-\mu$, $-\mu$ and $U - 2\mu$, respectively. The corresponding grand partition function, the average occupation number and the average energy are

$$\mathcal{Z} = \sum_i \exp(-\beta \mathcal{H}_i) = 1 + 2\exp(\beta\mu) + \exp(2\beta\mu - \beta U)$$

$$<n> = \frac{1}{\beta}\frac{\partial \ln \mathcal{Z}}{\partial \mu} = 2\frac{\exp(\beta\mu) + \exp(2\beta\mu - \beta U)}{1 + 2\exp(\beta\mu) + \exp(2\beta\mu - \beta U)}$$

$$<H> = <E - \mu n> = -\frac{\partial \ln \mathcal{Z}}{\partial \beta}$$

$$= -\frac{2\mu \exp(\beta\mu) + (2\mu - U)\exp(2\beta\mu - \beta U)}{1 + 2\exp(\beta\mu) + \exp(2\beta\mu - \beta U)}$$

$$<E> = <H> + \mu <n> = \frac{U\exp(2\beta\mu - \beta U)}{1 + 2\exp(\beta\mu) + \exp(2\beta\mu - \beta U)}$$

The half-filling case $<n> = 1$ corresponds to $2[\exp(\beta\mu) + \exp(2\beta\mu - \beta U)] = [1 + 2\exp(\beta\mu) + \exp(2\beta\mu - \beta U)]$, namely $\exp(2\beta\mu - \beta U) = 1$ or $\mu = U/2$, independent of T. The curve of $<n>$ versus μ is shown in Fig. D.2 where we see that $<n> = 1$ at $\mu = 2 = U/2$ for all T. We see in this figure that

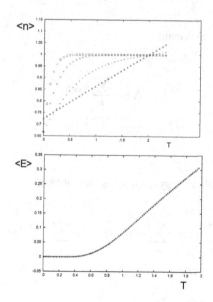

Fig. D.2 Top: Density of spin versus chemical potential μ for $U = 4$ at temperatures $T = 2$ (stars), 0.5 (vertical crosses), 0.2 (oblique crosses) and 0.1 (squares). Bottom: Average energy $<E>$ versus T at half-filling ($\mu = U/2 = 2$).

(i) for $\mu < 2$ (less than half-filling), as T increases the density decreases (ii) for $\mu > 2$, as T increases, the density of spin increases beyond 1.

Remark: (i) The action of U comes into play when $< n >$ crosses the half-filling (> 1): At $T > 0.5$ for example, adding one electron at a lattice site with an electron already there costs an energy U. This is seen in the curve $< E >$ in Fig. D.2 ($< E >$ becomes non zero) (ii) Imagine the situation at $T = 0$. The last filled energy level is the Fermi level: adding an electron to the system is to fill the next level. As long as the system is less than half-filled, this does not cost a U energy because one finds always an empty lattice to put the electron. But when the system reaches the half filling, the added electron comes on a site with another electron. This costs U. The energy discontinuity from 0 to U at half filling is called "Mott gap". This gap is not necessarily the band gap but it plays the same role: a gap is an indication of the insulating phase.

When $U = 0$ we have

$$\hat{\mathcal{H}} = -t \sum_{i,i',\sigma} b^+_{i'\sigma} b_{i\sigma} - \mu \sum_n \left(b^+_{i\uparrow} b_{i\uparrow} + b^+_{i\downarrow} b_{i\downarrow} \right) \tag{D.15}$$

Using the Fourier transforms

$$b_{\ell\sigma} = \frac{1}{N^{1/2}} \sum_k b_{k\sigma} e^{ik\ell} \tag{D.16}$$

$$b^+_{\ell\sigma} = \frac{1}{N^{1/2}} \sum_k b^+_{k\sigma} e^{-ik\ell} \tag{D.17}$$

where N is the number of system sites, we obtain

$$\hat{\mathcal{H}} = -t \sum_{k,\sigma} b^+_{k\sigma} b_{k\sigma} \sum_\rho \exp(ik\rho) - \mu \sum_k \left(b^+_{k\uparrow} b_{k\uparrow} + b^+_{k\downarrow} b_{k\downarrow} \right)$$
$$= \sum_{k,\sigma} [-2t \cos k - \mu] b^+_{k\sigma} b_{k\sigma} \tag{D.18}$$

where $\rho = \ell - \ell' = \pm 1$ $(a = 1)$ is the vector connecting two nearest sites and where we have used the sum rules on k and n.

Using (D.18), the grand partition function is written as

$$
\begin{aligned}
\mathcal{Z} &= \sum_{n_{k1,\sigma}, n_{k2,\sigma}, \ldots} \exp[-\beta \sum_{k,\sigma} (-2t\cos k - \mu)n_{k,\sigma}] \\
&= \sum_{n_{k1,\sigma}} \exp[-\beta(-2t\cos k_1 - \mu)n_{k1,\sigma}] \\
&\quad \times \sum_{n_{k2,\sigma}} \exp[-\beta(-2t\cos k_2 - \mu)n_{k2,\sigma}]\ldots \\
&= (1 + \exp[-\beta(-2t\cos k_1 - \mu)]) \, (1 + \exp[-\beta(-2t\cos k_2 - \mu)]) \times \ldots \\
&= \prod_k (1 + \exp[-\beta(-2t\cos k - \mu)])
\end{aligned}
\tag{D.19}
$$

where $n_{k,\sigma} = b_{k\sigma}^{+} b_{k\sigma}$ and we have used for the n sum (second line) $n_{k_i,\sigma} = 0, 1$.

The average energy and average occupation number are given by

$$
\begin{aligned}
<n> &= \frac{1}{\beta}\frac{\partial \ln \mathcal{Z}}{\partial \mu} \\
&= \sum_k \frac{1}{1 + \exp[\beta(\epsilon_k - \mu)]} \\
<H> &= <E - \mu n> = -\frac{\partial \ln \mathcal{Z}}{\partial \beta} \\
&= \sum_k \frac{\epsilon_k - \mu}{1 + \exp[\beta(\epsilon_k - \mu)]} \\
<E> &= <H> + \mu <n> = \sum_k \frac{\epsilon_k}{1 + \exp[\beta(\epsilon_k - \mu)]} \\
C &= \frac{d<E>}{dT} = -\frac{\partial^2 \ln \mathcal{Z}}{\partial \beta^2}\frac{d\beta}{dT} \\
&= \frac{1}{k_B T^2}\frac{\epsilon_k^2 \exp[\beta(\epsilon_k - \mu)]}{(1 + \exp[\beta(\epsilon_k - \mu)])^2}
\end{aligned}
$$

where $\epsilon_k = -2t\cos k$, C is the specific heat. The second equation is the Fermi-Dirac distribution. The plot of $<E>$ and C is shown in Fig. D.3 with $t = 1$ at half filling ($\mu = U/2 = 0$).

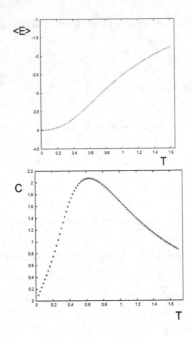

Fig. D.3   Average energy $< E >$ (top) and specific heat C (bottom) versus T for $t = 1$ and half-filling $\mu = 0$ $(U = 0)$.

Appendix E

Kosterlitz-Thouless Phase Transition

We consider the XY spins on a two-dimensional lattice with a ferromagnetic interaction between nearest neighbors. In the ground state, the spin configuration is a perfect ferromagnetic state, namely all spins are parallel. However, this system does not have a normal second-order transition at a finite temperature. There is no long-range ordering when the temperature is not zero, as indicated by the Mermin-Wagner theorem [129] for two-dimensional systems with spin continuous degrees of freedom (see the discussion in the conclusion of chapter 12). Kosterlitz and Thouless [102] have shown that the system has a special phase transition due to the unbinding of vortex-antivortex pairs at a finite temperature below (above) which the correlation function decays as a power law (exponential law) with increasing distance. This transition, called Kosterlitz-Thouless (KT) or Kosterlitz-Thouless-Berezinskii transition, is of infinite order. Let us outline in the following some important points which help us understand the mechanism behind the KT transition.

The Hamiltonian of the system is given by

$$H = -J \sum_{<i,j>} \vec{S}_i \cdot \vec{S}_j = -J \sum_{<i,j>} \cos(\theta_i - \theta_j) \qquad (E.1)$$

where the sum runs over all nearest neighbor spin pairs $\vec{S}_i$ and $\vec{S}_j$ in the lattice, and θ_i denotes the angle of the spin $\vec{S}_i$ with respect to some (arbitrary) polar direction in the two dimensional vector space containing the spins. At low temperatures $(T << J)$, the spins are nearly parallel on neighboring sites, we can expand the cosine to the second order to obtain the lattice gaussian model

$$H = E_0 + \frac{J}{2} \sum_{<i,j>} (\theta_i - \theta_j)^2 \qquad (E.2)$$

where $E_0 = -CJN/2$ (C: coordination number) is the energy per spin of the parallel spin configuration. The continuum limit is written as

$$H = E_0 + \frac{J}{2} \int d\vec{r} \left[\vec{\nabla}\theta(\vec{r}) \right]^2 \tag{E.3}$$

The partition function is given by

$$Z = e^{-\beta E_0} \int D[\theta] \exp\left\{ -\beta\frac{J}{2} \int d\vec{r}[\vec{\nabla}\theta(\vec{r})]^2 \right\} \tag{E.4}$$

where $D[\theta]$ is performed over all configurations of $\theta(\vec{r})$. Going back to the original Hamiltonian (E.1): hierarchically, one has $H = E_0(\text{ferro}) + H(\theta_v)(\text{vortices}) + \text{spin-waves excited on vortices}$. Using these in the exponential argument of Z one sees that the integration can be divided into two parts: a sum over the local minima θ_v of $H[\theta]$ (v: vortex) and a sum on the fluctuations θ_{sw} (sw: spin waves) around the minima, namely $\int D[\theta] = \sum_{\theta_v} \int D[\theta_{sw}]....$ We have

$$Z = e^{-\beta E_0} \sum_{\theta_v} \int D[\theta_{sw}] \exp\{-\beta[H(\theta_v) + \frac{J}{2} \int d\vec{r}_1 \int d\vec{r}_2$$

$$\times \theta_{sw}(\vec{r}_1) \frac{\partial^2 H}{\partial\theta(\vec{r}_1)\partial\theta(\vec{r}_2)} \theta_{sw}(\vec{r}_2)]\} \tag{E.5}$$

Note the absence of the term $\frac{\partial H}{\partial\theta(\vec{r})}$ because it is zero at the minimum. This corresponds to $|\vec{\nabla}\theta(\vec{r})| = 0$ which has two possible solutions

   i) $\theta(\vec{r})$=constant, namely the ferromagnetic ground state,

   ii) solutions of vortices of the director field around vortex centers. There are two cases

- For all closed curves encircling the position r_0 of the center of the vortex, we have

$$\oint \vec{\nabla}\theta(\vec{r}) \cdot d\vec{\ell} = 2\pi n \tag{E.6}$$

- For all paths that do not encircle the vortex position r_0

$$\oint \vec{\nabla}\theta(\vec{r}) \cdot d\vec{\ell} = 0 \tag{E.7}$$

At a given distance from r_0, we have $\theta(\vec{r}) = \theta(r)$, therefore the solution (E.6) yields $|\vec{\nabla}\theta(r)|2\pi r = 2\pi n$, namely $|\vec{\nabla}\theta(r)| = n/r$. Replacing this in

Eq. (E.3), we obtain

$$E - E_0 = \frac{J}{2} \int d\vec{r} \left[\vec{\nabla} \theta(\vec{r}) \right]^2$$

$$= \frac{Jn^2}{2} \int_0^{2\pi} d\theta \int_a^L r dr \frac{1}{r^2}$$

$$= \pi J n^2 \ln \frac{L}{a} \tag{E.8}$$

where a is the lattice constant (distance between nearest neighbors) and L is the system linear size. Note that n is called "charge" of the vortex. When $n > 1$, it is called a multi-charge. The energy of a large system is large even with a single charge. When we encircle a vortex, the integral (E.6) gives 2π, and when we encircle an antivortex, the integral is equal to -2π. If we encircle a pair of vortex-antivortex at a large enough distance, the integral is zero. So, the vortex-antivortex energy is just the energy of the vortex and antivortex at their distance R given by the following expression

$$E_{2\,vort} = 2E_c + E_1 \ln \frac{R}{a} \tag{E.9}$$

where E_c is the energy of the vortex core and E_1 is proportional to J. We emphasize here that $E_{2\,vort}$ does not go to infinity when $L \to \infty$, unlike the energy of a single vortex E given by Eq. (E.8).

Now, we consider the free energy $F = E - TS$ where S is the entropy. We imagine a system of independent vortices excited at T. The entropy is the number of ways to place these vortex centers at centers of the square lattice cells: the number of cell centers is L^2/a^2. Therefore $S = k_B \ln(L^2/a^2)$. We have

$$F = E_0 + (\pi J - 2k_B T) \ln \left(\frac{L}{a} \right) \tag{E.10}$$

We see here that when $L \to \infty$, F tends to ∞ if $\pi J - 2k_B T > 0$, namely $T < \pi J/2k_B$. We conclude that single vortices cannot be excited at low temperatures for large systems. However, pairs of vortices can be excited at low temperatures because their energy, Eq. (E.9), does not diverge. So, the only thing which can happen at temperature $T < \pi J/2k_B$ is the excitation of pairs of bound vortices. Of course, the energy of bound vortices depends on T through $< R >$ which can be calculated using the partition function. For $T > \pi J/2k_B$ pairs of bound vortices are unbound to become independent single vortices which lower the free energy ($F \to -\infty$ for large systems). The unbinding of pairs of vortices occurs at $T = T_{KT} = \pi J/2k_B$ which is called the KT transition temperature.

Let us show the absence of long-range ordering in the present two-dimensional XY-spin system at finite temperature. We calculate $< S_x >$ as follows:

$$< S_x > \; = \; < \cos\theta(\vec{r}) > \; = \; \frac{\int D[\theta]\cos\theta(\vec{r})e^{-\beta H}}{\int D[\theta]e^{-\beta H}} \qquad (E.11)$$

$$\theta(\vec{r}) = \sum_{\vec{k}} \theta(\vec{k})e^{i\vec{k}\cdot\vec{r}}$$

$$|\vec{\nabla}\theta(\vec{r})| = \sum_{\vec{k}} k\theta(\vec{k})e^{i\vec{k}\cdot\vec{r}}$$

$$|\vec{\nabla}\theta(\vec{r})|^2 = \sum_{\vec{k},\vec{k}'} kk'\theta(\vec{k})\theta(\vec{k}')e^{i(\vec{k}+\vec{k}')\cdot\vec{r}}$$

$$\int d\vec{r}|\vec{\nabla}\theta(\vec{r})|^2 = \sum_{\vec{k},\vec{k}'} kk'\theta(\vec{k})\theta(\vec{k}')\int d\vec{r}e^{i(\vec{k}+\vec{k}')\cdot\vec{r}}$$

$$= \sum_{\vec{k},\vec{k}'} kk'\theta(\vec{k})\theta(\vec{k}')\delta_{\vec{k},-\vec{k}'}$$

$$= \sum_{\vec{k}} k^2\theta(\vec{k})\theta(-\vec{k})$$

$$= \int \frac{d\vec{k}}{(2\pi)^d}k^2\theta(\vec{k})\theta(-\vec{k}) \qquad (E.12)$$

Replacing the last equality into the exponential argument of Eq. (E.11) [see the partition function Z given in Eq. (E.4)], we obtain after some calculations

$$< S_x > = \exp\left[-\frac{T}{2Ja^{2-d}}S_d\int_{\pi/L}^{\pi/a} dk k^{d-3}\right] \qquad (E.13)$$

where S_d is a constant. $< S_x >$ depends thus on the integral

$$I(L) = \int_{\pi/L}^{\pi/a} dk k^{d-3} \qquad (E.14)$$

For $d < 2$, $I(L) \propto L^{2-d} \to \infty$ as $L \to \infty$. Thus $< S_x > = 0$: there is no ordering for $d < 2$. For $d > 2$, we have

$$I(L) = \frac{1}{d-2}\left(\frac{\pi}{a}\right)^{d-2}$$

Therefore, $< S_x >$ is not zero at finite T: the ordering exists for this case. The case $d = 2$ is special: for $d = 2$ the integral $I(L)$ is logarithmically divergent $I(L) = \ln(L/a)$ which tends to ∞ for large L. $< S_x >$ thus goes to zero for any non-zero temperature. In spite of the absence of a long-range ordering at finite T, there is a special phase transition, called KT transition, of infinite order as shown above.

We examine now the behavior of the correlation function. We write

$$< \vec{S}(\vec{r}) \cdot \vec{S}(0) > = < \cos[\theta(\vec{r}) - \theta(0)] >$$
$$= \text{Re} < \exp\{i[\theta(\vec{r}) - \theta(0)]\} > \qquad (\text{E.15})$$

We expand in the Fourier modes

$$\theta(\vec{r}) - \theta(0) = \sum_{\vec{k}} \theta(\vec{k})[e^{i\vec{k}\cdot\vec{r}} - 1] = \int \frac{d\vec{k}}{(2\pi)^d} \theta(\vec{k})[e^{i\vec{k}\cdot\vec{r}} - 1] \qquad (\text{E.16})$$

Using Eqs. (E.4), (E.12) and (E.16), we rewrite Eq. (E.15) as

$$G(\vec{r}) = < \exp\{i[\theta(\vec{r}) - \theta(0)]\} >$$
$$= \frac{\int D[\theta] \exp\{i[\theta(\vec{r}) - \theta(0)]\} \exp\{-\beta\frac{J}{2} \int d\vec{r}[\vec{\nabla}\theta(\vec{r})]^2\}}{\int D[\theta] \exp\{-\beta\frac{J}{2} \int d\vec{r}[\vec{\nabla}\theta(\vec{r})]^2\}}$$
$$= \frac{\int D[\theta] \exp\{-\int \frac{d\vec{k}}{(2\pi)^d} \left[\beta\frac{J}{2} k^2\theta(\vec{k})\theta(-\vec{k}) - i\theta(\vec{k})[e^{i\vec{k}\cdot\vec{r}} - 1]\right]\}}{\int D[\theta] \exp\{-\beta\frac{J}{2} \int \frac{d\vec{k}}{(2\pi)^d} k^2\theta(\vec{k})\theta(-\vec{k})\}}$$
$$= \exp[-g(\vec{r})] \qquad (\text{E.17})$$

where

$$g(\vec{r}) = \frac{k_B T}{J} \sum_{\vec{k}} \frac{1 - \cos(\vec{k} \cdot \vec{r})}{k^2} \qquad (\text{E.18})$$

Proofs of Eq. (E.17): Using Eqs. (E.4), (E.12) and (E.16), we rewrite Eq. (E.15) as

$$G(\vec{r}) = < \exp\{i[\theta(\vec{r}) - \theta(0)]\} >$$
$$= \frac{\int D[\theta] \exp\{i[\theta(\vec{r}) - \theta(0)]\} \exp\left\{-\beta\frac{J}{2} \int d\vec{r}[\vec{\nabla}\theta(\vec{r})]^2\right\}}{\int D[\theta] \exp\left\{-\beta\frac{J}{2} \int d\vec{r}[\vec{\nabla}\theta(\vec{r})]^2\right\}}$$
$$= \frac{\int D[\theta] \exp\left\{-\int \frac{d\vec{k}}{(2\pi)^d} \left[\beta\frac{J}{2} k^2\theta(\vec{k})\theta(-\vec{k}) - i\theta(\vec{k})[e^{i\vec{k}\cdot\vec{r}} - 1]\right]\right\}}{\int D[\theta] \exp\left\{-\beta\frac{J}{2} \int \frac{d\vec{k}}{(2\pi)^d} k^2\theta(\vec{k})\theta(-\vec{k})\right\}}$$

To integrate this integral, let us separate the real and imaginary parts as follows: we put $\theta(\vec{k}) = R_{\vec{k}} + iI_{\vec{k}}$. Since $\theta(\vec{r})$ is real, we have $\theta(-\vec{k}) = \theta^*(\vec{k})$

so that $R_{\vec{k}}$ is even and $I_{\vec{k}}$ odd in $\vec{k}$. Then we can write the argument of the exponential of the numerator as

$$\int \frac{d\vec{k}}{(2\pi)^d} \beta \frac{J}{2} k^2 \left[\left(R_{\vec{k}} - i\frac{\cos(\vec{k}\cdot\vec{r}) - 1}{\beta J k^2} \right)^2 + \left(I_{\vec{k}} + i\frac{\sin(\vec{k}\cdot\vec{r})}{\beta J k^2} \right)^2 \right]$$

$$+ \int \frac{d\vec{k}}{(2\pi)^d} \frac{1 - \cos(\vec{k}\cdot\vec{r})}{\beta J k^2} \tag{E.19}$$

where we have used $[\cos(\vec{k}\cdot\vec{r}) - 1]^2 + \sin^2(\vec{k}\cdot\vec{r}) = 2[1 - \cos(\vec{k}\cdot\vec{r})]$. The integrals over $R_{\vec{k}}$ and $I_{\vec{k}}$ are gaussian, they give the same value as the gaussian integrals of the denominator although they do not have the same gaussian centers as the denominator. Hence, these gaussian integrals cancel out. Only the last term in (E.19) remains. We have then

$$G(\vec{r}) = \exp\left[-\frac{k_B T}{J} \int \frac{d\vec{k}}{(2\pi)^d} \frac{1 - \cos(\vec{k}\cdot\vec{r})}{k^2} \right] = \exp[-g(\vec{r})]$$

This is Eq. (E.17).

In $d = 2$, we have, in changing the sum into an integral,

$$g(\vec{r}) = \frac{k_B T}{J} \frac{1}{(2\pi)^2} \left[\int_0^{\pi/r} 2\pi k dk \frac{1 - \cos(\vec{k}\cdot\vec{r})}{k^2} + \int_{\pi/r}^{\pi/a} 2\pi k dk \frac{1 - \cos(\vec{k}\cdot\vec{r})}{k^2} \right]$$

$$= \frac{k_B T}{J} \frac{1}{2\pi} \left[\int_0^{\pi/r} 2\pi k dk \frac{1 - 1}{k^2} + \int_{\pi/r}^{\pi/a} 2\pi k dk \frac{1}{k^2} \right]$$

$$= \frac{k_B T}{J} \frac{1}{2\pi} \int_{\pi/r}^{\pi/a} dk \frac{1}{k}$$

$$= \frac{k_B T}{2\pi J} \ln(r/a) \tag{E.20}$$

where we have replaced the cosine term by 1 in the region where $0 < k < \pi/r$ (very small k indeed) canceling thus the first integral, and for $kr > \pi$ we considered that the cosine term oscillates so rapidly that it makes a zero contribution to the second integral. Equation (E.17) then gives

$$G(\vec{r}) \propto r^{-\eta} \quad \text{where } \eta = \frac{k_B T}{2\pi J} \tag{E.21}$$

The correlation function obeys thus a power law with the exponent η depending on T. Note that in the case of superfluid, J is replaced by the reduced superfluid density $\bar{\rho}_s = (\hbar/m)^2 \rho_s$. At the KT transition

$T = T_{KT} = \pi J/2k_B$, we have $\eta = 1/4$. In the ordinary second-order transition, the power law happens only at the "critical point", but in the two-dimensional XY case here, the power law is found for the whole temperature range from T_{KT} down to zero. This is the reason why we call the line below the KT transition a "critical line". Above T_{KT}, the correlation function obeys an exponential law. It is noted that in the calculation of the correlation, we did not take into account the vortices. This is because, as shown above, independent vortices cannot occur at $T < T_{KT}$, otherwise the free energy will go to infinity for large systems. There exist of course pairs of vortex-antivortex at short distances below T_{KT} as seen above but they do not affect long-distance correlation. Note that due to the absence of vortices in Eq. (E.21), this result cannot be used at temperature above T_{KT}. The renormalization group treats in a much more elegant manner the whole temperature range, but this advanced calculation is so lengthy to present in this book.

Appendix F

Low- and High-Temperature Expansions of the Ising Model

The low- and high-temperature expansions are useful not only for studying physical properties of a spin system in these temperature regions, but also for introducing a new concept called duality which allows to map a system of weak coupling into a system of strong coupling, as seen in the problem below.

F.1 The case of the square lattice

We consider the Ising model on a square lattice with the Hamiltonian

$$\mathcal{H} = -J \sum_{<i,j>} \sigma_i\, \sigma_j \tag{F.1}$$

where the sum is performed over nearest neighbors and $\sigma_{i(j)} = \pm 1$.

- The partition function:

$$
\begin{aligned}
Z &= \sum_{\sigma_1 = \pm 1, \ldots} \exp(K \sum_{<i,j>} \sigma_i\, \sigma_j) \\
&= \sum_{\sigma_1 = \pm 1, \ldots} \prod_{<ij>} \exp(K \sigma_i\, \sigma_j)
\end{aligned} \tag{F.2}
$$

where $K = J/(k_B T)$. In the ground state (GS), we have $E_0 = -2JN = -N_b J$ and $Z = 2 \exp(N_b K)$ where the factor 2 comes from the GS degeneracy (reversing all spins), $N_b = 2N$ is the total number of links. Let us construct the dual lattice by drawing the links (broken lines) perpendicular to the real links (solid lines) as shown in Fig. F.1. The case of the square lattice is special: its dual lattice formed by the broken lines is also a square lattice. This is not the case in general: for example the dual lattice of the triangular lattice is the honeycomb lattice.

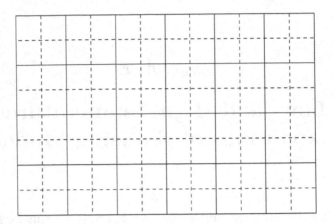

Fig. F.1   Real square lattice (solid lines) and its dual lattice (broken lines).

- Low-temperature expansion:
  - one reversed spin: there are 4 broken links around it, degeneracy=N (the number of choices of a spin among N spins), the energy is increased from -4J to +4J so that $E = E_0 + 8J$ (see Fig. F.2),
  - two reversed neighbors: 6 broken links, degeneracy=N_b (the number of choices of a link among N_b links), the energy is increased from -6J to +6J: $E = E_0 + 12J$,
  - three reversed neighboring spins: 8 broken links for both configurations (trimer with two links on a line, or trimer with two perpendicular links), degeneracy=$6N$ (there are N choices of the central site and 6 choices to form a trimer with the central site: one straight trimer along x axis and the second straight trimer along y axis, and four perpendicular trimers), $E = E_0 + 16J$.

Calculation of Z with the first excited states of energies $E_0 + 8J$, $E_0 + 12J$, $E_0 + 16J$: to do that, we have to find all excited states for each level. The first two levels concern one and two reversed neighboring spins as found above. However, the level $E_0 + 16J$ corresponds not only to the excited trimer shown above but also to two other cases: a cluster of four sites forming a square, two disconnected reversed spins. Both cases have 8 broken links. The first case has a degeneracy of N (the number of choices of the first site of the square), the second case has a degeneracy of $N(N - 5)$ (the number of choices of the first reversed spin is N, the number of choices of the second disconnected reversed spin is $N - 5$ where 5 is the number of spins concerned by the

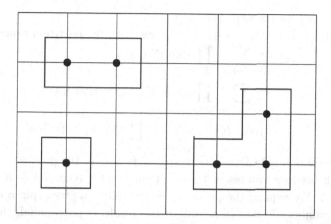

Fig. F.2　Graphs (heavy lines) crossing broken links for one, two three reversed spin clusters. The reversed spins are shown by black circles, other spins are not shown.

first reversed spins which are to be avoided). We finally have

$$Z = 2e^{2NK}[1 + Ne^{-8K} + 2Ne^{-12K}$$
$$+[6N + N + N(N-5)]e^{-16K} + ...] \qquad (F.3)$$

- Connected paths:
  We draw a path P which encircles each cluster of reversed spins: this path crosses the broken links around the cluster. For such a closed path, we can verify that it crosses an even number of broken links as follows. Imagine a rectangular path, the number of broken links is even. Now, including an additional site anywhere around the path will add two additional broken links. Excluding a site inside the path will reduce the number of broken links by 2. Such a construction shows clearly that any closed path crosses an even number of broken links. Let $\ell(P)$ be the number of broken links crossed by the path. Since the variation of energy when breaking a link is $\Delta E = J - (-J) = 2J$. A path crossing $\ell(P)$ broken links corresponds to $\Delta E = +2\ell(P) J$, so that the system energy is $E(P) = E_0 + \Delta E = E_0 + 2\ell(P) J$. The partition functions is thus

$$Z = \sum_P e^{-E(P)/k_B T} = 2e^{N_b K} \sum_P e^{-2K\ell(P)} \qquad (F.4)$$

- High-temperature expansion:
  Using Eq. (17.203) or Eq. (13.91) we write the partition function as

$$Z = \sum_{\sigma_1 = \pm 1, \dots} \prod_{<ij>} \exp(K\sigma_i \, \sigma_j)$$

$$= \sum_{\sigma_1 = \pm 1, \dots} \prod_{<ij>} (\cosh K + \sigma_i \, \sigma_j \sinh K)$$

$$= (\cosh K)^{N_b} \sum_{\sigma_1 = \pm 1, \dots} \prod_{<ij>} (1 + \sigma_i \, \sigma_j \tanh K) \qquad (F.5)$$

Since there are N_b links, there are N_b factors in the product. We draw a link between two nearest sites in each factor (this link is in the real space). We expand the product, we see that if a given spin appears an odd number of times in a term of the resulting polynomial, then when summing on its values, this term gives two opposite values yielding a zero contribution. Each nonzero term contains an even number of times of each spin, so that the spin factor in front of $\tanh K$ is equal to 1. The power of $\tanh K$ for this term is nothing but the number of links of spins in front of it. Consider for example the term $x = \sigma_1 \sigma_2 \, \sigma_2 \sigma_3 \, \sigma_3 \sigma_{10} \, \sigma_{10} \sigma_9 \, \sigma_9 \sigma_8 \, \sigma_8 \sigma_1 \tanh^6 K$ shown in Fig. F.3. This term has 6 links each of which connects two spins. Each spin appears twice because there are two links emanating from it. Hence, $x = \tanh^6 K$. The corresponding graph is shown by the lower left graph in Fig. F.3. Therefore, we can replace the polynomial by the sum

$$Z = 2^N (\cosh K)^{N_b} \sum_P (\tanh K)^{\ell(P)} \qquad (F.6)$$

where 2^N comes from the sum $\sum_{\sigma_1 = \pm 1, \dots}$ and $\ell(P)$ is the number of links in the closed graph P.

- Duality:
  The partition functions Z in (F.4) and (F.6) have the same structure: since the prefactors are non singular, the summations over the closed paths determine the singularity of Z. Note that $\ell(P)$ in (F.4) corresponds to the path drawn in the dual lattice (see Fig. F.2). While, $\ell(P)$ in (F.6) corresponds to the path drawn in the real lattice (see Fig. F.3). Nevertheless, these two kinds of path have the same structure with an even number of links in each path. Therefore, we can connect the two Z by fixing

$$e^{-2K^*} = \tanh K \qquad (F.7)$$

$$\text{Hence,} \quad K^* = -\frac{1}{2} \ln \tanh K \qquad (F.8)$$

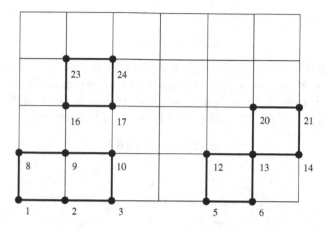

Fig. F.3 Graphs linking nearest sites: the lower left graph represents the term $\sigma_1\sigma_2\ \sigma_2\sigma_3\ \sigma_3\sigma_{10}\ \sigma_{10}\sigma_9\ \sigma_9\sigma_8\ \sigma_8\sigma_1\ \tanh^6 K = \tanh^6 K$, the right one represents the term $\sigma_5\sigma_6\ \sigma_6\sigma_{13}\ \sigma_{13}\sigma_{14}\ \sigma_{14}\sigma_{21}\ \sigma_{21}\sigma_{20}\ \sigma_{20}\sigma_{13}\ \sigma_{13}\sigma_{12}\ \sigma_{12}\sigma_5\ \tanh^8 K = \tanh^8 K$. The upper left graph represents the term $\sigma_{16}\sigma_{17}\ \sigma_{17}\sigma_{24}\ \sigma_{24}\sigma_{23}\ \sigma_{23}\sigma_{16}\ \tanh^4 K = \tanh^4 K$. The spins crossed by the graphs are shown by black circles, other spins are not shown.

where K^* corresponds to the low-T phase and K to the high-T phase. The above relation (F.8) is called the "duality" condition which connects the low- and the high-T phases. We deduce from (F.6) and (F.4) the following relation between the high-temperature $Z(K)$ and the low-temperature $Z(K^*)$:

$$Z(K) = 2^N (\cosh K)^{N_b} \sum_P (\tanh K)^{\ell(P)}$$

$$Z(K) = 2^N (\cosh K)^{N_b} e^{-N_b K^*} Z(K^*) \qquad (F.9)$$

where $\sum_P (\tanh K)^{\ell(P)}$ has been replaced by $\sum_P e^{-2K^*\ell(P)}$ and then by $Z(K^*)/(2e^{N_b K^*}) = Z(K^*)/(e^{N_b K^*})$ from (F.4) (the factor 2 in the denominator is neglected because N is large).

- Critical temperature:

  The critical temperature of the Ising model on the square lattice is obtained by using the duality. We have

$$\sinh 2K = 2 \sinh K \cosh K = 2 \tanh K \cosh^2 K = \frac{2 \tanh K}{1 - \tanh^2 K}$$

$$= \frac{2e^{-2K^*}}{1 - e^{-4K^*}} = \frac{2}{e^{2K^*} - e^{-2K^*}}$$

$$= \frac{1}{\sinh 2K^*} \qquad (F.10)$$

from which

$$\sinh 2K \, \sinh 2K^* = 1 \qquad (F.11)$$

This relation is symmetric with respect to K and K^*: when K^* increases, K decreases, and vice-versa. If the system undergoes a single phase transition, it should undergo at the same point of K and K^*, namely $K_c^* = K_c$. To satisfy (F.11), we should have

$$\sinh 2K_c = \sinh 2K_c^* = 1 \qquad (F.12)$$

Thus,

$$\frac{e^{2K_c} - e^{-2K_c}}{2} = 1$$

$$e^{2K_c} - 2e^{-2K_c} - 2 = 0$$

$$e^{4K_c} - 1 - 2e^{2K_c} = 0$$

$$X^2 - 2X - 1 = 0 \quad (X = e^{2K_c})$$

$$X = 1 + \sqrt{2} \quad \text{(positive solution)}$$

$$K_c = \frac{\ln(1 + \sqrt{2})}{2} \qquad (F.13)$$

Therefore, $k_B T_c / J = 1/K_c \simeq 2.27$ which is the exact Onsager's solution.

F.2   The case of the triangular and honeycomb lattices

We consider

$$\mathcal{H} = -J \sum_{<i,j>} \sigma_i \, \sigma_j \qquad (F.14)$$

where the sum is performed over nearest neighbors on the triangular lattice and $\sigma_{i(j)} = \pm 1$.

The dual lattice by construction is a honeycomb lattice shown by the broken lines in Fig. F.4.

First, we follow the same method as for the square lattice described above: writing the partition functions of the triangular lattice using low-temperature expansion with graphs on links of the dual lattice and high-temperature expansion with graphs on links of the real lattice, we obtain a relation similar to Eq. (F.9)

$$Z_t(K) = 2^N (\cosh K)^{N_b} e^{-N_b K} Z_h(K^*) \qquad (F.15)$$

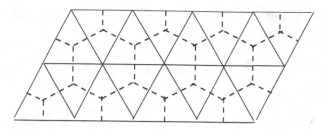

Fig. F.4 Triangular lattice (solid lines) and its dual honeycomb lattice (broken lines).

where $Z_t(K)$ denotes the low-temperature partition function of the triangular lattice, $Z_h(K^*)$ the high-temperature dual (honeycomb) lattice and $N_b = 3N$ is the total number of links.

Now, we can relate the two partition functions in another relation as follows. For convenience, the dual lattice is shifted as shown in Fig. F.5. We consider the plaquette defined by three sites 1,2 and 3 with a site at the center. We calculate the following quantity

$$
\begin{aligned}
A_p(K^*) &= \sum_{\sigma_0 = \pm 1} \exp[K^* \sigma_0(\sigma_1 + \sigma_2 + \sigma_3)] \\
&= \sum_{\sigma_0 = \pm 1} \cosh^3 K^* (1 + \sigma_0\, \sigma_1 \tanh K^*)(1 + \sigma_0\, \sigma_2 \tanh K^*) \\
&\quad \times (1 + \sigma_0\, \sigma_3 \tanh K^*) \\
&= 2 \cosh^3 K^* [1 + \tanh^2 K^* (\sigma_1\, \sigma_2 + \sigma_2\, \sigma_3 + \sigma_3\, \sigma_1)]
\end{aligned}
\tag{F.16}
$$

where we used the remark before Eq. (F.6) to expand the product of the first equality.

Now, we consider the plaquette defined by three sites 1, 2 and 3 of the triangular lattice. We calculate the following quantity

$$
\begin{aligned}
B_p(K^+) &= \exp[K^+(\sigma_1\, \sigma_2 + \sigma_2\, \sigma_3 + \sigma_3\, \sigma_1)] \\
&= \cosh^3 K^+ (1 + \sigma_1\, \sigma_2 \tanh K^+)(1 + \sigma_2\, \sigma_3 \tanh K^+) \\
&\quad (1 + \sigma_3\, \sigma_1 \tanh K^+) \\
&= \cosh^3 K^+ [1 + \tanh^3 K^+ + (\tanh K^+ + \tanh^2 K^+) \\
&\quad \times (\sigma_1\, \sigma_2 + \sigma_2\, \sigma_3 + \sigma_3\, \sigma_1)] \\
&= (\cosh^3 K^+ + \sinh^3 K^+)[1 + \frac{\sinh 2K^+}{2\cosh 2K^+ - \sinh 2K^+} \\
&\quad \times (\sigma_1\, \sigma_2 + \sigma_2\, \sigma_3 + \sigma_3\, \sigma_1)]
\end{aligned}
\tag{F.17}
$$

If we set the factors of $(\sigma_1 \, \sigma_2 + \sigma_2 \, \sigma_3 + \sigma_3 \, \sigma_1)$ in (F.16) and (F.17) to be equal, then we have

$$\tanh^2 K^* = \frac{\sinh 2K^+}{2 \cosh 2K^+ - \sinh 2K^+} \tag{F.18}$$

so that

$$\sum_{\sigma_0 = \pm 1} \exp[K^* \sigma_0 (\sigma_1 + \sigma_2 + \sigma_3)] = \frac{2 \cosh^3 K^*}{\cosh^3 K^+ + \sinh^3 K^+}$$
$$\times \exp[K^+ (\sigma_1 \, \sigma_2 + \sigma_2 \, \sigma_3 + \sigma_3 \, \sigma_1)] \tag{F.19}$$

This relation connects the two dual lattices. To find the full partition functions, it suffices to sum over all spins of the plaquette and to take the product over all plaquettes on each side of the above equation, we then have

$$Z_h(K^*) = \left(\frac{2 \cosh^3 K^*}{\cosh^3 K^+ + \sinh^3 K^+} \right)^N Z_t(K^+) \tag{F.20}$$

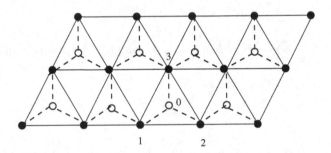

Fig. F.5　Triangular lattice (solid lines) and its dual honeycomb lattice (broken lines) shifted for convenience: the four sites are numbered from 0 to 3 for calculating the partition function.

Replacing $Z_h(K^*)$ given by (F.20) in (F.15), we have

$$Z_t(K) = 2^N (\cosh K)^{3N} e^{-3NK^*} \left(\frac{2 \cosh^3 K^*}{\cosh^3 K^+ + \sinh^3 K^+} \right)^N Z_t(K^+) \tag{F.21}$$

Using (F.18) to eliminate K^* in the above equation, and after some algebra, we get

$$\frac{Z_t(K)}{(\sinh 2K)^{N/2}} = \frac{Z_t(K^+)}{(\sinh 2K^+)^{N/2}} \tag{F.22}$$

Let us calculate the critical temperature of the triangular lattice. We note that by graph constructions for low- and high-temperatures, we obtained Eq. (F.7) which is very general, independent of the lattice structure: it was established using the square lattice, but all arguments leading to it are also valid for the triangular lattice. Using Eq. (F.7) to eliminate K^* in Eq. (F.18), and after some algebra, we obtain

$$(e^{4K} - 1)(e^{4K^+} - 1) = 4 \tag{F.23}$$

As before in the case of the square lattice, this relation shows that if K increases, K^+ decreases, and vice-versa. The transition should occur at the same critical temperature $K_c = K_c^+$. The solution of Eq. (F.23) at K_c is thus

$$e^{4K_c} - 1 = e^{4K_c^+} - 1 = 2, \quad \text{hence} \quad k_B T_c/J = 3.640$$

To find the critical temperature of the honeycomb lattice, we follow the same method as above: we obtain

$$\cosh 2K_c = 2, \quad \text{hence} \quad k_B T_c/J = 1.518.$$

Bibliography

[1] A. A. Abrikosov, Zh. Eksp. Teor. Fiz. **32**, 1442 (1957) (English translation: Sov. Phys. JETP **5**, 1174 (1957)].) In Abrikosov's original paper, the vortex structure of Type-II superconductors was derived as a solution of Ginzburg-Landau equations for $\kappa > 1/\sqrt{2}$.

[2] K. Akabli, H. T. Diep and S. Reynal, J. Phys.: Condens. Matter **19**, 356204 (2007).

[3] K. Akabli and H. T. Diep, J. Appl. Phys. **103**, 07F307 (2008).

[4] K. Akabli and H. T. Diep, Phys. Rev. B **77**, 165433 (2008).

[5] K. Akabli, Y. Magnin, M. Oko, I. Harada and H. T. Diep, Phys. Rev. B **84**, 024428 (2011).

[6] J. W. Allen, G. Locovsky, and J. C. Mikkelsen Jr., Solid State Commun. **24**, 367 (1977).

[7] S. Alexander, J. S. Helman, and I. Balberg, Phys. Rev. B **13**, 304 (1975).

[8] D. J. Amit, *Field Theory, Renormalization Group and Critical Phenomena*, World Scientific, Singapore (1984).

[9] P. W. Anderson, Science **235** (4793), 11961198 (1987).

[10] S. A. Antonenko and A.I. Sokolov, Phys. Rev. B **49**, 15901 (1994).

[11] S. S. Aplesnin, L. I. Ryabinkina, O. B. Romanova, D. A. Balaev, O. F. Demidenko, K. I. Yanushkevich and N. S. Miroshnichenko, Phys. Solid State **49**, Number 11, 2080-2085 (2007).

[12] N. W. Ashcroft and N. D. Mermin, *Solid State Physics*, Saunders College, Philadelphia (1976).

[13] P. Azaria, H. T. Diep and H. Giacomini, Phys. Rev. Lett. **59**, 1629 (1987).

[14] M. N. Baibich, J. M. Broto, A. Fert, F. Nguyen Van Dau, F. Petroff, P. Etienne, G. Creuzet, A. Friederich, and J. Chazelas, Phys. Rev. Lett. **61**, 2472 (1988).

[15] M. N. Barber, *Finite-size scaling*, in *Phase Transitions and Critical Phenomena*, vol. 8, Ed. C. Domb and J. L. Lebowitz, Academic Press, London (1983).

[16] J. Bardeen, L. N. Cooper and J. R. Schrieffer, Phys. Rev. **106**, 162164 (1957).

[17] J. Bardeen, L. N. Cooper and J. R. Schrieffer, Phys. Rev. **108**, 11751205

(1957).

[18] A. Barthélémy *et al.*, J. Magn. Magn. Mater. **242-245**, 68 (2002).

[19] See review on Oxide Spintronics by Manuel Bibes and Agnès Barthélémy, in a Special Issue of IEEE Transactions on Electron Devices on Spintronics, IEEE Trans. Electron. Devices **54**, 1003 (2007).

[20] R. J. Baxter, *Exactly Solved Models in Statistical Mechanics*, Academic, New York (1982).

[21] J. G. Bednorz and K. A. Müller, Z. Physik, B **64**, 189193 (1986).

[22] N. E. Bickers, D. J. Scalapino, R. T. Scalettar, Int. J. Mod. Phys. B **1**, 687 (1987).

[23] K. Binder, *Critical Hehaviour at Surfaces*, in *Phase Transitions and Critical Phenomena*, Vol. 8, Ed. C. Domb and J. L. Lebowitz, Academic Press, London (1983).

[24] K. Binder and D. W. Heermann, *Monte Carlo Simulation in Statistical Physics*, Springer-Verlag, Berlin (1992).

[25] J.A.C. Bland and B. Heinrich (Editors), *Ultrathin Magnetic Structures*, vol. I and II, Springer-Verlag (1994).

[26] N. N. Bogoliubov, J. Phys. (USSR) **11**, 23 (1947); Izv. Akad. Nauk USSR **11**, 77 (1947); Bull. Moscow State Univ. **7**, 43 (1947).

[27] N. N. Bogoliubov, Zhurnal Eksperimental'noi i Teoreticheskoi Fiziki **34**, 58 (1958).

[28] N. N. Bogolyubov and S. V. Tyablikov, Doklady Akad. Nauk S.S.S.R. **126**, 53 (1959) [translation: Soviet Phys.-Doklady **4**, 604 (1959)].

[29] M. Born and K. Huang, *Dynamical Theory of Crystal Lattices*, Clarendon Press, Oxford (1954).

[30] G. Brown and T.C. Schulhess, J. Appl. Phys. **97**, 10E303 (2005).

[31] A. Bunker, B. D. Gaulin, and C. Kallin, Phys. Rev. **B 48**, 15861 (1993).

[32] J. Cardy, *Scaling and Renormalization in Statistical Physics*, Cambridge Lecture Notes in Physics, Cambridge University Press (2000).

[33] C. Cercignani, *The Boltzmann Equation and Its Applications*, Springer-Verlag, New York (1988).

[34] P. M. Chaikin and T. C. Lubensky, *Principles of Condensed Matter Physics*, Cambridge University Press (1995).

[35] S. Chandra, L. K. Malhotra, S. Dhara and A. C. Rastogi, Phys. Rev. B **54**, 13694 (1996).

[36] L. N. Cooper, Phys. Rev. **104**, 1189 (1956).

[37] P. P. Craig *et al.*, Phys. Rev. Lett. 19, 1334 (1967).

[38] M. Debauche, H. T. Diep, H. Giacomini and P. Azaria, Phys. Rev. B **44**, 2369 (1991).

[39] M. Debauche and H. T. Diep, Phys. Rev. B **46**, 8214 (1992).

[40] P.-G. de Gennes and J. Friedel, J. Phys. Chem. Solids **4**, 71 (1958).

[41] H. W. Diehl, *Field-theoretic Approach to Critical Behaviour at Surfaces*, in *Phase Transitions and Critical Phenomena*, Vol. 10, Ed. C. Domb and J. L. Lebowitz, Academic Press, London (1986).

[42] H. W. Diehl, Int. J. Mod. Phys. B **11**, 3503 (1997).

[43] See *Review on Semiconductor Spintronics* by T. Dietl, in Lectures Notes,

vol. 712, Springer-Verlag, Berlin, pp. 1-46 (2007).

[44] Diep-The-Hung, J.C. S. Levy and O. Nagai, Phys. Stat. Solidi (b) **93**, 351 (1979).

[45] Diep-The-Hung, Phys. Status Solidi (b) **103**, 809 (1981).

[46] H. T. Diep, Phys. Rev. B **43**, 8509 (1991).

[47] H. T. Diep, M. Debauche and H. Giacomini, Phys. Rev. B (rapid communication) **43**, 8759 (1991).

[48] H. T. Diep (Editor), *Magnetic Systems with Competing Interactions* , World Scientific, Singapore (1994).

[49] *Frustrated Spin Systems*, 2nd Edition, Ed. H. T. Diep, World Scientific, Singapore (2013).

[50] Hung T. Diep, *Theory of Magnetism - Application to Surface Physics*, World Scientific, Singapore (2014).

[51] H. T. Diep and H. Giacomini, *Frustration - Exactly Solved Models*, p. 1-58, in *Frustrated Spin Systems*, Ed. H. T. Diep, World Scientific, Singapore (2013).

[52] B. Diu, C. Guttmann, D. Lederer, B. Roulet, *Physique Statistique*, Hermann, Paris (1989).

[53] See the series *Phase Transitions and Critical Phenomena*, Ed. C. Domb and J. L. Lebowitz, Academic Press, London.

[54] J. Du, D. Li, Y. B. Li, N. K. Sun, J. Li and Z. D. Zhang, Phys. Rev. B **76**, 094401 (2007).

[55] J. B. C. Efrem D'Sa, P. A. Bhobe, K. R. Priolkar, A. Das, S. K. Paranjpe, R. B. Prabhu, P. R. Sarode, J. Mag. Mag. Mater. **285** , 267 (2005).

[56] Fabian H. L. Essler, Holger Frahm, Frank Göhmann, Andreas Klümper and Vladimir E. Korepin, *The One-Dimensional Hubbard Model*, Cambridge University Press, London (2005).

[57] A. M. Ferrenberg and R. H. Swendsen, Phys. Rev. Let. **61**, 2635 (1988).

[58] A. M. Ferrenberg and R. H. Swendsen, Phys. Rev. Let. **63**, 1195 (1989).

[59] A. M. Ferrenberg and D. P. Landau, Phys. Rev. B **44**, 5081 (1991).

[60] A. Fert and I. A. Campbell, Phys. Rev. Lett. **21**, 1190 (1968); I. A. Campbell, Phys. Rev. Lett. **24**, 269 (1970).

[61] A. L. Fetter and J. D. Walecka, *Quantum Theory of Many-Particle Systems*, McGraw-Hill, New York (1971).

[62] M. E. Fisher and J.S. Langer, Phys. Rev. Lett. **20** , 665 (1968).

[63] M. E. Fisher and R.J. Burford, Phys. Rev. **156**, 583 (1967).

[64] P. J. Ford, *The Rise of the Superconductors*, CRC Press (2005).

[65] A. Gaff and J. Hijmann, Physica A**80**, 149 (1975).

[66] Jian-Feng Ge *et al.*, arXiv:1406.3435 (2014).

[67] A. Georges, G. Kotliar, W. Krauth and M. J. Rozenberg, Rev. Mod. Phys, **68**, 13 (1996).

[68] V. L. Ginzburg and L.D. Landau, Zhurnal Eksperimental'noi i Teoreticheskoi Fiziki **20**, 1064 (1950).

[69] L.P. Gor'kov, Sov. Phys. JETP **36**, 1364 (1959).

[70] C. J. Gorter and H. G. B. Casimir, Phys. Z. **35**, 963 (1934); C. J. Gorter and H. G. B. Casimir, Z. Tech. Phys. **15**, 539 (1934).

[71] W. Greiner, *Quantum Mechanics, an Introduction*, 3e Edition, Springer-Verlag (1994).

[72] R. B. Griffiths and M. Kaufman, Phys. Rev. B **26**, 5022 (1982).

[73] C. Gros, D. Poilblanc, T. M. Rice, F. C. Zhang, Physica C. 153155: 543548 (1988).

[74] P. Grunberg, R. Schreiber, Y. Pang, M. B. Brodsky, and H. Sowers, Phys. Rev. Lett. **57**, 2442 (1986); G. Binasch, P. Grunberg, F. Saurenbach, and W. Zinn, Phys. Rev. B **39**, 4828 (1989).

[75] M. C. Gutzwiller, Phys. Rev. Lett. **10**, 159 (1963); M. C. Gutzwiller, Phys. Rev. **137**, A1726 (1965).

[76] C. Haas, Phys. Rev. **168**, 531 (1968).

[77] A. Haug, *Theoretical Solid State Physics*, vol. II, p. 258, Pergamon Press (1972).

[78] X. He, Y.Q. Zhang and Z.D. Zhang, J. Mater. Sci. Technol. **27**, 64 (2011).

[79] Shaolong He *et al.*, Nat. Mat. **12**, 605 (2013); arXiv:1207.6823.

[80] C. Henley, Phys. Rev. Lett. **62**, 2056 (1989).

[81] B. Hennion, W. Szuszkiewicz, E. Dynowska, E. Janik, T. Wojtowicz, Phys. Rev. B **66**, 224426 (2002).

[82] A. Herpin, *Théorie du Magnétisme* (in French), Presse Universitaire de France (1968).

[83] Danh-Tai Hoang, Y. Magnin and H. T. Diep, Mod. Phys. Lett. B **25**, 937 (2011).

[84] Danh-Tai Hoang and H. T. Diep, Phys. Rev. E **85**, 041107 (2012).

[85] P. C. Hohenberg and B. I. Halperin, Rev. Mod. Phys. **49**, 435 (1977).

[86] J. Hoshen and R. Kopelman, Phys. Rev. B **14**, 3438 (1974).

[87] J. Hubbard, Proc. Roy. Soc. London A **276**, 238 (1963); J. Hubbard, Proc. Roy. Soc. London A **277**, 237 (1964).

[88] M. Inui, S. Doniach, P. J. Hirschfeld, A. E. Ruckenstein, Z. Zhao, Q. Yang, Y. Ni, G. Liu, Phys. Rev. B **37**, 5182 (1988).

[89] M. Itakura, J. Phys. Soc. Jpn **72**, 74 (2003).

[90] L. P. Kadanoff *et al.*, Reviews of Modern Physics **39**, 395 (1967).

[91] H. Kadowaki, K. Ubukoshi, K. Hirakawa, J. L. Martinez and G. Shirane, J. Phys. Soc. Jpn. **56**, 4027 (1987).

[92] J. Kanamori, Prog. of Theor. Phys. (Kyoto) **30**, 275 (1963).

[93] Kazuki Kanki, Damien Loison and Klaus-Dieter Schotte, J. Phys. Soc. Jpn. **75**, 015001 (2006).

[94] K. Kano and S. Naya, Prog. Theor. Phys. **10**, 158 (1953).

[95] T. Kasuya, Prog. Theor. Phys. **16**, 58 (1956).

[96] M. Kataoka, Phys. Rev. B **63**, 134435 (2001).

[97] M. Kaufman, Phys. Rev. **36**, 3697 (1987).

[98] M. P. Kawatra, J. A. Mydosh, and J. I. Budnick, Phys. Rev. B **2**, 665 (1970).

[99] C. Kittel, *Introduction to Solid State Physics*, 8-th edition, John Wiley & Sons, New York (2008).

[100] C. Kittel, *Quantum Theory of Solids*, John Wiley & Sons, New York (1987).

[101] J. Kondo, Prog. Theor. Phys. **32**, 37 (1964).

[102] J. M. Kosterlitz and D. J. Thouless, J. Phys. C **6**, 1181 (1973); J. M. Kosterlitz, J. Phys. C **7**, 1046 (1974).

[103] L. Kramer, Phys. Rev. B **3**, 3822, (1971).

[104] G. Kotliar, Jialin Liu, Phys. Review B **38**, 5182 (1988).

[105] D. P. Landau and K. Binder, *A Guide to Monte Carlo Simulations in Statistical Physics*, Cambridge University Press, Cambridge (2000).

[106] L. D. Landau, Sov. Phys. JETP **3**, 920 (1956).

[107] L. D. Landau, Sov. Phys. JETP **5**, 101 (1957).

[108] L. D. Landau, Sov. Phys. JETP **8**, 70 (1958).

[109] L.D. Landau, J. Phys. USSR, **5**, 71 (1941).

[110] L. Landau and E. Lifschitz, *Physique Statistique*, Mir, Moscow 1967.

[111] Review by Ph. Lecheminant, in *Frustrated Spin Systems*, Ed. H. T. Diep, 2nd edition, World Scientific (2013).

[112] S. Legvold, F. H. Spedding, F. Barson and J. F. Elliott, Rev. Mod. Phys. **25**, 129 (1953).

[113] Y. B. Li, Y. Q. Zhang, N. K. Sun, Q. Zhang, D. Li, J. Li and Z. D. Zhang, Phys. Rev. B **72**, 193308 (2005).

[114] E. H. Lieb and F. Y. Wu, Phys. Rev. Lett. **20**, 1445 (1967).

[115] R. Liebmann, *Statistical Mechanics of Periodic Frustrated Ising Systems*, Lecture Notes in Physics, vol. 251 Springer-Verlag, Berlin (1986).

[116] Defa Liu *et al.*, Nature Communications **3**, 931 (2012); arXiv:1202.5849.

[117] F. London and H. London, Proc. Roy Soc. London, Ser. A **149**, 71 (1935).

[118] F. London, Proc. Roy. Soc. **152**, 24-34, (1935).

[119] C. L. Lu, X. Chen, S. Dong, K. F. Wang, H. L. Cai, J.-M. Liu, D. Li and Z. D. Zhang, Phys. Rev. B **79**, 245105 (2009).

[120] Shang-Keng Ma *Statistical Mechanics*, World Scientific, Singapore (1985).

[121] Y. Magnin, K. Akabli, H. T. Diep and I. Harada, Computational Materials Science **49**, S204-S209 (2010).

[122] Y. Magnin, K. Akabli and H. T. Diep, Phys. Rev. B **83**, 144406 (2011).

[123] Y. Magnin, Danh-Tai Hoang and H. T. Diep, Mod. Phys. Lett. B **25**, 1029 (2011).

[124] Y. Magnin and H. T. Diep, Phys. Rev. B **85**, 184413 (2012).

[125] N. Majlis, *The Quantum Theory of Magnetism*, World Scientific, Singapore (2000).

[126] A. Malakis, S. S. Martinos, I. A. Hadjiagapiou, N. G. Fytas, and P. Kalozoumis, Phys. Rev. E **72**, 066120 (2005).

[127] F. Matsukura, H. Ohno, A. Shen and Y. Sugawara, Phys. Rev. B **57** , R2037 (1998).

[128] D. C. Mattis, *The Theory of Magnetism I: Statics and Dynamics*, 2nd ed., Springer-Verlag, Berlin (1988).

[129] N. D. Mermin and H. Wagner, Phys. Rev. Letters **17**, 1133 (1966).

[130] N. Metropolis, A. W. Rosenbluth, M. N. Rosenbluth, A. H. Teller, E. Teller, J. Chem. Phys. **21**, 1087 (1953).

[131] A. Montorsi, *The Hubbard Model*, World Scientific, Singapore (1992),

[132] G. Misguich and C. Lhuillier, *Two-dimensional Quantum Antiferromagnets*, in *Frustrated Spin Systems*, Ed. H. T. Diep, World Scientific, Singa-

pore (2013).

[133] S. R. Mobasser and T. R. Hart, Proceed. Society of Photo-Optical Instrumentation Engineers (SPIE), Conference Series **524**, 137 (1985).

[134] M.E.J. Newman and G. T. Barkema, *Monte Carlo Methods in Statistical Physics*, Clarendon Press, Oxford (2002).

[135] V. T. Ngo and H. T. Diep, J. Appl. Phys. **103**, 07C712 (2008).

[136] V. T. Ngo and H. T. Diep, Phys. Rev. E **78**, 031119 (2008).

[137] V. T. Ngo and H. T. Diep, Phys. Rev. **B75**, 035412 (2007).

[138] V. Thanh Ngo, D. Tien Hoang and H. T. Diep, J. Phys.: Cond. Matt. **23**, 226002 (2011).

[139] V. Thanh Ngo, D. Tien Hoang and H. T. Diep, Phys. Rev. E **82**, 041123 (2010).

[140] V. Thanh Ngo, D. Tien Hoang and H. T. Diep, Modern Phys. Letters B **25**, 929-936 (2011).

[141] T. Oguchi, H. Nishimori and Y. Taguchi, J . Phys. Jpn. **54**, 4494 (1985).

[142] P. Peczak and D. P. Landau, J. Appl. Phys. **67**, 5427 (1990); *ibid.*, Phys. Rev. B **47**, 14260 (1993).

[143] A. Peles, B. W. Southern, B. Delamotte, D. Mouhanna, and M. Tissier, Phys. Rev. B **69**, 220408 (2004).

[144] A. E. Petrova, E. D. Bauer, V. Krasnorussky and S. M. Stishov, Phys. Rev. B **74**, 092401 (2006).

[145] C. Pinettes and H. T. Diep, J. Appl. Phys. **83**, 6317 (1998).

[146] H. Puszkarski, Acta Physica Polonica A **38**, 217 (1970); *ibid.* A **38**, 899 (1970).

[147] See, for example, R. Quartu and H. T. Diep, Phys. Rev. B **55**, 2975 (1997).

[148] M. Rasetti (Ed.), *The Hubbard Model, Recent Results*, World Scientific, Singapore (1991).

[149] J. E. Sacco and F. Y. Wu, J. Phys. A **8**, 1780 (1975).

[150] S. Sachdev and K. Park, Ann. Phys. **298**, 58 (2002).

[151] Subir Sachdev, *Quantum Phase Transitions*, Handbook of Magnetism and Advanced Magnetic Materials, John Wiley & Sons, Ltd (2007).

[152] C. Santamaria, R. Quartu and H. T. Diep, J. Appl. Physics **84**, 1953 (1998).

[153] T. S. Santos, S. J. May, J. L. Robertson and A. Bhattacharya, Phys. Rev. B **80**, 155114 (2009).

[154] P. Sharma, Science **307** (5709), 531533 (2005).

[155] A. Schilling, M. Cantoni, J. D. Guo, H. R. Ott, Nature **363**, 6424 (1993).

[156] B. J. Schulz, K. Binder, M. Müller, and D. P. Landau, Phys. Rev. E **67**, 067102 (2003).

[157] F. C. Schwerer and L. J. Cuddy, Phys. Rev. **2**, 1575 (1970).

[158] K. Seeger, *Semiconductor Physics*, Springer-Verlag (1982).

[159] J. M. Speight, Phys. Rev D **55**, 3830 (1997).

[160] J. Stephenson, J. Math. Phys. **11**, 420 (1970); Can. J. Phys. **48**, 2118 (1970); Phys. Rev. B **1**, 4405, (1970).

[161] S. M. Stishov, A.E. Petrova, S. Khasanov, G. Kh. Panova, A.A.Shikov, J. C. Lashley, D. Wu, and T. A. Lograsso, Phys. Rev. B **76**, 052405 (2007).

[162] R. H. Swendsen and J.-S. Wang, Phys. Rev. Lett. **58**, 86 (1987).

[163] S. M. Sze and Kwok K. Ng, *Physics of Semiconductor Devices*, Third Edition, John Wiley & Sons (2006).

[164] W. Szuszkiewicz, E. Dynowska, B. Witkowska and B. Hennion, Phys. Rev. B **73**, 104403 (2006).

[165] R. A. Tahir-Kheli and D. Ter Haar, Phys. Rev. **127**, 88 (1962).

[166] M. Tissier, B. Delamotte and D. Mouhana, Phys. Rev. Lett. **84**, 5208 (2000).

[167] G. Toulouse, Commun. Phys. **2**, 115 (1977).

[168] See review by E. Y. Tsymbal and D. G. Pettifor, *Solid State Physics* (Academic Press, San Diego), Vol. 56, pp. 113-237 (2001).

[169] V. Vaks , A. Larkin and Y. Ovchinnikov, Sov. Phys. JEPT **22**, 820 (1966).

[170] J. Villain, J. Phys. C**10**, 1717 (1977).

[171] J. Villain, Phys. Chem. Solids **11**, 303 (1959).

[172] J. Villain, R. Bidaux, J.P. Carton, and R. Conte, J. Physique **41**, 1263 (1980).

[173] F. Wang and D. P. Landau, Phys. Rev. Lett. **86**, 2050 (2001); Phys. Rev. E **64**, 056101 (2001).

[174] X. F. Wang, T. Wu, G. Wu, H. Chen, Y. L. Xie, J. J. Ying, Y. J. Yan, R. H. Liu and X. H. Chen, Phys. Rev. Lett. **102**, 117005 (2009).

[175] Qing-Yan *et al.* (2012), Chin. Phys. Lett. **29**, 037402 (2012); arXiv:1201.5694.

[176] G. H. Wannier, Phys. Rev. **79**, 357 (1950); Phys. Rev. B **7**, 5017 (E) (1973).

[177] J. C. Wheatley, Progress in Low Temperature Physics **VI**, 77 (1970), Ed. C. J. Gortor, North-Holland Publishing Co.

[178] K. G. Wilson, Phys. Rev. B **4**, 3174 (1971).

[179] K. G. Wilson and M. E. Fisher, Phys. Rev. Lett. **28**, 240 (1972).

[180] K. G. Wilson, Rev. Mod. Phys. **47**, 773 (1975).

[181] U. Wolff, Phys. Rev. Lett. **62**, 361 (1989); Phys. Lett. B **228**, 379 (1989).

[182] J. Wosnitza, R. Deutschmann, H. von Löhneysen and R. K. Kremer, J. Cond. Matter. Phys. **6**, 8045 (1994).

[183] A. L. Wysocki, K. D. Belashchenko, J. P. Velev, and M. van Schilfgaarde, J. Appl. Phys. **101**, 09G506 (2007).

[184] J. Xia, W. Siemons, G. Koster, M. R. Beasley and A. Kapitulnik, Phys. Rev. B **79**, R140407 (2009).

[185] A. Yoshimori, J. Phys. Soc. Jpn. **14**, 807 (1959).

[186] A. Zangwill, *Physics at Surfaces*, Cambridge University Press (1988).

[187] G. Zarand, C. P. Moca and B. Janko, Phys. Rev. Lett. **94**, 247202 (2005).

[188] Y. Q. Zhang, Z. D. Zhang and J. Aarts, Phys. Rev. B **79**, 224422 (2009).

[189] J. Zinn-Justin, *Quantum Field Theory and Critical Phenomena*, Oxford Unversity Press (2002).

[190] D. N. Zubarev, Usp. Fiz. Nauk **187**, 71 (1960)[translation: Soviet Phys.-Uspekhi **3**, 320 (1960)].

[191] I. Žutić and S. Das Sarma, *Spintronics: Fundamentals and Applications*, Reviews of Modern Physics **76** (2), 323 (2004); arXiv:cond-mat/0405528.

Index

Printed in the United States
By Bookmasters